Book of Abstracts of the 61st Annual Meeting of the European Association for Animal Production

EAAP - European Federation of Animal Science

The European Association for Animal Production wishes to express its appreciation to the
Ministero delle Politiche Agricole e Forestali (Italy) and the
Associazione Italiana Allevatori (Italy)
for their valuable support of its activities.

Book of Abstracts of the 61st Annual Meeting of the European Association for Animal Production

Heraklion - Crete Island, Greece, August 23rd - 27th, 2010

ISBN 978-90-8686-152-1
e-ISBN: 978-90-8686-708-0
DOI: 10.3920/978-90-8686-708-0

ISSN 1382-6077

First published, 2010

© Wageningen Academic Publishers
The Netherlands, 2010

This work is subject to copyright. All rights are reserved, whether the whole or part of the material is concerned. Nothing from this publication may be translated, reproduced, stored in a computerised system or published in any form or in any manner, including electronic, mechanical, reprographic or photographic, without prior written permission from the publisher:
Wageningen Academic Publishers
P.O. Box 220
6700 AE Wageningen
The Netherlands
www.WageningenAcademic.com
copyright@WageningenAcademic.com

The individual contributions in this publication and any liabilities arising from them remain the responsibility of the authors.

The designations employed and the presentation of material in this publication do not imply the expression of any opinion whatsoever on the part of the European Association for Animal Production concerning the legal status of any country, territory, city or area or of its authorities, or concerning the delimitation of its frontiers or boundaries.

The publisher is not responsible for possible damages, which could be a result of content derived from this publication.

61st EAAP , President's Message

On behalf of the Greek Organizing Committee, I am very pleased to welcome you to the 61st Annual Meeting of the EAAP here in Crete. This is the second time (1985) that the EAAP meets in Greece. The theme of the Meeting is "**Impact of food demand, in quantity and quality wise, on Animal Production**" which is a hot topic in Europe and a very appropriate topic in view of the current worldwide demands for both human society and livestock industry. This Conference is also of great interest for the Greek livestock sector and gives a great opportunity to participants to share the experience of the wide range of Greek livestock farming systems which are related to physical and cultural features of Greece.

The participants will, also, have the opportunity to attend a very interesting scientific programme with a selected number of presentations from a great number of abstracts and to take part in workshops and discussions of the latest and most relevant research in the field of Animal Science, including Aquaculture. This would be very fruitful for young scientists and students in particular. During the meeting days, the scientific programme will be enriched with different technical visits, social events and a special programme for accompanying persons, all of them designed to showcase the extraordinary variety of the local culture, and of the architectural and gastronomic inheritance of our country, particularly in Crete.

George Zervas
Rector of the Agricultural University of Athens and
President of the Hellenic Society of Animal Production
President of the Greek Organising Committee

National organisers of the 61st eaap annual meeting

National Conference Organizers

- Agricultural University of Athens,Department of Animal Production
- Hellenic Society for Animal Production

Organizing Committee

President

- ***Prof. George Zervas*** - Rector of the AgriculturalUniversity of Athens and President of the Hellenic Society of Animal Production (gzervas@aua.gr)

Vice-President

- ***Mr. George Zacharopoulos*** - General Director of Animal Production Dept Ministry of Rural Development and Food (ve46u064@minagric.gr)

Executive Secretary

- ***Prof. Kostas Fegeros*** - Dept of Animal Sciences and Aquaculture of the Agricultural University of Athens (cfeg.@aua.gr)

Members

- ***Prof. Stelios Deligeorgis*** - Dept of Animal Sciences and Aquaculture, Agricultural University of Athens (sdel@aua.gr)
- ***Prof. Andreas Georgoudis*** - Dept of Animal Sciences, Aristotle University of Thessaloniki (andgeorg@agro.auth.gr)
- ***Assoc. Prof. Eleftheria Panopoulou*** - Dept of Animal Sciences and Aquaculture, Agricultural University of Athens (pae@aua.gr)
- ***Assist. Prof. John Hadjigeorgiou*** - Dept of Animal Sciences and Aquaculture, Agricultural University of Athens (ihadjig@aua.gr)
- ***Dr. George Papadomichelakis*** - Dept of Animal Sciences and Aquaculture, Agricultural University of Athens (gpapad@aua.gr)
- ***Dr. Christina Ligda*** - Nacional Agricultural Research Institution, Thessaloniki (chligda@otenet.gr)
- ***Dr. Eleni Tsiplakou*** - Dept of Animal Sciences and Aquaculture, Agricultural University of Athens (eltsiplakou@aua.gr)
- ***Dr. Katerina Tsolakidi*** - Dept of Animal Production, Ministry of Rural Development and Food (katerina1975@yahoo.com)

Greek Scientific Committee

President

- ***Prof. John Hadjiminaoglou*** - Dept of Animal Sciences, Aristotle University of Thessaloniki (ioahat@agro.auth.gr)

Vice-President

- ***Prof. Emmanouel Rogdakis*** - Dept of Animal Sciences and Aquaculture, Agricultural Univesrity of Athens (erog@aua.gr)

Secretary

- ***Assoc. Prof. Jean Boyazoglu*** - Dept of Animal Sciences, Aristotle University of Thessaloniki (jean.boyazoglu@wanadoo.fr)

Members

- ***Prof. Apostolos Karalazos*** - Dept of Animal Sciences, Aristotle University of Thessaloniki (karalazo@agro.auth.gr)
- ***Prof. Dimitris Dotas*** - Dept of Animal Sciences, Aristotle University of Thessaloniki (dotas@agro.auth.gr)
- ***Mr Angelos Baltas*** - Ministry of Rural Development and Food (eze@aua.gr)

Guest member

- ***Prof. Nicolas Zervas*** - Agricultural University of Athens (eze@aua.gr)

Coordinators of Study Commissions

Animal Genetics

- ***Prof. Emmanuel Rogdakis*** - Dept of Animal Sciences and Aquaculture, Agricultural University of Athens (erog@aua.gr)

Animal Physiology

- ***Assoc. Prof. Stella Chadio*** - Dept of Animal Sciences and Aquaculture, Agricultural University of Athens (shad@aua.gr)

Animal Nutrition

- ***Prof. Apostolos Karalazos*** - Dept of Animal Sciences, Aristotle University of Thessaloniki (karalazo@agro.auth.gr)

Animal Management Wellbeing and Health

- ***Assoc. Prof. Eftychia Xylouri*** - Dept of Animal Sciences and Aquaculture, Agricultural University of Athens (efxil@aua.gr)

Cattle Production

- ***Prof. Andreas Georgoudis*** - Dept of Animal Sciences, Aristotle University of Thessaloniki (andgeorg@agro.auth.gr)

Sheep and Goat Production

- ***Dr. Dimitris Papavasiliou*** - Technical Educational Institute of Epirous (papavas@teiep.gr)

Pig Production

- ***Mr Kostas Thessalos*** - Ministry of Rural Development and Food (kthessalos@gmail.com)

Horse Production

- ***Prof. John Menegatos*** - Dept of Animal Sciences and Aquaculture, Agricultural University of Athens (jmen@aua.gr)

Livestock Farming Systems

- ***Assist. Prof. John Hadjigeorgiou*** - Dept of Animal Sciences and Aquaculture, Agricultural University of Athens (ihadjig@aua.gr)

Greece

Greece is a rather small country situated on the south - eastern part of Europe with a population of about 11 million. The climate varies from north to south but in general is mild and favours the growth of a wide range of crops, cereals, roots, vegetables, cotton, tobacco, olive trees, vineyards, fruits and alfalfa, as well as forest.
The land area of Greece is 13.2 million hectares of which 30% is devoted to crops, 40% to pasture, and 20% to forestry. The average size of the agricultural holdings falls within a broad range of 2 to 10 hectares with a considerable number of mixed holdings (crops and livestock).
The population of Greece is distributed in lowlands (69%), hills (21%) and mountains (10%), with 16.3% of the active population involved in agriculture. Agriculture represents only 6.7% of the Gross National product, with 75% of its coming from crops and 25% from livestock production.
The seasonality of production of the natural vegetation, combined with low rainfall, high temperature and low organic content of the soil are some of the main constrains on the development of livestock production in relation to crops.

Agriculture in greece

Dairy sheep (8 millions) and goats (5.5 millions) are considered as the most significant livestock sector in Greece with a long continuity of the ancient tradition. Sheep (9 main dairy and 18 rare breeds) and goats (local native breeds) are kept mainly extensively in less favored areas, in rather small sized flocks, and produce milk which is transformed mainly into Feta cheese with special aroma, taste and flavor. This farming system is of great significance from an economic, cultural and environmental point of view.
Dairy cows, fattening cattle, pigs and poultry are kept mainly intensively indoors. The grasslands of Greece (83% state owned) are more suitable for sheep and goats grazing. They are composed mainly of annual species with great botanical interest (~6.500 species). These grazing areas of Greece receive no application of artificial fertilizers, nor agrochemicals and no agricultural management other than grazing which benefits a wide range of flora and fauna (e.g. birds, insects, snails, turtles, hedgehogs, hares, rabbits, foxes etc.).
The self-sufficiency of the country in products of animal origin is 28% in beef meat, 40% in pork, 82% in sheep and goat meat, 76% in poultry, 50% in cow's milk and milk products, 100% in sheep and goats milk, 97% in eggs and 87% in honey. Greece is also the European leader in aquaculture with high exports.

Crete

On the island of CRETE, goat rising has been reported since the Neolithic period. Homer, in his famous ancient Greek book, the Odyssey, describes the use of dairy sheep and goats during the Mycenaean times (about 1.200 B.C.), when the Cyclop Polyphemus in his cave sat down to milk his sheep and goats, then put aside half of the milk to be curdled in wicker baskets with a previous day's whey. This description of cheese making fits still today the practice by Greek mountain farmers making Feta cheese. Minoans and Mycenaeans loved cheese made from goat and sheep milk.
The surface area of Crete is 832.000 hectares. Nearly 43 % of this is devoted to agriculture (annual crops, olive trees, fruit trees, vineyards, vegetables etc.), 28 % to permanent meadows and pastures and 27 % forests and scrub woodland. Fifty percent of the green houses of Greece is located to Crete due to its mild climate. The livestock sector of Crete is of great importance with farms of high productivity, and consists of 2.200 bovines, 1.500.000 sheep, 627.000 goats, 61.000 sows, 406.000 rabbits and 1.430.000 poultry. Sheep and goat farming is the most important on its extensive or semi-intensive production system. Among the current sheep breeds the "Sfakia" has the largest population of autochthonous sheep, being subjected to a genetic improvement programme. Twenty five percent of the hard cheeses of Greece is produced in Crete from sheep and goat milk (Kefalograviera). A great proportion of the feedstuffs used by the livestock sector of Crete is imported mainly from the midland due to limited agriculture area for feedstuffs production.

Crete Island

Crete is located in the most southern part of Greece and is the biggest Greek island with 570.000 inhabitants. It separates the Aegean from the Libyan Sea and marks the boundary between Europe and Africa.

The town of Heraklion is the largest urban centre of Crete, the capital of the region of Crete and the economic centre of the island. During the Minoan period the town of Heraklion served as a port to Knossos. Currently the population of Heraklion is approximately 150.000. Since the Minoan period the town of Heraklion became the host of varied cultures and civilisations whose indelible marks are visible in the form of fountains, castles, walls, palaces, and other monuments.

Heraklion has a lot to offer to those who are in search of culture and entertainment. In particular, each summer the municipality of Heraklion organises cultural events ranging from theatrical performances, to music nights, to art exhibitions, and other events.

Host City "Hersonissos" is located 26 km east of Heraklion on the road to Agios Nikolaos. It keeps the name of the ancient city that was located in the place that the harbour of Hersonissos is today. The village of Old Hersonissos, located within a small distance to the south was built due to the fear of the pirates, that forced the inhabitants of the village to move in the inland.

Near Old Hersonissos there are two picturesque small villages, Piskopiano and Koutouloufari, overlooking the busy resort of Limin Hersonissos.

Conference centre

Creta Maris Conference Center
700 14 Hersonissos
Crete, Greece
Tel: +3028970 27000
Fax: +3028970 22130
E-mail: creta@maris.gr
Website: www.maris.gr

Sponsorships

Below the different types of Sponsorships

1. Meeting sponsor – From 3000 euro up

- Acknowledgements in the book of abstracts with contact address and logo.
- One page allowance in the final programme booklet of Barcelona.
- Advertising/information material inserted in the bags of delegates.
- Advertising/information material on a stand display.
- Acknowledgement in the EAAP Newsletter with possibility of a one page of publicity.
- Possibility to add session and speaker support (at additional cost to be negotiated).

2. Session sponsor – from 2000 euro up

- Acknowledgements in the book of abstracts with contact address and logo.
- One page allowance in the final programme booklet of Barcelona.
- Advertising/ information material in the delegate bag.
- ppt at beginning of session to acknowledge support and recognition by session chair.
- Acknowledgement in the EAAP Newsletter.

3. Speaker sponsor from 1000 euro up

- Half page allowance in the final programme booklet of Barcelona.
- Advertising / information material in the delegate bag.
- Recognition by speaker of the support at session.
- Acknowledgement in the EAAP Newsletter.

4. Registration Sponsor (equivalent to a full registration fee of the Annual Meeting)

- Acknowledgements in the book of abstracts with contact address and logo.
- Advertising/information material in the delegate bag.

Board of Trustees

Chair
Dr. Sandra Edwards
University of Newcastle
United Kingdom

Treasurer/Secretary
Prof. Andrea Rosati
EAAP Secretary General
Italy

Contact and further information

If you are interested to become a sponsor of the "EAAP Program Foundation" or want to have further information, please contact the Treasurer/Secretary Andrea Rosati:

rosati@eaap.org / eaap@eaap.org
Fax: +39 06 44266798
Phone +39 06 44202639

Acknowledgements

www.imv-technologies.com

www.dsm.com

www.dansire.dk

www.animal-journal.eu

ALCASDE

www.alcasde.eu

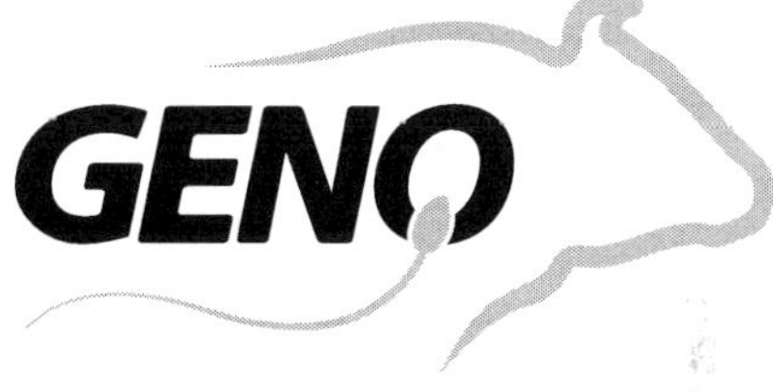

www.geno.no

crv@crvholding.com

EUROLACTIS.COM

Donkey's Milk & Dairy World leader

www.eurolactis.com

www.rednex-fp7.eu

European Association for Animal Production (EAAP)

President: Kris Sejrsen
Secretary General: Andrea Rosati
Address: Via G.Tomassetti 3, A/I
I-00161 Rome, Italy
Phone: +39 06 4420 2639
Fax: +39 06 4426 6798
E-mail: eaap@eaap.org
Web: www.eaap.org

62nd EAAP Annual meeting of the European Association for Animal Production

29th August to 2nd September 2011
Stavanger, Norway

Rica Forum Conference Centre
Website: http://www.eaap2011.com
E-mail: eaap2011@umb.no

Local Organising Committee:

President: Dr. Torbjørn Auran, Felleskjøpet Fôrutvikling

Secretary: Dr. Ina Andersen-Ranberg, Norsvin

Theme of the meeting:
"The importance of animal production for food supply, food quality and environment"

Mediterranean Symposium in Zadar, Croatia, 2010

October 27-29, 2010

Animal Farming and Environment Interactions in Mediterranean Region

Organized by:

University of Zadar, Dept of Ecology, Agronomy and Aquaculture

EAAP – European Association for Animal Production

CIHEAM - International Centre for Advanced. Mediterranean Agronomic Studies.

Secretariat: medit.zadar2010@unizd.hr
Website: www.unizd.hr/zadar2010

Scientific Programme EAAP 2009

Monday 23 August 8.30 - 12.30	Monday 23 August 14.00 - 18.00	Tuesday 24 August 8.30 - 12.30	Tuesday 24 August 14.00 - 18.00
Session 2 New phenotypes for new breeding goals Chair: Alfred de Vries ——— Session 3 Opportunities and challenges for grassland-based systems Chair: I. Hadjigeorgiou ——— Session 5 a: Innovation to support economic sustainability in management of small ruminants Chair: Eva Ugarte b: Animal Fibre Science Chair: C. Renieri ——— Session 6 Equine Milk for Human Health Chair: E. Salimei ——— Session 7 Intestinal health in monogastrics Chair: G. van Duinkerken ——— Session 8 New issues in meat quality in Cattle Chair: J.F. Hocquette	Session 9 New developments in animal health – from claw disorders to early detection of disease Chairs: E. Stassen and J. Krieter ——— Session 10 Feed Evaluation Chair J.E. Lindberg ——— Session 11 Evaluation and use of alternative pig nutrient sources Chair: S. Chadd ——— Session 12 Free Communications Genetics Chair: D. Wittenburg ——— Session 13 Animals in extreme environments Chairs: I. Casasus and A. Georgoudis ——— Session 14 ERFP/FAO/EAAP Symposium Strategies to add value to local breeds Chairs: I. Hoffmann, C. Ligda and S.J. Hiemstra ——— Session 15 Reproductive physiology of sheep and goats Chair: L. Bodin	8.30 - 9.30 Leroy Fellowship Award Lecture Michel Doreau ——— 9.30 - 12.30 Plenary Session Interactions between livestock industry and climate change	Session 4 Societal demands and policy instruments in relation to animal welfare and food production Chair: H. Spoolder ——— Session 16 Product quality (milk, meat) in small ruminants Chair: M. Gauly ——— Session 17 Symposium Alternatives to Castration in Pigs Chair: M. Angels Oliver ——— Session 18 Economic and animal aspects of mastitis & fertility in cattle Chair: A. Kuipers ——— Session 19 Horse Genetics – behaviour traits Chairs: M. Hausberger and B. Ducro ——— Session 20 Evolution of mammary gland and milk secretion: Consequences for lactation in farm animals Chair: R. Bruckmaier ——— Session 21 Industry Session: Innovations in feed evaluation Chair: B. Cooke
Session 1 The 10th International Workshop on Biology of Lactation in Farm Animals Chair: K. Stelwagen		**Tuesday 24 August 18.00 – 20.00**	
		Poster Session (Series A Posters)	

Wednesday 25 August 8.30 – 12.30	Wednesday 25 August 14.00 – 18.00	Thursday 26 August 8.30 – 12.30	Thursday 26 August 14.00 – 18.00
Session 22 Advances in quantitative genetics Chair: J. Szyda ——— Session 23 Foetal programming in farm animals Chair: M. Kuran ——— Session 24 Aquaculture Chair: S. Papoutsoglou ——— Session 25 Nutrition and management of lactating sows Chair: D. Torrallardona ——— Session 26 Robust cattle production systems Chairs: M. Coffey and M. Klopčič ——— Session 27 Health control and eradication programmes in small ruminants. Chair: N. Kemper ——— Session 28 Relationships between intensity of production and sustainability of Livestock Farming Systems Chair: B. Dedieu	Commission future programme and elections meeting (14.00 – 15.00) ——— Followed by free communications on (15.00 - 18.00) Session 29 Animal Genetics Chair: L. Dybdahl Pedersen Session 30 Animal Nutrition Chair: L. Bailoni Session 31 Animal Management & Health Chair: S. Edwards Session 32 Animal Physiology Chair: N. Scollan Session 33 Livestock Farming Systems Chair: V. Matlova Session 34 Improving cattle reproduction Chair: G. Thaller Session 35 Sheep & Goats Production Chair: M. Milerski Session 36 Pig Production Chair: P. Knap Session 37 Horse Production Chair: M. Saastamoinen	Session 38 Free Communications in Animal Genetics Chair: E. Strandberg ——— Session 43 Breeding and recording strategies in small ruminants Chair: M. Schneeberger	Session 44 Application of genomic selection Chair: T. Meuwissen ——— Session 45 Pigs and sheep as models for human medicine Chair: H. Quesnel ——— Session 46 Free Communications in Animal Nutrition Chair: A. Connolly
		Session 39 Symposium Cost Action: Feed for Health Chairs: I. Givens, L. Pinotti, C. Knight and A. Baldi ——— Session 40 Symposium Environmental impact of animal production: 1. Evaluating negative and positive externalities Chair: P. Lecomte 2. Designing more sustainable LFS and food chains Chair: T. Kristensen ——— Session 41 Seminar Consumer attitudes to food quality products in Southern Europe Chair: M. Klopčič ——— Session 42 Horse Network workshop: Role of Equids in socio economy and on human well being Chair W. Martin-Rosset, J. Flanagan and R. Kjaldman	
Wednesday 25 August 18.00 – 20.00			
Poster Session (Series B posters)			

Commission on Animal Genetics

Prof. Dr Simianer	President	University of Goettingen
	Germany	hsimian@gwdg.de
Dr Gandini	Vice-President	University of Milan
	Italy	gustavo.gandini@unimi.it
Dr Meuwissen	Vice-President	Norwegian University of Life Sciences
	Norway	theo.meuwissen@umb.no
Dr Strandberg	Secretary	SLU
	Sweden	Erling.Strandberg@hgen.slu.se
Dr Szyda	Secretary	Agricultural University of Wroclaw
	Poland	szyda@karnet.ar.wroc.pl
Dr De Vries	Industry rep.	CRV

Commission on Animal Nutrition

Dr Lindberg	President	Swedish University of Agriculture
	Sweden	jan-eric.lindberg@huv.slu.se
Dr Bailoni	Vice-President	University of Padova
	Italy	Lucia.bailoni@unipd.it
Dr Connolly	Vice-President/	
	Industry rep.	Alltech
	Ireland	aconnolly@alltech.com
Mrs Tsiplakou	Secretary	Agricultural University of Athens
	Greece	eltisplakou@aua.gr
Dr Cenkvàri	Secretary	Szent Istvan University
	Hungary	Czenkvari.Eva@aotk.szie.hu

Commission on Animal Management & Health

Dr Fourichon	President	Oniris INRA
	France	Christine.fourichon@oniris-nantes.fr
Dr Spoolder	Vice-President	ASG-WUR
	Netherlands	Hans.spoolder@wur.nl
Prof. Dr Krieter	Vice-President	University Kiel
	Germany	jkrieter@tierzucht.uni-kiel.de
Dr Edwards	Secretary	University of Newcastle Upon Tyne
	United Kingdom	Sandra.edwards@ncl.ac.uk
Mr Das	Secretary	University of Goettingen
	Germany	gdas@gwdg.de
Mr Pearce	Industry rep.	Pfizer
	United Kingdom	Michael.C.Pearce@pfizer.com

Commission on Animal Physiology

Dr Vestergaard	President	Aarhus University
	Denmark	Mogens.Vestergaard@agrsci.dk
Dr M. Kuran	Vice-President	Gaziosmanpasa University
	Turkey	mkuran@gop.edu.tr
Dr Driancourt	Vice president/	
	Industry rep.	Intervet
	France	Marc-antoine.driancourt@sp.intervet.com
Dr Bruckmaier	Vice-President	University of Bern
	Switzereland	Rupert.bruckmaier@physio.unibe.ch
Dr Quesnel	Secretary	INRA Saint Gilles
	France	Helene.quesnel@rennes.inra.fr
Dr Scollan	Secretary	Institute of Biological, Environmental and rural sciences
	UK	ngs@aber.ac.uk

Commission on Livestock Farming Systems

Dr Bernués Jal	President	CITA
	Spain	abernues@aragon.es
Dr Hermansen	Vice-President	DIAS
	Denmark	john.hermansen@agrsci.dk
Dr Leroyer	Vice president/	
	industry rep.	ITAB
	France	Joannie.leroyer@itab.asso.fr
Dr Matlova	Vice-President	Res. Institute for Animal Production
	Czech Republic	matlova.vera@vuzv.cz
Dr Eiler	Secretary	Wageningen University
	Netherlands	karen.eilers@wur.nl
Dr Ingrand	Secretary	INRA/SAD
	France	ingrand@clermond.inra.fr

Commission on Cattle Production

Dr Kuipers	President	Wageningen UR
	Netherlands	abele.kuipers@wur.nl
Dr Thaller	Vice-President	Animal Breeding and Husbandry
	Germany	Georg.Thaller@tierzucht.uni-kiel.de
Dr Lazzaroni	Vice-president	University of Torino
	Italy	carla.lazzaroni@unito.it
Dr Coffey	Vice president/	
	Industry rep.	SAC, Scotland
	UK	Karin.hendry@sac.ac.uk
Dr Hocquette	Secretary	INRA
	France	hocquet@clermont.inra.fr
Dr Klopcic	Secretary	University of Ljublijana
	Slovenia	Marija.Klopcic@bfro.uni-lj.si

Commission on Sheep and Goat Production

Name	Position / Country	Institution / E-mail
Dr Schneeberger	President	ETH Zentrum
	Switzerland	markus.schneeberger@inw.agrl.ethz.ch
Dr Ringdorfer	Vice-President	LFZ Raumberg-Gumpenstein
	Austria	
	ferdinand.ringdorfer@raumberg-gumpenstein.at	
Dr Bodin	Vice President	INRA-SAGA
	France	Loys.bodin@toulouse.inra.fr
Dr Papachristoforou	Vice President	Agricultural Research Institute
	Cyprus	Chr.Papachristoforou@arinet.ari.gov.cy
Dr Milerski	Secretary/	
	Industry rep.	Research Institute of Animal Science
	Czech Republic	m.milerski@seznam.cz
Dr Sagastizabal	Secretary/	
	Industry rep.	NEIKER-Tecnalia
		eugarte@neiker.net

Commission on Pig Production

Name	Position / Country	Institution / E-mail
Dr Knap	President	PIC International Group
	Germany	pieter.knap@pic.com
Dr Chadd	Vice-President	Royal Agric. College
	UK	steve.chadd@royagcol.ac.uk
Dr Torrallardona	Vice-President	IRTA
	Spain	David.Torrallardona@irta.es
Dr Pescovicova	Secretary	Research Institute of Animal Production
	Slovak Republic	peskovic@vuzv.sk
Dr Manteca	Secretary	Universitat Autònoma de Barcelona
	Spain	xavier.manteca@uab.es

Commission on Horse Production

Name	Position / Country	Institution / E-mail
Dr Miraglia	President	Molise University
	Italy	miraglia@unimol.it
Dr Burger	Vice president	Clinic Swiss National Stud
	Switzerland	
	Dominique.burger@mbox.haras.admin.ch	
Dr Janssen	Vice president	BIOSYST
	Belgium	Steven.janssens@biw.kuleuven.be
Dr Lewczuk	Vice president	IGABPAS
	Poland	d.lewczuk@ighz.pl
Dr Coenen	Vice president	University of Leipzig
	Germany	coenen@vetmed.uni-leipzig.de
Dr Palmer	Vice president/	
	Industry rep.	CRYOZOOTECH
	France	ericpalmer@cryozootech.com
Dr Holgersson	Secretary	Swedish University of Agriculture
	Sweden	Anna-Lena.holgersson@hipp.slu.se
Dr Hausberger	Secretary	CNRS University
	France	Martine.hausberger@univ-rennes1.fr

Session 01. The 10th International Workshop on Biology of Lactation in Farm Animals (BOLFA)

Date: 23 August 2010; 08:30 - 18:00 hours
Chairperson: Stelwagen (NZ)

Session 02. New phenotypes for new breeding goals

Date: 23 August 2010; 08:30 - 12:30 hours
Chairperson: De Vries (NL)

Theatre **Session 02 no. Page**

Session 03. Opportunities and challenges for grassland-based systems

Date: 23 August 2010; 08:30 - 12:30 hours
Chairperson: Hadjigeorgiou (GR)

Session 04. Societal demands and policy instruments in relation to animal welfare and food production

Date: 24 August 2010; 14:00 - 18:00 hours
Chairperson: Spoolder (NL)

Theatre **Session 04 no. Page**

Session 05. a: Innovation to support economic sustainability in management of small ruminants

Date: 23 August 2010; 08:30 - 10:15 hours
Chairperson: Ugarte Sagastizabal (ES)

Session 05. b: Animal fibre science

Date: 23 August 2010; 10:45 - 12:30 hours
Chairperson: Renieri (IT)

Session 06. Equine milk for human health

Date: 23 August 2010; 08:30 - 12:30 hours
Chairperson: Salimei (IT)

Session 07. Intestinal health in monogastrics

Date: 23 August 2010; 08:30 - 12:30 hours
Chairperson: Van Duinkerken (NL)

Theatre **Session 07 no. Page**

Poster **Session 07 no. Page**

Session 08. New issues in meat quality in cattle

Date: 23 August 2010; 08:30 - 12:30 hours
Chairperson: Hocquette (FR)

Session 09. New developments in animal health: from claw disorders to early detection of disease

Date: 23 August 2010; 14:00 - 18:00 hours
Chairperson: Stassen (NL) and Krieter (DE)

Theatre **Session 09 no. Page**

Poster **Session 09 no. Page**

Session 10. Feed evaluation

Date: 23 August 2010; 14:00 - 18:00 hours
Chairperson: Lindberg (SE)

Session 11. Evaluation and use of alternative pig nutrient sources

Date: 23 August 2010; 14:00 - 18:00 hours
Chairperson: Chadd (UK)

Poster **Session 11 no. Page**

Session 12. Free Communications in Animal Genetics

Date: 23 August 2010; 14:00 - 18:00 hours
Chairperson: Wittenburg (DE)

Theatre **Session 12 no. Page**

Poster **Session 12 no. Page**

Session 13. Animals in extreme environments

Date: 23 August 2010; 14:00 - 18:00 hours
Chairperson: Casasus (ES) and Georgoudis (GR)

Theatre | **Session 13 no.** | **Page**

Poster **Session 13 no. Page**

Session 14. Strategies to add value to local breeds

Date: 23 August 2010; 14:00 - 18:00 hours
Chairperson: Hoffmann (DE), Ligda (GR) and Hiemstra (NL)

Theatre **Session 14 no. Page**

Session 15. Reproductive physiology of sheep and goats

Date: 23 August 2010; 14:00 - 18:00 hours
Chairperson: Bodin (FR)

Theatre **Session 15 no. Page**

Session 16. Product quality (milk, meat) in small ruminants

Date: 24 August 2010; 14:00 - 18:00 hours
Chairperson: Gauly (DE)

Poster **Session 16 no. Page**

Session 17. Symposium: Alternatives to castration in pigs

Date: 24 August 2010; 14:00 - 18:00 hours
Chairperson: Oliver (ES)

Theatre **Session 17 no. Page**

Session 18. Economic and animal aspects of mastitis & fertility in cattle

Date: 24 August 2010; 14:00 - 18:00 hours
Chairperson: Kuipers (NL)

Session 19. Horse genetics: behaviour traits

Date: 24 August 2010; 14:00 - 18:00 hours
Chairperson: Hausberger (CH) and Ducro (NL)

Poster **Session 19 no. Page**

Session 20. Evolution of mammary gland and milk secretion: consequences for lactation in farm animals

Date: 24 August 2010; 14:00 - 18:00 hours
Chairperson: Bruckmaier (CH)

Theatre **Session 20 no. Page**

Session 21. Industry session

Date: 24 August 2010; 14:00 - 18:00 hours
Chairperson: Cooke (UK)

Session 22. Advances in quantitative genetics

Date: 25 August 2010; 08:30 - 12:30 hours
Chairperson: Szyda (PL)

Theatre **Session 22 no. Page**

Poster **Session 22 no. Page**

Session 23. Foetal programming in farm animals

Date: 25 August 2010; 08:30 - 12:30 hours
Chairperson: Kuran (TR)

Theatre **Session 23 no. Page**

Poster **Session 23 no. Page**

Session 24. Aquaculture

Date: 25 August 2010; 08:30 - 12:30 hours
Chairperson: Papoutsoglou (GR)

Session 25. Nutrition and management of lactating sows

Date: 25 August 2010; 08:30 - 12:30 hours
Chairperson: Torrallardona (ES)

Poster **Session 25 no. Page**

Session 26. Robust cattle production systems

Date: 25 August 2010; 08:30 - 12:30 hours
Chairperson: Coffey (UK) and Klopcic (SI)

Theatre **Session 26 no. Page**

Session 27. Health control and eradication programmes in small ruminants

Date: 25 August 2010; 08:30 - 12:30 hours
Chairperson: Kemper (DE)

Session 28. Relationships between intensity of production and sustainability of livestock farming systems

Date: 25 August 2010; 08:30 - 12:30 hours
Chairperson: Dedieu (FR)

Session 29. Free Communications in Animal Genetics

Date: 25 August 2010; 14:00 - 18:00 hours
Chairperson: Dybdahl Pedersen (DK)

Poster **Session 29 no. Page**

Session 30. Free Communications in Animal Nutrition

Date: 25 August 2010; 14:00 - 18:00 hours
Chairperson: Bailoni (IT)

Session 31. Free Communications in Animal Management and Health

Date: 25 August 2010; 14:00 - 18:00 hours
Chairperson: Edwards (UK)

Theatre **Session 31 no. Page**

Poster **Session 31 no. Page**

Session 32. Free Communications in Animal Physiology

Date: 25 August 2010; 14:00 - 18:00 hours
Chairperson: Scollan (UK)

Theatre **Session 32 no. Page**

Poster **Session 32 no. Page**

Session 33. Free Communications in Livestock Farming Systems

Date: 25 August 2010; 14:00 - 18:00 hours
Chairperson: Matlova (CZ)

Theatre **Session 33 no. Page**

Poster **Session 33 no. Page**

Session 34. Improving cattle reproduction

Date: 25 August 2010; 14:00 - 18:00 hours
Chairperson: Thaller (DE)

Theatre **Session 34 no. Page**

Session 35. Free Communications in Sheep and Goat Production

Date: 25 August 2010; 14:00 - 18:00 hours
Chairperson: Milerski (CZ)

Theatre **Session 35 no. Page**

Poster **Session 35 no. Page**

Session 36. Free Communications in Pig Production

Date: 25 August 2010; 14:00 - 18:00 hours
Chairperson: Knap (DE)

Theatre **Session 36 no. Page**

Poster **Session 36 no. Page**

Session 37. Free Communications in Horse Production

Date: 25 August 2010; 14:00 - 18:00 hours
Chairperson: Saastamoinen (FI)

Theatre **Session 37 no. Page**

Poster **Session 37 no. Page**

Session 38. Free Communications in Animal Genetics

Date: 26 August 2010; 08:30 - 12:30 hours
Chairperson: Strandberg (SE)

Theatre **Session 38 no. Page**

Poster **Session 38 no. Page**

Session 39. Symposium: Cost Action - Feed for health

Date: 26 August 2010; 08:30 - 18:00 hours
Chairperson: Givens (UK), Pinotti (IT), Knight (UK) and Baldi (IT)

Theatre **Session 39 no. Page**

Session 40. Symposium: Environmental impact of animal production - 1. Evaluating negative and positive externalities

Date: 26 August 2010; 08:30 - 12:30 hours
Chairperson: Lecomte (FR)

Session 40. Symposium: Environmental impact of animal production - 2. Designing more sustainable LFS and food chains

Date: 26 August 2010; 14:00 - 18:00 hours
Chairperson: Kristensen (DK)

Session 41. Seminar on consumer attitudes to food quality products in Southern Europe

Date: 26 August 2010; 08:30 - 18:00 hours
Chairperson: Klopcic (SI)

Poster **Session 41 no. Page**

Session 42. Horse network workshop: Role of equids in socio economy and on human well being

Date: 26 August 2010; 08:30 - 18:00 hours
Chairperson: Martin-Rosset (FR), Flanagan (IR) and Kjaldman (FI)

Theatre **Session 42 no. Page**

Poster **Session 42 no. Page**

Session 43. Breeding and recording strategies in small ruminants

Date: 26 August 2010; 08:30 - 12:30 hours
Chairperson: Schneeberger (CH)

Theatre **Session 43 no. Page**

Poster **Session 43 no. Page**

Session 44. Application of genomic selection

Date: 26 August 2010; 14:00 - 18:00 hours
Chairperson: Meuwissen (NO)

Theatre **Session 44 no. Page**

Session 45. Pigs and sheep as models for human medicine

Date: 26 August 2010; 14:00 - 18:00 hours
Chairperson: Quesnel (FR)

Theatre **Session 45 no. Page**

Session 46. Free Communications in Animal Nutrition

Date: 26 August 2010; 14:00 - 18:00 hours
Chairperson: Connolly (IE)

Poster **Session 46 no. Page**

Bovine mammary stem cells: cell biology meets production agriculture

Capuco, A.V., USDA-ARS, Bovine Functional Genomics Lab, BARC-E, Beltsville, MD 20705, USA; tony.capuco@ars.usda.gov

Mammary stem cells (MaSC) provide for net growth, renewal and turnover of mammary epithelial cells, and are therefore potential targets for strategies to increase production efficiency. Appropriate regulation of MaSC can potentially benefit milk yield, persistency, dry period management and tissue repair. Accordingly, we and others have attempted to characterize and alter the function of MaSC. Recent data indicate that MaSC retain labeled DNA for extended periods because of their selective segregation of template DNA strands during mitosis. Using long-term retention of bromodeoxyuridine-labeled DNA, we identified putative bovine MaSC. These label retaining epithelial cells (LREC) are in low abundance within mammary epithelium (<1%). They are predominantly estrogen receptor-negative and localized in the suprabasal layer of the epithelium throughout the gland. Thus, the response of MaSC to estrogen, the major mitogen in mammary gland, is likely mediated by paracrine factors released by cells that are estrogen receptor-positive. This is consistent with considerable evidence for cross-talk within and between epithelial cells and surrounding stromal cells. Microarray analyses suggest several likely paracrine factors including amphiregulin and insulin-like growth factor I. Excision of classes of cells by laser microdissection and subsequent microarray analysis will hopefully provide markers for MaSC and insights into their regulation. Preliminary analysis of gene expression in LREC and non-LREC are consistent with the concept that LREC represent a population of stem/progenitor cells. We have attempted to modulate MaSC number by infusing a solution of xanthosine through the teat canal and into the ductal network of the mammary gland of prepubertal heifers. The treatment increased the number of putative stem cells, as evidenced by an increase in the percentage of LREC and increased telomerase activity within the tissue. The exciting possibility that stem cell expansion can influence milk production is currently under investigation.

Epigenetic regulation of milk production in dairy cows

Singh, K.[1], Erdman, R.A.[2], Arias, J.A.[3], Molenaar, A.J.[1], Swanson, K.M.[1], Henderson, H.V.[1] and Stelwagen, K.[1], [1]AgResearch Limited, Ruakura Research Centre, Private Bag 3123, Hamilton 3240, New Zealand, [2]University of Maryland, Department of Animal and Avian Sciences, College Park, Maryland 20742, USA, [3]LIC, Private Bag 3016, Hamilton 3240, New Zealand; kuljeet.singh@agresearch.co.nz

A potential role of epigenetic mechanisms in manipulation of mammary function in the dairy cow is emerging. Epigenetics is the study of heritable changes in genome function that occur due to chemical changes rather than DNA sequence changes. DNA methylation is a stable epigenetic event which results in the silencing of gene expression. However, recent studies investigating different physiological states and changes in milk protein gene expression suggest that DNA methylation may also play an acute regulatory role in gene transcription. This overview will highlight the role of DNA methylation in the silencing of milk protein gene expression during mastitis and mammary involution. Moreover, environmental factors such as nutrition, physiological, and disease state (e.g. mastitis) may induce epigenetic modifications of gene expression. In dairy cows, it is well established that diverse environmental factors including nutrition influence milk production, with an adverse environment resulting in reduced milk production and energy stores. Results from a preliminary analysis indicate that changes in dam milk production and energy stores, which reflect her environment and nutritional status, are associated with changes in daughter milk production. It is difficult to determine if this is an exclusive epigenetic effect as these results require confirmation using independent dairy cattle populations. The current research investigating the possibility of in utero and cross-generational epigenetic modifications in dairy cows will be discussed. Understanding how the mammary gland responds to environmental cues provides a huge potential to not only enhance milk production of the dairy cow but also that of her daughter.

Epigenetic regulation of mammary gland function in rodents

Rijnkels, M.[1], Wang, L.[2], Freeman-Zadrowski, C.[1], Hernandez, J.[1], Potluri, V.[1], Admani, N.[1] and Li, W.[2], [1]Baylor College of Medicine, Pediatrics USDA/ARS CNRC, 1100 Bates street, Houston TX 77030, USA, [2]Baylor College of Medicine, Duncan Cancer Center, One Baylor Plaza, Houston TX 77030, USA; rijnkel@bcm.tmc.edu

Epigenetics can be defined as 'the stable alterations in gene expression potential that arise during development and proliferation' or as 'modifications of DNA or associated factors that have information content and are maintained during cell division, other than the DNA sequence itself'. The former is the consequence of the latter. Modifications of DNA and associated factors result in changes in chromatin organization. Chromatin regulates the access of regulatory complexes to the DNA and is involved in both acute transcriptional regulation and cell memory driving development and differentiation. The role of epigenetic regulation in mammary gland function has been studied in the past but only with recent new technologies are we starting to gather a more complete picture of epigenetic regulation of mammary gland function. A combination of cell culture and tissue analysis have shown that functional differentiation of MEC is associated with changes in epigenetic marks – nuclear organization, chromatin-nuclear matrix interactions, long-distant chromatin interactions, histone profiles, Dnase I hypersensitivity, and DNA methylation – in conjunction with the recruitment or displacement of trans-acting factors. In milk protein gene genomic regions these changes lead in general to a more open and permissive chromatin structure at these loci in the differentiated MEC. The fact that some of these marks are present at developmental stages when milk protein genes are not expressed yet suggests that they convey a cellular memory preparing these cells for their identity of fully functional mammary epithelial cells during lactation. We currently are expanding to genome wide analysis of epigenetic marks in the mammary gland, the preliminary analysis will be presented.

Milking-induced prolactin release has proliferative and survival roles on mammary epithelial cells in dairy cows

Boutinaud, M.[1], Angulo Arizala, J.[2], Bruckmaier, R.M.[3], Lacasse, P.[4] and Lollivier, V.[1], [1]INRA AGROCAMPUS OUEST, UMR1080 Production du lait, 35590 Rennes, France, [2]Universidad de Antioquia, Facultad de Ciencias Agrarias, medellin, Colombia, [3]University of Bern, Bremgartenstr. 109a, CH-3001 Bern, Switzerland, [4]AAFC, Dairy and Swine R&D centre, CP 90, J1M 1Z3, Sherbrooke, Canada; marion.boutinaud@rennes.inra.fr

Recently we demonstrated that the release of prolactin (PRL) induced at milking is galactopoietic in bovine. The role of PRL as a survival factor for mammary epithelial cells (MEC) has been clearly identified in rodents but has never been evidenced in ruminants. To assess the effect of the inhibition of PRL release in lactating dairy cows, nine Holstein cows were assigned randomly to treatments during 3 5-d periods: 1) daily i.m. injections of 2 mg of Quinagolide (a PRL release inhibitor), 2) daily i.m. injections of 2 mg of Quinagolide and twice a day (at milking time) i.v injections of PRL (2 µg/kg of body weight), 3) daily injection of vehicle as controls. MEC were purified from milk to assess their viability and mammary biopsies were harvested for immunohistological analyses of cell proliferation using PCNA staining. The data were analysed using the MIXED procedure of SAS where the cow and the treatment effects were treated. Quinagolide and PRL injection effect were tested using contrasts. Daily injections of Quinagolide reduced milking-induced PRL release ($P<0.05$) whereas the PRL injections mimicked the endogenous PRL release at milking. Quinagolide decreased milk production ($P<0.05$). Injections of PRL did not restore milk yield but tended ($P=0.09$) to increase protein yield. Injections of Quinagolide increased the number of MEC harvested from milk ($P<0.05$) whereas PRL injections tended to decrease it ($P=0.10$). PRL injections also increased the viability of MEC harvested from milk ($P<0.05$). Injections of Quinagolide decreased cell proliferation in mammary gland ($P<0.05$) whereas PRL injections increased it ($P<0.01$). In conclusion, PRL is a survival factor for bovine mammary epithelial cells.

Differences on sheep and goats milk fatty acid profile between conventional and organic farming systems

Tsiplakou, E., Kotrotsios, B., Hadjigeorgiou, I. and Zervas, G., Agricultural University of Athens, Animal Nutrition Physiology and Feeding, Iera Odos 75, GR 11855, Greece; eltsiplakou@aua.gr

The objective of this study was to investigate whether there is a difference in chemical composition and particularly in fatty acid (FA) profile, with emphasis on cis-9, trans-11 CLA, of milk obtained from conventional and organic dairy sheep and goat farms under the farming conditions practiced in Greece. Four dairy sheep (CS) and four dairy goat (CG) farms, representing common conventional production systems and another four dairy sheep (OS) and four dairy goat (OG) farms, organically certified, representing organic production and feeding systems were selected from all over Greece. One hundred and sixty two individual milk samples were collected from those farms in January-February 2009, about three months after parturition. The milk samples were analyzed for their main chemical constituents by IR spectrometry and their FA profile by gas chromatography. Farming system effects (CS vs. OS and CG vs. OG) on milk chemical composition and FA profile in each animal species was test by one-way ANOVA using SPSS statistical package (release 9.0.0). Post-hoc tests were performed using Duncan's multiple range test and significance was set at $P<0.05$. The results showed that the production system affected milk chemical composition: in particular fat content was lower in the organic sheep and goats milk compared with the corresponding conventional. Milk from organic sheep had a higher content in MUFA, PUFA, α-LNA, cis-9, trans-11 CLA and ω-3 FA, whereas in milk from organic goats α-LNA and ω-3 FA content was higher than that in the conventional ones. These differences are, mainly, attributed to different feeding practices used by the two production systems. In conclusion, the results of this study show that the organic milk produced under the farming conditions practiced in Greece has higher nutritional value, due to its FA profile, compared with the respective conventional.

Continuous lactation and mammary function

Collier, R.J., University of Arizona, Animal Sciences, 1650 E. Limberlost, Tucson, Arizona 85719, USA; rcollier@ag.arizona.edu

The dry period is required to facilitate cell turnover in the bovine mammary gland in order to optimize milk yield in the next lactation. Traditionally, an 8 week dry period has been a standard management practice for dairy cows based on retrospective analyses of milk yields following various dry period lengths. However, as milk production per cow has increased, transitioning cows from the non-lactating state to peak milk yield has grown more problematic. Currently, the transition period is known as the most dangerous period of the dairy cow's life-cycle. The majority of health problems in dairy cows occur in the first weeks of lactation and disease incidence in early lactation is associated with increased risk of culling. Given the size and extent of this health issue in the dairy industry it is proper to evaluate possible alternative management schemes for the dry and transition periods. Recent studies indicate a clear parity effect on dry period requirement. First parity animals require a 60 day dry period while lactations following later parities demonstrate no negative impact with 30 day dry periods or even eliminating the dry period when somatotropin is also used to maintain milk yields. Since greater parity animals are also more prone to transition problems they are candidates for reduced dry period management schemes. The majority of new intra-mammary infections occur during the dry period and persist during the following lactation. There is therefore, the possibility of altering mastitis incidence by modifying or eliminating the dry period in older parity animals. Since the composition of mammary secretions including immunoglobulin's may be reduced when the dry period is reduced or eliminated there is the possibility that immune status of cows during the peripartum period is influenced by the length of dry period.

The role of circadian rhythm in mammary function in the cow

Plaut, K. and Casey, T., Michigan State University, Animal Science, 1290 Anthony Hall, East Lansing, MI 28854, USA; kplaut@msu.edu

Environmental variables such as photoperiod, heat, stress, nutrition and other external factors have profound effects on quality and quantity of a dairy cow's milk. The way in which the environment interacts with genotype to impact milk production is unknown; however evidence from our lab suggests that circadian clocks play a major role. Daily and seasonal endocrine rhythms are coordinated in mammals by the master circadian clock in the hypothalamus. Peripheral clocks are distributed in every organ and coordinated by signals from the master clock. We and others have shown that there is a circadian clock in the mammary gland. Approximately 7% of the genes expressed during lactation had circadian patterns including core clock and metabolic genes. Amplitude changes occurred in the core mammary clock genes during the transition from pregnancy to lactation and were coordinated with changes in molecular clocks among multiple tissues. *In vitro* studies using a bovine mammary cell line showed that external stimulation synchronized mammary clocks, and expression of the core clock gene, BMAL1, was induced by lactogens. Female Clock mutant mice, which have disrupted circadian rhythms, have impaired mammary development and their offspring failed to thrive suggesting that the dam's milk production was not adequate enough to nourish their young. We envision that, in mammals, during the transition from pregnancy to lactation the master clock is modified by environmental and physiological cues that it receives, including photoperiod length. In turn, the master clocks coordinate changes in endocrine milieu which signals peripheral tissues. In dairy cows, it is clear that changes in photoperiod during the dry period and/or during lactation influences milk production. We believe the photoperiod effect on milk production is mediated, in part by the 'setting' of the master clock with light, which modifies peripheral circadian clocks including the mammary core clock and subsequently impacts milk yield and may impact milk composition.

Cell-based models to test bioactivity of milk-derived bioactives

Purup, S. and Nielsen, T.S., Faculty of Agricultural Sciences, Aarhus University, Animal Bioscience and Health, Blichers Allé 20, 8830 Tjele, Denmark; stig.purup@agrsci.dk

The life science industries have a strong interest in screening for novel bioactives in complex mixtures like milk and dairy products. Food bioactives are not only important for public health in general, but also have potential therapeutic applications for the treatment of a number of diseases. To identify these novel bioactives, establishment of robust screening assays is essential. The use of *in vitro* cell-based models for screening and testing have the advantage, that several concentrations of mixtures or specific compounds can be assayed at the same time in cells from specific tissues. Primary cell cultures from target organs or established cell lines can be used to identify the most sensitive cells. In addition, a large number of transfected cell lines with very specific sensitivities have been developed. Different endpoints inherent to basal or more sophisticated cellular functions can be investigated, such as cell viability, apoptosis, migration, intracellular signalling, regulation of gene expression, morphology and metabolic alterations. The gastro-intestinal tract is an obvious target for bioactive molecules delivered through milk and dairy products, because it lies at the interface between dietary components in the lumen and the internal processes of the host. Identification of bioactive factors that affect proliferation or migration of epithelial cells may have potential applications in promoting gastrointestinal health in both humans and animals. The mammary gland is another target organ of considerable interest in relation to milk bioactives since it has been estimated that dietary factors contribute to the etiology of 30-50% of all breast cancers. A large number of gastro-intestinal and mammary epithelial cell lines are commercially available, but in order to study some cellular functions, primary cultures of freshly isolated cells are often preferred, since established cell lines do not always express specialised properties in culture.

Host-defence related proteins in cows' milk

Wheeler, T.T., Haigh, B.J., Molenaar, A.J. and Stelwagen, K., AgResearch, Dairy Science and Technology, East Street, 3240, New Zealand; tom.wheeler@agresearch.co.nz

Milk is a source of bioactive molecules with wide ranging functions, relating to areas such as cancer, bone, cardiovascular and gut health, and immune function. Among these, the immune properties have been the best characterized. Traditionally, this function has been considered to be mediated predominantly by the immunoglobulins in milk. However, it is now realised that milk also contains a range of minor proteins that collectively form a significant first line of defence against pathogens, acting both within the mammary gland itself as well as in the digestive tract of the suckling neonate (e.g. lactoferrin, lactoperoxidase, complement proteins and antimicrobial peptides). We have used proteomics technologies to characterise the repertoire of host-defence related milk proteins in detail, revealing at least 100 distinct gene products in milk, of which at least 15 are known host-defence related proteins. Those having intrinsic antimicrobial activity likely function as effector proteins of the local mucosal immune defence (e.g defensins, cathelicidins, and the calgranulins). Among these, we have investigated the activities and biological role of the milk cathelicidins, as well as some antimicrobial proteins not previously described in milk, mammary serum amyloid A and histatherin. The function is much less clear for the immune-related milk proteins that do not have intrinsic antimicrobial activity. We hypothesise that at least some of these facilitate recognition of microbes, resulting in activation of innate immune signalling pathways in cells associated with the mammary and/or gut mucosal surface. These investigations are elucidating how an effective first line of defence is achieved in the bovine mammary gland and how milk contributes to optimal digestive function in the suckling calf. This work will contribute to understanding the balance between commesal and pathogenic microbes in the mammary gland and gut, and ultimately may lead to development of immune-related milk bioactives as high value ingredients from milk.

Udder morphometry in Gir cows and the subclinical mastitis prevalence

Porcionato, M.A.F., Reis, C.B.M., Soares, W.V.B., Cortinhas, C.S., Barreiro, J.R. and Santos, M.V., School of Veterinary Medicine and Animal Science, Department of Nutrition and Animal Production, Duque de Caxias Norte, 225, 13635-900 Pirassununga-SP, Brazil; mveiga@usp.br

The anatomical characteristics of the mammary gland of dairy cattle are not equal for all breeds, so that the udder and teats morphology may favor the individual or herd performance. For example, the depth udder measurements and teat size are closely related to the resistance to mastitis. Because of these differences, some anatomical measurements of mammary glands, obtained by ultrasound, have been used in research with dairy cattle. The ultrasound techniques used to study udder and teat cisterns appear as a minimally invasive tool and result in important information for genetic selection, the decision of management to be adopted or the prevention of mastitis. This trial aimed to evaluate the relation between morphological teat characteristics of Gir cows with ultrasound and subclinical mastitis prevalence. Eighty lactating Gir cows with 90 to 200 days of the 2nd or 3rd lactation were grouped in accordance with the milk flow: fast or slow and milked twice a day with a mechanical milker. Data were analyzed using SAS (version 8.2). Differences were considered significant at $P<0.05$. The ultrasonography images showed a higher ($P<0.05$) teat channel in flow slow (25.68 mm) than flow fast (22.31 mm) groups. No significant correlation ($P>0.05$) was observed between LogSCC and morphological teat characteristics. The infrared thermography technique was used to evaluate the udder temperature variation in cows with subclinical mastitis, but no differences ($P>0.05$) were observed to microorganism's type or LogSCC. The channel length and distance for the teat to floor had an influence on the prevalence of subclinical mastitis, as well as the mastitis causing pathogens in Gir cows.

Integrated immunological and metabolic responses of the mammary gland to LPS challenge in bovine

Silanikove, N., Rauch-Cohen, A. and Leitner, G., Agr. Res. Org., Biology of lactation Lab., P.O. Box 6, 50 250, Bet Dagan, Israel; nsilanik@agri.huji.ac.il

The aim of this study was to test the hypothesis that the diverse immune and metabolic responses of the mammary gland to LPS challenge express an integrated mechanism that fights against Gram negative bacteria. The present study confirmed that a LPS challenge is associated with a NO burst from surrounding mammary epithelial cells and that the consequential nitrosative stress was induced by formation of nitrogen dioxide from nitrite by lactoperoxidase. This response, most likely, plays a major role in fighting against bacterial invasion during acute mastitis, but is also associated with increased nitration of milk proteins. We confirmed the involvement of xanthine oxidase as a hydrogen peroxide generator and of catalase as a nitrosative stress modulator by oxidizing nitrite to nitrate in the mammary gland secretion. The plasmin system appears to play a pivotal role in inducing down-regulation of milk secretion and mammary gland epithelial cell metabolism. The acute shift of the mammary gland epithelial cell metabolism from mitochondrial oxidative respiration to cytosolic glycolysis, as reflected in a transient sharp increase in milk lactate concentration and a transient inverse decrease in milk citrate concentration, allows diversion of metabolic resources needed to produce milk to support the immune system. In turn, the acute increase in the concentration of lactate and malate in milk and reduction in lactose concentration were shown to be important in controlling *E. coli* growth. We found that a LPS challenge increases transiently lactate secretion and conversely decreases lactose secretion into the medium of mice HC11 cell line, which extent the general validity of our results.

Phytoestrogens found in milk inhibit COX enzyme activity in murine macrophages

Purup, S., Nielsen, T.S. and Hansen-Møller, J., Faculty of Agricultural Sciences, Department of Animal Health and Bioscience, Blichers Allé 20, 8830 Tjele, Denmark; stig.purup@agrsci.dk

Phytoestrogens are a large group of naturally occurring non-steroidal plant-derived compounds with potential beneficial effects on health, as indicated in epidemiological studies and experimental data from animal studies. We have previously documented that bovine milk contains various phytoestrogens, especially the isoflavone equol, when cows are fed high amounts of especially clover. Because long-term ingestion of foods rich in phytoestrogens and long-term ingestion of non-steroidal anti-inflammatory drugs (NSAIDs), have been associated with reduced cancer incidence in humans, the anti-inflammatory activity of individual phytoestrogens found in bovine milk were examined. Phytoestrogens included equol, formononetin, daidzein, biochanin A and genistein. Cells of the murine macrophage cell line RAW 264.7 were treated with lipopolysaccharide (LPS; 50 ng/ml) or peptidoglycan (PGN; 30 µg/ml) and either phytoestrogen at 1, 10 and 40 µM or vehicle (DMSO or DMSO/ethanol) in duplicate. After incubation for 16 hours thromboxane B2 (TrxB2) and prostaglandin E2 (PGE2) concentrations were measured in the culture media by ELISA. In a separate assay cell viability was measured by a resazurin metabolism assay. All phytoestrogens inhibited the LPS and PGN stimulated response of TrxB2 and PGE2. Formononetin was the most potent inhibitor. Maximal inhibition of all phytoestrogens at 40 µM corresponded to 74-100% inhibition of the increase in TrxB2 and PGE2. The viability of RAW 264.7 cells was not affected by any of the five phytoestrogens. In conclusion, the results demonstrate that the phytoestrogens that can all be found in cow´s milk, have the potential to contribute to a COX-inhibitory effect. Future studies will show if milk containing a full spectrum of different phytoestrogens can provide an even stronger anti-inflammatory effect than single phytoestrogens.

Session 01 Poster 13

Milk and blood cholesterol levels during the lactating cycle

Viturro, E., Schlamberger, G., Wiedemann, S., Kaske, M. and Meyer, H.H.D., Physiology Weihenstephan, Weihenstephaner Berg 3, 85354 Freising, Germany; viturro@wzw.tum.de

Milk and dairy products are the second principal source of cholesterol in the modern western diet. During the last years, we have initiated the in-depth study of milk cholesterol level regulation and transport into milk, aiming to establish the knowledge basis necessary for future attempts on reduction of the levels of this metabolite in dairy products. The present data focuses on the study of the physiological levels of cholesterol during the lactation cycle and the correlation between them. Milk samples were obtained weekly from 16 Brown Swiss cows (4.6±1.4 years) fed a standard diet and free of clinical problems. Milk samples were analyzed for major composition (% Fat, Protein, Lactose) and for cholesterol content. Blood samples were collected at week -4, -2, and -1 before expected calving, within 24 h postpartum (wk 0), and at week 1, 2, 4, 6, 8, 12, 16, 24, 36, 24 and 44 post-partum. Samples were analyzed for their total cholesterol, LDL- and HDL-cholesterol and triglyceride concentrations. During the lactation cycle, milk cholesterol content decreases gradually from the initial mean of 0.9 mM to a mean basal level of 0.4 mM during the final third of the cycle. However, blood cholesterol levels increase at a constant rate after week 4 post partum (from 3.5 to 6 mM at the end of the lactation). A huge interindividual variability on milk cholesterol levels was observed among animals with similar blood cholesterol concentration or at the same lactation week. Evenmore, milk cholesterol levels were not dependent on milk fat content. The presented data lead to establish the hypothesis that the transfer of cholesterol into milk might be a specific active transport process, as milk cholesterol levels are not directly related neither with blood levels of the metabolite nor with the milk fat content. The extensive interindividual variations observed also support this hypothesis and open wide intervention roads for a near future.

Session 01 Poster 14

Evaluation of the cytotoxic effects induced by ochratoxin A in a bovine mammary cell line

Baldi, A., Fusi, E., Rebucci, R., Pecorini, C., Pinotti, L., Saccone, F. and Cheli, F., Università degli Studi di Milano, Department of Veterinary Science and Technology for Food Safety, via Trentacoste 2, 20134 Milano, Italy; antonella.baldi@unimi.it

Ochratoxin A (OTA) is a mycotoxin that occurs in several agricultural products and causes diseases both in humans and animals. OTA can act in different ways by inducing cytotoxicity, oxidative cell damage and increased cell injury. The aim of the study was to evaluate the damage induced by OTA in a bovine mammary cell line (BME). The effects of OTA treatments on cell viability and membrane integrity were determined using MTT test and LDH release, respectively. BME cultures were exposed to increasing concentrations of OTA (0-10 µg/ml) up to 24 h. In order to detect the DNA damage the diphenylamine method, ladder formation by agarose gel electrophoresis and TUNEL assays were performed. BME cells were seeded in chamber slides and cultured for 24 hours in complete medium. Based on results obtained with previous assays, the appropriate dose was added to each chamber slide for the following 12 and 24 hours. Obtained data were analysed by one-way ANOVA, with $P \leq 0.05$ considered statistically significant. At all incubation times, the BME cells were sensitive to OTA cytotoxicity (LC_{50} = 0.8 µg/ml after 24 h). LDH release by mammary epithelial cells in the presence of several concentrations of OTA at 24 and 48 h of incubation were significantly increased ($P<0.01$). DNA fragmentation and DNA ladder formation were found to be dose dependent. The dark-brown apoptotic nuclei in BME cells, co-incubated for 12 h with 0.6 µg/ml of OTA, were only 4-6% of the observed cells. After 24 h of OTA incubation above LC_{50}, monolayers were completely destroyed and cell debris invaded all microscope fields. In BME cells OTA demonstrated to be able to induce cellular damage, LDH release and DNA fragmentation already at low concentration. The mechanisms by which OTA induces its toxicity, which process dominates (necrosis or apoptosis) depends the concentration/dosage and the time-exposure used.

Study of udder morphology in different stages of lactation in dairy ewes

Milerski, M., Research Institute of Animal Science, Přátelství 815, 104 00 Prague - Uhříněves, Czech Republic; milerski.michal@vuzv.cz

Good and homogenous udder morphology of dairy ewes is desirable for good milkability and udder health, especially if machine milking is in use. For that reason udder characteristics need to be included in recording and selection schemes for dairy sheep, beside the milk production and quality traits. The main aim of this study was to determine the optimal time during milking period for udder measurements and linear assessments. Investigation was carried out on 92 Lacaune ewes, which were examined five times during milking period in month intervals. Udder height (UH), udder width (UW) and teat length (TL) were measured, teat position (TP) was subjectively assessed by linear scoring using 5-point scale and cistern cross section areas were determined on the basis of ultrasound scanning. On the same day, the daily milk yield, fat and protein content were recorded. Average UH during the whole milking period was 14.24 cm (s.d. 2.42 cm), UW 13.60 cm (s.d. 1.83 cm), TL 2.92 cm (s.d. 0.53 cm) and average sum of areas of both cisterns was 52.16 cm^2 (s.d. 20.20 cm^2). Milk yield on the day of observation was significantly correlated with UH (r=0.62); UW (r=0.70); TL (r=0.36) and cistern areas scanned by ultrasound technique (r=0.81), while the correlation with linear scoring of TP was close to zero (r=0.04). It is advisable to perform udder measurements and linear assessments once in the first three months of milking period.

Session 01 Poster 16

Marker selection for udder health alters the cytokine secretion of bovine mammary epithelial cells after bacterial infection

Kliem, H.[1], Grandek, M.[1], Griesbeck-Zilch, B.[1], Kühn, C.[2] and Meyer, H.H.D.[1], [1]Physiology Weihenstephan, Technische Universität München, Weihenstephaner Berg 3, 85354 Freising, Germany, [2]Research Unit Molecular Biology, Research Institute for the Biology of Farm Animals (FBN), Wilhelm-Stahl-Allee 2, 18196 Dummerstorf, Germany; heike.kliem@wzw.tum.de

A new strategy to improve udder health is the selection of less susceptible cows using genetic markers for quantitative trait loci (QTL) on chromosome 18, which seems to influence the somatic cell count in milk. We established a cell culture model with bovine primary mammary epithelial cells from 28 Holstein-Friesian cows either marker or conventionally selected to investigate the activation of the innate immune system after an infection for 1, 6, 24 hours (h) with *Staphylococcus aureus* (*S. aureus*) and *Escherichia coli* (*E. coli*). We evaluated the protein concentration of bovine Interleukin-8 (IL-8) in cell culture supernatant. Results: A comparison between *E. coli* and *S. aureus* treatment revealed a significant increase of IL-8 after 24 h for both bacteria. Furthermore the IL-8 protein concentration at 24 h was significantly higher after *E. coli* compared to *S. aureus* treatment. A further comparison was conducted between high (q) or low (Q) susceptibility cows to mastitis in the marker assisted (MAS) group and the conventionally (CON) selected group. The results after *E. coli* treatment demonstrate a higher protein concentration of IL-8 in the MAS compared the CON group. A significant increase of IL-8 was found in the CON-Q compared the CON-q group. The protein concentration of IL-8 after S. aureus infection was also higher in the MAS group and revealed a significant higher amount in the MAS-Q compared to the MAS-q group. Conclusion: Breeding cows after marker assisted selection seems to be a promising method against mastitis, as these animals exhibit a higher IL-8 secretion and a better immune response than conventionally selected cows.

Session 01 Poster 17

Cortisol release after ACTH administration, weaning and first milking: relations between stress and behavior

Negrao, J.A.[1], Rodrigues, A.D.[2] and Stradiotto, M.M.[1], [1]USP/FZEA, University of São Paulo, Basic Science Department, Av. Duque de Caxias norte, 225, 13630-970, Pirassununga/SP, Brazil, [2]UNESP/ FCAV, University of São Paulo State, Animal Production Department, Via Prof. Paulo Donato Castellane, 14884-900, Jaboticabal/SP, Brazil; jnegrao@usp.br

Some studies have suggested that direct relationships exist between ACTH, CORT and discomfort levels caused by parturition. However maternal interactions between ewes and lamb following parturition can influence ewe behaviour in the milking parlor. The objective of this study was to investigate behavioural changes in Santa Ines ewes after imposition of two stressor agents (weaning and first milking) and exogenous ACTH administration as well as the relationships between behavioral and cortisol. Sixty ewes were submitted to ACTH administration, weaning and first milking machine. Ewes were in full view throughout the experiment and recordings started immediately after parturition and imposed stress, with general activity patterns of each cow videotaped for 72 h. Compared to baseline, cortisol increased significantly after ACTH administration and first milking to reach their maximal levels at 60 min; however cortisol levels measured at 300 min not differ from basal levels. In contrast, cortisol increased significantly after weaning to reach their maximal levels at 300 min. Cortisol response was significantly lower for weaning and first milking than after ACTH administration. Regarding ewes behavior in the milking parlor the ewes were classified as calm, agitated and very agitated in the milking parlor. Baseline cortisol concentration were higher ($P>0.05$) in the very agitated ewes, however area under the cortisol response curve were lower ($P>0.05$) in the very agitated cows when compared with calm cows. During experimental milking, very agitated ewes were more difficult to milk and spent more time in the milking parlor. This indicates that it is possible identify calm ewes to improve their adaptation to weaning.

Session 01 Poster 18

Responses of Holstein cows to ACTH administration: relations between behavior and cortisol levels

Negrao, J.A.[1], Porcionato, M.A.F.[1], Passillé, A.M.[2] and Rushen, J.[2], [1]USP/FZEA, University of São Paulo, Basic Science Department, Av. Duque de Caxias norte, 225, Pirassununga, SP 13630-970, Brazil, [2]Agriculture and Agri-Food Canada, Pacific Agri-Food Research Centre, 6947 Highway 7, Agassiz, BC, V0M 1A0, Canada; jnegrao@usp.br

In ruminants a variety of factors can cause stress and influence behaviour, and imposed stress in cattle has been associated with increased motor activity, low food intake, incomplete milk ejection, and high levels of CORT. The objective of this study was to investigate behavioural changes in Holstein cows caused by exogenous ACTH administration as well as the relationships between behavioral and cortisol. Forty multiparous cows were classified as calm, agitated and very agitated in the milking parlor, and subsequently submitted to one saline or ACTH injection. Cows were in full view throughout the experiment and recordings started immediately after ACTH and saline administration with general activity patterns of each cow videotaped for 24 h. Baseline cortisol concentration were higher ($P>0.05$) in the very agitated cows, however area under the cortisol response curve were lower ($P>0.05$) in the very agitated cows when compared with calm cows. Behavioural results during the first 2 experimental hours showed that cows were less active and spent more time lying following ACTH than after saline treatment ($P<0.05$). Also, cows spent significantly less time ruminating immediately following ACTH injection ($P<0.05$). However, feed intake measured after 4 and 24 h were similar for both treatments ($P>0.05$). There was no significant influence of ACTH treatment on frequency or duration of behaviours during the 4 and 24 h periods following injection ($P>0.05$). Consequently, ACTH can be used to measure reactivity of the adrenal cortex without involving biologically significant consequences, and the relations between behaviour and cortisol suggest that both can be used to study the cow temperament.

Gene expression of NPC1 is down regulated in the bovine mammary gland following dietary supplementation with sunflower seeds

Møller, J.W., Theil, P.K., Sørensen, M.T. and Sejrsen, K., Aarhus University, Agr. Science, Blichers alle, 8830 Tjele, Denmark; jeppew.moller@agrsci.dk

Supplementing cow's diet with unsaturated fatty acids can increase the content of unsaturated fatty acids in milk. One such supplement is sunflower seeds, which contain high proportions of linoleic acid (LA). LA increases the activity of sphingomyelinase in blood, which breaks down sphingomyelin to ceramide and phosphocholine. Ceramide is metabolized in the lysosomes to sphingosine, which subsequently is phosphorylated to sphingosine-1-phosphate, a potent activator of SREBP1, the key regulator of de novo fatty acid synthesis in the mammary gland. The efflux of sphingosine from lysosomes is mediated by NPC1 protein, and decreased NPC1 activity is associated with increased lysosomal sphingosine-1-phosphate levels. The objective was to investigate the effect of sunflower seed supplementation (SFS) on the mammary gene expression of NPC1, SREBP1, and SCAP, the activator of SREBP1. Twenty four lactating cows (186±20 days in milk (DIM), 25.3±2.5 kg/d) were blocked according to parity and DIM. Cows were allocated to one of four groups fed either control diet or diets supplemented with 5, 10 or 15% sunflower seed (% of DM) for five weeks. Content of fatty acids ≤C16 was reduced from 75% to 38% by increasing SFS, indicating reduced de novo synthesis of fatty acids. mRNA abundance was determined by RT-PCR on mammary biopsies. SFS decreased mRNA levels of NPC1 ($P<0.001$), SREBP1 ($P=0.034$), and SCAP ($P=0.0075$). The reduced expression of NPC1 may be a consequence of reduced sphingosine levels in the mammary gland. A reduced sphingosine level may via an inhibition of SREBP1 and SCAP also be involved in reducing de novo fatty acid synthesis.

Session 01 Poster 20

mRNA abundance of the components of the adiponectin system in adipose tissue and in liver of dairy cows supplemented with or without conjugated linoleic acids throughout lactation

Vorspohl, S.[1], Mielenz, M.[1], Pappritz, J.[2], Dänicke, S.[2] and Sauerwein, H.[1], [1]Insitute of Animal Science, Katzenburgweg 7-9, 53115 Bonn, Germany, [2]Federal Research Institute for Animal Health, Bundesallee 50, 38116 Braunschweig, Germany; sauerwein@uni-bonn.de

Adiponectin (Ad) is an adipokine related to lipid metabolism and insulin sensitivity. Activation of the Ad receptors AdR1 and AdR2 increases insulin sensitivity and decreases inflammation. Early lactation results in negative energy balance associated with reduced insulin sensitivity. Supplementing diets with conjugated linoleic acids (CLA) decreases milk-fat and might thereby reduce the energy requirements for milk production. We aimed to investigate the mRNA expression of Ad, AdR1 and AdR2 during an entire lactation thereby considering potential lactation stage effects of CLA. Holstein Frisian cows were divided into a control group (n=10) and a CLA group (n=11, receiving 10 g each of the cis-9,trans-11- and the trans-10,cis-12-CLA isomers per day from d 1 post partum until d 182). Biopsies were collected from subcutaneous fat (ScF) and from liver at d -21 and d 1, 21, 70, 105, 182, 196, 224, 252 relative to calving. The mRNAs of Ad, AdR1 and AdR2 were quantified by real-time RT-PCR (for the CLA group only d -21, d 21, 105, 196 and 252). Data were analysed with a general linear model or nonparametric tests ($P\leq0.05$). The mRNA abundance of Ad in scF and of its receptors in liver was not different between CTR and CLA. Ad mRNA was decreased in fat samples from d 21, 196, and 252 compared to d -21. AdR1 in liver was higher on d 105 and 169 than on d -21, and lower on d 252 than on d 105. AdR2 values were higher on 21 d in comparison to all other time points; on d -21, the AdR2 content was also lower than on d 105. We herein provide a comprehensive longitudinal study about lactation-related changes in mRNA abundance of the 3 components of the Ad system. The mRNA expression seems unaltered by CLA indicating that CLA does not affect the Ad system under the conditions investigated.

Effects of sunflower seed supplementation on mammary expression of genes regulating sphingomyelin synthesis

Møller, J.W., Theil, P.K., Sørensen, M.T. and Sejrsen, K., Aarhus University, Agr. Science, Blichers alle, 8830 Tjele, Denmark; jeppew.moller@agrsci.dk

Sphingomyelin (SM) is a phospholipid that stabilizes eukaryotic cell membranes. Dietary SM can inhibit cholesterol absorption in the gut and lower plasma triglyceride and cholesterol levels in humans. SM is a component of the milk fat globule membrane (MFGM), and dairy products are a major source of human dietary SM. De novo synthesis of SM in mammals is initiated by Serine C-palmitoyl Transferase (SPTLC) using serine and C18 fatty acids to produce dihydro-sphingosine (DS). LASS2 utilizes DS to produce ceramide, which is a substrate for SGMS2 (SM synthesis) and UGCG (glucosyl-ceramide synthesis). The fatty acid part of SM is characterized by a high degree of saturation, which is important for membrane stability and cholesterol binding. However, dietary unsaturated fatty acids can modify the fatty acid composition of SM. The objective was to examine the effect of increased sunflower seed supplementation (SFS) on mammary expression of genes associated with SM metabolism. Twenty four cows (186±20 DIM; 25.3±2.5 kg/d) were randomly assigned to four groups and fed a control diet or diets supplemented with 5, 10, or 15% sunflower seed for five weeks. Dry matter intake and milk yield was reduced in cows fed 10 and 15% SFS but not in cows fed 5% SFS, when compared to controls. All levels of SFS tended to increase milk fat content (P=0.08) and the level of unsaturated fatty acid in milk increased from 25% to 40%. Gene expression was analyzed by RT-PCR in mammary biopsies using the ΔΔCT method. 10% and 15% SFS reduced mRNA abundance of subunit 2 of SPTLC (SPTLC2, P=0.001), while expression of LASS2 tended to be reduced (P=0.07). SGMS2, UGCG, and SPTLC1 were not affected by treatment. Our results show that mammary expression of the SPTLC2 is reduced when cows are fed 10% and 15% SFS. This regulation may serve as a protective mechanism to prevent incorporation of unsaturated fatty acids into sphingomyelin which otherwise may destabilize the MFGM.

Identification of genes regulating milk β-carotene content

Berry, S.D.[1], Davis, S.R.[1], Beattie, E.M.[1], Thomas, N.L.[1], Burrett, A.K.[1], Ward, H.E.[1], Ankersmit-Udy, A.E.[1], Van Der Does, Y.[2], Macgibbon, A.H.K.[2], Spelman, R.J.[3], Lehnert, K.[1] and Snell, R.G.[1], [1]ViaLactia Biosciences, P.O. Box 109-185, Auckland 1149, New Zealand, [2]Fonterra Research Center, Private Bag 11029, Palmerston North, New Zealand, [3]LIC, Private Bag 3016, Hamilton 3240, New Zealand; sarah.berry@vialactia.com

Our aim was to discover genes regulating milk β-carotene content to allow the differentiation of milk using genetic selection. Using a Friesian-Jersey crossbred herd (n=850), founded by six F1 bulls, β -carotene was measured at peak, mid and late lactation. QTL were identified on BTA15, 17, and 18. Each QTL was significant at each lactation stage. Candidate genes were identified: β-carotene oxygenase 2 (BCO2, on BTA15); scavenger receptor class B, member 1 (SCARB1, on BTA17); and β-carotene oxygenase 1 (BCMO1, on BTA18). The coding regions were sequenced, along with 2kb of 5' sequence. Single nucleotide polymorphisms (SNPs) were genotyped across the trial pedigree to determine whether these explained the QTL. For BCO2, a G>A mutation, causing a premature stop codon, was discovered in exon three. BCO2 cows homozygous for this mutation produced milk with 78% more β-carotene than homozygous wild type animals. In addition, quantification of the BCO2 mRNA in liver identified an eQTL which collocated with the milk β-carotene QTL on BTA15. In BCMO1, three SNPs were discovered. The most striking of these, N341D, resulted in a 32% increase in milk β-carotene. The mRNA expression of BCMO1 was not affected by genotype. In SCARB1, one SNP was discovered, in the 5' regulatory region (-321 bp relative to the +1 translation start site), which resulted in a 10% increase in milk β-carotene content. The mRNA expression of SCARB1 was significantly affected by SCARB1 genotype. Thus, milk β-carotene may be decreased or increased using genetic selection, allowing the differentiation of milk for specific industrial applications, including the production of bovine milk enriched for β-carotene to alleviate vitamin A deficiency in humans.

Milking frequency and milk production in pasture-based lactating dairy cows

Rius, A.G., Kay, J.K., Phyn, C.V.C., Morgan, S.R. and Roche, J.R., DairyNZ, Animal Science, Private Bag 3221, Hamilton, New Zealand; john.roche@dairynz.co.nz

The objective of this study was to test the effect of modified milking frequency (MF) during early lactation on milk production in grazing dairy cattle. Multiparous Holstein-Friesian cows (n=150) were randomly assigned to one of five treatments at parturition: milked once daily (1X) for 21 d (1X21), milked 1X for 42 d (1X42), milked twice daily (2X), milked thrice daily (3X) for 21 d (3X21), and milked 3X for 42 d (3X42). All cows were milked 2X post treatment until wk 24 in lactation. Animals were offered a generous allowance of fresh pasture and supplemented with 4 kg DM/d of concentrate during the first 16 wk in milk and 2 kg DM/d for 8 wk thereafter. Effects of MF, duration of MF, and interactions during treatment and post treatment periods were tested using mixed models (GenStat 12.1). During the treatment period, a MF x duration interaction was detected for milk, protein, and fat yields. Relative to 3X21, 3X42 failed to increase milk production further. However, 1X42 had lower ($P<0.05$) milk (2.4 kg/d), protein (0.10 kg/d), and fat (0.12 kg/d) yields compared with 1X21 during the treatment period. Relative to 2X, 3X cows produced more milk (1.5 kg/d; $P<0.05$); however, protein and fat yields were not different during or after the treatment period. There was no MF x duration interaction post treatment. An adverse effect in production occurred for 1X in the post treatment period; however, 3X cows failed to sustain increased production compared with 2X. Relative to 2X, 1X cows had lower yields of fat (0.1 kg/d; $P<0.01$) and protein (0.05 kg/d; $P<0.05$) post treatment. Body weights were reduced in 2X cows compared with 1X during the treatment (476 vs. 484 kg; $P<0.05$) and post treatment periods (500 vs. 512 kg/d; $P<0.01$). In summary, 1X for the first 21 or 42 DIM impaired milk production and the losses continued for the remainder of the lactation. Relative to 2X, 3X in early lactation did not improve milk production beyond the period of increased milking frequency.

Haptoglobin mRNA expression in bovine adipose and liver tissue: physiological and conjugated linoleic acids (CLA)-induced changes throughout lactation

Saremi, B.[1], Mielenz, M.[1], Vorspohl, S.[1], Dänicke, S.[2], Pappritz, J.[2] and Sauerwein, H.[1], [1]Insitute of Animal Science, Katzenburgweg 7-9, 53115 Bonn, Germany, [2]Federal Research Institute for Animal Health, Bundesallee 50, 38116 Braunschweig, Germany; bsaremi@uni-bonn.de

Haptoglobin (Hp) is the most relevant acute phase protein in cattle, secreted mainly by the liver. Hp in mammary gland and neutrophiles is reported for cattle and has importance in local inflammatory processes. In rats, supplementation with CLA affects Hp in serum, but not Hp mRNA abundance in liver. For dairy cows, mRNA expression (exp) of Hp in liver has not yet been comprehensively characterized during an entire lactation period; for adipose tissue (AT) no data were available until now. We thus aimed to investigate Hp mRNA exp in subcutaneous fat (ScF) and liver during lactation, considering additionally potential effects of CLA. Holstein Frisian cows were grouped as control (n=10) or CLA (n=11, receiving 10 g each of the cis-9,trans-11- and the trans-10,cis-12-CLA isomers from d 1 in milk (DIM) until d 182). Biopsies were collected from ScF and liver at -21 and 1, 21, 70, 105, 182, 196, 224, 252 DIM. Hp mRNA was quantified by real-time RT-PCR (for CLA only d -21, d 21, 105, 196 and 252). Data were analyzed using repeated measurement analysis (SAS 9.2, $P<0.05$). At calving, Hp mRNA abundance in liver increased ($P<0.05$) from prepartal values by a factor of 9 and then decreased again to precalving values. Between DIM 182 and 196, Hp mRNA increased again [5.3 fold, $P<0.05$]. In AT, the amount of Hp mRNA was 2×10^6 fold lower than in liver and was detectable in 49% of the samples only. No time-related differences were observed throughout lactation for Hp mRNA in AT and CLA did neither affect liver nor AT Hp mRNA abundance. The peak in liver Hp mRNA after calving corresponds to Hp serum concentrations and confirms the liver as main source for systemic acute phase reactions. In contrast, the role of Hp exp in AT might rather be related to local inflammatory reactions and requires further investigation.

mRNA expression of genes related to lipogenesis and lipolysis in adipose tissue and plasma leptin in dairy cows during the dry period and in early lactation

Sadri, H.[1,2], Van Dorland, H.A.[2] and Bruckmaier, R.M.[2], [1]Department of Clinical Science, Faculty of Veterinary Medicine, University of Tabriz, 5166616471, Iran, [2]Veterinary Physiology, Vetsuisse Faculty, University of Bern, CH-3001, Switzerland; sadri.ha@gmail.com

Adipose tissue is a key player in the regulation of lipid metabolism according to the physiological status of the animal. Lipid metabolism is regulated by three major biochemical sites: fatty acid uptake, lipogenesis and lipolysis. The objective of this study was to investigate the changes in plasma leptin concentration in parallel to adipocyte regulation of genes related to lipogenesis and lipolysis in dairy cows from prepartum to early lactation. Subcutaneous fat biopsies were taken at the tail head from 27 dairy cows in week 8 antepartum (AP), on day 1 (D1) and in week 5 postpartum (PP). Blood samples were collected every two weeks during this period, and analyzed for concentrations of leptin. Adipose tissue samples were analyzed by real-time RT-PCR for mRNA encoding for leptin, hormone-sensitive lipase (HSL), perilipin (PLIN), lipoprotein lipase (LPL), acyl-CoA synthase long-chain family member 1 (ACSL1), acetyl-CoA carboxylase (ACC), and fatty acid synthase (FASN). Data were analyzed by ANOVA using the MIXED procedure of SAS. The mRNA abundance of ACC was lower on D1 than in AP ($P<0.05$). The mRNA encoding FASN was lowest in abundance on D1, followed by PP, and highest in AP ($P<0.05$). Mean plasma concentrations of leptin were lower on D1 and in PP compared with AP ($P<0.05$). There were no significant differences observed for mRNA abundance of Leptin, HSL, PLIN, LPL, and ACSL1 over time. The reduced mRNA abundance of key genes involved in the de novo synthesis of fatty acids in the adipose tissue confirms an attenuation of fatty acid synthesis around parturition. The observed hypoleptinemia during the postpartum period would be in favor of increasing metabolic efficiency and energy conservation, for mammary function and reconstitution of body reserves.

Stage of lactation alters metabolic profiles in blood and liver of cows subjected to feed restriction

Moyes, K.M., Bjerre-Harpøth, V., Damgaard, B.M. and Ingvartsen, K.L., Aarhus University, Blichers Allé 20, Postbox 50, Tjele, Denmark; kasey.moyes@agrsci.dk

How do cows in different stages of lactation adapt to feed restriction (FR)? Our objective was to determine the effect of stage of lactation on metabolic profiles of cows during FR. Twenty-nine healthy Holstein dairy cows in early (E;n=14;0-90 days in milk) and mid (M;n=15;91-220 days in milk) lactation were used. Of these, 13 cows were primiparous and 16 cows were multiparous. Prior to FR, all cows were fed a standard ration for ad libitum intake. After 8-d, all cows were FR to provide ~40% of net energy for lactation requirements based on body weight, milk production and composition by supplementing the standard ration with 60% wheat straw. After 4-d of FR, cows returned to full feed. Blood was collected every morning and plasma was analysed for concentration of non-esterified fatty acids (NEFA), beta-hydroxybutyric acid (BHBA), glucose and plasma urea nitrogen (PUN). Liver biopsies were collected once before and during FR and analyzed for triglycerides, glycogen, phospholipids, glucose and total lipid content. For each cow, the change in each variable before and during FR was calculated. The MIXED procedure of SAS was used to determine the effect of stage of lactation on metabolic profiles during FR. Changes in NEFA and PUN were greater ($P<0.05$) for M than E cows whereas changes in BHBA and glucose were greater ($P<0.01$) for E than M cows during FR. The changes in liver triglyceride, glycogen and total lipid content were not different ($P>0.07$) between E and M cows whereas changes in liver phospholipids and glucose were greater ($P<0.05$) in E than M cows during FR. Our results show that cows in early lactation respond differently to feed restriction than cows in mid lactation. These results provide insight into the homeorhetic mechanisms controlling the partitioning of nutrients of dairy cows during early lactation and should be beneficial for our understanding of how to maintain animal health and productivity throughout the lactation cycle.

Session 01 Poster 27

Nutritional and lactational responses of nulliparous ewes induced to lactate according to breed and treatment protocol

Ben Khedim, M., Caja, G., Salama, A.A.K., Schlageter, A., Albanell, E. and Carné, S., Universitat Autònoma de Barcelona, Grup de Recerca en Remugants (G2R), Campus universitari, Edifici V, 08193 Bellaterra, Barcelona, Spain; gerardo.caja@uab.es

Two treatments based on s.c. injections of estradiol and progesterone (d 1 to 7) at reduced doses of the standard protocol (a half dose: HD, 0.25 and 0.63 mg/kg BW; one third dose: TD, 0.17 and 0.42 mg/kg BW, respectively) and hydrocortisone (d 18 to 20; 50 mg/d) were used to induce lactation in 47 ewe-lambs of 9 mo of age (Manchega, n=24, 54 kg BW; Lacaune, n=23, 58 kg BW). Ewe-lambs were penned in 8 groups and fed ad libitum a total mixed ration. Machine milking (×2 daily) started at d 21 and lactation success (Manchega, >0.25 L/d; Lacaune >0.5 L/d) was evaluated at d 35. Low yielding ewe-lambs were dried-off after d 14 of lactation. Lactating ewe-lambs were treated with growth hormone (bST, 250 mg) at d 48 and 62 of lactation. Group dry matter intake (DMI) was recorded daily during the experiment (4 wk before and 10 wk after milking started) including individual estimation during lactation by using polyethylene glycol 6000 (PEG, 50 g/d for 14 d). Induction treatments decreased DMI (HD, -28%; TD, -18%; P<0.05) but recovered thereafter. Lactation success was 55% and did not vary by treatment. Onset of lactation increased DMI by 16% in both breeds. Lacaune ewe-lambs produced nearly 2 times more milk than Manchega at d 14 of lactation and varied by treatment (HD, 817 ml/d; TD, 458 ml/d; P<0.01), whereas Manchega did not vary (351 ml/d). No differences in milk composition were detected by breed or treatment during the first 14 d of lactation. Milk yield increased markedly by effect of bST (Manchega, 114%; Lacaune, 90%; P<0.01), but only a decrease in milk protein content (P<0.01) and a numerically greater DMI (P=0.16) were detected when bST and control lactating ewe-lambs were compared. In conclusion, marked differences were found in lactation induction response by breed which were related to their own hormonal environment.

Session 01 Poster 28

Effects of gestation and transition diets on composition and yield of sow colostrum and piglet survival

Theil, P.K., Hansen, A.V. and Sørensen, M.T., Aarhus University, Department of Animal Health and Bioscience, Blichers allé 20, DK-8830, Denmark; Peter.Theil@agrsci.dk

The experiment was carried out to evaluate whether gestation or transition diets (or both) affect colostrum synthesis of sows when transfer to transition diets occurs one week prior to expected farrowing. In total, 48 sows were fed one of four gestation diets until d 108 of gestation, and then sows were fed one of six transition diets. Three gestation diets were formulated to contain 35% dietary fiber (DF; mainly from sugar beet pulp, pectin residue or potato pulp) and a low fiber control diet contained 17.5% DF from mainly wheat and barley. Transition diets contained either 3% supplemented animal fat +/- 2.5 g/d of hydroxyl methyl butyrate or 8% supplemented fat originating from coconut oil, sunflower oil, fish oil or fish oil/octanoic acid mixture (4+4%). At parturition, colostrum samples were collected from unsuckled mammary glands no later than 3 h after birth of the first piglet. Colostrum intake of individual piglets was calculated based on their 24 h weight gain. The experiment was regarded as a complete balanced randomized factorial design and repeated measurements within a sow was taken into consideration in the statistical analysis. Sows fed pectin residue during gestation had a higher colostrum yield and secreted more energy via colostrum as compared to sows fed potato pulp. Sows fed coconut oil during the transition period had a higher secretion of colostral energy as compared to sows fed 3% animal fat and sows fed 8% fish oil. Colostral contents of protein (transition), lactose (gestation+transition) and dry matter (transition) were affected by the dietary treatments given in brackets. Secreted colostral energy was associated with 24 h piglet survivability. The study suggests that sow nutrition is important for colostrum synthesis and may be a promising way of improving neonatal piglet survival.

Differentially expressed proteins in mammary gland during lactation in sheep

Signorelli, F., Cifuni, G.F., Napolitano, F. and Miarelli, M., CRA-PCM, Proteomic, Via Salaria,31, 00015 Monterotondo (RM), Italy; maria.miarelli@entecra.it

The objective of this study was to evaluate changes in mammary protein profiles during lactation in sheep. We analyzed extracts of mammary gland, from 3 ewes of the Sarda breed, slaughtered at early, mid and late lactation stages, using 2D-SDS PAGE. Protein spots, showing differential expression, between one stage and the others, were identified by MALDI-MS or LC-MS/MS. We noted that the majority of these proteins are related to biochemical and physiological functions of the mammary gland: enzymes involved in carbohydrate and lipid metabolism, transport, cell communication and cellular processes, and apoptosis. The majority of the identified proteins are upregulated at mid lactation, compared to early and late stages, so following the same trend as milk yield, and belong to the pathway for lactose and lipid biosynthesis. At the end of lactation, among the upregulated proteins, we found k casein as well as proteins belonging to the pathway of cellular processes. The knowledge of the biochemical processes contributing to the functional and metabolic adaptation of the mammary gland to perform lactation will contribute to understand the variability of important economic traits, like milk yield and quality, and lactation length.

Effects of milk production and seasonal calving on some blood metabolite changes in dairy cows at postpartum period

Daghigh Kia, H., Ziaee, E., Moghaddam, G., Rafat, A. and Davarpoor, H., University of Tabriz, Department of Animal Science, 29 Bahman bolvard, 5166614766,Tabriz, Iran; hsz6955@yahoo.com

Negative energy balance in the early lactation period results in serious problems in health, milk production and reproduction of dairy cows. Stress (resulting from both high production or bad environmental conditions) can increase risk of metabolic disorders. The objective of this study was to evaluate the relationships between some serum metabolites and milk production in the postpartum period of dairy cattle. This study was conducted with 43 postpartum lactating Holstein cows from November 2007 to May 2008 in a commercial dairy farm located in Iran. Cows were divided into four groups on the basis of milk production and season of calving (cold and warm season). Daily milk yield in low and high groups were less and more than 35 kg/day respectively. Blood samples were collected once weekly from tail vein by venoject tubes. Plasma was analyzed for glucose, urea, phosphorus, BHBA, and total protein. Cows were housed in a free stall barn, milked three times daily, and fed a total mixed ration (TMR) according to NRC (2001) to meet production requirements. TMR included approximately 19% CP, 1.67 NEL (MJ/kg) and 23.6 DMI (kg/day) for lactating cow. Statistical procedures were performed using SAS software (SAS, 1999). Data for blood metabolites were analyzed by the general linear models (GLM). The results of the experiment showed that the concentration of serum glucose in low milk producing cows that calved in the cold season, was significantly higher than high milk production cows at postpartum in the warm season ($P<0.05$). BHBA levels in cold season-low milk producing cows were significantly lower than warm season-high producing cows ($P<0.05$). Serum phosphorus concentration in the cold season-high producing group was significantly lower than warm season-high producing cows ($P<0.05$). Results of the present study indicated that metabolite changes during a critical period of postpartum in dairy cows may be affected by levels of milk production and seasonal calving.

Hormonal (thyroxin, cortisol) and immunological (Leucocytes) responses to cistern size and heat stress in Tunisia

Ben Younes, R.[1], Ayadi, M.[2], Najar, T.[1], Caccamo, M.[3], Schadt, I.[3], Caja, G.[4] and Ben M'rad, M.[1], [1]Institut National Agronomique de Tunisie, Avenue Charles Nicolle 43, 1082 Tunis, Tunisia, [2]Institut Superieur de Biologie Appliquee, Route El Jorf – Km 22.5, 4119 Medenine, Tunisia, [3]CoRFiLaC, Regione Siciliana, S.P. 25 Km 5 Ragusa mare, 97100 Ragusa, Italy, [4]Universitat Autònoma de Barcelona, Plaça civica, 08193 Bellaterra, Spain; caccamo@corfilac.it

This study was carried out in 2006, in North Tunisia, using a block design per udder cistern size, using 60 Holstein cows (170±15 DIM; 18.0±5.7 L/cow/d milk) classified according to udder cistern size by ultrasonography as large-cisterned (44±13 cm^2; LC) and small-cisterned (21±8 cm^2; SC). The experiment was conducted at four times, April - 5 (D1), July -19 (D2), August -19 (D3) and September -19 (D4). On each test day, temperature and relative humidity data were registered hourly and cows' blood was sampled from the jugular vein to determine serum concentrates cortisol. Thyroxin (T4) serum concentrates were additionally analyzed in D2 – D4. Leucocytes (lymphocytes, eosinophils, neutrophils and monocytes) were counted differentially, and percentages of lymphocytes relative to total counted cells were calculated. Mean THI values were 62±2, 79±2, 84±2, and 77±1 in D1, D2, D3, and D4, respectively. Lymphocyte % relative to total cell counts and T4 concentrations were affected by test day ($P<0.001$). Lymphocytes % least square means (LSMeans) and standard error (SE) in D1, D2, D3, and D4, respectively, were as follows (superscripts differ by $P<0.05$): 73.6^{a}±2.7, 64.7^{b}±1.1, 65.6^{b}±1.3, 60.4^{c}±1.1. LSMeans and SE of T4 (nmol/L) in D2, D3, and D4, respectively, were as follows: 86.4^{a}±0.1, 46.5^{b}±0.1, 53.0^{b}±0.1. T4 concentrations were higher ($P<0.01$) in SC cows (67.7±0.1) compared to LC cows (52.7±0.1). Cortisol concentration was effected neither by test day nor by cistern size. However, the decrease of lymphocyte concentration during summer compared to spring evidences suppression of cows' immune system under hot environments.

New phenotypes for new breeding goals in cattle

Boichard, D., INRA, UMR1313 Animal Genetics and Integrative Biology, Bat 211, 78352 Jouy en Josas, France; didier.boichard@jouy.inra.fr

Cattle have to face new challenges for sustainable production with its three pillars, economic, societal and environmental. Three main factors will drive cattle selection in the future: 1) In dairy cattle, after a long period of selection on production, most functional traits have been deteriorated, sometimes up to a critical point, and need to be restored. This is particularly the case for fertility, mastitis resistance, longevity, metabolic diseases, i.e. traits with a low heritability and difficult to select. 2) Genomic selection offers new opportunities: selection pressure can be almost doubled and phenotype recording could be decoupled from selection and limited to several thousand animals; 3) New devices and sources could be used to generate new informations. Milk composition should be adapted to better meet human nutrition requirements. The corresponding phenotypes, mainly fatty acids, could be measured through a new interpretation of the usual medium infra red spectra. Milk composition can provide additional information about reproduction and health. Modern milk recorders also provide milking speed information. Electronic devices could be used to detect estrus, to record behavioral traits. Carcass traits could be effectively selected by using slaughterhouse data. A genetic solution has to be proposed as an alternative to dehorning for a better welfare. Cattle produce a lot of green house gas and contribute to global change. In the near future, the acceptability of cattle production could depend on its ability to decrease its ecological footprint. A first solution is to increase survival and longevity to decrease replacement needs and number of non productive animals. At the individual level, the rumen activity could probably be selected in order to limit the methane production, with a concomitant improvement of feed efficiency. A major effort should be dedicated to this new field of research. Finally, pathogens pressure will increase due to global change and disease resistance will become a major concern.

New phenotypes for new breeding goals in pigs

Knol, E.F., Duijvesteijn, N., Leenhouwers, J.I. and Merks, J.W.M., Institute for Pig Genetics BV, Postbox 43, 6640 AA Beuningen, Netherlands; jan.merks@ipg.nl

The generalized breeding goal for pigs is: quality pork against the lowest cost price. Historical selection was on the production efficiency (gain during finishing) and quality of the carcass (lack of backfat). These two traits take care of feed efficiency too. Ten to twenty years ago effective selection for increased litter size became feasible, reducing the cost price of a finishing piglet. These three phases, (1) fertility and litter size, (2) efficient finishing and (3) carcass and meat quality have further developed and include traits as easy cycling, number of teats, meat color, water binding capacity and marbling. New trait developments are on (a) livability; in early life, finishing phase and in terms of sow survival and longevity; (b) uniformity; in birth weight to decrease mortality, during finishing to increase protein efficiency and at shipment to increase slaughter plant efficiency; (c) society driven trends; non castration requires selection against boar taint, group housing and non-tail docking necessitate insight in behavioral mechanisms and worries about zoo-noses have started on lowering medication and increasing disease resistance; (d) environmental (non) sensitivity traits must allow animals to thrive in different regions of the world and/or adapt to climate change. Statistical and computational trends allow simultaneous estimation of more genetic effects for the same trait: (i) sow and service sire effect for litter size and gestation length, (ii) additive-maternal effects for stillborn, (iii) additive-nurse sow effects (note: different then additive maternal), estimating livability of piglets and mothering ability of sows, (iv) social interaction (pen mate effects). At the end of the day all genetic effects are tested against 'lowering cost price of quality pork' to accommodate interests of all stakeholders: wellbeing of farmers, animals, consumers and society. Simplicity and straightforwardness of the breeding goal has to be weighed against completeness and complexity.

Selecting layers on individually recorded behaviour and performance data in group housing systems

Icken, W. and Preisinger, R., Lohmann Tierzucht GmbH, Am Seedeich 9-11, 27472 Cuxhaven, Germany; icken@ltz.de

Consumers and politics demand group housing systems for laying hens, known for creating an appropriate environment for the birds. Breeding hens for such systems require objective parameters about their performance and behaviour. For this goal, innovative techniques, i.e. the electronic pop hole and funnel nest box, were developed to capture data in a cage free environment. In this study, single hen data captured from four consecutive flocks were analysed. In total, 1,214 full pedigreed hens of a brown layer line, under similar rearing conditions, were tested for a period of five months. All hens were tagged with a transponder on their leg which guaranteed consistent data recording of important production traits such as the duration of stay in the nest box, the exact oviposition time, frequency of free-range visits as well as the essential egg number. For statistical analysis, the observation period was divided into five 28-day laying periods. In the first laying period, the hens laid their eggs a half hour later than in the following production period. Irrespective of laying period, the average stay in a nest was 30 minutes. Differences in free-range behaviour were analysed with ongoing familiarisation to the system. The number of free-range visits increased from 9 to 13 passages as well as the time that one hen spent per day outside (4.0 to 6.7 hours). Estimated heritabilities for these behaviour traits and their correlations to the most important performance trait, egg number, clarified their genetic relevance as potential selection traits. The heritabilities for all behaviour traits ranged between 38% to 49%. Negative genetic correlations to the laying performance were estimated for the oviposition time (r_g = -0.46) and the time that hens spent in the free-range area per day (r_g = -0.75). For the nest occupation time, the estimated correlation was positive (r_g = +0.13). Hens with an optimal occupation time per nest visit of 30 to 40 minutes, showed the highest laying performance.

Genetic correlation between longevity and production traits in the Pirenaica beef cattle breed using bivariate models

Varona, L.[1], Moreno, C.[1], Van Melis, M.H.[2] and Altarriba, J.[1], [1]Universidad De Zaragoza, Unidad De Genetica Cuantitativa Y Mejora Animal, C/ Miguel Servet 177, 50013. Zaragoza, Spain, [2]University Of Sao Paulo, Facultade De Zootecnia, 13635-970, Brazil; lvarona@unizar.es

Survival or longevity is an economically relevant trait in beef cattle. The main inconvenience for its inclusion in the selection criteria is the delayed recording of phenotypic data. Thus, identification of a longevity-correlated trait early registered in lifetime would be very useful for selection purposes. Moreover, knowledge of the structure of genetic correlation with the usual criteria of selection will be very useful for predicting unexpected consequences on longevity. The objective of this study is to estimate the genetic correlation between survival (SURV) with cold carcass weight (CW), and subjective scales for conformation (CON), fatness (FAT) and colour (COL) measured on the carcass in the Pirenaica beef cattle breed. Here, survival was measured in discrete time intervals defined by the number of parities per cow and modelled through a sequential threshold model. We used data from 194090 parities from 48965 dams, and 29019, 24914, 22845 and 5278 records for CW, CON, FAT and COL, respectively. Four independent bivariate analyses were performed between cow survival and the analyzed traits. Posterior mean estimates (and posterior standard deviation) for heritabilities were 0.14 (0.01), 0.35(0.02), 0.21(0.02), 0.24(0.02) and 0.14(0.03) for SURV, CW, CON, FAT and CAL. Posterior mean estimates (and posterior standard deviation) for genetic correlations with SURV were 0.15(0.04), -0.19(0.05), 0.33(0.06) and 0.08(0.10), for CW, CON, FAT and COL, respectively.

Born to be a loser cow?

Pedersen, L.D.[1], Jørgensen, H.B.H.[1], Sørensen, M.K.[1,2], Thomsen, P.T.[3] and Norberg, E.[1], [1]Faculty of Agricultural Sciences, Aarhus University, Department of Genetics and Biotechnology, Blichers Allé P.O. Box 50, 8830 Tjele, Denmark, [2]Danish Agricultural Advisory Service, Udkærsvej 15, 8200 Aarhus N, Denmark, [3]Faculty of Agricultural Sciences, Aarhus University, Department of Health and Bioscience, Blichers Allé P.O. Box 50, 8830 Tjele, Denmark; Louise.DybdahlPedersen@agrsci.dk

A group of dairy cows referred to as 'loser cows' has received increasing awareness in Denmark within recent years. The loser cows are characterized by a generally lowered health and production status and a previous study, aimed at defining loser cows clinically, has estimated that the overall prevalence of loser cows in Danish Holstein herds is 3.2%. The clinical definition of a loser cow is based on an examination of the four categorical traits: lameness, hock lesions, other cutaneous lesions, and condition of the hair coat. These traits were chosen on the basis of practical considerations as they could be scored from a distance without fixating the cow. We will present results for the genetic and phenotypic parameters for the loser cow state and the underlying traits. The results are based on records on 6,098 cows which were analyzed with an animal model including fixed effects of herd, season of scoring and location of scoring, age at first calving, lactation stage and parity in addition to additive genetic effects and permanent environmental effects. Based on the results, we reason that some cows are born loser cows, as the clinically defined loser cow state is heritable with an estimated heritability of 0.08. Furthermore, we argue that the loser cow score or the traits included herein could be included in a total merit index, either directly or as correlated information traits, aimed at breeding for more robust cows. This will of course depend on whether the information provided by the loser cow traits is already incorporated in other traits in the total merit index. Accordingly, the association between the 'loser cow' traits and index traits will be examined and the results will be presented.

Precise phenotyping of 2,000 first lactation Holstein cows for claw disorders in a designed experiment

Weidling, S.[1], Schöpke, K.[1], Alkhoder, H.[1], Pijl, R.[2] and Swalve, H.H.[1], [1]Institute of Agricultural and Nutritional Sciences, Martin-Luther-University Halle-Wittenberg, Theodor-Lieser-Str. 11, 06120 Halle, Germany, [2]Claw Health GmbH, Fischershäuser 1, 26441 Jever, Germany; kati.schoepke@landw.uni-halle.de

Feet and leg problems are major reasons for a reduced lifespan, an impaired well-being and an inferior productivity in dairy cows. In dairy cattle, laminitis is the most important disease. Based on the knowledge from previous pilot projects, an experiment was designed for the collection of 2,000 records from seven large commercial herds. Herds and cows from these herds were selected to fulfil the following criteria: (1) Large herd size, herds have standard slatted flooring, (2) first lactation cows only, (3) lactation stage varying between 50 and 150 days in milk. Phenotyping included: (1) scores for claw disorders collected at trimming, (2) weight of the cow, (3) stature in cm, (4) body condition score, (5) ultrasonic backfat measurement. Blood samples were taken for later genotyping. Additionally, all records that commonly are available from milk recording schemes, classification of conformation, and insemination records are available. The sample is subject to quantitative as well as molecular SNP analysis as a part of the German FUGATO-Plus project on feet and legs in livestock. Collecting of phenotypes resulted in usable samples of 1,968 Holstein cows and first statistical results are available. Incidences for clinical and subclinical cases of laminitis (lam), digital dermatitis (dd), white line disease (wld), sole ulcer (su), and interdigital dermatitis (did) in rear legs were 57.3, 17.1, 12.8, 7.1 and 6.9%. Estimates of heritabilities were obtained applying threshold models and were 0.28, 0.25, 0.22, 0.35 and 0.02 for lam, dd, did, wld and su.

PhenoFinLait: French national program for high scale phenotyping and genotyping related to fine composition of ruminant milk

Faucon, F.[1,2], Barillet, F.[3], Boichard, D.[3], Brunschwig, P.[2], Duhem, K.[1], Ferrand, M.[2], Fritz, S.[2], Gastinel, P.L.[2], Journaux, L.[2], Lagriffoul, G.[2], Larroque, H.[3], Lecomte, C.[1], Leray, O.[4], Leverrier, S.[5], Martin, P.[3], Mattalia, S.[2], Palhière, I.[3], Peyraud, J.L.[3] and Brochard, M.[2], [1]CNIEL-FCL, Paris, 75009, France, [2]Institut Elevage-UNCEIA, Paris, 75012, France, [3]INRA, Paris, 75007, France, [4]Actilait, Poligny, 39800, France, [5]LILANO, Saint-Lo, 50000, France; ffaucon@cniel.com

Thanks to genomic selection, it is possible to evaluate a sire, at birth, on different criteria such as milk yield, fat content… Nowadays dairy factories become more concerned by the nutritional quality of milk products mostly reflected in the fine composition of milk for fat and protein (individual protein (IP) and fatty acid (FA)). Some of these new criteria are already considered by some dairies in paying milk to farmers. In this context, all scientific and economic stakeholders, from milk production to milk processing have formed the PhenoFinlait consortium to carry out a R&D phenotyping project on fine milk composition. The aim is to develop a cheap and large scale phenotyping system for individual milk components (FA and IP) and apply this procedure in commercial farms allowing the analysis of genetic and environmental factors (feeding) involved in milk composition. First results were the development of calibration equations to estimate FA content by Mid Infra-Red (MIR) spectrometry. A reference method for IP identification and quantification has been chosen. Collection of milk and blood samples (for MIR analysis and DNA mapping) and data record on animal feeding is ongoing on 20 000 cows, ewes and goats located in 1500 farms with different herd management systems. Large scale genotyping will complement the design of this ambitious project aimed to provide new interlinked tools for selection and farm management to quickly react to human nutrition demand for the benefit of the dairy industry and consumers. This program is funded by ANR, Apis-Gène, CASDAR, CNIEL, FGE, FranceAgriMer, Ministry of Agriculture

Towards novel selection strategies to improve production and fertility by integrating expression profiles and high throughput genotyping in cattle

Pimentel, E.C.G.[1], Tietze, M.[1], Simianer, H.[1], Reinhardt, F.[2], Bauersachs, S.[3], Wolf, E.[3] and König, S.[1], [1]Georg-August University Göttingen, Department of Animal Sciences, Albrecht-Thaer-Weg 3, 37075 Göttingen, Germany, [2]Vereinigte Informationssysteme Tierhaltung w.V., Heideweg 1, 27283 Verden, Germany, [3]Ludwig-Maximilians-Universität München, Feodor-Lynen-Str. 25, D-81377 Munich, Germany; epiment@gwdg.de

The objective of this work was to integrate findings from functional genomics studies with genome-wide association studies for fertility and production traits in dairy cattle. Association analyses of production and fertility traits with SNPs located within candidate genes revealed in a series of gene expression studies were performed. A set of 2294 Holstein-Friesian bulls genotyped for 39557 SNPs was used. Fifty nine SNPs were located within chromosomal segments covered by a candidate gene. Allele substitution effects for each SNP were estimated using a mixed model with a fixed effect of marker and a random polygenic effect. For the random part, two alternative covariance matrices were assumed, derived either from marker or from pedigree information. Correlation between marker and pedigree-derived relationship coefficients was 0.78. Almost all SNP effect estimates that were significant in the analysis with the kinship matrix built from marker genotypes were also significant in the analysis with the pedigree-derived relationship matrix. Many of the SNP effects provided evidence of the antagonistic relationship between production and fertility at the molecular level. We found several SNP alleles having favorable effects on yield traits (milk, fat, and protein), as well as on at least one fertility trait though. While most of quantitative genetic studies proved genetic antagonisms between yield traits and functional traits, improvements in both production and functionality may be possible when focusing on a few relevant SNPs. Investigations combining input from quantitative genetics and functional genomics with association analysis may be applied for the identification of such SNPs.

Genetic parameters for litter traits including farrowing duration and piglet survival up to weaning in the French Large White and Landrace sows

Merour, I.[1], Bernard, E.[1], Bidanel, J.P.[2] and Canario, L.[2], [1]IFIP, BP 35104, 35651 Le Rheu, France, [2]INRA, UMR 1313 GABI, 78352 Jouy-en-Josas, France; isabelle.merour@ifip.asso.fr

Genetic parameters were estimated for number of piglets born in total (NBT), born alive (NBA), and stillborn (NSB) per litter as well as birth to weaning survival rate (SR), farrowing duration (FD), litter weight and within-litter standard deviation of piglet body weight at birth (SDBW). Since 2002, data were collected on 29,153 French Large White (LW) and 16,354 French Landrace (LR) litters, in 29 LW and 17 LR nucleus herds. The FD was assessed according to four classes: <2 hours, 2 to 3 hours or 3 to 4 hours, and >= 4 hours. The traits were analyzed using the REML methodology in a multi-trait animal model including the fixed effects of herd, year x month and parity, and the random effects of the sire of the litter as well as the additive genetic value and permanent environment of the sow. Heritability estimates ranged from 0.10 to 0.13 for litter size, from 0.06 to 0.11 for mortality traits and were around 0.08 for FD. In LW and LR sows, NSB was unfavorably correlated with NBT (rg=0.46±0.06 and 0.56±0.05), but lowly correlated with NBA (rg=0.05±0.08 and 0.18±0.07). Both NBT and NBA showed a genetic antagonism with SR (-0.24±0.09 and -0.32±0.07). The FD was genetically associated with NBT (0.23±0.09 and 0.33±0.05) but not with NBA. The NSB was positively correlated with FD (rg=0.64±0.08 and 0.41±0.09). Litter weight at birth showed low (LR) to moderate (LW) genetic correlations with SR. Higher SDBW was associated with lower SR (rg= -0.21±0.18 and -0.28±0.33). These results indicate that peripartum survival could be improved by selecting for NBA instead of NBT, with possible use of FD which is recorded routinely by French breeders. Selecting against SDBW could also improve piglet survival during lactation. Other innovative traits based on the scoring of sow maternal ability, farrowing ease and ability to switch to a catabolic state at farrowing were registered on these sows and will also be analyzed.

Session 02 Theatre 10

Identification of pigs with increased overall disease resistance using haematological and immunological tests

Maignel, L.[1]*, Wyss, S.*[1]*, Fortin, F.*[2]*, Hurnik, D.*[3] *and Sullivan, B.*[1]*,* [1]*Canadian Centre for Swine Improvement inc, Central Experimental Farm, Building 54, 960 Carling Avenue, Ottawa (Ontario) K1A 0C6, Canada,* [2]*Centre de développement du porc du Québec inc, 2795 boul. Laurier, bureau 340, Québec (Québec) G1V 4M7, Canada,* [3]*Atlantic Veterinary College, 550 University Avenue, Charlottetown (Prince Edward Island) C1A 4P3, Canada; laurence@ccsi.ca*

A study was conducted on potential methods to identify sire families with increased overall disease resistance in Yorkshire, Landrace and Duroc pigs using haematological and immunological parameters as well as performance and mortality information from the Canadian Centre for Swine Improvement national database. Blood samples were collected on 893 pigs from 63 targeted sire families in 13 herds across Canada and tested for haematological and immune response tests. Mortality and cause of death were also recorded in participating herds from birth to market weight. A total of 73,845 piglets were tracked from birth to weaning in 5 participating farms and 55,683 of these were tracked through to market weight. Breed differences were observed for certain haematological parameters, transformed animal blastogenic indices and for a lymphocyte proliferation index. Most haematological and immunological parameters were found to be low to moderately heritable. Residual correlations between blood parameters and on-farm performance (growth, backfat thickness and lean depth) were found to be very low. Relationships between blood parameters and mortality data in sire families were explored. The results were very encouraging overall, showing that the different parameters studied are heritable and relatively independent from performance traits. More data is required to clearly study the relationship between survival rates and in-vitro indicators of immune response.

Session 03 Theatre 1

Grazing systems in Mediterranean Europe with focus on dairy sheep and goats: understanding and driving the forage-animal-livestock product continuum

Molle, G.[1,2]*, Decandia, M.*[1]*, Cabiddu, A.*[1] *and Sitzia, M.*[1]*,* [1]*Agris Sardegna, Department of Animal Production Research, loc. Bonassai, 07040 Olmedo, Italy,* [2]*Agris Sardegna, Department of Research on Horses, P. D. Borgia, 4, 07014 Ozieri, Italy; gmolle@agrisricerca.it*

After a long history of system simplification where the trend has been to increase milk production based on soil tillage, grass-dominated forage crops, N fertilization and concentrate supplementation, small ruminant grazing systems of Mediterranean Europe have been recently challenged by a prudent retro-gradation towards more diverse lower-input systems. Science has generally shown that small ruminants, for their moderate genetic value and their size, are not particularly efficient at incorporating high level of food (either fibrous or starchy) into products. High level of tilling and supplementation are energy-consuming, expensive operations and can be hazardous to the environment. Furthermore dairy and meat products sourced by these 'simplified' systems have no specific fingerprints and fall into the large market of highly standardized food, where dairy cow products are dominant. This review will focus on advances in the understanding of the effect of retro-gradation towards more self-sustainable grazing systems where different pasture and browse sources are used not only to cope with animals requirements, but also: to satisfy their preference and welfare, support environment preservation and landscape beauty, and offer a range of animal products with potential attractive nutracines and flavours. The mechanics of some key trade-offs between plant and grazer/browser and through it up to dairy or meat product quality will be highlighted as well as the way of driving these systems to targeted objectives at the farm level. To conclude an overview of key research and technology transfer topics to be urgently addressed is provided.

Session 03 Theatre 2

Modeling heterogeneous ecosystems with large herbivores

Coughenour, M.B., Colorado State University, Natural Resource Ecology Laboratory, Fort Collins, Colorado, 80523, USA; mikec@nrel.colostate.edu

A spatially explicit landscape ecosystem model called SAVANNA has been used, further developed and updated for research and decision making in heterogeneous ecosystems with native and domestic large herbivores for over two decades. It is a generalized model that has been applied in a wide variety of habitats and vegetation types. Applications in the U.S. have included Rocky Mountain and Yellowstone National Parks, and the Pryor Mountain Wild Horse Range. Applications have also been developed in Eastern and Southern Africa, Australia, and Inner Mongolia. The most common uses have been assessments of sustainable ecological carrying capacity for large herbivores and assessments of ecosystem responses to climatic variability and change. A process based modeling approach is used to predict plant growth, vegetation dynamics, nutrient cycling, water budgets, and herbivore foraging, energetic status, and population dynamics. Spatial data are used to initialize the model and spatial-dynamic outputs can be viewed with the 'Savanna Modeling System' or a GIS. Multiple plant and animal species or functional types can be simulated, enabling assessments of interactions between woody and herbaceous vegetation, effects of herbivory on vegetation composition and function, and interactions between wild and domestic herbivores. This presentation will provide an overview of this approach to spatially explicit ecosystem modeling.

Session 03 Theatre 3

Local challenges for the future of extensive livestock farming systems in contrasted regional environments

Gibon, A.[1], Ickowicz, A.[2] and Tourrand, J.F.[2], [1]INRA, Centre de Recherche de Toulouse Auzeville, BP 52627, 31326 Castanet-Tolosan cedex, France, [2]CIRAD, Campus Montpellier SupAgro-INRA, 2 place Viala, 34060 Montpellier cedex 1, France; annick.gibon@toulouse.inra.fr

To consider the future of extensive livestock farming systems (LFS) at a local scale, we use here the results of a set of case studies (CS) in Africa, South America and Europe made in the ADD-TRANS project (2005-09). The CSs were carried out in selected gerographical areas in Urugayan pampa, Bresilian Amazonia, sub-sahelian steppes in Senegal, and mountains and hills in Southern France. They primilary aimed at supporting sustainabilty of local LFSs and land use (LU). A same method was applied in all of them: the participatory building of an agent-based model with farmers and other local actors for a spatially-explicit simulation of scenarios of change in local LFS and LU. The results stress out contrasts but also similarities in trends and pressures for change among the the CS areas. In particular, globalisation and liberalisation of agricultural markets challenge the future of local LFS and the perenniality of semi-natural grasslands is at risk in all of the CS areas with a long past of grassland-based systems. Competition of cash crops appears as a most common threat for local LFS and LU sustainability, its expansion logics and processes differing according to natural and socio-economical environment specificities. A main finding of the study is the strong evidence of the importance to attach to long-term behaviours of individual farmers for improving the understanding of local changes in extensive LFS and assessing opportunities and challenges for their future. In each CS area, participatory groups regarded the modelling of the local variety in farmers' long-term objectives and coping strategies used to adapt to their environment, as a requisite for a sound simulation of scenarios. Some similarities in farmer behaviours amongst the CSAs lead us to put a new perspective on some basic contradictions in the current search for sustainable development of extensive LFS.

Livestock farms in mountainous grassland area: outlook '2015'

Veysset, P., Rozière, B., Benoit, M. and Laignel, G., INRA, UR1213 Herbivores, F-63122 Saint-Genès-Champanelle, France; veysset@clermont.inra.fr

Grass is the most valued resource recognised by cattle and sheep farmers in mountainous areas. Consistency of production systems is the result of an interactive management of environmental constraints, production targets, environmental and socio-economic policies. Health Check of the CAP, abolition of milk quotas, increased input prices, uncertain prices of products, specifications for PDO cheeses, are all issues that farmers will have to cope with in the short and medium term. This study proposes to understand and analyze, through simulation / optimization, possible changes and predictable strategies for cattle and sheep farmers to face these changes announced for 2015. A group of regional experts validated the choice of 10 farm-types studied in Auvergne, a mountainous grassland region, and the anticipated changes in policies and economics. Overall, the impact of these changes on the farm income appears to be positive for most grassland farmers in Auvergne. The high subsidies revaluation allocated to grassland largely explains this outcome. The economic results of suckler cattle systems should not change, but they should increase for the dairy cattle and sheep meat systems. Only the low grazing cattle systems should experience a fall in their income. Regarding the evolution of technical systems, beef cattle systems should evolve towards the production of younger animals in order to save inputs, leading to lower live weight production. Concerning the dairy cattle systems, milk production seems driven by the abolition of quotas and the integration of PDO cheese market chain (restrictive technical specifications with premium price for the milk) also seems interesting. Sheep meat systems have the most to gain from an economic standpoint and the simplification of flock management practices could lead to highly increased income, provided that reproductive performance will be maintained. For all systems, in an increasing input prices context, improved grazing management should have a marked positive impact on farm income.

Maximising the use of rangelands in a meat sheep farming system (South of France): impact on technical, economic and environmental performance

Benoit, M.[1] and Jouven, M.[2], [1]INRA, Unité Recherche Herbivores 1213, Centre de Theix, 63122 St Genes-Champanelle, France, [2]INRA-SupAgro, UMR868, 2 place Viala, Montpellier, 34060, France; marc.benoit@clermont.inra.fr

An experimental meat sheep flock of 330 ewes with high potential productivity (Romane breed, prolificacy 220%) is reared outdoors in a harsh environment: 300 ha of French Causse rangelands (Larzac plateau, southern Massif Central), of which 18 ha have been fertilized for 20 to 30 years. In the current farming system, 40 to 50% of the energy requirements of the flock are met with conserved feed (forage, concentrates) produced in adjacent fields or purchased. A detailed analysis of the diversity of rangeland vegetation types in the paddocks, in relation to the past grazing management, suggested a new way to organize grazing on non-fertilized rangelands. This new management, which requires a slight reduction in flock numbers, is based on a flexible seasonal organization of grazing. Low-potential lands are considered as major forage resources in specific periods of the year, instead of 'complementary resources' or 'buffer resources' in case of climatic hazards. The new feeding and grazing management proposed should decrease by half the consumption of conserved feed. Utilising a modelling study, we compare the former and the new system at three levels: 1) technical performance of the flocks, 2) economic returns depending on the political and economic context and 3) environmental performance (greenhouse gas emission and energy consumption) of the systems. Opportunities for sustainable high-performance meat sheep systems based on rangeland utilization, within the European context, are discussed.

Goat's grassland-based systems in the Western Mediterranean area: present situation and future challenges: case study of Spain

Mena, Y.[1], Ruiz, F.A.[2] and Castel, J.M.[1], [1]Universidad de Sevilla, Ciencias Agroforestales, Carretera de Utrera, km 1, 41013, Spain, [2]Instituto Andaluz de Investigación y Formación Agraria, Economía y Sociología Agraria, Camino de Purchil. Apartado de correos 2027, 18004, Spain; yomena@us.es

Goat production systems in the Mediterranean basin tend to intensification. Nevertheless, goat's grassland-based systems are still important for a variety of reasons. Milk production in pastoral systems of Western Mediterranean countries(France, Italy and Spain) range from 214 to 482 l of milk sold /goat/ year and differences between milk income and feed cost range from 55.8 to 189.7 €/goat. The best farm economic results are related to the greater goat milk production and higher milk prices. However, research on the nutritional utilization of pastures and appropriate feed supplementation should be encouraged. Milk productivity should be improved, without neglecting the balance between hardiness and general productivity. In Spain, a 50% of goat's grassland-based farms still exist as compared to the 90's. The analysis of sustainability (using MESMIS methodology) of dairy goat systems in mountainous areas of South Spain shows a similar behaviour for the attribute productivity among semi-extensive (76.0%), semi-intensive (71.5%) and intensive (73.4%) systems. However, the most intensive goat farms demonstrate less capacity for self-management: semi-extensive (60.1%), semi-intensive (44.3%) and intensive (47.2%) systems. Multiple functions and values of these systems like the social importance and environmental benefits are commented.E.U. Common Agricultural Policy frameworks and its influence on the sustainability of dairy goat systems in Spain, are also analyzed. Some of the strategies to improve the sustainability of these systems are: appropriate feeding and reproductive management, labour organization, valorisation of products, participation of the farmer in the added value of their primary production and social and economical valorisation of their multiple functions and values.

Intensive dairy farming in Denmark and the role of grazing

Kristensen, T., Aarhus University, Faculty of Agricultural Sciences, Dept. of Agroecology and Environment, Blicher Alle 20, DK 8830 Tjele, Denmark; troels.kristensen@agrsci.dk

This paper describes briefly the development in herd size, feeding and farming structure in Danish dairy farming during the last 10 years. The position of grazing in this development is described based on information from an questioner, sent to 800 Danish dairy farmers with more than 100 cows. Grazing by the dairy cow was used at 34% of the farms, but only on 25% of the conventional farms. Herd size has increased more on the non grazing farms during the last five years and also investment in housing was higher in this group. The non grazing farms had a higher proportion of AMS milking. Stocking rate, cows/ha, was identical between the conventional non grazing herds and the farms using grazing, but access to land was more difficult at the non grazing farms defined as distance to at least 0.3 ha/cow and the number of roads to be crossed. Problems with walking lanes was the single most important physical obstacle for grazing, followed by access to land and also a higher work load was identified as problematic. There was a clear difference between the grazing and non grazing farms in their judgments of the management problems in relation to grazing. Feeding management was identified as being very problematic by 75% of the non grazing farms, but only by 28% of the grazing farms. A 27% of the grazing farms had no problems in relation to maintaining milk yield, while 65% of the non grazing claimed it to be a very problematic area. Besides this more than half of non grazing farms expected a lower productivity of roughage, which was only the case for 16% of the grazing farms. These results show that there is a strong correlation between the way of farming and the attitude and expectation to problems and possibilities in a system. Nevertheless, looking 10 years ahead, more than half of the non grazing farmers expect that grazing would increase based on regulations, as well as half of the non grazing farmers expect an increasing demand for grasslands based milk.

Effect of post-grazing sward height on grass production and performance of four beef heifer genotypes

Minchin, W. and Mcgee, M., Teagasc, Beef Production Research Centre, Grange, Dunsany, Co. Meath, Ireland; william.minchin@teagasc.ie

An experiment was conducted to evaluate the effects of two contrasting grassland management systems on grass production and performance of four late-maturing crossbred breeding heifer genotypes. One hundred and thirty-six heifers comprising of 4 genotypes: Limousin × Holstein-Friesian, Limousin × Simmental, Charolais × Limousin, and Charolais × Simmental were blocked by genotype, live weight, body condition score (BCS), age and suckler beef value, randomly assigned to one of two grassland management systems: grazing to a post-grazing sward height of either 4.0 or 6.0 cm. There were two replications of each grazing system resulting in four groups of 34 animals on 40 ha. The stocking rate was 2.5 LU/ha. Herbage surplus to grazing requirements was removed from the rotation by harvesting relevant paddocks for silage. Animal live weight, BCS and ultrasonic fat and muscle depths were measured. Grassland measurements included: pre and post-grazing sward heights, herbage mass (> 4 cm), sward density, herbage production, farm cover and grazing days. An additional 20 tonnes DM of surplus grazed grass (26 v. 6 t DM) was removed from the 4 cm system compared to the 6 cm system. There was no difference (P>0.05) in animal performance between the grazing systems. Potential exists to increase herbage production by grazing to a lower post grazing height without sacrificing animal performance.

Naturally occurring phytoestrogen effects on twinning rate in Maremmana cattle breeding

Primi, V., Danieli, P.P., Marchitelli, C. and Nardone, A., University of Tuscia, Animal Science, Via S. Camillo de Lellis, snc, 01100 Viterbo, Italy; primival@unitus.it

Twinning in cattle occurs globaly at a frequency of 1-2%. Twinning rate is known to be affected by age and parity of the dam, season, feeding and genetic factors. Little is known about the effects of phytoestrogens on twinning rate in cattle. The aim of this study was to explore the role of naturally occurring phytoestrogens on twinning rate in a Maremmana cattle herd. The Maremmana cattle breed in Castelporziano farm (Rome) was used as 'case study' because of the high twinning rate (11%) recorded in the period 1996-2008. In the period 2006-2007, during mating season, a total of 186 cows were divided into 3 groups which were assigned to different grazing parcels (A, B and C). Forage samples were collected in all parcels in spring. After species identification, samples were dried and ground to pass a 1 mm sieve. Phytohormones (Daidzein, Formononetin, Genistein, Biochanin A and Coumestrol) were extracted by acid hydrolysis and quantified by reverse phase HPLC with UV and fluorimetric detection. Data were analyzed by Statistica 7 software package (StatSoft Inc., Tulsa, OK, USA). Significance was declared at P<0.05. Total phytohormones found in parcel A (10.11±4.34 g/kg DM) were higher (P<0.05) than those found in parcel B and C (2.34±1.56 and 1.18±0.51 g/kg DM, respectively). Parcel A had a high proportion of subterranean clover, characterized by high levels of phytoestrogens compared to other forages. During experiment, the average twinning rates of cows reared on parcels A, B and C were 12.3%, 3.7% and 0%, respectively. Twining rate in parcel A was higher than those in parcels B and C considered together (chi square= 7.21±5.91; P<0.01), although total calving per parcels were not different. Results reported herein would suggest a possible effect of naturally phytoestrogens on twin birth rate in cattle, although several studies refer about negative effects of them on animal fertility.

Nutrient utilization and growth performance in sheep receiving different silages with or without concentrate

Nisa, M., Shahzad, M.A., Khan, S.H. and Sarwar, M., University of Agriculture, Faisalabad, Institute of Animal Nutrition and Feed Technology, Assistant Professor,Institute of Animal Nutrition and Feed Technology,University of Agriculture, 38040, Faisalabad, Pakistan; Linknisa@gmail.com

The experiment was conducted to investigate the influence of maize (*Zea mays*), sorghum (*Sorghum bicolor*) and millet (*Pennisetum americannum*) silages with or without concentrate on nutrients intake, digestibility, nitrogen balance and weight gain in Sipli sheep. Six experimental diets were formulated having 100% maize silage (MS), maize silage and concentrate at 50:50 (MSC), 100% sorghum silage (SS), sorghum silage and concentrate at 50:50 (SSC), 100% millet silage (MiS) and millet silage and concentrate at 50:50 (MiSC), respectively. Twenty four Sipli lambs were allotted to the six experimental diets in a completely randomized design to evaluate nutrition parameters in a ninety days trial. The results indicated that among silage diets, lambs fed MS consumed more dry matter (DM) than those fed SS and MiS diets. Likewise, lambs offered MSC had higher DM intake than those fed SSC and MiSC diets. Crude protein (CP) and neutral detergent fiber (NDF) consumed by the lambs also followed similar trend. Higher DM, CP and NDF digestibilities were also observed in lambs fed MS and MSC diets than those fed SS, SSC, MiS and MiSC diets. Overall digestibilities of DM, CP and NDF were higher in experimental diets containing silage with concentrate. Lambs fed MS diet had 2.79 and 4.45 g/ day higher N retention than those fed SS and MiS, respectively. Similarly, lambs fed MSC diet had 2.24 and 5.12 g/day higher N retention than those fed SSC and MiSC diets, respectively. The results showed that lambs fed MSC had higher daily weight gain and better feed conversion ratio than those fed MS, SS, SSC, MiS and MiSC diets. Moreover, lambs fed MSC diet had higher nutrients intake, digestibility, nitrogen balance and weight gain.

Effect of grazing time on milk fatty acids composition in sustainable dairy cattle systems of humid areas

Roca-Fernández, A.I.[1], González-Rodríguez, A.[1], Vázquez-Yáñez, O.P.[1] and Fernández-Casado, J.A.[2], [1]Agrarian Research Centre of Mabegondo, P.O. Box 10, 15080, Spain, [2]Agrarian and Fitopathologic Laboratory of Galicia, P.O. Box 365, 15640, Spain; anairf@ciam.es

Milk from grazing dairy cows show an improved milk FA profile with lower saturated fatty acids (SFA) and higher unsaturated fatty acids (UFA), mainly conjugated linoleic acid (CLA), than milk from silage-fed cows. Sixty-one autumn calving Holstein-Friesian dairy cows were balanced and randomly assigned to one of three treatments. The aim of this study was to investigate the effect on milk FA composition of different proportions of grazing (G0, zero; G12, 12-hr and G24, 24-hr) in cows supplemented with concentrate containing oilseeds (average, 6.3 kg DM cow^{-1} day^{-1}) and its variation according to the grazing season. Daily milk yield and weekly milk composition (fat, protein and urea) were individualy recorded. Individual body weight and body condition score were also recorded regularly. Herbage quality of each paddock was determined by NIRS and milk FA composition was analyzed by gas chromatography. Cows grazing rotationally pastures of perennial ryegrass and white clover, in spring, were stocked at high and low grazing pressure in G12 and G24, respectively. Grazing 24-hr caused a significant ($P<0.05$) decrease in short and medium chain fatty acids (SCFA, 8.34 and MCFA, 39.24 g 100 g^{-1} of FA, respectively) and a significant ($P<0.05$) increase in long chain fatty acids (LCFA, 42.29 g 100 g^{-1} of FA). The lowest ratio saturated/unsaturated FA was obtained in G24 (2.08). The monounsaturated and polyunsaturated fatty acids (MUFA, 25.20 and PUFA, 4.24 g 100 g^{-1} of FA, respectively) were significantly ($P<0.05$) higher in G24. The CLA content showed a significant ($P<0.05$) increase with grazing time, from 0.47, 0.72 and 1.23 g 100 g^{-1} of FA for G0, G12 and G24, respectively. From April to July, CLA levels in milk from cows grazing grass all day were three times higher than those fed silage while at the end of summer and in autumn these differences were reduced at half.

The effect of linseed supplemention on growth, carcass traits, fatty acid profile, retail shelf life, and sensory attributes of beef from steers finished on grassland of the northern Great Plains

Kronberg, S.L.[1], Scholljegerdes, E.J.[1], Lepper, A.N.[2] and Berg, E.P.[2], [1]USDA-ARS, P.O. Box 459, 58554 Mandan, ND, USA, [2]NDSU, Animal Sciences, P.O. Box 6050, 58108 Fargo, ND, USA; scott.kronberg@ars.usda.gov

Trial objective was to determine if daily supplementation of linseed for 85 days to steers finished on grassland would influence growth rates, carcass traits, or fatty acid profile, tenderness, and sensory attributes of ribeye steaks. Eighteen yearling steers (initial BW 399±21 kg) were randomly divided into 3 equal-sized groups. Treatment 1 steers (LN) received a daily supplement of ground flaxseed (0.20% of body weight) while Treatment 2 steers (CS) received a daily supplement of ground corn and soybean meal (0.28% of BW) that had levels of CP and TDN similar to the supplement for LN steers. Control steers (CT) were not supplemented. Treatments were individually fed from mid-August to November 7th, the day before slaughter. All steers grazed growing forage from early-May to slaughter. Growth rate of LN steers was 25% greater ($P<0.01$) than that of CT steers, but similar ($P=0.45$) to that of CS steers. No differences were observed for carcass traits ($P\geq0.14$), tenderness (Warner-Bratzler shear force; $P\geq0.24$), or sensory attributes ($P\geq0.40$) except for a slight off flavor detected in steaks from LN and CT steers ($P=0.06$). The n-3 fatty acids α-linolenic acid and eicosapentaenoic acid were 62 and 22% higher, respectively, in beef from LN vs. CT ($P<0.001$). The ratio of n-6 to n-3 fatty acids was lower ($P<0.001$) in beef from LN compared to CT and CS. Results also indicate that daily supplementation of linseed to steers grazing growing grasses on the northern Great Plains may improve growth rate and enhance the n-3 fatty acid profile of their steaks. However, changes in color measurements of steaks presented in retail display indicated that supplementing flaxseed to forage-finished cattle may reduce the acceptability of fresh beef to consumers when purchased from retail display cases.

Session 03 Poster 13

Rearing of dairy cattle herd replacements at pasture

Roca-Fernández, A.I., González-Rodríguez, A. and Salvatierra-Rico, J.A., Agrarian Research Centre of Mabegondo, Animal Production, P.O. Box 10, 15080 La Coruña, Spain; anairf@ciam.es

Feeding and management of dairy cattle herd replacements during rearing period are major factors to consider on animal performance, reproduction, health and welfare of dairy cattle. Rearing on the farm, especially in those where grazing is the main source of feed for heifers, should be based on concentrating calving and achieving the highest lifetime production of cows calved at two years. The purpose of this study was to identify appropriated grazing strategies to apply in dairy herd replacements for achieving acceptable daily live-weight gains at pasture. As a first approximation to the factors that control rearing of heifers at pasture, a trial was conducted at CIAM involving Holstein-Friesian heifers (n=39). Animals grazing pastures of perennial ryegrass and white clover rotationally, were supplemented with silage (grass and maize), when pasture production and/or sward quality was not appropriate to achieve predefined growth. Factors related to grassland management (stocking rate, sward utilization and quality) were controlled and parameters for an adequate development of heifers performance (bodyweight, body condition score, rump height and daily live-weight gains) were determined aiming for calving at around two years of age. Average stocking rate was 3.85 heifers/ha, with 65% sward utilization and grass had 14.5 g/kg of crude protein and digestibility of 78.3%. The results of this study show that mean daily live-weight gains of up to 0.773 kg/head are possible at pasture when quality grass is offered to heifers. Average body condition score of 2.86 (over 5) was maintained and 440 kg/head with a rump height of 137 cm was reached at insemination. Age at insemination was at 18 months and the percentage of pregnancy was 67%, whereas calving was synchronized in spring at 27 months of age. Increased reliance on grazed grass for dairy heifers might be a successful strategy for Galician dairy farms to minimize costs.

Determination of relationship between botanical composition and biomass quantity of grasslands and NIR reflectance by using NDVI derived from Remote Sensing data

Bozkurt, Y.[1], Basayigit, L.[2] and Kaya, I.[3], [1]Suleyman Demirel University, Faculty of Agriculture, Department of Animal Science, Isparta, 32260, Turkey, [2]Suleyman Demirel University, Faculty of Agriculture, Department of Soil Science, Isparta, 32260, Turkey, [3]Kafkas University, Faculty of Veterinary Medicine, Department of Animal Science, Kars, 36100, Turkey; ybozkurt@ziraat.sdu.edu.tr

In this study, the possible relationship between gross botanical composition and biomass quantity and NIR (Near Infra Red) reflectance was investigated by using NDVI (Normalised Difference Vegetation Index) derived from remote sensing data. For this purpose, red (0.45-0.52 µm), near infra-red (0.52-0.60 µm) and infra-red (0.63-0.69 µm) spectral bands of the LANDSAT 5 TM satellite images taken in 2005 were used and NDVI was calculated as a ratio between measured reflectivity in the red and NIR portions of the electromagnetic spectrum. Field measurements were carried out in North Eastern part of Turkey in 23 test stations, during vegetation growth period, to determine botanical composition and fresh biomass quantity in order to compare with the satelite data. Regression analysis showed that there was a poor relationship between botanical composition and NDVI (R2=0.22) while a good coefficient of determination was observed between fresh biomass quantity and NDVI (R2=0.67). It can be concluded that field spectral measurements and satellite derived information can provide a valuable tool towards the evaluation of grasslands.

Session 03 Poster 15

Effect of indoor alfalfa hay supplementation on pasture intake in a rationed dairy sheep grazing system

Arranz, J.[1], García-Rodríguez, A.[1], Ruiz, R.[1], De Renobales, M.[2] and Mandaluniz, N.[1], [1]NEIKER, P.O. Box 46, E-01080 Vitoria, Spain, [2]University Basque Country, P.O. Box 450, Vitoria, Spain; rruiz@neiker.net

Dairy sheep systems in the Basque Country, during spring, are based on part-time grazing. During this period forage resources are managed by matching grazing and indoor supplementation with production requirements. The aim of this study was to evaluate the effect of indoor alfalfa hay supplementation on fresh grass intake. Moreover, grazing time, milk yield and milk quality were monitored. The experiment was conducted with 36 multiparous Latxa dairy ewes for 4 weeks. Sheep were blocked into 3 homogeneous groups of 12, and each group was randomly assigned to different alfalfa rates: (A) 300 gDM/d, (B) 600 gDM/d and (C) 900 gDM/d. Each ewe received 500 g DM of concentrates per day and had access to a poliphyte pasture 4 hours/d after morning milking. Grass intake was estimated by n-alkanes technique comparing faecal recuperation of C_{32} vs C_{33}. Pellets with external alkanes were administered orally to 6 ewes/group during morning milking for 10 days. Faecal samples were collected the last 3 days and faecal n-alkanes were determined by gas chromatography. Grass intake increased as indoor alfalfa supplementation was reduced (1.46, 1.09 and 0.48 kg DM grass/d/ewe in Group A, B and C, respectively), therefore fresh grass contribution to diet was 65, 50 and 26%, respectively. All groups produced the same amount of milk (1.21, 1.18 and 1.21 l/d, respectively), whereas milk from group A had significantly higher unsaturated fatty acid content (1186, 918 and 954 µmol/g fat, respectively), mainly the isomer c9t11 of the conjugated linoleic acid (50.3, 34.4 and 32.9 µmol/g fat, respectively). These results show the possibility to increase the use, during spring, of locally available resources in Latxa dairy system (which could compose as much as 65% of the diet), by reducing indoor feeding. This increase did not compromise milk yield, while the resulting milk had a 'healthier' fat profile.

Use of a combination of n-alkanes and their carbon isotope enrichments as diet composition markers

Derseh, M.B.[1,2], Pellikaan, W.[1] and Hendriks, W.[1], [1]Wageningen University, Animal Sciences, P.O. Box 338, 6700 AH Wageningen, Netherlands, [2]Hawassa University, Animal and Range Sciences, P.O. Box 5, Hawassa, Ethiopia; melkamu.derseh@wur.nl

Plant cuticular n-alkanes have been successfully used as markers to estimate diet composition and intake of grazing herbivores. In botanically diverse vegetation types, however, additional markers are required to accurately estimate diet composition. This study was conducted to evaluate the potential of using a combination of n-alkane profiles and their carbon isotope enrichments to increase discriminatory power of plant cuticular compounds. A total of 23 plant species were collected from the Mid Rift Valley rangelands of Ethiopia and analyzed for long chain n-alkanes ranging from heptacosane to pentatriacontane (C27-C35), as well as carbon isotope enrichment (13C) of these n-alkanes. The analysis was conducted by gas chromatography/isotope ratio mass spectrometry following saponification, extraction, and purification. The dominant n-alkanes in the plant species were C31 (283±246 mg/kg dry matter) and C33 (149±98 mg/kg dry matter). The carbon isotopic enrichment of the n-alkanes ranged from -19.5 to -37.4. Principal component analysis was used to examine the pattern of interspecies differences in n-alkane profiles and their isotope composition. Large variability among the pasture species was observed. The first three principal components explained most of the interspecies variance (n-alkanes 0.91; 13C values of n-alkanes 0.74). Comparison of the principal component scores using orthogonal procrustes rotation indicated that about 0.84 of the interspecies variance, explained by the two types of data sets, were independent of each other, suggesting that use of a combination of the two markers could improve diet composition estimations. It was concluded that, while the n-alkane profile of the pasture species remains a useful marker for use in the study region, the 13C values of n-alkanes could provide additional information in discriminating diet components of grazing animals.

Session 03 Poster 17

Growth, botanical composition, yield and quality of annual and perennial grass-legume swards under rainfed conditions in Greece

Economou, G.[1], Theodorou, E.[1], Tsiplakou, E.[2], Kotoulas, V.[1] and Hadjigeorgiou, I.[2], [1]Agricultural University of Athens, Department of Agronomy, Faculty of Plant Science, 75 Iera Odos, 11855, Athens, Greece, [2]Agricultural University of Athens, Department of Nutrition Physiology & Feeding, Faculty of Animal Science & Aquaculture, 75 Iera Odos, 11855, Athens, Greece; ihadjig@aua.gr

Grass-legume swards were compared on the basis of forage yield and quality under rainfed conditions in Greece. The mixtures comprised of annuals: a) Hordeum vulgare + Pisum sativum (H-P) and b) Hordeum vulgare + Vicia sativa (H-V) and perrenials: a) Lolium perenne + Festuca arundinancea, (L-F), b) Dactylis glomerata + Festuca arundinancea + Trifolium repens, (D-T), c) Lolium perenne + Festuca arundinancea + Trifolium repens, (L-F-T), d) Lolium perenne + Festuca arundinancea + Dactylis glomerata + Trifolium repens, (L-F-D-T), each one at three different proportions of species. Mixtures were sown in mid October 2008, at a completely randomised design with four replicates. Herbage production, botanical composition and the fatty acid (FA) composition of forage produced were determined in the 6 performed harvests. Over the production period, 430 mm of rain and mild winter-spring temperatures were recorded. Results indicate differences between mixtures, where the highest yields produced by the annuals. H-V mixture on average produced the highest (7 tnDM/ha), while on perennial mixtures L-F performed well (5,5 tnDM/ha) and F-L-T produced the lowest (4tnDM/ha). However, the denser plot and suppression of weeds was produced by F-L-D-T mixture, in all combinations. Linolenic acid followed by linoleic acid were the most abundant FA in all treatments, while the richest was F-L-D-T mixture, in all possible combinations. The survey constitutes the baseline of sward establishing in Greece, that will allow increase in production of a quality feed.

Comparison of the eating behaviour of two cow types on pasture

Kunz, P., Wetter, A., Roth, N. and Thomet, P., Swiss College of Agriculture, Department of Animal Science, Laenggasse 85, 3052 Zollikofen, Switzerland; peterkunz@bfh.ch

Various studies have shown that New Zealand Holstein Friesian (NZ HF) cows produce more milk per kg of body weight (BW) than Swiss (CH) or North American type cows. The authors of these studies agree that a principal reason must be the higher intake of grazed grass by the NZ HF cows. The question remains whether there is a genetic influence on the eating behaviour of cows. The present study examined whether NZ HF cows, compared to CH cows, eat more grass at places where the grass has grown higher (e.g. rank patches), which would therefore result in a higher intake of dry matter and, with it, a higher intake of energy. On eight dairy farms, 28 cow pairs (28 NZ HF and 28 CH cows) were observed for 2.5 hours during eating time after milking. Bite frequency (number of bites per minute) and eating time in rank patches (seconds per minute of grazing time) were measured eight times per cow within the 2.5 hours. The two cow types differ significantly (Wilcoxon signed-rank test) in their eating behaviour. NZ HF cows ate more in rank patches (16.3 sec/min; P=0.004) and had lower bite frequency (60.5 bites/min; P=0.027) than CH cows (6.9 sec/min; 62.4 bites/min). The energy-corrected milk yield of the NZ HF cows was higher during the survey (3.6 kg/100 kg of BW) than that of the CH cows (3.1 kg/100 kg of BW). It is concluded that grass intake on pasture is influenced by, among other things, the differing behaviour of cows differing genetically.

Will new regulation focusing on outcomes meet society's demand for improved animal welfare?

Sandøe, P.[1], Jensen, K.K.[1] and Sørensen, J.T.[2], [1]University of Copenhagen, Faculty of Life Sciences, 1958 Frederiksberg C, Denmark, [2]University of Aarhus, Faculty of Agricultural Sciences, 8830 Tjele, Denmark; pes@life.ku.dk

European livestock farming faces a growing public call for animal welfare to be improved. It has also become clear that legally based minimum requirements governing the housing of farm animals are insufficient to avoid revelations of serious welfare problems. This raises the question of how the regulation could be designed to better meet public expectation. Currently, the public regulation of farm animal production is undergoing transformation. Rather than detailed rules for procedures and bureaucratic controls, regulation is increasingly governed by principles of 'own control' and product responsibility. The authorities set up conditions the final products must meet, leaving the choice of means open for the producers. Intervention is resorted to only when there is reason to believe a producer does not have the will, or the competence, to control production. The present paper analyses the application of these principles in animal welfare regulation. To date such regulation has been designed mainly to ensure minimal standards of housing and handling are met, but there are now examples in Europe of regulation set up with the objective of minimizing a specific welfare problem. When the prevalence of the problem on a farm is higher than deemed acceptable, the farmer can be required to work out a plan to improve matters. Underlying these developments there seems to be a change of focus: minimum standards for the protection of individual animal were once the key issue; standards determining when problems in the herd reach an unacceptable level have now become more prominent. Also, there is a change from merely using negative sanctions to the creation of positive incentives. Besides outlining these developments the paper seeks to assess the likely success of the principles in the light of competing criteria, particularly whether the principles promote the central goal of improving animal welfare.

Session 04 Theatre 2

Animal welfare initiatives in Europe: goals, instruments, actors and success factors analyzed in the EconWelfare project

Schmid, O. and Kilchsperger, R., Research Institute of Organic Agriculture (FiBL), Socio-economic division, Ackerstrasse, CH-5070 Frick, Switzerland; otto.schmid@fibl.org

In the EU research project EconWelfare an on-line survey was conducted on 84 Animal Welfare Initiatives (AWIs) in DE, ES, IT, NL, PL SE, UK and Macedonia. A special grouping and assessment system has been developed. 40 regulatory initiatives (legislation or voluntary organic and non-organic standards) and 44 non-regulatory initiatives (e.g. education/information, research, quality assurance and cross-compliance) were assessed. The main goals of AWIs were, besides improving AW, also the creation of awareness amongst target groups and response to consumer concerns. Main instruments were regulatory instruments (legislation and private combined with penalties); labelling (mostly private); financial incentives (private and public); codes of practice and mostly private information/education campaigns. Farmers and farmers groups, major retailers, processors and abattoirs, certification bodies and national governments were main actors in the regulatory initiatives, and in the non-regulatory initiatives AW organisations and researchers. The analysis showed also country specific differences regarding goals, instruments and actors. Few AWIs have already been quite successful in reaching multiple goals, using different policy instruments and involving broader networks. But many of the AWIs still have too narrow goals, do not combine enough instruments and neglect important actors (e.g. farmers). Reflections have to be made how different policy instrument could be best combined for achieving multiple goals, with which actor networks. More dynamic governance models are needed. As countries are in different states/levels of AW development, we will need varying policy instruments to realise improvements.

Session 04 Theatre 3

Strengths and weaknesses of animal welfare standards

De Roest, K. and Ferrari, P., Research Centre for Animal Production, Corso Garibaldi, Reggio Emilia, Italy; k.de.roest@crpa.it

In the last years public and private bodies in Europe have launched and promoted voluntary welfare standards which have the aim to comply with a societal demand for increased animal welfare on farms, during transport and in slaughterhouses. In March the Council of Ministers has backed these standards as a useful instrument to enhance animal welfare beyond the EU legal minimum opening the road for a labelling system analogous to the system of organic production. In this paper, produced with the framework of the EconWelfare project (www.econwelfare.eu), strengths and weaknesses of voluntary animal welfare standards are analysed according to view of different types of stakeholders: scientists, multiple retailers, animal protection and consumers' organisations, farmers' organisations, transport companies and processing industry. This stakeholder consultation has been carried out by means of literature studies and by involving directly the different categories of stakeholders in workshops. Strengths and weaknesses have been analysed from different view angles. Scientists expressed their opinion on the different aspects of the standards from the 'animal' point of view assessing the advantages and disadvantages for the welfare of the animal. Multiple retailers and animal protection organisations analysed the standards according to their efficacy to raise animal welfare, whereas the representatives of the animal production chains gave an assessment of the technical and economic feasibility of the introduction of the standards in the main stream of animal production systems. Finally the different points of view on the animal welfare standards have been compared in a synoptic overview emphasizing those aspects of animal welfare where an agreement among the different stakeholders easily can be been reached and other aspects with a strong disagreement among stakeholders which will have more difficulty to advance. The research has interested fattening bulls, veal calves, dairy cows, sows and piglets, fattening pigs, laying hens and broilers.

Econwelfare: designing policy instruments and indicators for good animal welfare in a socio-economic context

Immink, V.[1], Ingenbleek, P.[1], Hubbard, C.[2], Garrod, G.[2], Guy, J.[2] and Keeling, L.[3], [1]Social Sciences Group of Wageningen Univ and Research Centre, Wageningen, Box 29703, Netherlands, [2]Newcastle Univ, Newcastle Upon Tyne, NE1 7RU, United Kingdom, [3]Swedish Univ of Agricultural Sciences, Uppsala, 750 07, Sweden; linda.keeling@hmh.slu.se

This paper describes the part of the research performed in the EU-funded project Econwelfare that aims to develop policy instruments to support implementation of the Community Action Plan on the Protection and Welfare of Animals as well as to identify indicators to document the relative effectiveness of these policy instruments. The word 'policy instrument' is used in the broadest sense to mean instruments used by government departments or agencies, private enterprises, academic bodies or other non-governmental organizations who formulate standards for animal welfare (AW). 'Indicators' means any quantitative or qualitative measure that can be used to monitor progress at the level of the animal, the food chain and society. As a first step an assessment of strategic AW issues at country level was made. Analyses included stakeholder consultations to validate the findings. A SWOT methodology was used and differences between countries were depicted by their relative position on a matrix representing the relative level of AW, as perceived by stakeholders in the society and chain, on one axis, and the extent to which AW is regulated by legislation or whether higher AW is coordinated in market arrangements by private and non-governmental parties, on the other axis. In the second step, a Delphi Policy exercise was carried out with experts from nine European countries. The aim of this exercise was to verify and refine the set of potential 'policy' instruments and indicators identified in the first step that will allow for the identification of cultural, socio-economic and structural differences between countries. The results highlight the opportunities for different strategies to improve AW in different member states of the EU.

Animal welfare policy: directions for solutions

Backus, G., Immink, V. and Ingenbleek, P., Wageningen UR, Agricultural Economics Institute, P.O. Box 29703, 2502 LS The Hague, Netherlands; gé.backus@wur.nl

Marketing and law are the primary classes of strategic tools for the management of animal welfare. These tools will be considered with respect to specific targets and specific animal welfare issues for which the targets may or may not have any motivation, opportunity, and/or ability to cooperate. The improvement of animal welfare up to a level that is higher than the legal minimum is seen as a task for market actors rather than the government. Because an improvement of animal welfare seems directly related to an increase of costs, price is an important barrier to the improvement of animal welfare in both consumer and business-to-business markets. As compared to promotion, price has for example a much stronger effect on sales. The central question is therefore how the additional costs of animal welfare can be managed in such a way that they are no longer a barrier to welfare improvements. In order to overcome the barriers to improve animal welfare, directions for solutions are: (1) the removal of products that are not animal friendly from the super market shelves, (2) connecting animal welfare to parallel cost decreases, (3) connecting animal welfare to customer value, (4) price increase to cover animal welfare costs, (5) separate payments for animal welfare (separating customer and societal value).

Promoting animal welfare: an example in french cattle slaughterhouses

Mirabito, L.[1], Marzin, V.[1], Morlevat, S.[1], Alleyrangues, X.[2], Frencia, J.P.[2] and Vialter, S.[2], [1]Institut de l Elevage, 149 Rue de Bercy, 75012 Paris, France, [2]ADIV, 10 rue Jacqueline Auriol, ZAC des Gravanches, 63039 Clermont-Ferrand cedex 02, France; luc.mirabito@inst-elevage.asso.fr

The new regulation on the welfare of animals during slaughter, adopted in 2009, by the European Union give to the industry a new template for the development of higher standard of animal care in the abattoirs. Anticipating these changes, french organizations of slaughterhouses by collaborating with research institutes have carried out since 2008 a working program to promote and implement these new rules. The first version of the final guide has been released in 2010 and is based on three main tools. The first one is a hazard analysis realized step by step from the unloading of the animals until the death. For each step, the hazard have been identified in relation with 5 factors (environment, equipment, method, operators, animal) and recommendations about conception and management have been proposed. The second tool consist in flowcharts and instructions. For each step, the standard operating procedures are described and conditional/ decision points are identified in relation with risk for animal welfare. At each node, in case of failure, operators have to take corrective measures in order to solve the problem before going further in the process. The third tool is a self monitoring form, which will be used by the Animal Welfare Officers, in order to check the efficiency of the different procedures. This form is based on a list of ressource-based measures (for example, availability of water, stocking density, etc.) and animal-based measures (for example, slips, falls, handling, unconsciousness which have also been checked during the program on repeatability between observers). This guide is already partly implemented in some slaughterhouses and ongoing activities are development of training for Animal Welfare Officers and operators in order to develop its use in routine process.

Religious slaughter: improving knowledge and expertise through dialogue and debate on issues of welfare, legislation and socio-economic aspects

Miele, M. and Anil, H., Cardiff University, King Edward VII Avenue, CF10 3WA Cardiff, France; haluk_anil@hotmail.com

Religious slaughter, a controversial and emotive issue due to animal welfare considerations as well as cultural and human rights issues, has been subject to much debate recently in Europe. There are two main types of religious slaughter: Halal slaughter for meat intended for Muslims and kosher slaughter for meat intended for Jewish consumers. While the market for kosher products is very small and stable in Europe, in the last decade the demand for Halal meat has increased significantly and market forecasts predict further growth, especially for non-stunned meat. There is considerable variation in current practices in Europe and the rules regarding religious requirements are still unclear. The Dialrel project (www.dialrel.eu, 2006-2010), funded under the EU 6th Framework Programme, was undertaken by 17 partners aimed at providing information relating to slaughter techniques as well as Halal and Kosher consumers' expectations, market shares and the organization of the supply chains for Halal and Kosher products and general public concerns about animal welfare and market transparency. More crucially the project aimed at creating a dialogue with the religious minorities and the industry in Europe in order to promote and disseminate the best practices that will address the welfare of animals, whilst operating within the legal frame of the new EU regulation. This legislation, protection of animals at time of killing (1099/2009) grants exemption from stunning in the case of religious slaughter. To this end, national and international consultations and other dissemination activities were undertaken in order to agree on a list of practical recommendations for improving the welfare of animals at religious slaughter that would be compatible with religious rules. This paper will illustrate how the dialogue between the religious minorities and the researchers working in the project was articulated and how final practical recommendations were produced.

Farm animal welfare and purchasing behaviour

Sossidou, E.N.[1], Szücs, E.[2], Konrád, S.[2], Cziszter, L.[3], Peneva, M.[4], Venglovsky, J.[5] and Bozkurt, Z.[6], [1]National Agricultural Research Foundation, Veterinary Research Institute, N.AG.RE.F. Campus, 57001 Thermi, Thessaloniki, Greece, [2]University of West-Hungary, Var 2. Mosonmagyaróvár, H-9200, Hungary, [3]Banat University, Calea Aradului 119, 300645 Timişoara, Romania, [4]University of National and World Economy, Studenski Grad Hristo Botev, 1700 Sofia, Bulgaria, [5]University of Veterinary Medicine, Komenskeho 73, 041 81 Koşice, Slovakia (Slovak Republic), [6]Afyon Kocatepe University, Ahmet Necdet Sezer Campus, 03200 Afyon, Turkey; Szucs.Endre@mkk.szie.hu

In this paper consumer purchasing behaviour is analyzed in relation to their attitudes towards farm animal welfare, environment and food quality. The survey was part of the activities through the WELANIMAL project and it was undertaken by the face to face questionnaire interview method and at the same time, electronically through WELANIMAL homepage. A random of 465 consumers was interviewed in Greece, Hungary, Romania, Bulgaria, Slovakia and Turkey. Overall, consumers do not find it easy to find farm animal welfare information from labeling. Responses do not differ between different ages and gender ($P \geq 0.05$). On the contrary, nationality together with political and religion positions significantly affect consumers' responses ($P \leq 0.05$). When consumers asked if they are aware of animal welfare issues when purchasing meat, the 43% of them claimed 'yes, most of the time' or 'yes, some of the time'. However, the 53% of them 'very rarely' or 'never' consider these issues. This purchasing behaviour is significantly depended on the nationality and politic positions ($P \leq 0.05$). It is notable that only 3% of respondents stated that there is certainly no positive impact to animal welfare by their purchasing behaviour. The willing to pay more for animal welfare products differs between consumers from countries under study ($P \leq 0.05$). Political positions, marital and educational status, internet facilities at home are important factors that affect this behaviour ($P \leq 0.05$).

Innovative technologies for sustainable management of small ruminants sheep

Caja, G.[1], Carné, S.[1], Rojas-Olivares, M.A.[1], Salama, A.A.K.[1], Ait-Saidi, A.[1], Mocket, J.H.[1], Costa, A.[1,2] and Aguiló, J.[2], [1]Universitat Autònoma de Barcelona, Grup de Recerca en Remugants, G2R, 08193 Bellaterra, Spain, [2]Universitat Autònoma de Barcelona, Microelectrònica i Sistemes Electrònics, MSE, 08193 Bellaterra, Spain; gerardo.caja@uab.es

Diversity, poor development and low income are traits of current small ruminant scenario which limit adoption of new technologies. Traditional practices are not sustainable in modern sheep and goats industry when human labour is scarce. Innovating systems may offer support for the economic sustainability in management. Artificial vision and imaging have been implemented for body composition evaluation (i.e. carcass grading and body condition scoring), identification and traceability (i.e. retinal imaging, detection of diseases and evaluation of morphological traits). Ultrasound scanning is used for *in vivo* carcass and udder evaluation. Infrared imaging for body condition scoring and mastitis detection have been recently used. Moreover, electronic identification (e-ID), adopted as mandatory for sheep and goat in the EU, may be a key tool for the implementation of all systems that use individual data. Radiofrequency by passive low-frequency transponders is the most common technology used in the current automated e-ID equipment for livestock. Use of e-ID allows reading the animals without visual contact and, combined with sensors and positioning equipment, provides a reliable method for monitoring, performance recording, and traceability of animals and most of the animal products in on-field conditions. In sheep and goat farms, e-ID is being implemented to collect information for individual feeding, performance (e.g. weight and milk) and intake recording (e.g. feed and water), for oestrus and het-stress detection and to register movements within the pasture. Dynamic reading using panel readers in conventional race-ways is the faster way for sorting groups and to maintain updated flock or herd inventories. Recent studies on e-ID implementation concluded that it is cost-effective and recommended its use in practice.

Combination of retinography and RFID (Rtn+EID) for an advanced imaging identification in mouflons (*Ovis orientalis musimon, Gmelin* 1774)

V. Petruzzi, V.[1], Cappai, M.G.[2], Muzzeddu, M.[3], Macciotta, L.[1], Nieddu, G.[2] and Pinna, W.[2], [1]University of Sassari, Veterinary Pathology and Clinic Dept., Via Vienna 2, 07100 Sassari, Italy, [2]University of Sassari, Animal Biology Dept., Via Vienna 2, 07100 Sassari, Italy, [3]Institution for Foreste Sardegna, S.S., Sassari-Fertilia, 07040 Olmedo, Italy; mgcappai@uniss.it

Reference about retinographic peculiarities in mouflons lack important descriptions of specific pictures normally used nowadays as a fine tool in neurophtalmic semiology in many animal species. A total of 30 mouflons, firstly electronically identified by means of transponders HDX 134.2 kHz, ISO 11784-11785 contained in endoruminal ceramic boluses (75 g. 70x21 mm RUMITAG bolus®), were investigated by 12 retinographies (Kowa RC–2 retinograph fundus camera con Kodak Elitechrome film, 100 asa) on both eyes, after 2 - 3 drops of tropicamide (Visumidriatic 0,5%) to allow adequate pupil dilation. The detection and record of transponder's electronic code (EIC) were performed by a handy reader (static reading) and by a fixed reader (dynamic reading), together with real time download on a notebook of transponder's electronic codes. As far as vascularization, retinographic pictures were characterized by the presence of 3 main arteries: one emerges from the papilla following a dorsal direction; 2 follow lateral directions and veins behave as satellites of arteries, often twisted to the correspondent artery. The retinographic pictures observed, as emergence, number, pathways and distribution, according to the respective topographic relationships of vessels in the fundus of the eye, allowed to recognize for each single mouflon one own individual retinographic picture as a biologically originated fingerprint for the identification of each animal. The integrated system (retinography & RFID) opens interesting fields of investigation for future chances of development of new softwares in the Advanced Imaging Technology that is increasing for human applications and that will presumably involve also the animal sector in the next years.

The profitability of early silage harvesting on Norwegian dairy goats farms

Flaten, O.[1], Asheim, L.J.[1] and Dønnem, I.[2], [1]NILF, Box 8024 Dep., 0030 Oslo, Norway, [2]Animal and Aquacultural Sciences, Box 5033, 1432 Ås, Norway; ola.flaten@nilf.no

This study aims to evaluate how harvesting time of grass silage influences economic result and optimal use of inputs, in particular fertilizers and concentrate in dairy goat farming. Due to a cold climate, goats in mountain areas of Norway are fed indoors for up to 9 months. Early harvesting improves silage quality and the need for purchased concentrates can be reduced. A linear programming model was developed to examine three harvesting regimes: very early (HR1), early (HR2) and normal (HR3), producing forage containing 7.2/6.2, 6.2/6.2, and 5.3/5.3 MJ of net energy lactation/kg DM, in primary growth/regrowth, respectively. Response curves in the model were derived from field experiments with two levels of fertilization and animal trials with two levels of concentrate for each harvesting time. The model maximises profit of a goat farm with 70,000 l milk quota, and stalling capacity for 100 goats. By assuming 6 ha of land, optimal fertilization level, concentrate supplementation and milk production per goat were highest when fed HR1 and HR2 silage. However, HR3 was most profitable as benefits such as higher milk yield and better paid milk due to higher total solids content, could not offset the higher opportunity costs of producing silage following lower yields and poorer winter survival of the swards for HR1 and 2. HR1 was particularly unfavourable as it was impossible to fully produce the quota. With more farmland available, intensity in silage and milk production was lowered. HR2 became most profitable with more land, whereas HR1 never performed best. By abolishing the quota, a high intensity in milk production was beneficial. HR3 was still profitable at limited availability of land, but the gains of HT2 increased with more land, and with plenty of land HR1 was most competitive. The study demonstrates that farmland availability profoundly influence the choice of production intensity and the profitability of producing and feeding high quality grass silage to dairy goats.

Types of milking machines-parlors for dairy sheep and goats in Greece

Kiritsi, S., Katanos, I., Skapetas, B. and Laga, V., Alexander Technological Educational Institute, Department of Animal Production, P.O. Box 141, 57400 Thessaloniki, Greece; ikatanos@ap.teithe.gr

The different types of milking machines-parlors for milking of small ruminants in Greece could be divided into four groups: milking machines with buckets, mobile milking machines with trolley, milking parlors with milk pipelines, and rotary milking parlors. The bucket milking machine system is used in a small number of farms with few animals and the throughput is 70 ewes or 50 goats/h/milker. The trolley mobile milking machine systems have the same problems with the bucket milking systems and are used in small farms. Milking machines with milk pipelines (fixed parlor installations) exist in different type variations that differentiate according to the position of the animal in relation with the milker's position and/or the method of entrance and exit of the animals. The various types could be classified into two groups: the fixed 'casse', tunnel and herringbone, and the dynamic system of continuous movement (rotary and fast exit). The majority of the milking machines are of the fixed 'Casse' type. These milking parlors have one pit and 1-2 ramps. The total number of stalls (animal positions) is varying (6-66). The ratio of the milking units to the milking stalls could be 1:1, 1:2, 1:3 or 1:4, with the most frequent ratio being 1:2. The throughput of this system is 160-170 ewes/h/milker. In the rotary milking parlors (Carrousel), the animals and milking units are moving in a circular way on a rotating platform. This system is used in farms with large number of animals and the throughput is from 300 to 400 animals/h/milker, when automatic milking cluster removers are used. The mobile milking parlors are used in a very small percentage of farms and only in cases where flocks of the same farm are in different areas or in extensive nomadic or semi-nomadic flocks. All the equipment of the milking system is on a 4 wheel platform and the throughput can reach 100-120 ewes or 60-80 goats /h/milker.

Milking machines-parlors' throughput for dairy goats in Central Macedonia, Greece

Katanos, I., Kiritsi, S., Skapetas, B., Laga, V., Sentas, A. and Bampidis, V.A., Alexander Technological Educational Institute, Department of Animal Production, P.O. Box 141, 57400 Thessaloniki, Greece; ikatanos@ap.teithe.gr

The aim of this work was to study the efficiency of different types of milking machines for dairy goats in the region of Central Macedonia, Greece. Thirty-two dairy goat farms were used totally. The average size of the farms was above 200-300 goats in milk. In each farm, five visits were carried out during the first 5 months of the milking period of dams. All the visits were conducted in the evening milking. Two questionnaires were used for the registering of different technical data in the milking parlors such as the time of starting and finishing of milking procedure, the time of entering and exit of dams in the parlors, the time of milking of each animal group, the time of hand stripping, the quantity of milk per hour, the farm size, the animal's breed, the milking machines type and size etc. From these data, the average throughput of the milking machines-parlors and the average time of the milking procedure were calculated. The average throughput of the Carrousel type parlors ranged from 245 to 427 dams/h, according to their size and the number of the milkers during milking. In the traditional «Casse» type parlors this parameter ranged from 138 to 400 dams/h, while in the modern «Casse» type (fast exit) parlors from 148 to 466 dams/h. The lowest throughput was found for the bucket system milking parlors (45 dams/h). The variable milked dams/milker/h ranged from 55 to 129 because of the low milkability of the animals and the inappropriate labor organization in the milking parlors. The average milking time of flocks ranged from 1 to 2.3 hours. This parameter may be considered in normal limits having in mind that the farmers are applying time-consuming milking routines. It's concluded that the average throughput of milking machines-parlors for dairy goats ranged in medium or low levels in Greece.

Nutritional and physiologically competitive factors affecting fine fibre production in animals

Galbraith, H.[1,2], [1]*University of Aberdeen, Institute of Biological and Environmental Sciences, 23 St Machar Drive, Aberdeen AB24 3RY, United Kingdom,* [2]*University of Camerino, Department of Environmental and Natural Sciences, Via Pontoni 5, 62032 Camerino, Italy; h.galbraith@abdn.ac.uk*

Hair fibre is produced by specialist hair follicles embedded in the skin of animals. These follicles are composed of epidermal and dermal components in general commonality with other tissues of the mammalian integument. The major components of these tissues are proteins which include low-sulphur (cysteine)-containing keratins and higher sulphur-containing keratin-associated proteins produced during proliferation and differentiation of epidermal keratinocytes. This study tested the hypothesis that mohair fibre production is affected by supply of amino acids from the digestive tract and endogenously and apparent competition, physiologically, for these between keratinising tissues. The effect of dietary supplementation of the rumen protected sulphur-containing amino acid, methionine, was studied in ten Angora goats given dietary supplementation of either 0 or 2.5 g/day for 102 days. Methionine supplementation significantly ($P<0.05$) increased clean mohair yield (7.9 vs 15.6±1.02 g/day), and deposition of cysteine (9.70 vs ±10.5 g/100 g). Non-fibre effects included increased fleece-free liveweight gain, initial (left front hoof) dorsal horn growth and hardness associated with greater concentration of cysteine. Examination of extracts of mohair fibre and hoof horn by one-dimensional electrophoresis and scanning densitometry demonstrated commonalty in presence of individual proteins except for one band at 64-65kDa which was absent in fibre. Certain significant differences were recorded in relative quantitative expression of proteins according to molecular mass. The results provide quantitative information relating amino acid nutrition to mohair fibre production. They also inform on competitivity between fibre and non-fibre products and show that this is associated with commonality in chemical composition of synthesised products.

A two-locus model for the Suri/Huacaya phenotype in the alpaca (*Vicugna pacos*)

Presciuttini, S.[1], *Valbonesi, A.*[2], *Apaza Castillo, N.*[3], *Antonini, M.*[2], *Huanca Mamani, T.*[3] *and Renieri, C.*[2], [1]*Università di Pisa, Dipartimento di Scienze fisiologiche, Via San Zeno 31, 56123 Pisa, Italy,* [2]*Università di Camerino, SARRF di Scienze Ambientali, Via Gentile III da Varano, 62032 Camerino, Italy,* [3]*INIA, ILLPA Puno, Rinconada Salcedo, Puno, Peru; carlo.renieri@unicam.it*

Genetic improvement of fiber-producing animal species has often induced transition from double coated to single coated fleece, accompanied by dramatic changes in skin follicles and hair composition, likely implying variation at multiple loci. Huacaya, the more common fleece phenotype in alpaca (Vicugna pacos), is characterized by a thick dense coat growing perpendicularly from the body, whereas the alternative rare and more prized single-coated Suri phenotype is distinguished by long silky fiber that grows parallel to the body and hangs in separate, distinctive pencil locks. A single-locus genetic model has been proposed for the Suri-Huacaya phenotype, where Huacaya is recessive. We show that this model is rejected in a set of controlled test-crosses using a maximum likelihood approach. In addition, we present two surprising observations: 1) a large proportion (about 3/4) of the Suri animals is hybrid (with at least one Huacaya offspring), even in rearing conditions where the Huacaya trait would have been almost eliminated; 2) a model with two different values of the segregation ratio fit the data significantly better than a model with a single parameter. These facts can be parsimoniously explained by a genetic model in which two linked loci must simultaneously be homozygous for recessive alleles in order to produce the Huacaya phenotype. The estimated recombination rate between these loci was 9.9%.

The effect of weight loss on wool protein expression profiles in Australian Merino Lambs: a proteomic study using 1DE and Mass Spectrometry

Almeida, A.M.[1,2,3], Ariike, L.[3,4], Kilminster, T.[5], Scanlon, T.[5], Greeff, J.[5], Cardoso, L.A.[1,2], Oldham, C.[5] and Coelho, A.V.[3], [1]IICT - Instituto de Investigação Científica Tropical, CVZ, FMV, Av. Univ. Técnica, 1300-477 Lisboa, Portugal, [2]CIISA - Centro Interdisciplinar de Investigação em Sanidade Animal, FMV, Av. Univ. Técnica, 1300-477 Lisboa, Portugal, [3]Instituto de Tecnologia Quimica e Biologica, MS laboratory, Av. Republica EAN, 2781 Oeiras, Portugal, [4]Competence Center of Food and Fermentation Technology & Tallinn University of Technology, Ehitajate tee 5, 19086 Tallinn, Estonia, [5]Dep. of Agriculture and Food Western Australia, 3 Baron-Hay Court South Perth, WA 6151, Portugal; aalmeida@fmv.utl.pt

Commercial wool production is based on merino breeds under extensive management practices, frequently in tropical dry regions. Seasonal weight loss (SWL) is the most important constraint to animal production in tropical countries. The influence of SWL on wool quality parameters has been studied, however, there is a lack of information on how does it affects protein expression. Twenty four Australian merino (Peppin) ram lambs were divided into two groups: control (C) and underfed (U). C animals were fed on maintenance diets; U animals on a restricted diet. Experiment lasted 42 days until U animals reached 85% of their initial body weight. Wool present in a 100 cm^2 area over the left scapulae was sampled. Protein was extracted using the Shindai method. One dimension electrophoresis (SDS-PAGE) stained with colloidal coomassie were conducted. Gels were digitized and analyzed calculating relative intensities for each band present that was identified using mass spectrometry techniques. Two bands that showed higher expression in the U animals, that were identified as Keratin I and Glycine tyrosine proteins. Results suggest the existence of alterations in wool fibril structure at the level where those proteins are located as a consequence of weight loss.

Analysis of KIT (mast/stem cell growth factor receptor) mutations in the skin of merino sheep

Saravanaperumal, S.A., Pediconi, D., Renieri, C. and La Terza, A., University of Camerino, Dept. of Environmental and Natural Sciences, Via Gentile III da Varano, 62032 Camerino (MC), Italy; carlo.renieri@unicam.it

KIT, a type III receptor protein tyrosine kinase also referred to as steel factor, plays an essential role in melanocyte development, migration and survival. Mutations in the KIT gene have previously been shown to cause white coat colour phenotypes in pigs, mice and humans. To investigate the role of KIT in dominant white merino sheep, the full length cDNA was cloned, sequenced and the 5' & 3'UTRs were characterized by RACE experiments from the skin biopsies. RT-PCR analysis revealed two basic splice variants with (+) and without (-) a 12bp insert in the coding region for the amino acids 'GNSK' (i.e., GTAACAGCAAAG pos.1532-1543). The cDNA sequence analysis revealed a 'C' to 'T' transition, 11bp just before the insertion site corresponding to the amino acid S507F in (+) to (-) form. Semi-quantitative RT-PCR of the white, black and brown animal exhibited the predominant expression of the (-) form over the (+) form in skin. Two different 5'UTRs with distinct N-terminus sequences were determined by 5'RACE in black against white and brown animals. In contrast, a single product of 795bp was obtained by 3'RACE. In our study, we indentified 15 point mutations in the coding region. Twelve SNPs (T22P, L57S, K167R, M195I, N261I, S468P, I609T, N649S, V665A, V750A, T815I, P862S) were found to be missense and three (242V, 725V, 739S) were silent mutations. Eight SNPs were located on the extracellular IgG-like domain involved in the receptor dimerization and seven SNPs were found in the intracellular catalytic domain responsible for trans-phosphorylation, activation, and intracellular signaling. Molecular dynamics simulations in combination with gene expression studies are in progress to better unveil the relevance of KIT receptor-ligand (SCF) interaction linked to coat color genetics.

Session 06 Theatre 1

Chemical and physico-chemical properties of equid milk

Fox, P.F. and Uniacke, T., University College Cork, Food and Nutritional Sciences, College Road, NA, Ireland; pff@ucc.ie

The milk of equids is dilute, containing ~2% fat, ~1.5% protein, ~7% lactose and ~0.3% ash. The chemistry and properties of lactose have been studied extensively. Milk also contains oligosaccharides (OS) containing 3-10 monosaccharides; they contain lactose at the reducing end and also fucose, galactosamine and sialic acid. Human milk contains 15 g OS/L; the milk of monotremes, marsupials, bears and elephants also contain high levels of OSs. Equine colostrum contains ~25 mg OSs /L; there are no reports on the OSs in equine milk. Equid milk contains ~2% lipids, ~80% of which are triglycerides, ~5% phospholipids and ~9% fatty acids (FAs). The lipids contain relatively high levels of middle-chain FAs and polyunsaturated FAs. The lipids occur as globules (2-3 mm), stabilized by a membrane acquired during expression of the globules from the mammocytes. Long (0.5-1 mm) filaments, comprised of mucins, extend from the surface of the globules. The filaments facilitate the adhesion of fat globules to the intestinal epithelium, prevent bacterial adhesion and may protect mammary tissue against tumours. The ratio of caseins and whey proteins in equid milk is ~1:1. The principal whey proteins are a-lactalbumin, b-lactoglobulin, lysozyme, and immunoglobulins. The principal casein is b-casein with a lesser amount of a_s-casein and a very low level of k-casein. The primary structure and other molecular properties of the caseins are known. The caseins exist as micelles, diameter 250 nm, which are sedimented at low gravitational forces, are quite heat stable > pH 7.0 but are very unstable < pH 6.8. Equine milk is destabilized by chymosin but does not form a gel, partly because of the low casein concentration; if the pH of milk is reduced, the caseins aggregate and sediment on renneting without forming a gel. Asinine milk, also, does not form a gel on renneting; cheese can not be made from equine or asinine milk.

Session 06 Theatre 2

History, ethnology and social importance of horse milking in central Asia

Langlois, B., INRA, Animal genetics, Bige-INRA-CRJ, 78 352-Jouy-en-Josas, France; bertrand.langlois@jouy.inra.fr

It is now known that horse milking can be traced back to the early domestication time of the Botai culture, in the north of Kazakhstan near 3500 y. B.C. Then after 1000 y. B.C the horse cultures spread in the steppian regions because horses are particularly adapted to these countries characterized by draught in the summer, very cold winters and where the vegetal primary production is sparse. Horses are very narrow grazers but are also able to browse woody plants. Besides their ability to move, they can eat during14 hours of a 24 h day, which is a considerable advantage in these conditions. Their ability to depose and to mobilize fat makes them a kind of Camel of the cold desert (mountain, steppe and even taiga).It is therefore not surprising that they are used as the main tool of shepherds to farm this environment. The use of horses has also evolved becoming a formidable weapon with the development of the saddle (400 B.C) and that of the composite bow. Following the Scythians, today nations perpetuate this heritage in Central Asia and Asia. These are the Kazakhs, the Kirghizes, the Baschkirs, the Yakutes, the Kalmuks, and the Mongols. The latter is certainly the most emblematic of the common culture and we will consider them as an example in this paper to illustrate the place of horse milk in the core of the turco- mongol societies.

Advances on equine milk and derivatives for human consumption

Martuzzi, F. and Vaccari Simonini, F., Università di Parma, Dipartimento di Produzioni Animali, via del Taglio 8, 43100 Parma, Italy; francesca.martuzzi@unipr.it

Recent advances in research concerning mare's milk are reviewed. It is estimated that around 30 million people consume equine milk regularly throughout the world. In many Countries of Asia and East Europe mare's milk is consumed as koumiss, airag or chigee, alcoholic dairy products obtained by means of a mixed culture of fermenting yeasts and bacteria. A considerable variability in the microorganisms isolated from fermented mare's milk is observed. Isolation and characterization of Lactobacillus spp. from koumiss are addressed in numerous publications from East Asia: 240 strains have been identified since 2004. Several studies take in account mare's milk potential probiotic aspects. Many biologically active factors are present in equine colostrum and milk, in particular lysozyme in high amount. Especially the tissue repair function of some bioactive molecules has been tested on patients with several pathologies and positive responses were documented in clinical studies. Equine milk has a whey protein:casein ratio close to 50:50 and while biochemistry and structure of whey proteins are well known, characterization of the casein fraction is quite recent and considered in some studies with different methodological approaches. Some researches are aimed to study alternative milk for children with cow milk protein allergy. Mare's milk immunoblottings obtained with anti-beta-lactoglobulin and anti-total casein polyclonal antibodies, produced against cow milk proteins, showed very mild immunoreactivity, while both antibodies strongly react with proteins from other dairy species. Fat content in equine milk is lower if compared to other species, ranging between 10-20 g/kg. Recent studies confirmed that fat composition is characterized by high content in polyunsaturated fatty acids: linoleic and alfa-linolenic fatty acids, essential for human nutrition, are present in very variable proportion that can exceed 20%, while conjugated linoleic acid content is around 0.1%, much lower than in cow or ewe milk.

Mare's milk composition in two breeds during different stages of lactation

Lehtola, K.[1] and Saastamoinen, M.T.[2], [1]University of Vaasa, Marketing, PL 700, 65101 Vaasa, Finland, [2]MTT Agrifood Research Finland, Equine Research, Opistontie 10, 32100 Ypäjä, Finland; katariina.lehtola@uwasa.fi

The composition of mare's milk resembles more human milk than cow's milk. The milk composition varies due to many factors, e.g. the breed of the horse, stage of lactation, age of the mare and nutrition. The aim of this study was to determine how breed, stage of lactation, number of previous lactations and mare's age influence the basic composition of mare's milk as well as its fatty acid, Ca, P, Mg, Zn and Fe contents. Milk samples were collected from 22 and 12 Finnhorse (FH) mares in two consecutive years, respectively, and from 8 Standardbred trotters (STB). The age of the mares ranged from 5 to 21 yrs. The milk samples (100 ml) were hand-milked and frozen (-20 °C) before analyses. The breed, the stage of lactation and the number of parturitions had an impact on the milk composition. The dry matter content was highest at foaling, especially due to its high protein content - then the content decreased, and increased again towards the end of the lactation. Also the ash (mineral) content was highest at foaling, and after that it lowered evenly until the end of the lactation. For the STB, their milk had lower fat content during the first lactation week compared to the FH. The protein content of the STB mare's milk was higher during the first 12 hours compared to the FH milk. The total solids, fat and protein contents decreased as the number of lactations increased, but the lactose and the ash contents did not vary, or they decreased only slightly. The effect of the lactation period on the milk composition was similar for both breeds. The milk of the STB mares contained more unsaturated fatty acids (C18:1; C18:2; C18:3) compared to the FH milk. The percentage of palmitic- and the oleicacids and the total unsaturated fatty acids were directly related to mares' parity in both breeds. The fatty acid composition in mare's milk is comparable with human milk. Due to its composition, mare's milk is suitable for infant nutrition when produced and handled in excellent hygiene conditions.

Thermal inactivation kinetics of alkaline phosphatase in equine milk

Marchand, S., Coudijzer, K. and De Block, J., Institute for Agricultural and Fisheries Research (ILVO), Technology and Food Sciences Unit, Brusselsesteenweg 370, 9090 Melle, Belgium; sophie.marchand@ilvo.vlaanderen.be

Alkaline phosphatase (ALP) is widely used as an indicator of proper pasteurization in bovine milk. Due to interest in the use of equine ALP as a time/temperature integrator (TTI) for evaluation of the efficacy of thermal processing of equine milk, its inactivation kinetics were evaluated in whole and skimmed equine milk. Experimentally determined decimal reduction times showed that equine ALP is more readily inactivated in equine milk than its bovine counterpart. Thus, considering the required 6 D reduction of pathogens and the rather low enzyme level present in equine milk, equine ALP will not be suitable as indicator for correct pasteurization of equine milk under the conditions currently used in the reference method for the determination of ALP in milk-based products.

Session 06 Theatre 6

Mechanical milking of equines, the european proposal

Simoni, A.[1], Drogoul, C.[2] and Salimei, E.[3], [1]University of Molise, S.A.V.A., Via De Sanctis, 86100 Campobasso, Italy, [2]LEGTA Semur Chatillon, Chatillon sur Seine, 21400 Chatillon sur Seine, France, [3]University of Molise, S.T.A.A.M., Via De Sanctis, 86100 Campobasso, Italy; salimei@unimol.it

Equine milk utilisation as an hypoallergenic food for infant requires the adoption of adequate milking equipments and routines in order to achieve a high standard quality product in a economically feasible production. The aim of the paper is to review the studies on mechanical milking technique for mares, started in France in 1990's, and for donkeys, as the breeding of this equine is developing in Italy since 2000. Different prototypes of milking facility for dairy jennies are described as related to an innovative milking routine. In dairy jennies, average time for milk ejection is about 60 seconds. In specific facility, the milking routine time lasts in average 3 minutes per head; mechanical milking is carried out in a environment projected to preserve donkey's milk safety and minimize the stress for animals and operators.

Session 06 Theatre 7

Advances on *Equus asinus* as a dairy species

Salimei, E., Università del Molise, di. S.T.A.A.M., via De Sanctis, 86100 Campobasso, Italy; salimei@unimol.it

The ancient utilization of donkey as dairy animal has recently achieved a relevant scientific interest in human nutrition since clinical evidences show that donkey's milk is well tolerated by infants with cow's milk allergy and its use is reported to be useful in both the treatment of human immune-related diseases and the prevention of atherosclerosis. Donkey's milk production differs greatly from that of traditional dairy species in terms of both anatomical and economical characteristics. Jenny's milk production is in fact low (350 - 850 ml per milking) and its gross composition differs significantly from that of cow's milk, showing similarity to human milk except for the low and variable average total solid and fat content. Current data on donkey's milk show in average a good hygiene standard, according to the European Community legislation for raw milk. Among the most studied donkey's milk components as related to a antigenic role, the protein fraction shows lower content of casein and β-lactoglobulin and higher content of lysozyme than in bovine milk. Moreover, proteomic studies clearly illustrate the similarity of equine and human milk when compared to conventional dairy species. As the most complete dietary source for growing newborns, milk is not only a source of antigens but also of bioactive and functional compounds, like metabolites, enzymes, hormones, trophic and protective factors. Some nitrogenous and lipids functional components as well as peptides with human-like activity have been determined in donkey's milk, suggesting the potential health promoting value of this innovative dairy food. Aiming to highlight some unique characteristics of the dairy donkey enterprise that influence milk yield and quality, genetic and environmental factors such as diet, stage of lactation, management of milking and thermal milk processing are reviewed.

Session 06 Theatre 8

***In vitro* digestion of donkey milk with human gastrointestinal enzymes**

Tidona, F.[1], Criscione, A.[1], Jacobsen, M.B.[2], Devold, T.[3], Bordonaro, S.[1], Vegarud, G.E.[3] and Marletta, D.[1], [1]University of Catania, DACPA, via Valdisavoia, 5, 95123 Catania, Italy, [2]University of Oslo, Dpt of Research and Development, Oestfold Hospital Trust, 1601 Fredrikstad, Norway, [3]Norwegian University of Life Sciences, Dpt of Chemistry, Biotechnology and Food Science, P.O. Box 5003, Aas, N-1432, Norway; d.marletta@unict.it

Tolerance and therapeutic properties of donkey milk are well known. The characteristic protein profiles, with a low casein content, the peculiar fatty acids composition and the high lactose content confer to this milk good tolerability, high nutritional and functional properties. In this study individual milk samples from 14 Ragusano donkeys were analysed in their qualitative and quantitative protein profiles. Isoelectrofocusing analysis revealed three IEF protein patterns. Individual variability was related both to the caseins (one defective sample, without αs1-casein) and the whey proteins (two defective samples, without β-lactoglobulin II); the remaining 11 samples showed a normal protein pattern. The distribution of nitrogen fractions, size and zeta potential of the casein micelles were also investigated in order to verify the possible correlation between protein polymorphism, nitrogen fractions content, casein micelles and protein digestibility in donkey milk. A simulated *in vitro* digestion study, carried out in a two-steps with human gastric juice (pH 2 and pH 4) and duodenal juice (pH 7.5), was performed. During digestion, the degradation profiles of donkeys' milk proteins and fatty acids were assessed to provide direct evidence of differences in the digestion of this milk. The observed polymorphism led to different protein profiles and a modified casein to whey protein ratio but did not show any appreciable difference in the size and the zeta potential of casein micelles. Individual variability affected the digestion of milk proteins only in the two defective samples lacking the β-lactoglobulin II, revealing a more rapid degradation in particular during the gastric digestion.

Thyroid hormones in blood and milk of lactating donkey: preliminary results

Todini, L.[1], Fantuz, F.[1], Malfatti, A.[1], Brunetti, V.[1] and Salimei, E.[2], [1]Università di Camerino, Dipartimento Scienze Ambientali, V Circonvallazione, 62024 Matelica (MC), Italy, [2]Università del Molise, Di STAAM, V De Sanctis, 86100 CB, Italy; francesco.fantuz@unicam.it

Thyroid hormones in milk could stimulate lactation in the mother and play physiological roles for the suckling offspring. Donkey milk is valuable for human infants with cow's milk allergy as well as for individuals with immune-related diseases. The aim was to assay thyroid hormone concentrations in milk and blood of 16 lactating jennies, 32-58 days postpartum at the beginning of the study, stabled with their foals and fed 8 kg of coarse hay and 2.5 kg of mixed feed (12.8% cp) daily. Samplings were carried out at 15 d-interval from 9th March to 1st June, at 11.00 am. Animals were machine milked after a 3 h-separation from their offspring and milk samples were immediately processed for iodothyronines extraction with alkaline ethanol at -20 °C. Blood samples were collected by jugular venipuncture in evacuated tubes (K3-EDTA), immediately centrifuged and the plasmas stored at -20 °C until assayed. Total concentrations of T3 in milk and T3 and T4 in plasma were assayed using EIA kits (Radim, Rome, Italy), expressly validated for donkey species (intra- and inter-assay CVs for T3 and T4 and for milk and plasma: 2-8%). High variability of individual mean plasma concentrations was observed, expecially for plasma T3 values (range from 5.1±0.4 to 38.1±5.3 ng/ml). Overall mean plasma T4 concentration was 76.9±8.8 ng/ml (individual means ranging from 38.9±10.1 to 166.4±33 ng/ml). Mean T3 in milk was less variable among individuals (mean 4±0.2; range 2.8±0.1 to 6±0.2 ng/ml). Plasma T4 was affected by time ($P<0.001$), showing a rise from April forward. The highest mean of plasma T3 was observed in June. Milk T3 concentrations were rather stable throughout lactation. To our knowledge, this is the first time that bioactive T3 in milk has been assayed by ELISA and in donkey throughout lactation, as well: further research will address its potential biological actions.

Session 06 Poster 10

Grana padano cheese making with lysozyme from ass's milk: first results

Galassi, L.[1], Salimei, E.[2] and Zanazzi, M.[1], [1]Ente Regionale Servizi Agricoltura Foreste-Regione Lombardia, Via Pilla, 25/b, 46100 Mantova, Italy, [2]Università del Molise, STAAM, Via De Sanctis, 86100 Campobasso, Italy; laura.galassi@ersaf.lombardia.it

Maize silage is one of the principal livestock forage in dairy farm where the PDO Grana Padano cheese is made. Unwelcome microorganisms can therefore move to milk by external contamination, with negative effects on cheese ripening. In order to prevent swelling in Grana Padano cheese ripening, lysozyme from hen egg white has to be used in Grana Padano cheese making. On the other hand, sensitive consumers can show allergy to hen eggs and their components. Since lyzsozyme content is reported to be high in ass's milk, a dairy food for consumers with food allergies, nine traditional Grana Padano cheeses (tGP) were compared to nine ass's milk added cheeses (amGP), in order to study the ripening process under the effect of lysozyme from the two different sources. Chemical, microbiological, radioscopic and sensory analyses were performed for evaluating the quality of milk, whey and cheese. Although the lowest defect score assigned by radioscopic analysis was observed for the tGP thesis, such difference didn't increase, as ripening was carried out. Moreover, pH values at 3 hours were significantly lower in amGP than in tGP thesis (5.367±0.013 vs. 5.562±0.014, $P<0.0001$), suggesting that ass's milk could make lactic fermentations easier.

Session 06 Poster 11

Microbiological characteristics of raw ass's milk: manual vs. machine milking

Sorrentino, E.[1], Di Renzo, T.[1], Succi, M.[1], Reale, A.[1], Tremonte, P.[1], Coppola, R.[1,2], Salimei, E.[1] and Colavita, G.[1], [1]Università del Molise, Di. S.T.A.A.M., via De Sanctis, 86100 Campobasso, Italy, [2]CNR - ISA, via Roma, 64, 83100 Avellino, Italy; colavita@unimol.it

On the basis of the advantages recognised in infant nutrition of the use of ass's milk for its dietary and therapeutic properties, the aim of this study was to evaluated the influence of the dairy jennies' milking on the microbiological characteristics (total mesophilic bacterial count, psicrotrophic bacteria, yeast and molds, lactic acid bacteria, *Enterococcus* spp., Enterobacteriaceae, total coliform, faecal coliform) of milk, raw and after storage. Moreover the effect of manual vs machine milking procedure was also evaluated on jennies' milk yield. The study was carried out in farm A (600 donkeys) where jennies were machine milked and in farm B (20 donkeys) where dams were manually milked. The raw milk bulk samples from the farm A showed excellent microbiological characteristics (mean total bacterial count ± s.d., log CFU/ml: A - 3.50±0.3; B - 4.70±0.2): the good microbiological quality did not change during the storage at 4 °C for 3 days. The presence of undesired microorganisms found after storage at 4 °C for 3 days in the bulk donkey's milk from farm B cannot be only attributed to the modalities of manual milking, but also to hygiene management of the animals. Results recommend special care in both management of dairy jennies and mechanical milking in order to achieve raw ass's milk with elevated hygienic-sanitary standard.

Session 06 Poster 12

Hygiene and health parameters of donkey's milk

Colavita, G.[1], Amadoro, C.[1], Maglieri, C.[1,2], Sorrentino, E.[1], Varisco, G.[3] and Salimei, E.[1], [1]Università del Molise, Di STAAM, via De Sanctis, 86100 Campobasso, Italy, [2]IPSSAR G. Giolitti, P.zza IV Novembre, 12084 Mondovì (CN), Italy, [3]Ist Zooprofilattico Sperim della Lombardia e Emilia, via Bianchi, 9, IT-25124 Brescia, Italy; colavita@unimol.it

The growing interest toward donkey's milk as food for infants with cow's milk protein allergy implies unexceptionable hygiene and health requirements. A biennial study was carried out on some microbiological parameters of bulk milk samples (per milking) from 10 Martina Franca-derived dairy jennies machine milked from April to September each year. A total of 39 sample of raw milk were collected and underwent microbiological evaluation of Salmonella spp, *Listeria monocytogenes*, *Listeria* spp. by PCR while *S. aureus*, *E. coli* and total coliform enumeration was carried out by plate count. Pathogens like Salmonella spp. and L. monocytogenes were not found in all the samples. *S. aureus* has been isolated only once (10CFU/ml). Among the hygiene indices, *E. coli* was observed in 2 samples (10^2CFU/ml), while total coliform counts varied from <10 CFU/ml to 3.7 x 10^6 CFU/ml. *Listeria* spp. has been isolated in 20 samples out of 39. Although the results have to be considered as preliminar, they allow a substantially positive judgement on health quality of donkey's milk machine milked due to the absence of pathogens like *Salmonella* spp. and *L. monocytogenes*. However, both the *E. coli* occasional isolation and the ample variability of the coliform contamination suggest that great attention has to be paid to the management of milk since milk hygiene can be compromised as soon as correct procedures of hygiene during milking and collection are disregarded. The presence of *Listeria* spp. in donkey's milk shows an ineffective sanitization of equipments and milk facilities.

Trace elements supplementation to dairy donkey's diet: milk yield per milking and major milk components

Fantuz, F.[1], Todini, L.[1], El Jeddad, A.[2], Bolzoni, G.[3], Lebboroni, G.[1], Ferraro, S.[1] and Salimei, E.[4], [1]Università di Camerino, Dip. di Scienze Ambientali, via Circonvallazione 93, 62024 Matelica (MC), Italy, [2]Az. Agr. Montebaducco, via Boiardo 26, Salvarano di Quattro Castella (RE), Italy, [3]Ist. Zooprofilattico Sperimentale della Lombardia e dell'Emilia, via Bianchi 9, Brescia, Italy, [4]Università del Molise, Di STAAM, via De Sanctis, 86100 Campobasso, Italy; francesco.fantuz@unicam.it

Clinical evidences show that ass's milk is well tolerated by infants with cow's milk protein allergy. Although minerals are crucial in infants nutrition little is know about the mineral content of donkey's milk and about minerals requirements of lactating jennies. As a part of a larger research on mineral nutrition of dairy donkeys, the aim of this study was to determine the effect of dietary trace elements supplementation on donkey's milk yield and major milk components. Sixteen lactating jennies (32-58 days from foaling) were used to provide milk samples obtained by milking machine at 11:00 am. Animals were divided into 2 groups (CTL and TE): jennies in CTL group were daily fed 8 kg of coarse hay and 2.5 kg of mixed feed. Jennies in TE group were fed the same diet as CTL group but mixed feed was added with 140 mgFe, 24 mg Cu, 148 mg Zn, 70 mg Mn, 3.2 mg I and 0.24 mg Se/kg concentrate. Jennies were housed with the foals that were separated from the dam 3 hours before milking. The study lasted 3 months and every 2 weeks milk yield was recorded and milk samples were collected to be analysed for major milk components. Average (±SE) milk yield per milking and milk gross chemical composition were in the range reported for donkey's milk. Results on ash are based on 3 sampling times. Milk yield (CTL=561.25±24.63 Vs TE=598.98±25.15), total solids (CTL=9.81±0.04 Vs TE=9.82±0.04), protein (CTL=1.76±0.01 Vs TE=1.75±0.01), fat (CTL=0.28±0.04 Vs TE=0.27±0.04), lactose (CTL=7.04±0.01 Vs TE=7.07±0.01) and ash (CTL 0.43±0.01 Vs TE=0.43±0.01) milk contents were not significantly affected by the dietary treatment.

Modulation of the human aged immune response by donkey milk intake

Jirillo, E.[1], D'alessandro, A.G.[2], Amati, L.[3], Tafaro, A.[1], Jirillo, F.[1], Pugliese, V.[4] and Martemucci, G.[2], [1]University of Bari, MIDIM, P.le Giulio Cesare, 70100 Bari, Italy, [2]University of Bari, PRO.GE.S.A., Via G. Amendola, 70126 Bari, Italy, [3]National Institute of Gastroenterology, Via Turi, 27, 70013 Castellana Grotte, Italy, [4]Geriatric Center, Via Cisterna, 70010 Turi, Italy; dalex@agr.uniba.it

Immunosenescence is characterized by a progressive decline of immune functions with age and both innate and adaptive immune responses are severely impaired and increases the susceptibility of old people to infections, tumours and autoimmune diseases. Nutrition seems to be fundamental in the mechanism of immune recovery in the elderly. A series of reports has emphasized the beneficial effects of donkey milk in the immuno- compromised host for the inhibition of bacterial activity and prevention of atherosclerosis because of its antioxidant properties. The aim of the study was the evaluation of pro-inflammatory and anti-inflammatory effect of donkey milk in aged individuals. In a group of 14 healthy aged subjects, donkey milk was administered for a period of one month. Cytokine profile -interleukin (IL)-12, IL-10, IL-1β, IL-8, IL-6 and Tumor Necrosis Factor (TNF)-α - was assessed before and after milk intake by means of a cytometric bead array test. Data demonstrated that IL-12 was undetectable, while IL-10, IL-1β and TNF-α were released in very low amounts. IL-8 was increased by donkey milk administration. Same pattern of response was noted with IL-6. These findings indicate that administration of donkey milk in the aged host is able to upregulate the acute immune response.

Possibility of small scale production of sour mare's milk product kumis

Koro, P.[1] and Saastamoinen, M.T.[2], [1]Equine College, Opistontie 10, 32100 Ypäjä, Finland, [2]MTT Agrifood Research Finland, Equine Research, Opistontie 10, 32100 Ypäjä, Finland; paivi.koro@hevosopisto.fi

Mare's has been shown to be suitable to people suffering cow's milk protein and other food intolerances. Sour mare's milk product called kumis has been used for centuries in Asia by many native people. It is made by fermenting mare's milk while stirring or churning. During the fermentation, Lactobacillus bulgaricum acidifies the milk, and Saccharomyces lactis turn it into a carbonated and mildly alcoholic drink. Kumis is the only form of mare's milk to preserve it as good food at least few weeks in the primitive nomadic life. Kumis is considered a very nutritious and rehabilitative drink. The first kumis sanatorium was founded in Russia in 1858. Particularly, in recovering from tuberculosis kumis has played a significant role. The biological value of kumis is based e.g. on its high-quality protein, easily absorbable fat and lactic acid and vitamin content. The aim of this study was to find out if a small scale kumis production is possible on horse farms, which can sell it e.g. to their customers of 'bed & breakfast' and green care services. Milk in the present study was milked from Finnhorse mares by hand, but also a milking machine and human breast pumps were tested. Finnhorses were shown to have gentle nature for milk production. The souring agents used were imported from Russia. Fresh kumis was stored for 1-76 days to study how its quality remained. The preservation quality of kumis was estimated by a panel, which stated kumis to remain fresh, aromatic and sparkling for 3 weeks. Even after 3 months of storage kumis was still drinkable, but the taste was clearly changed. In conclusion, it is possible to produce high-quality and sufficiently long-perishable kumis, which remain fresh at 3 weeks, in ordinary domestic kitchen conditions. Handling milk with high hygiene standards is important. When milking just one or two mares hand milking is the best way to get the milk. Milk production is one potential new use of many domestic and draft horse breeds in maintaining their populations.

Effect of breed on free fatty acids and cholesterol level in mare milk

Pieszka, M. and Łuszczyński, J., Agricultural University, Horse Breeding Department, Al. Mickiewicza 24/28, 30-059 Kraków, Poland; mpieszka@ar.krakow.pl

Cholesterol is important as food component according to its role in causing disease of cardiovascular system. It is endogenous sterol contained in each animal cell, mainly in plasma membrane but also in mitochondria, Golgi apparatus and in nuclei membrane. It is also important as a precursor of many steroids as hormones, bile acids, vitamin D and glycosides. Also fatty acids play important role in human health. Except they are nutritive component they also can help to prevent some diseases. The aim of this study was to evaluate the effect of breed on free fatty acids and cholesterol content in mare milk. Material for the study was collected from 15 Arabian and 13 Anglo-Arabian mares from Janów Podlaski Stud. All mares were kept in the same stable and fed according to Horse Feeding Requirements (1991) using good meadow hay, oats, wheat bran and carrot. Samples of milk were collected by hand-milking in 1, 3, 6 and 9 day after parturition. Cholesterol level was evaluated according to the enzymatic method using diagnostic test for total cholesterol Liquick Cor-CHOL Cormay (Lublin, Poland). Content of free fatty acids level was evaluated using photometric apparatus MILKOSCAN. Statistical differences were evaluated using one-factor variance and Tukey's test (Statistica 8 for Windows). It was observed that level of free fatty acids increased during first 9 day of lactation in both Arabian and Anglo-Arabian mares milk (respectively 1.01, 1.14, 1.28 and 1,54% in Arabian and 0.71, 0.74, 0.89 and 1.06% in Anglo-Arabian mares milk) whereas cholesterol level decreased (respectively 30.72, 26.63, 25.65 and 19.92 mg/ml in Arabian and 23.10, 19.55, 16.26 and 15.40 mg/ml in Anglo-Arabian mares milk). Arabian mares produced milk with significantly higher free fatty acids content ($P \leq 0.05$) and highly significantly higher level of cholesterol ($P \leq 0.01$).

Session 06 Poster 17

Concentration of estradiol and progesteron in Arabian mares milk

Pieszka, M. and Łuszczyński, J., Agricultural University, Horse Breeding Department, Al. Mickiewicza 24/28, 30-059 Kraków, Poland; mpieszka@ar.krakow.pl

Concentration of different hormones in milk is the subject which is not properly recognised yet. The role of sex hormones in the body is very well known whereas their role in milk is not clearly understood. Some authors suggests that activity of sex hormones can be the cause of heat diarrhoea in foals which is not proved yet. The object of the study was to evaluate the beta-estradiol and progesterone concentration in Arabian mare milk. Material for the study was collected from 30Arabian mares. All mares were kept in the same stable and fed according to Horse Feeding Requirements (1991). Samples of milk were collected by hand-milking every two days starting from 2 and finishing at 30 day after parturition. The activity of sex hormones was evaluated according to RIA method. Statistical analysis included the differences between successive days of lactation and also the effect of mares age on sex hormones activity in their milk. For that reason the one-factor variance analysis and Tukey's test were used. Average concentration of beta-estradiol in Arabian mares milk was 0,14 nmol/l and ranged between 0.123 in 2 day and 0.159 nmol/l in 12 day after parturition. The curve was not regular but the slightly increasing tendency could be observed. The study also showed the effect of mares age on estradiol concentration – older mares (more than 12 years) produced milk with significantly higher ($P \leq 0.05$) activity of estradiol than young mares (<8 years) and in average age (8-12 years), especially during first 15 days. Average concentration of progesterone was 2.7 nmol/l during the 30-days lactation with range between 0.4 at 8 day and 6.34 at16 day of lactation. From 8 to 16 days the sudden increase of this hormone activity was observed which could be connected with early pregnancy and then the concentration was decreasing. During the analysis the lower activity was observed in younger mares, especially from 16 to 24 days after parturition but the differences were not statistically significant.

Session 07 Theatre 1

Protection at the intestinal level; the role of nutrition - lessons from human and animal research

Spring, P.[1,2] and Moran, C.[1,2], [1]Swiss College of Agriculture, Länggasse 85, 3052 Zollikofen, Switzerland, [2]Alltech France, SARL 14, Place Marie-Jeanne Bassot, 93200, Place Marie-Jeanne Bassot, 93200, France; peter.spring@bfh.ch

The gastrointestinal tract is a complex interface between the host and its environment, and is continuously challenged with a diverse array of dietary and microbial compounds. These compound host interactions play key roles in molding both its structure and function. To maximize efficient transfer of nutrients from the gastrointestinal tract, the system depends on a large surface area and the mucosal epithelium varies along the gastrointestinal tract to meet its functional requirements. Many requirements for maximal nutrient digestion and absorption and microbial barrier function, create a conflict in, and thus necessitate a complex system of physical, biochemical and cellular mechanisms for protecting the intestinal mucosa from invading pathogens. Those mechanisms are constantly challenged and adapted depending on the interactions with compounds present in the digesta. Changes in the lumen particularly affect villi structure and function, leading to alterations of their number in the wall, modulation of surface exposure, synchronization of digestive enzymes with digesta, and surface protectants, such as mucus. In addition, changes in the lumen trigger the responses from the gut associated lymphoid tissue located at the mucosal surface and in the lamina propria. The complexity of the systems makes it challenging to study and interpret the effect of the diet or particular dietary ingredients on intestinal protection. Novel research approaches such as nutrigenomics do allow us to more holistically define changes that occur with dietary modifications, however, the interpretation of their short and particular long term effects on host health remains often vague. The paper will give an overview of interactions between the diet and the host at the intestinal interface and will discuss some specific feed ingredients, such as yeast cell wall and particular fractions from it in more detail.

Session 07 Theatre 2

Nutritional strategies to combat *Salmonella* in EU food animal production

Berge, A.C., Faculty of Veterinary Medicine, Ghent University, Department of Reproduction, Obstetrics and Herd Health, Salisburylaan 133, 9820 Merelbeke, Belgium; acbberge@hotmail.com

Nutritional strategies to minimize Salmonella in food animal production are a key component in producing safer food for European consumers. The current European approach is to use a farm to fork strategy where each sector has to implement measures to minimize and reduce Salmonella contamination. In the pre-harvest phase all tools available need to be employed to reduce Salmonella contamination such as; biosecurity, controls in animal trade, the animals, housing, management, cleaning, disinfection as well as the feed. It is important to prevent Salmonella introduction onto the farm through feed, and therefore heat treatment, acidification and Salmonella controls have been used. The feed can also modify the gastro-intestinal tract microflora and influence the immune system and thereby minimize Salmonella colonization and shedding. Physical feed characteristics such as course ground meal can delay gastric emptying and thereby raise the acidity of the gut. Feed additives such as organic acids, short and medium chain fatty acids, essential oils, probiotics, prebiotics and certain specific carbohydrates such as mannan-based compounds have potential to reduce Salmonella levels when added to the feed. These nutritional strategies could be evaluated, integrated and used in holistic control programmes to succesfully minimize Salmonella on farms.

Session 07 Theatre 3

The role of intestinal microflora in gut health in pre- and post-weaning piglets

Jansman, A.J.M.[1], Koopmans, S.J.[2] and Smidt, H.[3], [1]Wageningen UR Livestock Research, P.O. Box 65, 8200 AB Lelystad, Netherlands, [2]Wageningen UR, Biomedical Research, P.O. Box 65, 8200 AB Lelystad, Netherlands, [3]Wageningen University, Laboratory of Microbiology, Dreijenplein 10, 6703 HB Wageningen, Netherlands; alfons.jansman@wur.nl

Traditionally the gut is mainly seen as the site for the digestion and absorption of nutrients from the diet. More recently, other functions of the gut got much more attention in farm animals. The gut has an important barrier function between the intestinal lumen and the systemic circulation and prevents potential harmful constituents to pass the gut lining. The gut is also a main site for the host's immune system and hosts a complex microflora. The gut microflora, the local immune system and the intestinal mucosa show large interactions which together fulfill an important role in maintaining intestinal health. These interactions are largely influenced by the postnatal environment, diet composition in the pre- and post-weaning period (e.g. nutrients and (functional) ingredients). So far, a large focus was given towards the possibilities to influence gut health and function via the diet in the post weaning period. A variety of measures, especially in the area of feed additives, has already been implemented in practice. In the EU funded project INTERPLAY 'Interplay of microbiota and gut function in the developing pig – innovative avenues towards sustainable animal production' the focus is on the kinetics of colonization by commensal as well as pathogenic microbiota along the gastro-intestinal tract of young pigs and on the concomitant impact of the microflora on gastro-intestinal function throughout life and to provide a better understanding of the host-microbe interactions that drive gut development. A combination of different approaches (e.g. influencing of intestinal microflora by use of specific feed additives (e.g. pre- or probiotics) or selective ingredients and stimulation of the natural defense of the piglet) will be needed to further optimize gut health and function and minimize the use of antibiotics in pig production.

Session 07 — Theatre 4

Quantification of the reduction in voluntary feed intake and growth after a digestive challenge in post-weaned piglets

Pastorelli, H.[1,2], Van Milgen, J.[1,2], Lovatto, P.[3] and Montagne, L.[1,2], [1]INRA, UMR 1079 Systèmes d'Elevage Nutrition Animale et Humaine, Saint-Gilles, 35590, France, [2]Agrocampus Ouest, UMR1079 SENAH, Rennes, 35000, France, [3]UFSM, Campus Camobi, Santa Maria, RS, 97105-900, Brazil; helene.pastorelli@rennes.inra.fr

Health perturbations can affect technical and economical results of pig production. Their consequence on voluntary feed intake and growth of pigs has been extensively described but no general framework exists to quantify these. A database was developed including literature data reporting the effect of bacterial or viral infections, parasites, mycotoxins or poor sanitary housing conditions on performance in piglets and growing pigs. The aim of the present study is to analyse a subset of the database and to quantify, by a meta-analysis, the impact of a digestive challenge in post-weaning and during early growth on feed intake and growth. The challenge was obtained either by experimental infection with *E. coli* or by housing the pigs under poor sanitary conditions. To account for variation between studies, results were expressed relative to that of the control group. The response immediately following the challenge was distinguished from the response during the overall experimental period. On average, voluntary feed intake was reduced by 6%, independently of the measurement period. Consequently, daily gain was also reduced but covariance analysis indicated that the reduction in growth was greater immediately following the challenge compared with the overall response. A reduction of 10% in feed intake resulted in an immediate reduction in daily gain of 26%, whereas the reduction in the total experimental period was 14%. The observation that growth was affected to a greater extent than feed intake (especially for the immediate response) may suggest that nutrients are deviated from growth towards other functions (e.g., maintenance include the animal's defence system). These results are a step towards quantifying the impact of health perturbations in predictive growth models.

Session 07 — Theatre 5

Impact of feeding diets with inclusion of chicory on the faecal bacteria in growing pigs

Ivarsson, E. and Lindberg, J.E., Swedish University of Agricultural Science, Department of Animal Nutrition and Management, P.O. Box 7024, SE-750 07 Uppsala, Sweden; Emma.Ivarsson@huv.slu.se

Dietary fibre has the potential to improve animal production by increase the microbial diversity and/or stimulate health beneficial bacteria. Chicory (Cicorium intybus L), is a perennial herb, the root has a high inulin content, whereas the vegetative part has a high content of soluble uronic acid, the building block of pectins. Both inulin and pectins have been shown to beneficially affect the gut microflora. The aim of this study was to evaluate the effect of feeding growing pigs chicory forage (leaves and stem) or inulin on faecal bacteria and pH. Three experimental diets were formulated, a cereal-based control diet (C) and two diets with inclusion of 80 g/kg chicory forage or inulin (CH80 and INU80). Eighteen seven weeks old pigs (castrated males and females) with an initial weight of 17.2 + 0.8 kg were used in an 18-days experiment. The piglets had ad libitum access to feed and water. Faeces samples were collected at days 0, 7 and 18. Measurements of faecal pH and cultivation of Enterobacteriace and Lactobacillus were performed on fresh samples. Terminal restriction fragment length polymorphism (T-RFLP) and sequence analysis were performed on frozen samples. The statistical analyses were performed with procedure Mixed in SAS, cultivation data and faecal pH were analysed with repeated statement. Pigs fed diet INU80 had lower ($P<0.05$) counts (log cfu g^{-1}) of Enterobacteriaceae than pigs fed diet C and diet CH80, higher ($P<0.05$) counts of Lactobacillus and a lower ($P<0.05$) faecal pH than pigs fed diet CH80. Enterobacteriaceae tended ($P=0.06$) to decrease with time, whereas Lactobacillus and faecal pH were stable. The T-RFLP analysis showed highest relative abundance for unknown Prevotella among all diets, this group was also higher ($P<0.05$) in pigs fed diet CH80 compared to pigs fed diet C and INU80. This study shows that inulin and chicory forage stimulates different bacterial groups, and confirms the prebiotic properties of inulin.

Dietary glycerol level effects on performance traits, glycerol kinase gene expression and gut microbiota in broilers

Zoidis, E., Papadomichelakis, G., Mountzouris, K.C., Pappas, A.C., Arvaniti, K. and Fegeros, K., Agricultural University of Athens, Department of Nutritional Physiology and Feeding, Iera Odos 75, 11855, Athens, Greece; kmountzouris@aua.gr

Glycerol produced by biodiesel industry, is currently considered as an alternative to energy feedstuffs in monogastric animals. The study aimed to determine dietary glycerol effects on several performance traits, glycerol kinase (GK) gene expression and gut microbiota in broilers. Four hundred one-day old Cobb chicks were allocated in 4 groups of 5 replicates each. They were fed 4 isocaloric and isonitrogenous diets C, G7, G14 and G21, containing 0, 7, 14 and 21% of crude glycerol, respectively. Feed and water intake, weight gain, excreta moisture, litter score and mortality were recorded weekly. Liver GK gene expression was determined by RT-PCR analysis. In addition, the levels of bacteria belonging to total aerobes, total anaerobes, coliforms, *E. coli*, *Clostridium* spp, *Clostridium perfringens*, *Lactobacillus* spp, *Bifidobacterium* spp and gram-positive cocci were determined in the broiler ileal and caecal digesta. Data were analysed by one-way ANOVA using SPSS v.16.0. Feed intake, weight gain and final body weight were significantly higher ($P<0.05$) in G7, compared to C, G14 and particularly G21 broilers. Dietary glycerol level increased linearly water intake ($P<0.05$), excreta moisture ($P<0.05$), litter score ($P<0.05$) and mortality ($P<0.05$). GK mRNA levels increased for glycerol inclusion up to 14% and reached a plateau afterwards, suggesting a limiting capacity for dietary glycerol utilization. Groups C and G7 had higher *Bifidobacterium* spp ($P<0.05$) and lower coliform ($P<0.001$) and *C. perfringens* ($P<0.05$) levels in broiler caeca. In conclusion, dietary glycerol at 7% had beneficial effects on broiler performance, whereas high levels (>14%) impaired broiler growth, increased mortality and affected negatively litter condition and caecal gut microbiota composition.

Liquid feed fermented with a *Lactobacillus* strain with probiotic properties, reduces ileal and caecal *Salmonella* counts in broiler chickens

Savvidou, S.[1], Beal, J.D.[1], La Ragione, R.M.[2] and Brooks, P.H.[1], [1]University of Plymouth, Drake Circus, PL4 8AA, Plymouth, United Kingdom, [2]Veterinary Laboratories Agency, New Haw, KT15 3NB, Surrey, United Kingdom; savidousumela@googlemail.com

Lactobacillus Salivarius ss *Salivarius* NCIMB 41606 (*L. salivarius*) was assessed for its efficacy in reducing *Salmonella enterica Typhimurium* Sal 1344 nal^r carriage in broiler chickens. A total of 68 hatchlings were randomly divided into four groups. One group was provided with a daily dose of 10^7 CFU ml^{-1} of *L. salivarius*, delivered via the drinking water (WAT) from day old. The second group was provided with 10^9 CFU gr^{-1} of *L. salivarius* via fermented moist feed (FMF). The third group was provided with feed acidified (AMF) with 30.3 ml of lactic acid/ kg of wet feed from day old and the last group acted as the control (CON) without any addition of *L. salivarius* or acid. At two weeks of age chicks were challenged with circa10^6 ml^{-1} *Salmonella enterica* serovar *Typhimurium* Sal 1344 nal^r, by oral gavage. At 26 and 40 days post-inoculation, six birds from each group were randomly selected, euthanased and the ileum, caecum, liver and spleen sampled at post-mortem and subjected to bacteriological analysis. Post-mortem enumeration of *S. Typhimurium* and LAB was conducted according to the method described by La Ragione *et al.* (2004). At 26 days post *Salmonella* challenge the FMF group had significantly ($P<0.05$) lower ileal *Salmonella* counts (1.08 log_{10} CFU ml^{-1}) compared with the CON (6.61 log_{10} CFU ml^{-1}) and AMF groups (6.10 log_{10} CFU ml^{-1}) and FMF and WAT group had significantly ($P<0.05$) lower caecal *Salmonella* counts than the CON group (1.05, 1.20 and 5.03 log_{10} CFU ml^{-1}, respectively). There were no significant differences between the treatments, at 40 days post *Salmonella* challenge. In conclusion these studies demonstrated that FMF, produced using *L. salivarius*, is an effective means of controlling *S. Typhimurium* infection in poultry.

Session 07 Poster 8

Chicory (*Cichorium intybus* L.) in weaned pig diets: effects on ileum and colon

Liu, H.Y. and Lindberg, J.E., SLU (Swedish University of Agricultural Sciences), Department of Animal Nutrition and Management, P.O. Box 7024, SE-75007 Uppsala, Sweden; Liu.Haoyu@huv.slu.se

Chicory (*Cichorium intybus* L.) forage has a high content of uronic acids (the building block in pectin) and has potential as a fibre source in pig diets. However, very limited information is available on the impact on pigs. In contrast, the chicory root is rich in inulin, and has been extensively studied as a prebiotic. In the current work, the effect of dietary inclusion of chicory forage or inulin on organic acids, digesta pH and gut microflora composition of weaned piglet was studied. Eighteen 7 weeks old weaned piglets (male and female) were fed with one cereal-based control diet (C) and 2 experimental diets with either 80 g/kg chicory forage (C80) or inulin inclusion (I80) in an 18-day experiment. Ileal and colon digesta samples were collected to determine the organic acid concentration and pH. The microflora composition was assessed by terminal restriction fragment length polymorphism (T-RFLP). Statistical analysis was performed with procedure Mixed in SAS (SAS Institute, Cary, NC, USA, version 9.1). Bray Curtis index was used for cluster analysis. The T-RFLP data revealed 70 terminal restriction fragments (TRFs) in ileum and 79 TRFs in colon, indicating a more diversified microbiota in hindgut than in small intestine. Moreover, the composition of the ileal microbiota was highly individual. However, when colonic samples from pigs fed diet C80 and I80 were clustered according to their TRF patterns, 4 out of 5 pigs on diet C80 grouped together, against all 5 pigs on diet I80. The concentration of organic acids in colon, as well as the pH in ileum and colon were unaffected ($P>0.05$) by diet. The concentration of lactic acid was low in colon, but was numerically higher ($P=0.07$) in pigs fed diet I80 than in pigs fed diets C and C80. The molar proportion of butyric acid in colon was higher on diet C ($P=0.05$) than on diets C80 and I80. In conclusion, chicory forage and inulin affected intestinal organic acids, pH value and microflora composition of pigs differently.

Session 07 Poster 9

Effect of using prebiotic BioMos® on performance of broiler chickens

Rakhshan, M., Shivazad, M., Mousavi, S.N. and Zaghari, M., Tehran University, Animal Science, Karaj, Tehran, 89187, Iran; maryamrakhshan@ut.ac.ir

A total of 1920 one-day-old Ross 308 broiler chicks was assigned to four treatments in order to evaluate the effects of Bio-Mos® as growth promoter on the performance, jejunum histology and cecal microbial contents of the male and female birds. Each treatment contained 24 replicates of 20 chicks each. Treatments were: 1) diet without MOS for male chicks; 2) diet without MOS for female chicks; 3) diet with MOS for male chicks; 4) diet with MOS for female chicks; The chick performances were evaluated by recording the body weight gain, feed intake, feed conversion ratio (FCR), and mortality at 10, 24 and 47 days of age. At 47 days of age, the cecal contents from each bird were aseptically emptied into sterile plastic bags and stored at -20 °C for later microbiological analysis. At 47 days of age, 1 bird from each pen was slaughtered and a 2 cm segment of the midpoint of the jejunum was excised, washed in physiological saline solution, and fixed in 10% buffered formalin. Serial sections were cut at 5 µm and placed on glass slides. Sections were deparaffinized in xylene, rehydrated in a graded alcohol series, stained with hematoxylin and eosin, and examined by light microscopy. Data were analyzed using the GLM procedure of SAS. Differences among treatments were compared using a Duncan's multiple range tests. No difference in mortality was observed among the four treatments ($P>0.05$). Using MOS significantly improved FCR and increased body weight gain in 1-47 days in males and females ($P<0.05$). Body weight gain and feed intake were significantly higher in males than females in 1-47 d ($P<0.05$). FCR in males was lower than females. Bio-Mos® significantly increased villi height and decreased crypt depth ($P<0.05$). The counts of lactobacillus and coliform bacteria were not affected by the supplementation of Bio-Mos® ($P>0.05$).

Intact brown seaweed (*Ascophyllum nodosum*) in diets of weaned piglets: effect on performance and gut bacteria

Michiels, J.[1,2], Skrivanova, E.[3], Missotten, J.[2], Ovyn, A.[2], Mrazek, J.[4], De Smet, S.[2] and Dierick, N.[2], [1]University College Ghent, Schoonmeersstraat 52, 9000 Gent, Belgium, [2]Ghent University, Proefhoevestraat 10, 9090 Melle, Belgium, [3]Institute of Animal Science, Pratelstvi 815, 10400 Prague, Czech Republic, [4]Institute of Animal Physiology and Genetics, Pratelstvi 815, 10400 Prague, Czech Republic; skrivanova.eva@vuzv.cz

Previous research showed the antimicrobial properties of dried intact brown seaweed *Ascophyllum nodosum* against coliforms in the pig foregut. Here, the aim was to determine its effects on animal performances and to further elaborate its effect on gut bacteria. A total of 160 weaned piglets (21 d, 6.59±0.91 kg) were allocated to 32 pens of 5 animals each. A pre-starter diet was given to all animals until day 4, followed by 4 different starter diets until day 28 (8 pen replicates per diet): a control diet: cereals-soybean-milk products based and the control diet + 2.5, 5 or 10 g dried seaweed per kg. The diets did not contain organic acids and supplemental Cu and Zn were limited to 10 and 100 mg per kg respectively. At day 12/13 one piglet from each pen was sacrificed and gastric and small intestinal digesta were sampled for plate countings onto selective media (total anaerobic bacteria, coliforms, *E. coli*, streptococci, lactobacilli) and caecal digesta for molecular analysis using the PCR-DGGE method with universal bacterial primers targeting fragments of the 16S rRNA. Data were analysed by a linear model with treatment as fixed effect. Dietary A. nodosum supplementation had no effect on daily weight gain, nor did it alter feed conversion. Plate countings were not different between treatments. Dendrograms representing relationships between PCR amplified 16S rRNA fragment banding patterns from caecal digesta did not indicate that individuals grouped according to treatment. Thus, the addition of *A. nodosum* seaweed did not enhance performances of piglets nor altered the gut microbiota in contrast to previous experiments with corn-soybean basal diets.

Using odd and branched chain fatty acids to investigate dietary induced changes in the caecum of rabbits

Papadomichelakis, G., Mountzouris, K.C., Paraskevakis, N. and Fegeros, K., Agricultural University of Athens, Nutritional Physiology and Feeding, 75 Iera Odos Str., 118 55, Athens, Greece; gpapad@aua.gr

The description of the interactions between nutrition and caecal bacteria abundance using odd-numbered and branched-chain fatty acids (OBCFA) has not been studied in rabbits. Thus, the effect of diet on OBCFA pattern in rabbit caecal contents was investigated. Forty-eight weaned rabbits were fed 4 isolignocellulosic (ADF, 180 g/kg) diets with 2 levels of digestible fibre (LDF, 165 vs. HDF, 240 g/kg) supplemented with soybean oil (SO, 20 g/kg) or not, following a 2×2 factorial design. One rabbit from LDF group died during the experiment, and therefore at 77 days of age, 47 rabbits were sacrificed, caeca were removed and contents were freeze-dried to determine fatty acid (FA) composition. GLM procedures (two-way ANOVA) and principal component analysis (PCA) were used to examine the effects of diets and the relationships between caecal FA, respectively (SPSS v.16.0). The microbial origin of OBCFA vs. dietary FA in caecal contents was illustrated by clustering of these FA in the loading plots of PCA. The contribution of odd-numbered FA in the OBCFA pattern was increased ($P<0.05$) for HDF diets, whereas that of branched-chain FA decreased ($P<0.05$), indicating potential shifts in the relative abundance of gram-negative and gram-positive fibrolytic bacteria, respectively. The relative importance of OBCFA in total microbial FA was reduced by SO addition, as indicated by the lower total OBCFA ($P<0.001$). However, SO effect on the variations in OBCFA pattern appeared to depend on the readily available fermentable substrate, as indicated by the DF × SO interactions ($P<0.001$) for odd-numbered and branched-chain FA. In conclusion, different dietary treatments were associated with changes in OBCFA, which in turn could be related to potential shifts in caecal microbial populations that need to be determined. It is proposed that the use of OBCFA in rabbit nutrition research should be further assessed.

Future research priorities for animal production in a changing world

Scollan, N.D.[1], Greenwood, P.L.[2], Newbold, C.J.[1], Yanez Ruiz, D.R.[3], Shingfield, K.J.[4], Wallace, R.J.[5] and Hocquette, J.F.[6], [1]IBERS, Aberystwyth University, Wales, SY23 3EB, United Kingdom, [2]Beef CRC, University of New England, Armidale, NSW 2351, Australia, Australia, [3]CSIC, Profesor Albareda, 1, 18008 Granada, Spain, [4]Agricultural Research Centre of Finland, Kirkkotie, Jokioinen FIN-31600, Finland, [5]Rowett Institute of Food and Health, Aberdeen University, Aberdeen, United Kingdom, [6]INRA, Theix, 63122 Saint-Genes Champanelle, France; ngs@aber.ac.uk

This paper reports the outcomes from a Workshop on 'Animal Production in a Changing World' held in Theix (INRA, France) on Sept. 9-10, 2009, in which 35 participants from 15 countries took part. The objective was to discuss how to address the main challenges within the livestock sector: its environmental impact and role in global climate change and how to increase production of animal products coupled with lower footprint, and how to meet society needs in terms of product quality for the consumer. Key lectures presented the main drivers of animal agriculture: population growth, environmental impact, mitigation options, animal efficiency and animals' products quality. They highlighted the synergies between research needs and strategies dedicated to improving food quality and safety and those devoted to decreasing environmental impact of ruminants' livestock. Then, two discussion groups were set up. The remarks on product quality were that the existing knowledge is not fully applied, the priorities with regards to quality clearly differ between developing and developed countries and that an environmental index needs to be established taking into account carbon footprint, water and energy use, etc. The discussion on environmental issues highlighted the importance of focusing on whole life cycle analysis in the mitigation area, while the adaptation strategy should be based on selection for profitable animals under different production systems. In summary, a fundamental shift in designing our production systems is required, to meeat present needs without compromising future generations.

Improving the quality of red meat: the Australian lamb industry as an example

Pethick, D.W.[1], Banks, R.[2] and Ball, A.[2], [1]Murdoch University, 90 South St, 6150 Murdoch WA, Australia, [2]Meat & Livestock Australia, University of New England, 2350 Armidale NSW, Australia; d.pethick@murdoch.edu.au

Research undertaken by Meat and Livestock Australia recommends the future progression of red meat products to ideally follow 5 pillars of consumer interest, (i) integrity and tractability (ii) eating quality (iii) human nutritional value (iv) ethical and sustainable production systems and (v) value and efficiency. This paper describes the approach taken by the Australian lamb industry with respect to increasing lean meat yield of the carcase while simultaneously improving eating quality and the nutritive value of lamb to humans. The key proposition is that genetics and meat science research is undertaken simultaneously within a large synchronised project, as this will deliver the most efficient and powerful results. The outcomes are to underpin genetic and nongenetic (i.e. production systems) progress, which are then linked to clear delivery mechanisms. In the case for Australia the delivery systems revolve around the Sheep Genetics and Meat Standards Australia programs, which underpin industry mechanisms for progress on genetics and eating quality. The final project was launched as the Cooperative Research Centre for Sheep Industry Innovation in 2007 whereby a large breeding program called the Information Nucleus was established. This utilises approximately 100 new sires mated per year to 5000 ewes over 8 diverse production sites. Each year, initially for 5 years, 2000 slaughter lambs will undergo detailed measurement of lean meat yield, eating quality (eg taste tests, intramuscular fat, shear force) and human health attributes (Fe, Zn, fatty acid profile). The early results show considerable scope for managing the complex interactions between yield, eating quality and human health attributes through genetic means combined with the appropriate production knowledge to deliver the genetic outcomes.

Effect of lipid-rich plant extract on the fatty acid composition of the phospholipid fraction of *longissimus thoracis et lumborum*

Kim, E.J.[1], Richardson, R.I.[2], Lee, M.R.F.[1], Gibson, K.[2] and Scollan, N.D.[1], [1]Aberystwyth University, IBERS, Plas Gogerddan, Aberystwyth, SY233EB, United Kingdom, [2]University of Bristol, Division of Farm Animal Science, Langford, BS40 5DU, United Kingdom; nigel.scollan@aber.ac.uk

Nutrition is a key factor influencing the content of n-3 polyunsaturated fatty acid (PUFA) in beef lipids. This study examined effect of incremental inclusion of lipid-rich plant extract in steers fed on grass silage v. concentrate on the fatty acid composition of muscle phospholipids. The plant extract (PX) was developed from the liquid fraction extracted from fresh lucerne (Désialis, France), and then heat-treated and dried. Forty Belgian-Blue × Holstein steers (~400 kg liveweight) were allocated to one of five dietary treatments: 1) grass silage ad libitum, 2) grass silage ad libitum plus 75 g PX/dry matter intake (DMI), 3) grass silage ad libitum plus 150 g PX/DMI, 4) restricted barley straw and control concentrate (40:60 on a DM basis), and 5) restricted barley straw and concentrate with PX (25% in concentrate on a DM basis) (40:60 on a DM basis. Growth rate was controlled on all diets to achieve a similar growth rate to those fed on ad libitum forage. Animals were slaughtered when they achieved fat class 3 and samples of longissimus thoracis et lumborum were taken at 48 h post-mortem for fatty acid analysis. An analysis of variance was conducted with diet as the main factor using GenStat. Total fatty acids and amounts of the major saturated fatty acids in phospholipids were not different, averaging 487, 70.2, 48.2 mg/100 g meat for total fatty acids, 16:0 and 18:0, respectively. Feeding grass silage v. concentrate increased ($P<0.001$) deposition of n-3 relative to n-6 PUFA. Incremental PX on grass silage resulted in additional deposition of 18:3n-3 (and 18:2n-6) and longer chain derivatives EPA and DHA resulting in improvements in P:S and n-6:n-3 ratio. In conclusion, forage increased n-3 PUFA in beef muscle and further beneficial responses in n-3 PUFA were noted when PX was included in diet.

Effect of total replacement of sodium selenite with an organic source of selenium on quality of meat from double-muscled Belgian Blue bulls

De Boever, M.[1], De Smet, S.[1], Nollet, L.[2] and Warren, H.[3], [1]Ghent University, Lanupro, Proefhoevestraat 10, 9090 Melle, Belgium, [2]Alltech Netherlands and BV, Genteseteenweg 190 B1, 9090 Melle, Belgium, [3]Alltech Biotechnology Centre, Summerhill Road, Dunboyne, Co. Meath, Ireland; hwarren@alltech.com

Thirty seven Belgian Blue young fattening bulls were allocated, according to liveweight, to one of two dietary treatments: Control (n=18; All-Mash, 0.3 ppm selenium as sodium selenite) or Treatment (n=19; All-Mash, 0.3 ppm selenium as organic selenium (Sel-Plex®, Alltech Inc., Nicholasville, KY)). Both diets were formulated to contain 150 ppm Vitamin E. Animals were on trial for 193 and 174 days for Control and Treatment, respectively. Animals were slaughtered in the same slaughterhouse and refrigerated prior to sampling. Average age at slaughter was 21±1.9 and 21±1.5 for Control and Treatment, respectively. Drip loss was measured by weight difference after hanging the meat for 48 h at 4 °C in a plastic bag. At 24 h post mortem, samples of loin and bottom round were taken to measure L*, a* and b* values during 11 days cold storage (2-4 °C) under 1000 lux. Readings were taken on the surface and at 2 cm depth of the bottom round. Estimates of metmyoglobin were derived from the L*, a* and b* values. Organic selenium numerically reduced drip loss from 3.30 to 2.88%. Total replacement of sodium selenite with an organic source of selenium (Sel-Plex®) significantly improved a* values in both loin ($P<0.05$) and bottom round ($P<0.01$) resulting in an improved shelf-life under retail conditions. Organic selenium also significantly ($P<0.001$) reduced the formation of metmyoglobin (a measure of oxidation) after day 6 of storage. Total replacement of sodium selenite with an organic selenium source (Sel-Pex®) significantly improved meat quality with regards to redness of the meat and undesirable products of oxidation.

The possibilities of *in vivo* predicting of intramuscular fat content in the lean cattle

Tomka, J.[1], Polák, P.[1], Peškovičová, D.[1], Krupa, E.[1], Bartoň, L.[2] and Bureš, D.[2], [1]Animal Production Research Centre Nitra, Hlohovecká 2, 95141 Lužianky, Slovakia (Slovak Republic), [2]Institute of Animal Science, Přátelství 815, 10400 Praha Uhříněves, Czech Republic; tomka@cvzv.sk

The effect of measurement location along the musculus longissimus thoracis et lumborum on *in vivo* intramuscular fat content prediction was evaluated in our study. Attention was also paid to effect of echocoupler usage on the prediction accuracy. There were 57 bulls (Charolais, Beef Simmental, Czech Fleckvieh, crossbreds) used in the experiment. The cross sectional scanograms from musculus longissimus thoracis et lumborum were obtained from the site between 8th and 9th rib and between 12th and 13th rib using Sonovet 2000. The animals were measured at the age of 14 to 20 months. Within computer image analysis the mean gray value of the whole muscle area (GRAY8, GRAY12) and area fraction of intramuscular fat (INT8, INT12) were calculated. The actual intramuscular fat content was assessed within laboratory analyses. Pearson`s correlations between ultrasound measurements and laboratory determined intramuscular fat content were calculated. When data on whole dataset were considered, the correlations between ultrasound measurements and laboratory determined intramuscular fat content ranged from -0.03 to 0.59. When data on beef bulls were considered, the highest correlations were calculated ranging from 0.55 to 0.73. When data on crossbreds were considered the correlations between ultrasound measurements and laboratory determined intramuscular fat content ranged from -0.31 to 0.63. When data on Czech Fleckvieh bulls were considered the correlations between ultrasound measurements and laboratory determined intramuscular fat content ranged from -0.51 to 0.69. Linear regression models were designed for IMF prediction including single ultrasound measurement or the combination of two ultrasound measurements.

Quality characteristics of meats of podolian bulls slaughtered at different ages

Tarricone, S., Marsico, G., Celi, R., Colangelo, D. and Karatosidi, D., University of Bari, Animal Production, Via Amendola 165/A, 70126 Bari, Italy; despinakaratosi@yahoo.com

Podolian cattle, autochthonous breed of south Italy, is reared prevalently with a wild system or a semi wild one. Systems that permit a big utilization of natural grazing to produce quality and genuine meats. The aim of this study was to evaluate the quality of meat from Longissimus dorsi, so we analyzed the physical parameters, the chemical and the fatty acid compositions of sample cuts of Podolian bulls slaughtered at three different ages. This trial was carried out on 18 Podolian steers, reared in wild system and divided into 3 homogeneous groups of 6 animals each. The steers were slaughtered at the age of 14 months (A group), 16 months (B group) and 18 months (C group), according to veterinary police rules. The determination of meat physical and chemical characteristics was effected according to ASPA official methodologies. The data was analysed for variance (ANOVA) using the GLM procedure of SAS. The meats of B group show, even if with different levels of statistical significance, the highest pH at 1, 24 and 72 hours; and the lowest shear force of cooked meats. On the other half, the A group presents the highest value of L* and b* index and the major cooking loss. The chemical composition of raw meats from C group demonstrates that these meats have the lowest incidence of moisture and indeterminate ($P<0.01$), they have high value of fat ($P<0.01$); instead, the cooked meat from the same group shows the major incidences of moisture and fat, while the incidences of proteins and indeterminate are the lowest ($P<0.01$). Fatty acid composition of raw meat of B group demonstrates that these meats have the highest quantity of SFA, and the lower of PUFA and UFA, ω3, ω6 and the highest value of A.I. and T.I ($P<0.01$ and $P<0.05$). The same trend is showed on cooked meat of B group. The B group meat has, in the complex, the better physical and chemical characteristics, even if acidic quality of fat is not particularly useful for human feeding.

Using stable isotope ratio analysis to discriminate between beef from different production systems

Osorio, M.T.[1], Monahan, F.J.[1], Schmidt, O.[1], Black, A.[2] and Moloney, A.P.[2], [1]University College Dublin, School of Agriculture, Food Science and Veterinary Medicine, Belfield, Dublin 4, Ireland, [2]Teagasc, Grange Beef Research Centre, Dunsany, County Meath, Ireland; aidan.moloney@teagasc.ie

Consumers are increasingly interested in the provenance of their food. Light element stable isotope ratio analysis (SIRA) can distinguish between beef from cattle fed rations rich in C3 and C4 plant material such as barley and maize, respectively. The objective of this study was to determine the potential of SIRA in distinguishing between beef from cattle fed different C3-rich rations. Heifers (n=25/group) received for eleven months (November to October): ad libitum grazed grass (G), a barley-based concentrate (C), winter grass silage followed by grazed grass (SG), winter grass silage followed by grazed grass with concentrate (SGC). Longissimus dorsi muscle samples were collected at slaughter and stable isotope ratios of C ($^{13}C/^{12}C$) and N ($^{15}N/^{14}N$) were measured by isotope ratio mass spectrometry. Isotope ratios are expressed in delta notation [δ per mille (‰)]. The carbon isotopic ratio was -27.97‰, -29.15‰ and -30.95‰ in the concentrate, silage and grass, respectively. The corresponding nitrogen isotopic ratio was 6.96‰, 4.86‰ and 5.86‰. Beef from the G and SG cattle was grouped together due to their similar $\delta^{13}C$ (-27.72‰ and -27.56‰, respectively) and $\delta^{15}N$ (9.22‰ and 8.93‰, respectively) isotopic values which reflected the values in the feed. Beef from the C cattle ($\delta^{13}C$ = -25.02‰, $\delta^{15}N$ = 6.27‰) was clearly differentiated from G and SG while beef from the SGC cattle ($\delta^{13}C$ = -26.38‰, $\delta^{15}N$ = 7.92‰) was intermediate and separated from G/SG and C. It is concluded that SIRA of carbon and nitrogen can be used to authenticate grass-fed and concentrate-fed beef but cannot distinguish between long-term grass-fed beef and beef from cattle fed conserved grass prior to grazing.

Session 08 Theatre 8

Use of rotating box and turned-back position of cattle at the time of slaughter

Warin-Ramette, A.[1] and Mirabito, L.[2], [1]OABA, 10 place Léon Blum, 75011 Paris, France, [2]Institut de l Elevage, 149 Rue de Bercy, 75012 Paris, France; luc.mirabito@inst-elevage.asso.fr

Rotating box is a restraining method which was developed for ritual slaughter. The use of turned-back position of animals in the rotating box is however criticized in the EFSA report of 2004. Dunn et al. (1990) showed more stress of cattle in an old and rudimentary rotating box than when animals are slaughtered in an upright position.. The objective of this study was to have an overview of the use of modern rotating box in France. Use of rotating box is described in three commercial slaughterhouses and 259 adult animals were observed in their bleeding-out. Slaughter process was relatively similar amongst slaughterhouses, with the median time being one minute and a half, but varying from twenty seconds to four minutes. Cattle were restrained 22 seconds in average before their bleed-out but they were turned-back only during 3 seconds. Few animals vocalised (6%) and/or struggle (2%) when they were in inverted position. Frequency of vocalisations however increased with the time spent in inverted position. The proportion of bovine who vocalised when they were turned-back was three times higher when animals have already vocalised before rotation than when animals did not. Analyses of animal's struggles have brought similar results. To conclude, in modern rotating boxes, time spent in the turned-back position by cattle is hugely shorter than reported previously. Furthermore, inverted position seems to induce few reactions, probably in relation with time spent in this position and emotional state of the cattle. In practice, as long as this position improves working condition and consequently bleeding, it could be advised to perform bleeding as soon as possible after the end of the rotation (as a guideline, less then 3-5 seconds). The use of the 90° position should also be considered in order to reduce the manipulation of the animals.

Polish beef consumers: emerging or declining market?

Gutkowska, K., Żakowska-Biemans, S., Kosicka-Gębska, M., Sajdakowska, M. and Wierzbicki, J., Warsaw University of Life Science, Department of Organization and Economics, ul. Nowoursynowska 166, 02-787 Warszawa, Poland; jerzy.wierzbicki@pzpbm.pl

In the last two decades, the consumption of beef in Poland decreased from 17 kg per capita to 3.8 kg which seems to be strange as compared to other EU countries and worldwide. To get insight into factors that determine this tendency, there was a five years project developed 'Optimization of beef production in Poland in accordance with the strategy from fork to farm'. Here are the results of the consumer research, carried out within the first stage of the project, in order to identify consumers' perception of beef, habits, culinary preferences and factors hampering its consumption. The survey was conducted on a sample of 3195 respondents, selected from the database of PERSONICX developed by a market research agency ACXIOM, using computer assisted telephone interviews (CATI) technique from Dec. 2009 to Feb. 2010. The results confirmed results from Europe, Asia and South America, that diversification of consumers' behavior towards beef is due to different variables such as gender, income or place of living. Respondents attributed to beef positive features such as 'lean meat', 'tasteful meat' but negative aspects of beef also emerged since consumers claimed that beef is 'tough', 'difficult to prepare' and 'expensive'. The most frequently consumed beef type was for the purpose of cooking boullion, minced meat, entrecote and roast. Polish consumers most often use such culinary techniques like cooking, stewing, roasting. Using the segmentation developed by ACXIOM there were 7 segments of consumers identified that tend to consume more than on average beef and their socio-demographic profile is as follows: inhabitants of large cities with favorable income position, white collar workers, consumers with higher education, older couples without children and younger couples with children. The data analysis show that these consumers are willing to pay more for better quality beef what seems promising in terms of introduction of quality systems.

Session 08 Poster 10

Genomic profiling during myogenesis of fœtal clones

Hue, I.[1], Liszewska, E.[1], Valour, D.[1], De La Foye, A.[2], Meunier, B.[2], Picard, B.[2], Sandra, O.[1], Heyman, Y.[1] and Cassar-Malek, I.[2], [1]INRA, UMR1198, Centre de Jouy, Domaine de Vilvert, 78350 Jouy-en-Josas, France, [2]INRA, UR1213, Centre de Theix, 63122 Saint-Genès-Champanelle, France; isabelle.cassar-malek@clermont.inra.fr

Recently, we have shown that disturbances in both primary and secondary myogenesis occur in bovine cloned foetuses. This may impact on the meat quality derived from clones and their offspring. To further identify molecular pathways that may underly early disturbances in myogenesis, we analysed the transcriptome of cloned and control bovine muscle at 30 days of gestation (dpc). Using a bovine oligo-array (22K, INRA-AGENAE), 340 oligonucleotides corresponding to 215 genes were found to be differential between clones and controls ($P<0.015$). We also performed a comparative proteomic analysis (using 2D-gel electrophoresis) of the Semitendinosus muscle at 60 and 260 dpc and found changes in the muscle proteome in clones vs controls. As indicated by data mining of the transcriptomic and proteomic data, few functions were affected, mainly lipid metabolism and angiogenesis at 30 dpc, regulation of cell cycle/apoptosis at 60 dpc, and energy metabolism and chaperone activity at 260 dpc. At 30 dpc, the Ingenuity Pathway software identified 4 interconnected networks related to lipids. Moreover, differential expression of homeobox transcription factors and their target genes was also detected, as well as genes involved in myogenesis and angiogenesis. qPCR and western-blot validations are now in progress.

A potential genetic marker for meat tenderness in the calpastatin gene in Nellore (*Bos indicus*) cattle

Rosa, A.F.[1], Carvalho, M.E.[1,2], Silva, S.L.E.[1], Gomes, R.C.[1], Bonin, M.N.[1], Eler, J.P.[1], Ferraz, J.B.E.[1], Poleti, M.D.[1], Caritá, A.G.[1], Campos, N.L.[1], Oliveira, E.C.M.[1] and Balieiro, J.C.C.[1], [1]University of São Paulo, Basic Sciences, Av. Duque de Caxias Norte, 225, 13635-900 Pirassununga/SP, Brazil, [2]University of São Paulo, Animal Sciences, Av. Pádua Dias, 11, 13418-900 Piracicaba/SP, Brazil; afrosa@usp.br

The aim of this study was to analyze the association between a single nucleotide polymorphism (SNP) in the calpastatin gene and meat tenderness in Nellore cattle and evaluate the allelic frequency of the SNP. For this purpose, 76 animals (38 steers and 38 young bulls, 19-month old) from the beef cattle herd of University of São Paulo were feedlot finished and then slaughtered. Cattle were genotyped for a single nucleotide polymorphism (SNP) in the calpastatin gene (CAST; GeneBank accession number: AY008267, position 282) and Warner-Bratzler Shear Force (WBSF) was determined in Longissimus dorsi muscle samples aged for 0, 7 and 14 days post mortem. Effects of CAST on WBSF was evaluated using a mixed model (Mixed procedure, SAS Inst., Inc., Cary, NC) with repeated measures in animal effect and shear force and sex as the dependent variables. The results showed that the allele C was a little more frequent (0.51) than allele G (0.49) in the evaluated population and 22.4% of the population had GG genotype, whereas the frequencies for CC and CG genotypes were 23.7 and 53.9%, respectively. It also was observed an interaction between CAST and aging time for Warner-Bratzler shear force (WBSF; kgf). As expected, aging increased meat tenderness for all genotypes. There were no differences among genotypes for WBSF at 0 and 7 days of aging. However, animals that presented the CC genotype had lower WBSF (softer meat) than those with GG. WBSF did not differ between CC and CG, and CG and GG as well. The SNP in the calpastatin gene can be used as a marker for meat tenderness in Nellore (Bos indicus) cattle in genetic selection programs.

The GENOTEND chip: an oligochip based on genomic markers for the prediction of beef quality

Cassar-Malek, I.[1], Capel, C.[2], Vidal, V.[3], Jesson, B.[3], Levéziel, H.[4] and Hocquette, J.-F.[1], [1]INRA, UR1213, Theix, 63122 St-Genès-Champanelle, France, [2]Institut de l Elevage, 149 rue de Bercy, 75595 Paris Cedex 12, France, [3]IMAXIO, Biopôle Clermont-Limagne, 63360 St-Beauzire, France, [4]INRA, UGMA, Faculté des Sciences et Techniques, 87060 LIMOGES, France; isabelle.cassar-malek@clermont.inra.fr

Beef quality depends on factors associated with production, processing and meat preparation. Today, information on quality is only obtainable after slaughter, which is a limitation to the delivery of a consistent quality meat. Recently, muscle gene expression profiling revealed that unsuspected genes may be potential 'genomic markers' of sensory attributes, especially tenderness. The GENOTEND program aims to confirm the relationship between these genes and intrinsic quality attributes of meat quality (e.g. tenderness, flavour, juiciness). We developed an Agilent chip with specific probes of more than 3000 genes involved in muscle biology or meat quality. RNA from Longissimus thoracis muscle samples of Limousin young bulls or Charolais young bulls or steers was hybridised on the chip. The negative correlation between DNAJA1 expression and the initial or global tenderness (Genomic marker for meat tenderness, Patent EP06300943.5) was confirmed based on 18 specific probes (r=-0.40 to -0.60) in the Charolais young bulls or steers. However, the study showed that conditions of production must be taken into account for the prediction of beef quality. In the muscles of the Limousin bulls, we found a negative correlation (r=-0.30) between DNAJA1 expression and the level of muscle calpastatin (involved in post-mortem proteolysis and meat ageing). The expression of other members of the DNAJ family was associated with beef tenderness. In conclusion, the genomic markers of beef tenderness can be specific of an animal type (steer or young bull), of a breed or of production conditions linked to the environment. Some gene families (including that of DNAJA1) are associated with beef quality. The IMAXIO Company will soon propose in service the transcriptomic analysis of bovine muscles.

Session 08 Poster 13

A particular myosin heavy chain isoform variable between beef breeds

Picard, B.[1], Allais, S.[2], Jurie, C.[1], Cassar-Malek, I.[1], Leveziel, H.[3], Journaux, L.[4] and Renand, G.[2], [1]INRA, Unité de Recherches sur les Herbivores UR1213, Theix, 63122 Saint-Genès-Champanelle, France, [2]INRA, Unité de Génétique Animales et Biologie Intégrative, UR1313, 78352 Jouy-en-Josas, France, [3]INRA, UMR1061 Unité de Génétique Moléculaire Animale, Faculté des Sciences et Techniques, 87060 Limoges Cedex, France, [4]UNCEIA, MNE, 149 rue de Bercy, 75595 Paris Cedex 12, France; brigitte.picard@clermont.inra.fr

The myosin heavy chain (MyHC) isoform IIb, considered as not expressed in bovine muscle, has been detected by electrophoresis in muscles of some Blonde d'Aquitaine young bulls descendant of the same sire called 'Hiver'. Interestingly, the animals with this isoform had a higher tenderness (estimated by sensory analysis) and juiciness of their meat. The objective of this study was to search for the presence of this isoform in large samples of animals of the three French beef breeds: Blonde d'Aquitaine (BA), Charolaise (Ch) and Limousine (Li) and to validate its relation with beef sensory qualities. MyHC isoforms of the Longissimus Thoracis muscle were separated by electrophoresis and their proportions were evaluated by densitometry on 958 young bulls (roughly 1/3 of each breed). The BTA19 chromosome, where is localised the gene synthesizing MyHC isoform (MYH4), was screened for a QTL associated with the presence/absence of MyHC IIb in the 'Hiver' sire family. The co-segregation of 4 micro-satellites and MyHC IIb phenotypes was analysed using the QTLMAP software. MyHC IIb was observed with very different frequencies according to the breed: 25% in BA, 6% in Ch, 41% in Li. It is the first time that we observed such differences between breeds for a muscle characteristic. The relation between MyHC IIb and tenderness and juiciness was confirmed only in the BA breed and more particularly in the lineage of the bull 'Hiver'. A highly significant QTL for this isoform was detected, but no QTL of tenderness could be detected on BTA19. Further analyses are in progress to study thoroughly the role of this isoform in beef tenderness.

Session 08 Poster 14

Effect of genotype on fatty acid composition of several muscles (LT, ST, PM) of young fattening bulls

Somogyi, T.[1], Hollo, G.[1], Anton, I.[2] and Hollo, I.[1], [1]Kaposvár University, Guba S. 40., 7400 Kaposvár, Hungary, [2]Reseach Institute for Animal Breeding and Nutrition, Gesztenyés s 1., 2053 Herceghalom, Hungary; hollo.gabriella@sic.hu

We examined the ability of n-3 fatty acids of linseed supplementation (LS) in finishing diet to increase the n-3 fatty acid content of bovine muscles (longissimus thoracis-LT, semitendinosus-ST, psoas major-PM) from six different genotypes. For this study 10 Angus (A), 8 Charolais (CH), 11 Holstein (H), 10 Hungarian Grey (HG), 15 Hungarian Simmental (HS), 9 Charolais x Hungarian Grey F1 (CH X HG) fattening bulls were used. In the last period the bulls were given for 163 d long 25% LS in concentrate. The final live weight was 600 kg. A bulls had the highest intramuscular fat (IM) level (3.76), particularly in PM (4.74), followed by HG (3.40), crossbred HG (3.05), CH (2.33), HS (2.21), with the lowest level for H (2.11) bulls. IM for LT (2.87) and ST (1.77) muscles were significantly lower than PM (3.60) for all groups. SFA was significantly higher for the PM than other two, due to the high level of IM. The highest MUFA was detected in LT, significantly differed from others. ST contained the highest PUFA (12.99), followed by LT (10.30) and PM (10.19). The n-6 fatty acids were affected either by breed, or by muscle type. The linoleic acid was the highest in Holstein bulls except for PM, whilst the lowest was in all cases in A bulls. The same tendency can be seen in all cases for long chain n-6 fatty acids. The overall mean of n-3 fatty acids differed among genotypes from 1.36 to 1.80%. The highest level measured in ST of CH and H (2.19), whereas the lowest one LT of HG (1.22). Significant differences among genotypes were shown only for PM. In the same muscle the lowest n-3 fatty acids level had HG and the highest one HS. The level of n-3 series long chain fatty acids was the highest for ST, for LT and PM in H, CH and HS, respectively. LS resulted in beneficially lower n-6/n-3 ratios in muscles in all groups. The ratio of n-6 /n-3 was ranged from 4.48 to 6.22 and significantly less favourable in H than in A bulls.

Effects of FASN and SCD gene polymorphism on carcass traits and muscle fatty acid compositions in Japanese Black cattle

Katoh, K., Tohoku University, Animal Physiology, 1-1 Amamiyamachi, Aoba-ku, Sendai, 981-8555, Japan; kato@bios.tohoku.ac.jp

We recently reported that GH gene polymorphism has a crucial role for carcass traits and muscle fatty acid compositions in Japanese Black cattle. Cattle with AA-type allele produced a greater carcass weight while cattle with CC-type allele, which is a specific allele for Japanese Black cattle and shows reduced GH secretion, produced unsaturated fatty acid-rich beef. These characteristics are much more marked in heifers than in bulls. On the other hand, the significant role of the enzymes, fatty acid synthase (FASN) and stearoyl-CoA desaturase (SCD) in relation to synthesis and unsaturation of fatty acids is well documented. In the present study we assessed effects of FASN and SCD gene polymorphisms on carcass traits and fatty acid compositions in Japanese Black cattle. Muscle and adipose tissues (n=382) sampled from M. longissimus thoracis around the age of 30 months in Sendai region were classified by FASN and SCD gene polymorphisms, and were subjected for statistical analysis in relation to carcass traits and fatty acid compositions. PCR-RFLP and gas chromatography were employed to analyze polymorphisms and fatty acid compositions. SAS package program was used for statistical analysis. In FASN genotypes, WW-, WR- and RR-types were 11, 48 and 41%, respectively. In SCD genotypes VV-, VA- and AA-types were 10, 52 and 38%, respectively. WW-type FASN genotype showed a tendency for increasing muscular C18:1, but significantly lower C14:0 and C16:1 contents. On the other hand, AA-type SCD genotype showed a tendency for increasing C18:1 content. However, sex-dependency was not demonstrated for these effects of FASN and SCD polymorphisms. Taken together, the significance of FASN and SCD polymorphisms on beef production is much less than that of GH polymorphism.

The use of EUROP classification traits and X-ray computed tomography (CT) measurements of rib samples from six cattle breeds to predict slaughter value

Hollo, G.[1], Somogyi, T.[1], Anton, I.[2], Repa, I.[1] and Hollo, I.[1], [1]Kaposvár University, Faculty of Animal Science, Guba S. street 40., 7400 Kaposvár, Hungary, [2]Research Institute for Animal Breeding and Nutrition, Gesztenyés street 1., 2053 Herceghalom, Hungary; hollo.gabriella@sic.hu

The EUROP classification scores and the tissue composition of rib samples determined by CT-scanning were used to investigate the best method for prediction of carcass composition of young fattening bulls from different genotypes: Angus (A), Charolais (CH), Holstein (H), Hungarian Simmental (HS), purebred Hungarian Grey (HG) and crossbred Hungarian Grey (CHxHG). The animals were kept and fed under same condition. The target final live weight of all genotypes was determinate at 600 kg. The carcasses were evaluated according to EUROP grading system. For the X-ray computer tomography analysis rib samples were taken between the 11-13th rib. The ranking of breeds according to EUROP conformation class is the following: CH, HS, A, CHxHG, HG and H. At the same time the dissected lean meat yield of carcass and CT-muscle proportion of rib samples was significant higher in all genotypes (HS: 74 and 72%, CH: 72 and 70%, HG, CH x HG: 71 and 68%) than that of A (67 and 61%). Carcass fatness class changed together with CT determined fat content. Data showed the lean % and fat % in carcass correlated closer with CT-muscle and CT-fat (r=0,92 and r=0,85) than EUROP conformation and fat scores (r=0,34 and r=0,69). Predictors derived from CT alone accounted for a high proportion of the variance in dissected fat proportion (R^2 =0.8), but lower proportions for dissected lean meat yield (R^2 =0.7). Adding traits such as EUROP conformation score and carcass weight into equation increased prediction accuracy by up to 0.26. Bone content could be lower accuracy (R^2 =0.6) predicted using CT data and EUROP conformation score. Findings confirmed that the prediction of carcass composition can be achieved more objectively with the CT data inclusion into EUROP carcass classification system.

Effect of ageing time on meat characteristics of castrated or uncastrated brahman calves

Pargas, H.L.[1], Colmenarez, D.[1], Ciria, J.[2], Asenjo, B.[2] and Miguel, J.A.[2], [1]Office of the Dean of Veterinary Sciences, University Centroccidental Lisandro Alvarado. Barquisimeto, The State of Lara, Venezuela, 000, Venezuela, [2]Universidad de Valladolid, campus uniersitario s/n, 42004 Soria, Spain; jangel@agro.uva.es

This paper compares the shear force, the capacity of water retention and some sensory attributes in meat of the beef calves Brahman uncastrated and castrated at birth, depending on the time of ageing (1, 7 and 14 days). We used 25 each type animals raised on commercial terms in Venezuela. For the preparation of displays, muscle Longissimus thoracis was extracted from the left side of the carcass, and cut into fillets of 2.5 cm, was later pressurized vacuum and assigned at random one of the three periods of maturation, being stored in refrigeration at 20 °C so far from being scanned. In the analysis, swatches charters cooked in an electric Grill opened up an internal temperature of 350 °C. A Warner-Bratzler cell was used for the calculation of the shear force and losses by cooking were calculated for water retention capacity. Sensory analysis was done by a trained assessed the juiciness, myofibril tenderness and general panel, the amount of connective tissue, and intensity of flavor. Compared to the entire Brahman, the ageing time significantly influenced the sensory characteristics of meat from castrated animals as well the instrumental tenderness of the meat of both groups of animals. However, no effect of ageing on cooking water loss was seen for both types of animals. Shear force values were lower for castrated, corresponding to higher ratings assigned by the panel of tasters for tenderness. Shear force and the cooking water loss, except the intensity of flavor, mapped negatively with sensory characteristics.

Inclusion of maize silage in beef fattening diets: effects on performance and meat quality

Casasús, I.[1], Ripoll, G.[1], Revilla, R.[2] and Albertí, P.[1], [1]CITA, Avda. Montañana, 930, 50059 Zaragoza, Spain, [2]CTA, Avda. Movera, 50194 Zaragoza, Spain; icasasus@aragon.es

In order to reduce feeding costs in Spanish beef fattening enterprises, an alternative to conventional feeding on concentrates and straw was tested. Thirty-two female calves received in a feedlot either of two fattening diets ad libitum from weaning to slaughter at 450 kg. One group received concentrates (13.2%CP, 4.3% CF) and barley straw (CONV), and the other was fed a total mixed ration (80% maize silage: 20% high-protein concentrate, 14.8% CP, 17.4% CF) (TMR). Animals were weighed monthly and feed intake was registered daily per group. At slaughter, carcass characteristics were assessed and samples of Longissimus thoracis were taken for the determination of meat quality (colour determined with a spectrophotometer, instrumental texture as Warner-Bratzler shear force). Animals from both treatments had similar growth rates (ADG 1.119 vs. 1.181 kg/d in TMR and CONV, respectively, NS), and therefore took a similar period to reach target slaughter weight (273 vs. 258 d, NS). Dry matter intake was slightly higher in TMR animals (6.53 vs. 6.0 kg DM/d, not statistically tested) and so was feed conversion ratio (5.83 vs. 5.08 kg/kg). Carcass characteristics were similar between both groups (271 vs. 275 kg in TMR and CONV, respectively, NS; 59.2 vs. 59.9% dressing percentage, NS; 10.5 vs. 10.6 points conformation score (U), NS; 5.3 vs. 5.5 points fat cover, NS), and therefore income perceived per carcass was similar (1122 vs. 1160 €, NS). There were no statistical differences in subcutaneous fat colour (although TMR animals tended to have higher a* and b* values in, indicating a higher pigment content) or meat tenderness (with a trend towers more tender meat in TMR animals). Meat colour was similar at slaughter or after 13 d in either conventional or modified atmosphere packaging. These results show that, with similar economic revenue per carcass, feeding costs per kg gain were 13% lower in TMR animals, while other costs were kept constant and carcass and meat quality did not differ.

Effect of genotype on mineral content of different muscles (LT, ST, PM)

Somogyi, T., Hollo, G. and Hollo, I., Kaposvár University, Guba S. 40., 7400 Kaposvár, Hungary; hollo.gabriella@sic.hu

The aim of this study was to determine the effect of genotype on the mineral content of three different muscles. Altogether, 62 young bulls from six genotypes including Angus (A), Charolais (CH), Holstein (H), purebred and crossbred Hungarian Grey (HG, CHxHG) and Hungarian Simmental (HS) were used. Animals were kept and fed under the same conditions. Bulls were slaughtered in commercial slaughterhouse at the same live weight (600 kg). 24 h *post mortem* samples were taken for mineral content analysis from right half carcass longissimus thoracis (LT), semitendinosus (ST) psoas major (PM) muscles. The Na, Ca, K, Fe, Mg, Cu, Zn and Mn content (mg/1000 g tissue) were determined by atomic absorption or spectrophotomethry. The genotype influenced P, Mg, K, Na, Fe content, whilst the effect of muscle on all minerals except for Cu was significant. Significant genotype and muscle interaction can be seen only for Ca. ST muscle of H has the highest content of P, differed from the lowest one measured in LT muscle of A animals. LT muscle of A was the poorest for Mg, on the other hand it was the richest in PM of H and HG bulls. About K, significant differences among genotypes for PM muscle can be detected, CH had higher level of K than A and CH x HG. It is well known that high level of Na is not beneficial. Na content of ST in CH x HG was lower than H bulls. The overall Na level of H bulls was the highest, and it was the lowest one in crossbred HG bulls. Our data showed that the richest source of iron is the beef from HG, especially PM muscle. The iron content of HG bulls in same muscle differed from level of A, CH, HS bulls. Contrary to LT muscle, the highest Fe content had H bulls, the poorest one CH bulls. The highest Mn and Cu level were detected for LT muscle of A and ST muscle of H, resp. Referring to zinc, higher level was in LT especially for H animals. Our results confirmed that the beef mineral richness depends on genotype, and it is related to muscle type, too.

Influence of dietary crude protein content and source on Friesian young bulls performances and meat quality

Iacurto, M., Palomba, A., Ballico, S. and Vincenti, F., Agricultural Research Council, Research centre for meat production and genetic improvement, Via Salaria, 31, 00015 Monterotondo (RM), Italy; federico.vincenti@entecra.it

The concentration and ruminal degradability of crude protein in beef cattle diets may affect excretion of Nitrogen and consequently gases emission. The aim of this study was to evaluate the effect of dietary crude protein on in vita performances, slaughter and dissection performances and meat quality. The Study was carry out on 27 Frisian young bulls, when bulls were 13 months old, they were divided in three groups and fed with three different diets: C - Control group (7 animals) with 0.90 UFV/ kg DM and with 15.42% of CP; Cg - Chickpea group (10 animals) with 0.97 UFV/kg DM and CP from 15.61% to 11.34%; Bg - Broad bean group (10 animals) with 0.97 UFV/kg DM and CP from 15.23% to 11.34%. Live weights were recorded every 20 days in order to calculate growth curves. Bulls were slaughtered at about 544 kg of body weight. Carcass weights, carcass yields were evaluated. After 7 days of ageing pH, water losses, WBS on cooked meat and colour parameters were performed. No differences were found on *in vivo* performances (ADG: C= 1.16 kg/d; Cg= 0.97 kg/d; Bg= 1.02 kg/d). Crude protein content and source influenced some slaughter performances and some meat quality traits in fact statistical differences were found on carcass weight, fat score, lean meat, bone and fat percentage, pH, water losses, WBS, Lightness, Redness index and Hue values.

Session 08 Poster 21

***In vivo* assessment of carcass composition using muscle area and fat thickness measured by ultrasound**

Polák, P.[1], Tomka, J.[1], Bartoň, L.[2], Bureš, D.[2], Krupa, E.[1] and Oravcová, M.[1], [1]Animal Production Research Centre Nitra, Hlohovecká 2, 949 11 Nitra, Slovakia (Slovak Republic), [2]Research Institute for Animal Production, Přátelství 815, 104 00 Prah - Uhříněves, Czech Republic; polak@cvzv.sk

The aim of this study was to analyze possibility to assess carcass composition *in vivo* using muscle area and fat thickness measured on live animals by ultrasound. Fifty four young bulls either purebred Czech Fleckvieh and Polled Simmental or crosses of Czech Fleckvieh, Charolais and Piedmont were sonographically scanned on the back. Musculus longissimus thoracis et lumborum (MLTL) area and fat layer were measured on 8th (AREA8, FAT8) and 12th (AREA12, FAT12) thoracic vertebra. Correlation coefficients between selected carcass quality variables (weight of hot carcass - WHC, weight of meat in carcass - WMC and weight of valuable cuts in carcass - WVCC) and muscle areas were higher for AREA8 then AREA12 vertebra. FAT8 and FAT12 showed low negative correlations with selected carcass quality characteristics. Linear regression models using by area8 and FAT8 had not significantly higher coefficient of determination than those with using AREA12 and FAT12. Coefficients of determination for models including live weight before slaughter (LWBS), AREA8 and FAT8 were 0.86, 0.70 and 0.60 when WHC, WMC and WVCC were considered as the dependent variables. LWBS had the highest impact on prediction ability of each model. Sonographically obtained measurements (AREA8 and FAT8) increased R2 by 0.06 to 0.12. Both were significant. This means that they improve predicting ability of single live weight.

Session 08 Poster 22

Relationship of Magnetic Resonance Imaging (MRI) measurements with some meat quality data and biochemical properties of blood

Hollo, G.[1], Zsarnóczay, G.[2], Hollo, I.[1], Somogyi, T.[1], Anton, I.[3], Takács, I.[1], Farkas, A.[1] and Repa, I.[1], [1]Kaposvár University, Guba S. street 40., 7400 Kaposvár, Hungary, [2]Hungarian Meat Research Institute, Gubacsi street 6/b, 1097 Budapest, Hungary, [3]Reseacrh Institute for Animal Breeding and Nutrition, Gesztenyés street1., 2053 Herceghalom, Hungary; holo.gabriella@sic.hu

The relationship between MRI – data (T_2 relaxation time) and meat quality parameters (pH, colour, intramuscular fat, Warner-Bratzler shear force-WBSF) as well as blood parameters (ADT, ALT, CK, Glucose, Cholesterol, Urea, Fe) was investigated. For this purpose, 30 Hungarian Simmental bulls were used. The slaughter weight of animals was determinate at 600 kg. Blood samples were collected during bleeding. The m. longissimus thoracis between the 11-13th ribs was sampled 24 hours post mortem (pm) for MRI examination and for chemical analysis as well as for the determination of pH, L*, a*, b* and WBSF. The T_2 relaxation time on two occasion post mortem was investigated. Pearson correlation of coefficients was calculated among parameters. Significant longer T_2 relaxation time was measured in rib samples at pm 6. days (T_{22}) in opposite to those measured at pm 24 hours(T_{21}). The T_{21} relaxation time correlated to AST (r=-0.66), Cholesterol (r=0.51), Glucose (r=-0.56), WBSF (r=0.57) and pH (r=-0.59). The T_{22} relaxation time showed positive relationship with Lightness (r=0.54), fat content (r=0.50) and WBSF(r=0.77), whilst negative related to pH (r=-0.77). Significant negative correlations were seen between WBSF and AST, CK, Glucose, Urea and a positive correlation was observed between WBSF and Fe. It seems that MRI can be used for the evaluation of beef quality traits; further research is needed in this area.

Effect of different nonprotein nitrogen sources on meat characteristics of feedlot finished Nelore steers

Corte, R.R.P.S., Nogueira, J.C.M., Brito, F.O., Leme, P.R., Pereira, A.S.C., Aferri, G. and Silva, S.L., Universidade de São Paulo, A Duque de Caxias Norte, 13635900, Brazil; jocamano@usp.br

To assess the effects of soybean meal replacement by different nonprotein nitrogen sources (NNP) on the meat quality of finishing cattle, 46 Nelore steers with a mean initial weight and age of, respectively, 333 kg and 20 months, were fed one of four diets: 1) CTL (control diet): 12% soybean meal and 1% Urea, 2) O: 6% soybean meal and 1.8% Optigen®II, 3) U: 6% soybean meal and 1.66% Urea, and 4)U+O: 6% soybean meal, 1.0% Urea and 0.72% Optigen®II. Diets had 78.5% concentrate and were isoproteic (15.5% CP) and had the same energy (77.4% TDN) and rumen degradable protein (10.4%) values. Steers were allotted to four pens according to initial body weight and the experimental design was a randomized block. After 75 days, steers were slaughtered and 24 after, Longissimus muscle samples were taken for meat characteristics analyses. Shear force (SF), meat color (MC), and marbling score (MARB) were determined in samples aged 1, 14 and 21 days and ether extract (EE) was in samples aged 1 day. SF was obtained with a Warner Bratzler (AMSA, 1995) equipment, MC with MiniScan XE, Hunter Lab and MARB with USDA QUALITY GRADE (1999). EE analyses were realized according with AOAC (1997) in lyophilized meat samples. Effects of treatments were evaluated using SAS software. Meat characteristics were not affected by treatments ($P>0.05$). The general mean of SF was 4.89 kg, indicating a tenderness meat. MC means of L*, a*, b* were 38.24, 17.95 and 15.24. MARB was classified as select for all treatments, with a general mean of 4.74. EE mean was 2.56%, indicating a lean meat. These results indicated that replacement of soybean meal by NNP sources used in this study did not affect the meat characteristics of feedlot finished steers. Thus, in these conditions these nitrogen sources can be used to reduce the costs of feeding while still achieving similar results to traditional soybean meal-based diets.

An *in vitro* culture model of intramuscular bovine preadipocyte

Soret, B., Tiberio, P., Mendizábal, J.A. and Arana, A., Universidad Pública de Navarra, Producción Agraria, Los Olivos, Campus Arrosadia, 31006 Pamplona, Navarra, Spain; paula.tiberio@unavarra.es

The main objective of this work was to develop an *in vitro* cell culture system using a chemically defined medium in order to differentiate primary intramuscular bovine preadipocytes obtained from a local breed (Pirenaica), which shows low intramuscular fat deposition. Immediately after exanguination of the steers, samples (intramuscular and subcutaneous) from the left side of the carcass from the Longisimus thoracis muscle were collected and placed on a solution at 37 °C. Cells from the stromovascular fraction of the adipose tissues were isolated by collagenase digestion and plated on a medium containing M199, acetate, antibiotics and 10% new born calf serum. After confluence, the cells were challenged with differentiation inducer media which contained DMEM/F12, acetate and antibiotics, being this the 'basal medium', and different combinations of: insulin, tri-iodothyronine, dexamethasone, isometilbutilxantine, rosiglitazone, biotine, transferrine, selenium, bovine serum albumin, octanoate and oleic acid. Differentiation was assessed with Red O oil staining. Finally, a chemically defined differentiation media composed by basal medium plus a combination of insulin 1.6 mg/ml, dexametasone 10 nM, rosiglitazone 10 mM, isometilbutilxantine 0.5 mM, biotine 0.5 mM, bovine serum albumin 0,12% w/v and octanoate 1 mM was found to induce differentiation of both intramuscular and subcutaneous preadipocytes, the later used as a control adipose tissue. This model will be used in further assays examining the ability of the preadipocytes to differentiate and to study the differentiation of bovine intramuscular adipocytes at molecular level, aiming to gain a better understanding of the process of lipid accretion in economically important adipose tissue depots.

Quality of beef from different production systems marked in Portugal

Monteiro, A.C.G., Fontes, M.A.A., Prates, J.A.M. and Lemos, J.P.C., FMV TULisbon, CIISA, Av Universidade Técnica, 1300-477, Portugal; amonteiro@fmv.utl.pt

In Portugal there are several meat products with Protected Denomination of Origin label, originated from national beef breeds raised in traditional production systems. The promotion of certified products can be of great benefit to the rural economy, by improving farmers' incomes and by retaining the rural population. Portuguese certified beef consumption is increasing due to public perception of its higher nutritional value and safety. However, Commercial crossbred cattle produced under intensive regimens provide the main supply of beef at competitive prices. Moreover, import of Brazilian beef is increasing. Brazilian beef is obtained from local breeds raised in semi-extensive production systems. Carnalentejana-PDO beef is the Portuguese certified beef with the higher economic impact, being obtained from Alentejana purebred young bulls produced in semi-extensive production systems. These meat products have higher production costs and therefore higher sale prices, but the less intensive production systems often affect negatively beef colour and tenderness. So, the aim of this study was to compare the quality characteristics of 3 relevant types of beef marked in Portugal: Commercial, Brazilian and PDO beef. Fifteen samples of L. lumborum muscle of each meat type were collected in a hypermarket and used for pH, colour, intramuscular fat, pigments, collagen content and its solubility, myofibrilhar fragmentation index, cooking losses and WBSF determinations. Despite the lower Commercial beef pH, colour did not differ between meat types. Brazilian meat had higher MFI, but this higher maturation index was not reflected in a lower WBSF value. Commercial beef showed higher variability in colour and pigments content, whilst PDO had higher variability in the parameters related to tenderness. We concluded that despite the different origin, genetics and production systems, all beef types had similar physiochemical characteristics. Also meats had a reasonable value in WBSF, indicating to be medium tender.

Session 08 Poster 26

The effects of polymorphisms in the calpastatin and µ-calpain genes on beef tenderness are breed specific

Allais, S.[1,2], Hocquette, J.F.[3], Levéziel, H.[4], Payet-Duprat, N.[4], Lepetit, J.[5], Rousset, S.[6], Denoyelle, C.[7], Bernard-Capel, C.[7], Journaux, L.[1] and Renand, G.[2], [1]UNCEIA, 149, rue de Bercy, Paris, France, [2]INRA, UMR 1313, 78350 Jouy en Josas, France, [3]INRA, UR 1213, 63122 Theix, France, [4]INRA, UMR 1061/ Université de Limoges, 87060 Limoges, France, [5]INRA, UR 370, 63122 Theix, France, [6]INRA, UMR 1019, 63122 Theix, France, [7]Institut de l Elevage, 149, rue de Bercy, 75595 Paris, France; carine.capel@inst-elevage.asso.fr

The objectives of the study were to test the association of polymorphisms in the calpastatin (CAST) and µ-calpain (CAPN) genes with meat tenderness in three French beef breeds. A total of 1,114 Charolais, 1,254 Limousin and 981 Blond d'Aquitaine purebred young bulls were genotyped for three SNP in the CAST gene and four SNPs in the CAPN gene. Two of these markers can be found in Australian or American commercial genetic tests. Quantitative traits studied were Warner-Bratzler shear force and tenderness score evaluated by trained sensory panels. All the SNPs were informative in the three breeds. Analyses of individual marker and haplotype associations with traits were performed. Results were different between the three breeds. We found a significant effect of the G allele of a CAST marker (positions 97574679 on Btau4.0) on shear force (+0.18 RSD) and tenderness score (-0.20 RSD) in the Blond d'Aquitaine breed. In this breed, this marker was associated with another CAST SNP such as the GA haplotype appeared to be associated with a tougher meat. Two CAPN markers (positions 45221250 and 45241089 on Btau4.0) had a significant effect on both traits in the Charolais breed (from |0.11| to |0.25| RSD). In this breed, the haplotype analysis showed that two haplotypes made up three CAPN markers, ACA and AGG, were associated with a tender meat and a tougher meat respectively. Consequently, the current work suggests that effects of studied markers are breed-specific and can not be extended to all *Bos taurus* breeds.

Transcriptomic markers of beef meat tenderness

Bernard-Capel, C.[1], Cassar-Malek, I.[2], Renand, G.[3], Lepetit, J.[4] and Hocquette, J.F.[2], [1]Institut de l Elevage, 149 rue de Bercy, 75595 Paris Cedex 12, France, Metropolitan, [2]INRA, URH, Theix, 63122 Saint-Genes Champanelle, France, Metropolitan, [3]INRA, GABIE, Domaine de Vilvert, 78352 Jouy-en-Josas cedex, France, Metropolitan, [4]INRA, QuaPa, Theix, 63122 Saint-Genes Champanelle, France, Metropolitan; carine.capel@inst-elevage.asso.fr

In Livestock species and more particularly in bovine, meat tenderness is a major factor not only to help producers to give value to their animals but also to satisfy the consumers. Although there is no genomic selection yet in beef cattle to meet these expectations, many research studies have been recently conducted in order to better understand the physiological processes underlying meat tenderness and so to identify genes related to this trait. We analysed gene expression level in Longissimus thoracis and Semitendinosus muscles from 15- and 19-month-old Charolais bull calves using microarray technology. Comparison of the transcriptome of animals which produced hard or tender beef according to the Warner-Brätlzer shear force measurement allowed the identification of 160 and 276 genes differentially expressed ($P<0.05$) in Longissimus thoracis and Semitendinosus muscle, respectively. Among these genes, 22 were significantly related to tenderness in the two muscles, 9 of which were up-regulated and 13 were down-regulated in the muscles giving tender beef. However, gene expression level was correlated with the shear force value either in the Longissimus thoracis muscle or the Semitendinosus muscle but not in both muscles, suggesting that these transcriptomic markers are muscle specific. Interestingly, we found some down-regulated genes (DNAJA1, HSPB1) which were involved in the apoptosis pathway and which had already been shown to be related with beef sensory tenderness. In conclusion, these genes could be good genomic markers of tenderness evaluated by sensory or mechanical analyses.

Digital dermatitis in cows: current state of knowledge

Relun, A.[1,2], Guatteo, R.[1], Roussel, P.[2] and Bareille, N.[1], [1]ONIRIS, INRA, UMR1300 BioEpAR, BP40706, F-44307 Nantes, France, [2]Livestock Institute, 149 rue de Bercy, F-75012 Paris, France; anne.relun@oniris-nantes.fr

Digital dermatitis (DD) is a multifactorial contagious foot disease of cattle, associated with management practices, environmental and microbial factors. While not described before 1973, it is now worldwide one of the most common causes of lameness in cows. It is an increasing concern for farmers. Besides adverse effects of the disease on animal welfare and production, control schemes seem to be ineffective for its eradication, and sometimes even for decreasing its prevalence. Moreover, two of the most effective compounds used in footbath for herd treatment (formaldehyde and copper sulphate) could be forbidden in Europe because of human health or environmental hazards. This presentation aims at providing an overview on the current state of knowledge on this disease and to stress on questions and perspectives for its future control.

Relationships between estimated breeding values for claw health and production as well as functional traits in dairy cattle

Alkhoder, H.[1], Pijl, R.[2] and Swalve, H.H.[1], [1]Institute of Agricultural and Nutritional Sciences, Martin-Luther-University Halle-Wittenberg, Theodor-Lieser-Str. 11, 06120 Halle, Germany, [2]Claw Health GmbH, Fischershäuser 1, 26441 Jever, Germany; hermann.swalve@landw.uni-halle.de

The status of claw health in dairy cows can be examined at the time of hoof trimming. Since 2000, the professional hoof trimmer René Pijl has recorded sub-clinical and clinical disorders of claws of dairy cows during his professional work in Northern Germany. Data is recorded electronically and stored in a data base which in turn is linked to the central milk recording and pedigree data base located at VIT, Verden, Germany. At present, the data base contains 79,181 observations from repeated trimmings of 26,112 Holstein cows. Based on this data and prior analyses, heritabilities were estimated applying a threshold animal model. For laminitis (0.19), dermatitis digitalis (0.09), dermatitis interdigitalis (0.19), white line disease (0.13), and sole ulcer (0.14) were found. Using these results, breeding values for bulls were estimated and correlated with the official breeding values from the national genetic evaluation. For claw health breeding values, a minimum of 70% reliability was required. This requirement resulted in between 126 and 334 usable EBV of sires. For the most prominent disease, laminitis, favorable correlations between disease resistance and total merit index, production index, overall conformation, and functional longevity were obtained in a range of 0.04 to 0.34. Correlations with feet and leg scores also were correlated favorably, although small of magnitude (0.14 to 0.23). For laminitis, correlations with EBV for somatic cells were non-significant. For dermatitis digitalis (Mortellaro's disease), correlations were approaching the significance threshold and were unfavorable. In contrast, for dermatitis interdigitalis, again favorable and significant correlations (0.13 to 0.20) were found in relation with somatic cell score EBV.

Relationships of individual animal traits and sole hemorrhage scores in fresh heifers

Ouweltjes, W., Wageningen UR, Livestock Research, P.O. Box 65, 8200 AB Lelystad, Netherlands; wijbrand.ouweltjes@wur.nl

Sole hemorrhages occur frequently in dairy cattle. Epidemiological studies have identified housing, parturition and feeding as important risk factors. Mechanical overload is regarded as an underlying cause for effects of housing. However, even under the same keeping conditions individual animals vary considerably in the degree of sole hemorrhages. It is investigated if individual differences in body weight and behavior for a group of heifers around calving are related to differences in development of sole hemorrhages. At Waiboerhoeve experimental farm an experiment was performed with 44 heifers kept under 1 of 4 different housing conditions for 2 months after first calving and under the same conditions during the third month after first calving. The animals were monitored with IceTag sensors from at least 2 weeks before calving until the end of the experiment. Feed intake was monitored with a RIC system. From calving onwards the animals were weighed twice daily, before entering the parlor. Claw disorders and claw diagonals were scored in weeks 1, 9 and 14 after calving. Data were statistically analyzed with ASreml. Behavior before calving was positively related to behavior after calving, i.e. animals that were relatively active before calving tended to be relatively active after calving. Overall activity levels were low however, most of the non-lying time was spent standing still. In all housing groups the average sole hemorrhage score increased after calving. The animals were categorized in three classes for sole hemorrhage (none or minor, moderate and above moderate) as observed in week 14. Patterns for %standing, #steps/hour and #standing bouts for each category per day after calving were remarkably similar. Similar results were obtained for body weight change and feed intake. Under the circumstances of the study susceptibility for sole hemorrhages was not related to behavioral differences or body weight. This suggests that heavier heifers or more active heifers are not at increased risk for overload.

Session 09 Theatre 4

Transforming Growth Factor Beta 1 (TGFβ1) and Connective Tissue Growth Factor (CTGF) in the functionally important laminar region of the bovine claw

La Manna, V.[1,2], Di Lucca, A.[1] and Galbraith, H.[1,2], [1]University of Aberdeen, School of Biological Sciences, 23 St Machar Drive, Aberdeen AB24 3RY, United Kingdom, [2]University of Camerino, Environmental and Natural Sciences, Via Pontoni 5, 62032 Camerino, Italy; h.galbraith@abdn.ac.uk

The major source of bovine lameness arises from lesions of the claw. These have been associated with impairment of the weight-bearing dermal laminar suspensory system and damage to the white line and epidermal horn production of the sole. Causes may include parturition/lactational hormonally-induced breakdown of vascularised laminar tissue and/or aseptic inflammatory 'laminitis' attributed to ruminal acidosis. In considering mechanisms responsible, this study investigated TGFβ1 and CTGF. TGFβ1 is known to mediate (i) immune response, (ii) epidermal cell function and (iii) turnover of structural macromolecules in dermis (eg collagens, elastin) upstream of CTGF, in other anatomical locations. Samples of laminar tissue were collected post mortem from the wall region of claws of healthy non-pregnant female cattle. Expression of TGFβ1 by RT-PCR was detected in 0.75 of the analysed samples and similar to that of the housekeeping gene (18S rRNA). The expression of both TGFβ1 and CTGF was greater in samples collected in May than February ($P=0.03$ and $P=0.002$ respectively). Immunohistochemical signal for TGFβ1 active protein, was detected in both epidermis and dermis although with different intensity. TGFβ1 mRNA transcripts were detected by in situ hybridisation solely in the dermis, particularly around blood vessels and stromal cells proximal to the axial laminar tissue. For CTGF, mRNA transcripts localised in dermal lamellae and the abaxial reticular dermis with immunohistochemical detection in dermis and epidermis. The results provide evidence of presence and sites of synthesis of both TGFβ1 and CTGF and suggest the possibility of the generally conserved roles in immune response mediation and maintenance of integrity of structural molecules extending to epidermal and dermal tissue of the claw.

Session 09 Theatre 5

The early detection of disease in beef cattle through changes in behaviour

Szyszka, O., Edwards, S.A. and Kyriazakis, I., Newcastle University, School of Agriculture Food and Rural Development, Agriculture Building, NE1 7RU, United Kingdom; ollie.szyszka@ncl.ac.uk

In this study an attempt was made to quantify the changes in behaviour that could be used as disease indicators in beef cattle, using a mild challenge to the animal's immune system through vaccination. With the identification of specific changes in behaviour, early diagnosis of disease and detection of subclinical disease may be made possible. The behaviours focused on were overall activity, standing, lying, feeding and drinking and were observed in 24 Holstein/Friesian cross beef bulls aged between 5 and 11 months. At the start of the experiment, day -9, the animals were fitted with a pedometer (Icetag) to record activity. Feeding and drinking were monitored with the use of video recordings. Eight animals were randomly allocated to one of three treatments, whilst balancing for age and weight. On day 0 the animals were vaccinated with either 5 ml Rispoval 4 (Pfizer), a live attenuated vaccine, or 5 ml Bovipast (Intervet), an inactive vaccine, or 5 ml of a placebo for the controls. The experiment ran for 15 days post vaccination. The results showed no treatment effect on duration or frequency of recorded behaviours. However, the amount of time spent lying was significantly greater for the first treatment group compared to the second. Also, injection per se caused a persistent increase in animal activity across treatments. There could be various reasons as to why there were no other changes, the most obvious being that the vaccine did not constitute a strong enough challenge. Alternatively, the individual variation in response might have obscured any effects. As behaviour is known to change after a health challenge, it is reasonable to assume that with a slightly stronger or more prolonged challenge this may be detectable. It can be concluded that more research is needed to identify specific changes in behaviour of beef cattle faced with a health challenge, so that early detection of disease, and in particular of subclinical disease, can be made possible.

Session 09 Theatre 6

Improved detection of bovine respiratory disease in the young bull with a rumen temperature bolus

Timsit, E.[1], Assié, S.[1], Quiniou, R.[2], Seegers, H.[1], Fourichon, C.[1] and Bareille, N.[1], [1]ONIRIS, INRA, UMR1300 BioEpAR, BP40706, F-44307 Nantes, France, [2]INRIA, Campus de Beaulieu, F-35042 Rennes, France; christine.fourichon@oniris-nantes.fr

Bovine respiratory diseases are the most frequent health disorder in fattening bulls, whatever the farming system. Early antibiotic treatment of a case is recommended to improve the welfare of the animal, to limit losses due to impaired growth, and to prevent severe consequences of the disease and death. In farms, effectiveness of the treatment depends on the ability to early detect the occurrence of new cases of respiratory disease. The objective was to evaluate the use of rumen temperature boluses to detect respiratory diseases in young bulls. Twenty four young bulls received boluses at entry in the fattening unit and were observed 40 days. When rumen hyperthermia was detected, or when bovine respiratory disease signs were suspected by the farmer, a clinical examination was performed by a veterinarian and repeated until the end of the rumen hyperthermia episode. During rumen hyperthermia episodes, high rectal temperature were systematically observed (40.1+/-0.6 °C). Thirty-six bovine respiratory disease cases were detected in 21 animals. Diseased bulls showed rumen hyperthermia prior to the onset of respiratory disease signs detectable by visual appraisal. Furthermore, six respiratory disease cases would have been missed if rumen temperature was not monitored. The use of rumen temperature boluses enabled an earlier and more sensitive detection of bovine respiratory disease in young bulls in comparison to visual appraisal.

Session 09 Theatre 7

The use of a wireless ruminal bolus to detect subacute ruminal acidosis and other health challenges in cattle

Alzahal, O.[1], Alzahal, H.[1], Steele, M.A.[1], Kyriazakis, I.[2], Duffield, T.F.[1] and Mcbride, B.W.[1], [1]University of Guelph, Guelph, N1G 2W1, Canada, [2]Newcastle University, Newcastle upon Tyne, NE1 7RU, United Kingdom; oalzahal@uoguleph.ca

The objective of this study was to differentiate between changes in ruminal temperature (RT) due to subacute ruminal acidosis (SARA) vs. changes in RT that result from systemic changes, such as febrile responses by using a wireless ruminal bolus (WRB). Eight rumen-cannulated Holstein lactating dairy cows (586±37 of BW, 106±18 DIM) were used in a 4X4 repeated Latin Square design in a 2X2 factorial arrangement. Cows were randomly assigned to 1 of 2 diets, a low-grain (LG, 10% of DMI) or a high-grain (HG, 28% of DMI) diet, plus an intra-mammary infusion containing either lipopolysaccharide (LPS, 100 ug derived from *E. coli* 0111:B4) or sterile saline. Cows received the intra-mammary infusion at 900 h. Cows were fed at 900 and 1400 h daily. Each period consisted of 21 days. During the last day of each period, ruminal pH and RT was measured every minute per day via an indwelling logging system (INS) that resided in the ventral sac of the rumen and via WRB that resided in the reticulum. Statistical analysis was conducted on daily measures using PROC MIXED of SAS using grain level, LPS level, and their interaction, column (cow), and row (period) as fixed effects. Data obtained from INS indicated that HG cows had lower ($P<0.05$) ruminal pH compared with LG cows (278 vs. 120 minutes/day below pH 5.8), which indicated that HG cows were under SARA. Temperature recorded by INS showed an increase with both grain and LPS treatments ($P<0.05$) with cows receiving HG+LPS having the highest RT as indicated by time (min/day) spent above 39.2 °C. The WRB captured the increase in RT that was due to LPS infusion ($P<0.05$)and had a tendency ($P<0.1$) to detect the change in RT that was due to SARA. WRB can detect health challenges associated with febrile responses, and may have the potential to detect SARA.

Relationships between body surface thermographs with rumen pH, rumen and vaginal temperature in lactating dairy cows fed a control or an acidogenic diet and challenged with *E. coli* lipopolysaccharide

Montanholi, Y.R., Alzahal, O., Miller, S.P., Swanson, K.C., Schenkel, F.S. and Mcbride, B.W., University of Guelph, Department of Animal and Poultry Science, Guelph, ON, N1G 2W1, Canada; ymontanh@uoguelph.ca

Infrared thermography is a non-invasive technology that has been applied in the assessment of physiological processes and pathological conditions associated with productive performance in livestock. The objectives of this study were to compare telemetry readings of and rumen temperature and also records of the rumen pH and vaginal temperature to thermographs concomitantly taken from different body locations (e.g. flank, teat, forehead and eyeball) in healthy or febrile cows under dietary treatments. Eight lactating Holstein cows (586±37 of BW, 106±18 DIM). were arranged in a latin-rectangle design with four cows receiving either a control or a acidogenic diet and two cows in each dietary group receiving an intramammary injection of *E. coli* lipopolysaccharide (LPS) in each of the four 21d periods. Devices were placed into the rumen through a ruminal fistula for continuously measuring pH and temperature. Vaginal temperature was monitored using a data logger. Thermal images were taken every 1 h between 08.30 and 21.30 h. The complete dataset is being summarized in order to obtain the average of the telemetry and vaginal temperature readings around the time (+2 min.) that each of the thermographs was taken. The summarized dataset will be analyzed as a latin-rectangle design with repeated measures over time and the least square means for each trait within time of the day will be used for performing regression and correlation analysis. Preliminary analysis of the thermographs showed an increase of 1 to 2 °C in the rear area and a decrease of 3 to 4 °C in feet 5 to 6 h after the LPS injection (fever peak response). The complete evaluation of our dataset will help to clarify the strengths and potential complementarity of these technologies to ensure health and optimal productive performance in ruminant production systems.

Influence of rubber flooring on claw growth and health of lactating dairy cows in southeastern Sicily using infrared thermal imaging

Licitra, G.[1,2], Azzaro, G.[1], Scollo, C.[1], Currieri, E.[1], Caccamo, M.[1], D'emilio, A.[3], Ben Younes, R.[4] and Petriglieri, R.[1], [1]CoRFiLaC, Regione Siciliana, S.P. 25 Km5 Ragusa mare, 97100 Ragusa, Italy, [2]Catania University, DACPA, via Valdisavoia, 5, 95123 Catania, Italy, [3]Catania University, DIA, via Santa Sofia, 100, 95123 Catania, Italy, [4]Institut National Agronomique de Tunisie, Avenue Charles Nicolle 43, 1082 Tunis, Tunisia; azzaro@corfilac.it

Thirty lactating dairy cows (137±60 DIM; 38.26±6.8 kg/d milk) in a Sicilian herd were grouped based on parity number, milk yield level, and lactation stage. Cows in each group were randomly assigned to 2-level treatment of floor (concrete vs rubber) in the barn and observed in March, July and November 2009. Hoof trimming and claw length measurement were performed at each control day. Foot lesions and claw palmar surface were classified according to American Association of Bovine Practioners. Thermal images of the dorsal front hoof and palmar surface of each claw for all cows were acquired before and after trimming. The database comprised 801 trimming hoof records and 360 images. Claws growth was significantly higher ($P<0.05$) in rubber compared to concrete flooring group (65 vs 29 mm in July and 12 vs 0 mm in November). The incidence of foot lesions for rubber and concrete flooring, respectively, was as follows: laminitis lesions, 0.3%, 1.2%; digital dermatitis, 6.5%, 9.8%; deformity of claw, 15.1%, 13.3%; sole ulcers, 0.3%, 0%; interdigital dermatitis, 5.4%, 3.7%; heel erosion, 16.1%, 13.3%; sole hemorrhage, 28.0%, 34.2%; white line separation, 28.5%, 24.4%. Although no significant difference was found between rubber and concrete flooring, probably due to a high management level, a higher incidence of sole hemorrhage, digital dermatitis and laminitis was found in concrete compared to rubber flooring. Thermal images were processed through binarization using ImageJ software. Preliminary results showed that surface temperature was higher in presence of foot pathology, indicating that thermal imaging could be recommended as a non-invasive detection tool for dairy industrial use.

Improved sow longevity and welfare with a chelated mineral blend

Zhao, J.[1], Greiner, L.[2], Keith, N.[3], Vazquez-Anon, M.[1], Knight, C.D.[1], Stebbens, H.[4] and Harrell, R.J.[1], [1]Novus International, 20 Research Park Drive, 63304 St. Charles, MO, USA, [2]Innovative Swine Solutions, Carthage Veterinary Service P.O. Box 220, 62321 Carthage IL, USA, [3]Keith Association, P.O. Box 220, 62702 Springfield MO, USA, [4]Novus Europe, Ave Marcel Thiry 200, 1200 Brussels, Belgium; helen.stebbens@novusint.com

The objective of this study was to examine the benefits of feeding sows a chelated trace mineral blend (OTM, Mintrex®, Novus International Inc.) on sow longevity and welfare. Two sister sow farms with a common grandparent farm were fed either an inorganic control (ITM) or an OTM blend (Zn, Mn, and Cu), which replaced 50% of the ITM, with target levels of Zn, 165ppm, Cu, 16.5ppm, and Mn, 38.5ppm supplemented to the diets. Treatment was initiated at weaning and continued into the breeding herd. Gilts fed OTM had lower removal rates (8.4% vs. 7.5%, P=0.03) than gilts fed ITM from first service to first farrowing. Higher numbers of replacement females fed OTM remained within the herds to Parity 4 compared to those fed ITM (68.4% vs. 61.2%; respectively, P<0.001). The involuntary removal rate and relative removal rate due to locomotion (leg problems) were significantly reduced with OTM supplementation. In gilts, removal rate due to locomotion and involuntary removal rate were 7.4% vs. 14.3%, and 17.5% vs. 26.5% for the OTM and ITM groups, respectively (P<0.001). Similar results were observed in sows in that removal rates due to locomotion (12.6% vs. 19.4%, P<0.01) were reduced in sows fed OTM compared to sows fed ITM, and the observation was consistent across each parity. Results suggest OTM is beneficial for maintaining sow skeletal health and improving welfare assessed by higher survival rates to parity 4 and lower removal rates due to locomotion.

Effect of feed quality on particle size in faeces from different ruminant species fed high and low quality forage

Jalali, A.R.[1], Nørgaard, P.[1], Weisbjerg, M.R.[2] and Nielsen, M.O.[1], [1]Dept. of Animal and Veterinary Basic Sci., Faculty of Life Sci., Univ. of Copenhagen, 1870, Frederiksberg C, Denmark, [2]Faculty of Agricultural Sci., Univ. of Århus, 8830, Tjele, Denmark; pen@life.ku.dk

The aim of this study was to compare particle size distribution in faeces from different small ruminant species. The experimental design included ad libitum feeding with two forage types to three ruminating species during two periods in a balanced block design. The species included six adult non pregnant female llamas (L), Danish Landrace goats (G) and Shropshire ewes (E) with mean body weights of 140, 45 and 75 kg, respectively. The forage included green hay (GH) and grass seed straw (GSS). The content of CP, NDF and ADL in the DM was 15, 58 and 4% for the GH, and 7, 81 and 8% for the GSS, respectively. The content of in sacco un-degradable NDF in the GH and GSS was 14 and 30% of the NDF, respectively. Daily mean DM intakes were 1.8, 1.1, 0.7 kg and 1.0, 1.3, 0.7 kg for L, E and G in the GH and GSS, respectively. Faeces was collected over four days, washed in nylon bags with a pore size of 10 μm and freeze dried before being sorted into six sieving fractions with square holes of 2.36 (O), 1.0 (M), 0.5 (S), 0.212 (D), 0.106 (C) mm and a bottom bowl (B). The area, length and width of particles in sub-samples from each fraction were measured using image analysis software, ImageProPlus. The overall faecal arithmetic mean particle length (APL), the 95 percentile of particle length (95PL) and width values (95PW) were 1.1, 1.0, 3.2, 2.5 and 0.6, 0.6 mm at the GH and GSS, respectively. The proportions of particles in the B, C, M and M+O fractions were significantly affected by forage type, species and their interactions (P<0.001). The 95% of faecal particles was shorter than 0.45, 0.42 and 0.30 mm in L, E and G, respectively. In conclusion, feeding higher digestible forage greatly affects the proportion of large and small particles in washed faeces from ewes, goats and llamas, and this correlation is greater in llamas than in ewes and especially when compared with goats.

Nutritional and Methanogenic values of some tropical plants for ruminants

Chaudhry, A.S., Virk, M.R. and Khan, M.M.H., Newcastle University, Agriculture, Food and Rural Development, Agriculture Building, NE1 7RU, United Kingdom; a.s.chaudhry@ncl.ac.uk

The forage availability for livestock is unpredictable in many tropical regions. The unstable availability of forages causes poor livestock production as animals rely on poor quality and nutrient deficient feeds. Tree and shrub plants can potentially alleviate some problems of feed shortages and nutritional deficiencies for livestock during forage scarcity seasons. However, it would be necessary to nutritionally evaluate these plants before their use for feeding animals. This replicated 5x7 factorial study compared the nutrient contents and *in vitro* dry matter degradability (IVD) and methanogenic properties of 5 plants comprising grass nuts (grass) as a standard control and 4 shrubs (Baker or Adhatoda vasica= AV; Sanatha or Dodonea viscose=DV; Acacia or Acacia ampliceps= AA; and saltbush or Atriplex lentiformis= AL) at 7 incubation hours (0 to 96). The plants were statistically compared for their nutrients and IVD for each incubation time at $P<0.05$. These plants differed significantly for most nutrients, IVD at each incubation time and methane ($P<0.05$). A vasica contained significantly more CP but less NDF than other plants including grass nuts (($P<0.05$). The patterns of IVD of these plants with increasing times compared well with that of grass nuts. However, the extent of difference between mean IVD of plants depended upon the plant type and the incubation time ($P<0.001$). In fact, IVD of AV and AL were closer to grass nuts but greater than those of DV and AA at most incubations times. D viscose contained highest TP but lowest CP and IVD than other shrubs ($P<0.001$). Although these plants showed variable nutrients, IVD and methane as compared to grass, they can be potentially used to formulate ruminant diets during the feed shortage seasons of relevant regions of the tropical world. Further studies will look at the suitability of these plants as methane reducing supplements for forage consuming livestock especially in tropical countries to overcome food security issues.

Nutritional composition of spineless safflower (*Carthamus tinctorius* L. var. *inermis* Schweinf.) grown at different levels of N-fertilization

Primi, R.[1]*, Danieli, P.P.*[1]*, Ruggeri, R.*[2]*, Del Puglia, S.*[2]*, Rossini, F.*[2] *and Ronchi, B.*[1]*,* [1]*University of Tuscia, Animal Science, Via San Camilllo de Lellis, snc, 01100 Viterbo, Italy,* [2]*University of Tuscia, Crop Production, Via San Camilllo de Lellis, snc, 01100 Viterbo, Italy; danieli@unitus.it*

The aim of the present study was to assess how N-fertilization affects nutritional parameters of spineless safflower grown under Mediterranean conditions. Spineless safflower, sown in November 2008, was grown in 3 m x 18 m plot (45 plants m^{-2}). A randomized complete block design (RCBD) one cultivar x four winter N-fertilization levels (ammonium nitrate 26%; N_0=0 kg/ha, land allocation, N_1=35 kg/ha, N_2=70 kg/ha, N_3=105 kg/ha) (three replicates per each level) was used. Crop biomass production was recorded on May 2009 before blooming. 3 kg of fresh material for each plot was collected and pH was measured immediately. Dried and grounded samples (1 mm mesh) were analyzed for Crude Protein (CP), lipids (FAT), Crude Fiber (CFom) and ash (ASH) content. All data (n=24) were analyzed by one-way ANOVA; difference between means was declared significant at $P<0.05$. Biomass production showed a clear nitrogen-dose relationship (from N_0=1.80±0.29 to N_3=2.71±0.20 t ha^{-1} DM, $P<0.001$). N-fertilization significantly affected the plant CP content, with high values (6.13±0.68% DM) observed for the N_3 fertilization level, as well as accumulation of CP in aboveground biomass (from N_0=80.2±10.4 to N_3=165.3±10.2 CP kg ha^{-1}, $P<0.001$). FAT (1.36±0.23% DM), CFom (41.70±3.65% DM) and ASH (12.74±0.74% DM) were not significantly affected by N-fertilization. At harvest, the highest pH (6.0±0.1) was found for N_2 treatment and the lowest was recorded for the N_0 level (5.8±0.1) ($P<0.001$). Spineless safflower shows an interesting potential to be grown for ruminants feeding purposes in Mediterranean area. N-fertilization plays an important role in the productivity of safflower and, particularly, on its CP content. Further studies are in progress to fully characterize safflower both as hay and silage.

The anthelmintic effect of sainfoin (silage, hay, fresh) and the role of flavonoid glycosides

Ojeda-Robertos, N.[1], Manolaraki, F.[1], Theodoridou, K.[2], Aufrère, J.[2], Halbwirth, H.[3], Stich, K.[3], Regos, I.[4], Treutter, D.[4], Mueller-Harvey, I.[5] and Hoste, H.[1], [1]INRA/ENVT, 23 Chemin des Capelles, 31300, Toulouse, France, [2]INRA/Theix, Clermont-Ferrand, 63122 Saint-Genès-Champanelle, France, [3]InstituteVerfahrenstechnik,Umwelttechnik&TechnischeBiowissenschaften, Getreidemarkt 9, A-1060 Vienna, Austria, [4]TUM, Dürnast 2, 85354 Freising, Germany, [5]Agriculture,NutritionalSciencesResearch, Whiteknights, RG6 6AH Reading, United Kingdom; fofivet@yahoo.gr

Studies on the use of tannin-rich forages as nutraceuticals because of their anthelmintic properties are expanding. The aim was to examine whether methods of conservation influence the AH activity. 17 sainfoin extracts (silage, hay, fresh) were tested plus Lucerne (control). AH effects of 4 concentrations, were measured on Haemonchus contortus larvae by using the larval exsheathment inhibition asssay. In a second trial, the hypothesis tested was that hay or silage processes might change the ratio of flavonol/flavonol glycosides, the former compounds having higher AH effects. 3 flavonoids (quercetin, kaempferol, isorhamnetin) and the dihydrochalcone phloretin were compared to their respective glycoside molecules. AH effect were tested on 4 concentrations on H.contortus. Differences were detected between glycoside and aglycone molecules for the group quercetin(P=0.001), kaempferol (P=0.03) and isorhamnetin (P=0.04) with higher AH activity found for the aglycone molecules. No differences were detected between the dihydrochalcone phloretin and its glycoside phlorizin. Dose response effect was found for isorhamnetin and phlorizin groups. A higher AH activity was noticed in hay and silage samples compared to fresh. These differences seem partly explained with the presence of aglycone molecules with higher AH properties than the corresponding flavonol glycosides.

The evaluation of lagged and non-lagged versions of a zero-order model to describe ruminal DM and CP degradability

Fathi Nasri, M.H.[1], Danesh Mesgaran, M.[2] and Abbasi Daloii, T.[2], [1]University of Birjand, Department of Animal Science, Birjand, Iran, [2]Ferdowsi University of Mashad, Department of Animal Science, Mashad, Iran; tabbassid@yahoo.com

The lagged and non-lagged versions of a segmented model with three spline-lines delimited by two nodes or break points, constraining splines 1 and 3 to be horizontal asymptotes, and follows zero-order degradation kinetics were evaluated when fitting the ruminal DM and CP disappearance of raw and roasted whole soybeans from data obtained using the in situ technique. The model was run with or without the lag for DM and CP disappearance curves and some statistics including the mean square prediction error (MSPE), root of MSPE (rMSPE), R2 and plot of residuals against predicted values were used to compare between two versions of model. The F-values indicated that the more complex version (lagged) did not fit the data significantly better than the simpler (non-lagged) version. Moreover, decomposition of MSPE gave similar values of error in central tendency (ECT), error due to regression (ER), and error due to disturbance (ED) for both versions of model, and was mainly dominated by the disturbance component, which indicated that ruminal degradation of DM and CP of samples was well represented by both lagged and non-lagged forms. Therefore, the difference between parameter L and zero was tested by t-test ($P<0.05$). No significant difference was found suggesting that the parameter was unnecessary and data were best described by non-lagged form.

Comparing nutritive value of Almond hull of several Almond varieties by gas production technique

Jafari, S.[1], *Alizadeh, A.R.*[1] *and Imani, A.*[2], [1]*Islamic Azad University, Saveh branch, Animal Science Department, Saveh, 39187/366, Iran,* [2]*Seed and plant Improvement Institute (SPII), Horticulture Research Department,Karaj, 31375-764, Iran; jafarisaeid@rocketmail.com*

In vitro cumulative gas production techniques were developed to predict fermentation of ruminant feedstuffs and the close association between rumen fermentation and gas production has been recognized for over a century. Little information exists on the effective use of Almond hull as a feed for ruminant. The aim of this study was to compare nutritive value of hull of four varieties of Almond provided in Iran and to the sugar beet pulp (B) and alfalfa hay (A). Samples of Almond hull of Almond varieties including Rabbi (R), Mamaii (M), Shahrud15 (SH15) and Shokufe (SH) were used in this experiment. Ruminally fistulated steer were used for obtained rumen liquor sample and media solution mixed with rumen liquor (1:2 v:v). Each grinded feed samples were placed into three glass syringes and aforesaid solution were added to them. The gas production parameters measured by F-curve and data were analyzed by SAS. Among Almond hulls, a+b (ml/0.2 gDM) parameter was highest in R and M (respectively 97.2, 79.5, 79, 70.1, 63.4 and 63.1 for B, R, M, SH, A and SH15; $P<0.01$). Gas production in several times were significantly affected by Almond hull varieties and in initial incubation times (<8 h) R and M varieties had produced more gas volume than B (50.1, 49 vs. 46.1 ml for R, M and B, respectively; $P<0.01$).However, total gas production of all hulls was lower than B. Interestingly, R, M and SH produced gas more than alfalfa (98, 81, 80.2, 72.3, 65.2 and 64.2 ml for B, R, M, SH, A and SH15, respectively; $P<0.01$). The lack of change in gas production parameters between A and SH15 alongside increasing in gas production volume of R, M and SH varieties suggest that part of alfalfa hay can be substitute by Iranian Almond hulls in ruminant ration. Also, increased gas production in R and M varieties was consisted with high NFC content of this valuable by-product.

Effect of polyethylene glycol and polyvinylpyrrolidone on *in vitro* gas production of grape pomace

Besharati, M.[1,2] *and Taghizadeh, A.*[1], [1]*Dep. of Animal Science, Faculty of Agriculture, University of Tabriz, Tabriz, 51664, Iran,* [2]*Peyame noor University of Benis, Shabestar, 51665, Iran; ataghius@yahoo.com*

The aim of this study was to determine the effect of PEG and PVP on *in vitro* gas production kinetics of grape pomace. The chemical composition of grape pomace was determined using the methods recommended by AOAC (1999). (1991). Total phenolics (TP) were measured using the Folin Ciocalteau method (Makkar, 2000). Total tannin (TT) was determined after adding insoluble polyvinylpyrrolidone and reacting with Folin Ciocalteau reagent (Makkar, 2000). Tannic acid was used as the standard to express the amount of TP and TT. Gas production was measured by Fedorak and Hrudy (1983) method. Approximately 300 mg of dried and ground (2 mm) grape pomace sample was weighed and placed into serum bottles in the presence (300 mg) and in the absence of PEG or PVP. The gas production was recorded after 2, 4, 6, 8, 12, 16, 24, 36, and 48 h of incubation. The gas production profiles in triplicate fitted with equation of $Y=A(1 - e^{-ct})$. Data was subjected to ANOVA as a completely randomized design with 4 replicates by the GLM procedure (SAS, 2002), and treatment means were compared by the Duncan test. The DM, CP, ADF, NDF, Crude fat, Total phenols, Total tannins and OM in grape pomace were 93.3, 6.62, 18.4, 18.7, 1.41, 3.01, 2.27 and 87.7%. At the early incubation times (2, 4 and 6 h), the control treatment (treatment without PEG or PVP) had the highest gas production volume among treatments, but after 6 h the gas production volume in experimental treatments (treatments with PEG or PVP) was increased ($P<0.05$). There was considerable increase in gas production when the grape pomace was incubated in the presence of PEG or PVP. The increase in the gas production in the presence of PEG and PVP is possibly due to an increase in the available nutrients to rumen micro-organisms, especially the available nitrogen.

Effects of cry1ab gene on the rumen protozoa in lambs

Balieiro, G.N.[1], Bueno, M.S.[1], Nogueira, J.C.F.[2], Ferrari, E.J.[1], Nogueira, J.R.[1] and Moreira, H.L.[1], [1]Agência Paulista de Tecnologia dos Agronegócios APTA/SAA, Bandeirantes Avenue, 2419, Ribeirão Preto, São Paulo, 14030670, Brazil, [2]Faculdade de Zootecnia e Engenharia de Alimentos FZEA/USP, Duque de Caxias Avenue, 255, Pirassununga, São Paulo, 13635900, Brazil; geraldobalieiro@apta.sp.gov.br

This study aimed evaluating the effect of gene cry1Ab on the ruminal protozoa population in lambs fed silage maize hybrids with the cry1Ab trait versus its nonbiotech counterpart. The hybrids DKB 390 from Dekalb and AG 8088 from Agroceres that contains cry1Ab gene from Bacillus thuringiensis and yours isogenics that not contains the cry1Ab gene were used. Twenty Ile de France male lambs, with 33 to 50 kg live-weight and around 10 months old were fed with the experimental silages it for 28 days. Subsequently, the ruminal fluid was extracted through of a nasal-esophagic tube at two times, before feeding and 40 minutes after feeding. The experimental design was the randomized block with five replications, in factorial arrangement 2 x 2. There was significant interaction between the effects of cry1Ab gene and hybrids on the rumen protozoa. The ruminal protozoa population did not differed between lambs fed silages from hybrids AG 8088. However the ruminal protozoa population in lambs fed silage from hybrids DKB 390 that contains cry1Ab was smaller than in lambs fed silage from DKB 390 that not contains cry1Ab. The total ruminal protozoa population, and counts of Entodinium, Diplodinium, Epidinium and Dasytricha from lambs fed silage from DKB 390 that contains cry1Ab versus lambs fed silage from DKB that not contains cry1Ab were 32.8 vs 26.7; 25.5 vs 22.0; 1.9 vs 1.4; 1.52 vs 1.15 and 1.35 vs 0.85 multiplied by 10^4/ml of fluid ruminal, respectively. The changes on the protozoa population were attributed the higher percentage of crude fat from silage DKB 390 that contains cry1Ab than DKB 390 that not contains the cry1Ab (4.27 vs 3.53).

Session 10 Poster 9

Evaluation nutritive value of processed barley grain with different methods using in situ technique in sheep

Parand, E., Taghizadeh, A., Moghadam, G.A. and Janmohamadi, H., Dep. of Animal Science,Faculty of Agriculture, University of Tabriz, Tbariz, 51664, Iran; ataghius@yahoo.com

Several processing methods are developed to increase the digestibility of barley. In this study effects of four different barley processing methods on the degradation parameters of dry matter and crude protein were evaluated in a completely random design using in situ technique. Grains received three different processing (roasting (RBG), steam rolling (SRBG) and microwave processing (MBG)) and were compared with control group (grinded not processed barley GBG). Rumen degradation characteristics of feeds (Ørskov et al., 1980) were calculated after the incubation of 5 g sample of DM (grounded at 2 mm) in 12 cm×6 cm nylon bags with a pore size of 50 µm incubated in the rumen of 2 cannulated sheep for 0, 2, 4, 6, 8, 12, 16, 24, 36 and 48 h. The degradability data of DM and CP were fitted by equation of Y = a + b (1−e−ct). Results showed that processing methods had significant effect on the degradability of grains in rumen. MBG and GBG had more ED of DM and CP than others did and microwave processing increased the digestibility of barley grain numerically. SRBG had the least Degradability of DM and CP followed by RBG. SRBG decreased ruminal digestibility of barley grain significantly. processing methods differed in rate of degradation and RBG had the least rate of degradation for DM and CP. It seems that despite the heat during these processing methods turns starch to a more resistant form to digestion, denaturation of protein matrix reduces the starch degradability in the rumen. Use of processing methods like steam rolling and roasting in order to decrease ruminal degradation of starch can lower risk of incidence to disorders related to rapid starch degradation in barley grain, like acidosis

The effects of microbial inoculants and forage species on *in vitro* gas production of some hays and their silages

Kilic, U., Garipoglu, A.V. and Onder, H., Ondokuz Mayis University, Department of Animal Science, Faculty of Agriculture, 55139, Turkey; unalk@omu.edu.tr

This study was conducted to determine the effects of forage species (leguminious and graminious), storage types (drying and ensiling) and microbial inoculants (with Maize-All and without Maize-All) on *in vitro* gas productions and gas production kinetics of some forages used in ruminant feeding. In this study, corn hay-CH, corn silage-CS, grass hay-GH, grass silage-GS, alfalfa hay-AH, alfalfa silage-AS, vetch hay-VH, vetch silage-VS were used. Two SakızxKarayaka rams aged 2 with ruminal cannulas (average live weight 40 kg) were used in gas production technique. All of the feedstuffs were incubated for 3, 6, 9, 12, 24 and 48 hours. Data were evaluated in nested factorial design. While the leguminious forages have higher *in vitro* gas production values compared to the graminious forages for incubations up to 24 hours ($P<0.01$), this difference was found insignificant for 24 and 48 hours incubation. GH and GS had lowest *in vitro* gas production values for all incubation periods ($P<0.01$). The highest *in vitro* gas production values were found for CH and AH ($P<0.01$). Leguminious forages had higher gas production rates (c values) compared to the graminious forages either or not in the presence of microbial inoculants ($P<0.05$). Silage making and Maize All supplementation lowered the *in vitro* gas production for all the forages used in the experiment ($P<0.01$).

A comparison of rumen fluid sampling techniques for ammonia and volatile fatty acid determination

Rowntree, J.D.[1], Pierce, K.P.[1], Kenny, D.A.[2], Kelly, A.K.[1] and Boland, T.M.[1], [1]University College Dublin, School of Agriculture, Food Science and Veterinary Medicine, Belfield, Dublin 4, Dublin, Ireland, [2]Teagasc, Grange, Dunsany, Co. Meath, Ireland; jason.rowntree@ucd.ie

In order to obtain a clear picture of rumen fermentation profiles it is necessary to routinely sample the rumen contents. Ruminally cannulated animals allow for the frequent sampling of rumen liquid and solid fractions; however there are increased cost and animal care implications of such an approach. Oesophageal sampling allows for a less invasive, but more limited sampling of rumen contents to take place. The objective of thisstudy was to compare the volatile fatty acid (VFA) and ammonia concentrations of rumen fluid samples collected per rumen cannula or via the use of an oesophageal sampler (OS). Eleven ruminally cannulated, lactating Holstein Friesian cows were sampled on four separate dates over a two week period. Rumen fluid was sampled using a FLORA rumen scoop, oesophageal sampler (OS) or via the rumen cannula (C). These samples were subsequently analysed for VFA and ammonia concentration. Statistical analysis was performed using PROC MIXED in SAS. OS showed a reduction in the total VFA concentration ($P<0.001$) compared to the C method. Individual VFA proportions and acetate: propionate ratio were unaffected by sampling method ($P>0.05$). Similarly sampling method had no impact on rumen ammonia concentration ($P>0.05$). Furthermore correlation analysis showed a moderate to strong association (ranging from $R=0.58-0.89$) between OS and C methods for all variables measured except for total VFA's. These findings suggest that the OS technique gives an accurate representation of VFA proportions and ammonia concentration when compared to per cannula sampling, but it can underestimate total VFA concentration.

Evaluation and use of co-products from the biofuel industry in pigs

Zijlstra, R.T.[1] and Beltranena, E.[1,2], [1]University of Alberta, Agricultural, Food and Nutritional Science, 4-10 Ag/For Ctr, Edmonton, AB T6G 2P5, Canada, [2]Alberta Agriculture and Rural Development, #307, 7000-113 Str, Edmonton, AB T6H 5T6, Canada; ruurd.zijlstra@ualberta.ca

Recent price increases for traditional feedstuffs such as cereal grains, fats, and oilseeds have forced the pork industry to look for alternative feedstuffs. One contributing factor has been the rapid expansion of the biofuel industry. This industry also produces several co-products such as distillers dried grains with solubles (DDGS), canola meal (expeller-pressed, solvent-extracted), canola cake, and crude glycerol. These co-products can serve as alternative feedstuffs for swine feeding partially offsetting increases in feed costs, but also present several feeding challenges: First, processing of co-products adds variability in macronutrient profile beyond the variability intrinsic to the feedstock. Therefore, feed quality evaluation for energy, amino acids, and phosphorus content and digestibility is important, as is the system selected for evaluation. Second, fermentation and heat processing impact the availability of amino acids and phosphorus. Overheating reduces lysine availability due to Maillard reactions, reduces heat-labile anti-nutritional factors, but combined with fermentation may increase mineral availability due to partial breakdown of phytate. Third, co-products may contain chemical residues and mycotoxins that reduce voluntary feed intake and affect reproductive performance. Deoxynivalenol survives processing and will concentrate in DDGS. Finally, co-product use may impact not only carcass characteristics, but also pork quality. The high fibre content of co-products reduces dressing percentage. The high oil content of some co-products provides unsaturated fatty acids that soften pork fat. In conclusion, use of co-products from the biofuel industry may reduce feed costs per unit of pork produced, but also provides challenges to achieve cost-effective, predictable growth performance, carcass characteristics, and pork quality.

Effects of dietary glycerol on glycerol kinase gene expression and gut microbiota in growing piglets

Papadomichelakis, G., Zoidis, E., Mountzouris, K.C., Lippas, T. and Fegeros, K., Agricultural University of Athens, Nutritional Physiology and Feeding, 75 Iera Odos Str., 118 55, Athens, Greece; gpapad@aua.gr

Earlier studies have proposed that high dietary levels of glycerol cannot be efficiently utilized by pigs, due probably to the limiting role of glycerol kinase (GK), which regulates the metabolic activation of glycerol. In addition, a potential influence of glycerol on swine gut microbiota has not been previously assessed. Thus, the effects of 2 diets, G1 and G2 (with 7.5 and 15% of crude glycerol, respectively) vs. a control (C) were assessed in 18 weaned (aged 30 days) Large White×Pietrain piglets. Feed intake and weight gain were recorded weekly. At the end of growing period (72 days of age), the liver of the piglets was removed and frozen for further RNA isolation and GK gene expression analysis. Selected constituents of the gut microbiota composition, monitored at ileal and caecal level were assessed. Data were analysed by one-way (diet) ANOVA using SPSS v.16.0. Feed intake, weight gain and feed conversion were not affected by glycerol addition; however, G1 tended to be heavier (by 10%) compared to control piglets, at the end of the growing period. There was no effect of the glycerol addition levels on the examined ileal and caecal gut microbiota composition. Glycerol kinase gene expression, assayed by RT-PCR, increased linearly ($P<0.001$) with dietary glycerol content. This later evidence in combination with growth data, suggests that post-translational effects could primarily affect the efficiency of glycerol utilization. In conclusion, dietary glycerol level did not have any negative effect on piglet growth and gut microbiota composition. The present data need to be reconfirmed with larger number of pigs to eliminate the influence of inter-individual variations. The project (Ref. No. 200023) was funded by the Research Committee of the Agricultural University of Athens.

Session 11 Theatre 3

Preliminary study on the effects of crude glycerol on the intramuscular fatty acid composition in growing pigs

Lippas, T., Papadomichelakis, G., Zoidis, E., Mountzouris, K.C. and Fegeros, K., Agricultural University of Athens, Nutritional Physiology and Feeding, 75 Iera Odos Str., 118 55, Athens, Greece; fanis712@yahoo.gr

The effects of dietary glycerol content on the intramuscular fatty acid (FA) profile were investigated in 18 growing Large White×Pietrain piglets. Piglets, aged 30 days at weaning, were allotted in 3 groups and were fed either a control (C) or glycerol supplemented diets G1 and G2, with 7.5 and 15% of crude glycerol, respectively. Glycerol derived from a mixture of different oils (palm, soybean, cotton and rape seed oil) and had a 0.5% total FA content with 13.0% palmitic, 42.6% oleic, 31.6% linoleic and 5.9% α-linolenic acid. At 72 days of age, piglets were weighed and slaughtered. After chilling at 4 °C for 24 h, carcass weight was recorded and a chop of the loin muscle was excised and stored at -20oC, until analyzed for FA. Data were analysed by one-way (diet) ANOVA using SPSS v.16.0. Body weight at slaughter and carcass weight were not affected by the dietary glycerol level. Muscle fatness was similar between treatments, as indicated by the total weights of FA (mg/100 g fresh tissue). Muscle polyunsaturated (PUFA) to saturated (SFA) ratio was linearly increased (P<0.05), reflecting the decline (Plinear<0.001) of SFA and the increasing trend (P=0.097) of PUFA, with dietary glycerol level. The trend to higher PUFA was associated with increasing levels of α-linolenic acid (Plinear<0.01), in correspondence to the dietary FA profile, and long chain PUFA, mainly EPA (Plinear<0.05) and DHA (Plinear<0.01). In conclusion, changes in the intramuscular fatty acids could be related to the FA residues found in the supplementary crude glycerol. However, this influence is expected to vary, depending on the origin of crude glycerol (oils with different saturation degree of FA) and the level and regime of supplementation. The project (Ref. No. 200023) was funded by the Research Committee of the Agricultural University of Athens.

Session 11 Theatre 4

Digestibility of amino acids in organically cultivated white-flowering faba bean (*Vicia faba*) and cake from cold-pressed rapeseed (*Brassica napus*), linseed (*Linum usitatissimum*) and hemp seed (*Cannabis sativa*) in growing pigs

Høøk Presto, M., Lyberg, K. and Lindberg, J.E., Swedish University of Agricultural Sciences, Department of Animal Nutrition and Management, Box 7024, SE- 750 07 Uppsala, Sweden; Magdalena.Presto@huv.slu.se

According to the EU-regulations for organic production, diets used for livestock must not include synthetically produced amino acids or feedstuffs that have been treated with chemical solvents, such as oil seed meals from oil extraction. Therefore, there is a need to evaluate potential protein-rich feed resources in the geographical region where the pig meat production is located. In Northern Europe, this implies the use of a limited number of legume and oil seeds. White flowered faba beans (Vicia faba) are increasingly used in both conventional and organic pig feeds. However, there is limited information on their nutritional value. Cold-pressed rapeseed (Brassica napus) and linseed (Linum usitatissimum) cakes are rarely used in conventional pig feeds and as far as we know hempseed (Cannabis sativa) cake has not been used in conventional pig diets. The study aimed at determining the ileal apparent (IAD) and standardized ileal digestibility (SID) of crude protein (CP) and amino acids (AA) in organically cultivated white-flowering faba beans, and cakes from hempseed, linseed and rapeseed. The experiment was designed as a four period change-over trial with six castrated male Yorkshire pigs fitted with post valve t-caecum (PVTC) cannulas. The IAD and SID of CP for the feed ingredients were affected by dietary treatment (P=0.029 and P=0.027, respectively). The IAD and SID of most AA in the feed ingredients were also affected by dietary treatment (P<0.05) but without any significant consistent trend. The overall digestibilities were however in general comparable with conventional protein feed ingredients. Thus, these alternative protein feed ingredients have the potential to be used to a greater extent when formulating organic pig diets.

Consequence of extruded linseed incorporation in sows and/or pigs' diets on performance

Quiniou, N.[1], Goues, T.[1], Vautier, A.[1], Nassy, G.[1], Chesneau, G.[2], Weill, P.[2], Etienne, M.[3] and Mourot, J.[3], [1]IFIP, BP35104, 35650 LeRheu, France, [2]Valorex/Association Bleu-Blanc-Coeur, La Messayais, 35210 Combourtillé, France, [3]INRA, UMR1079 SENAH, 35590 S-Gilles, France; nathalie.quiniou@ifip.asso.fr

Extruded linseed (exLIN) is used most often in pigs' diets in order to improve the ratio between ω6 and ω3 PUFA in pork meat. The aim of this trial was to investigate the effect on performance of exLIN incorporation on a net energy basis in sows' and/or pigs' diets. At each stage, different diets were formulated on the same net energy bases (gestation: 9.3, lactation: 9.6, growing/ finishing periods: 9.3 MJ/kg) and amino acid bases (5.0, 8.5, 8.3 and 7.4 g digestible lysine, respectively). During gestation and lactation, diets with 3.5% exLIN (L group, n=29) were compared to diets enriched with an equivalent amount of lipids through 1.4% palm oil (P group, n=29) or without addition of lipid (C group, n=29). Pigs from L and C sows were studied during the fattening period in order to study different durations of exLIN utilisation. Pigs from C sows received diets containing either 0% (CC group, n=92) or 2% exLIN (CL group, n=93); pigs from L sows were fed a diet containing 2% exLIN (LL group, n=92). No significant difference in sows' body condition at farrowing, prolificacy and weaning to oestrus interval was observed. In largest litters, the farrowing progress was significantly faster in the L than in the C and P sows. The survival rate of piglets weighing between 1.0 and 1.4 kg at birth was significantly improved in the L group. From 24 to 110 kg BW, neither pigs' spontaneous daily feed intake, feed conversion ratio, or carcass fatness were significantly influenced by the treatment. Higher ADG between 24 and 70 kg and higher pH24 in Semimembranosus were obtained in LL than in CC ones. The effect of ω3 PUFA on reproductive hormone synthesis could partly explain differences in farrowing progress. Investigations on growing pigs bred in poorer sanitary conditions would be interesting taking into account the anti-inflammatory properties of ω3 PUFA.

Effect of short-term feeding of genetically modified (Bt MON 810) maize on weanling pig growth performance, organ weights and organ histopathology

Buzoianu, S.G.[1,2], Walsh, M.C.[1], Gardiner, G.E.[2], Cassidy, J.P.[3], Rea, M.C.[1], Ross, R.P.[1] and Lawlor, P.G.[1], [1]Teagasc, Moorepark Research Centre, Fermoy, Ireland, [2]Waterford Institute of Technology, Cork Road, Waterford, Ireland, [3]University College Dublin, Faculty of Veterinary Medicine, Dublin, Belfield 4, Ireland; stefan.buzoianu@teagasc.ie

Genetically modified (GM) maize is the second most cultivated transgenic crop worldwide. The objective of this study was to investigate the effect of feeding diets containing GM maize for 31 days on weanling pig growth performance, organ weights and organ histopathology.Thirty two male pigs were weaned at 28 days of age and following a 6 day basal period were blocked by weight and ancestry and randomly assigned as individuals to one of two treatments: GM Bt maize- (MON 810) or non-GM isogenic maize-based diets (388 g maize/kg). The pigs were individually penned in four identical climate controlled rooms (8 pigs/ room). Each treatment group was equally represented in each room to avoid a room effect. Body weight and feed disappearance were recorded on days 0, 7, 14, 21, 28 and 30. Pigs were sacrificed on day 31 and the heart, kidneys, spleen and liver, devoid of blood clots and fat, were removed, weighed and sampled for histopathological analysis. Statistical analysis was performed using the GLM procedure of SAS. Overall, pigs fed GM maize had a higher daily feed intake ($P<0.05$), tended to be heavier ($P=0.11$) and to have higher average daily gain ($P=0.11$) than the pigs fed the non-GM maize. Heart, liver and spleen weights did not differ between treatments ($P>0.05$) and no histopathological differences were observed for these organs. The kidneys of pigs fed GM maize were heavier than for non-GM fed pigs ($P<0.05$). However, no indicators of kidney damage were found on histopathological examination. These data demonstrate that short-term feeding of GM maize had no observed adverse effects on piglet growth or the health indicators investigated.

Effect of phytase supplementation to sorghum-soybean meal diets on the ileal digestibility of phosphorus, phytate, and energy in growing pigs

Morales, A.[1]*, Cervantes, M.*[1]*, Sánchez, E.*[1]*, Araiza, B.A.*[1]*, Carrillo, G.*[1] *and Zijlstra, R.*[2]*,* [1]*Universidad Autónoma de Baja California, ICA, MEXICALI, BC, 21100, Mexico,* [2]*University of Alberta, AFNS, Edmonton, AB, T6G 2P5, Canada; amorales24@yahoo.com*

Previous reports suggest that phytates may affect the digestibility of energy, but the effect of phytase supplementation on the apparent ileal digestibility (AID) of energy has been inconsistent. An experiment was conducted to evaluate the effect of supplementing phytase to a sorghum-soybean meal diet for growing pigs on the AID of amino acids, energy and phosphorus (P), as well as the ileal digestibility (ID) of phytates in growing pigs. Eight pigs (BW 22.1±1.3 kg) were fitted with a T-cannula at the distal ileum. Each period consisted of 9 d; 7 d for diet adaptation, and 2 d for digesta collection. Treatments (T) were: 1) basal sorghum-soybean meal diet, 2) basal diet plus phytase (1,050 FTU/kg diet). The AID of amino acids data was previously reported. The digestibility results for phosphorus, phytate, and energy were: phosphorus, 39.9, 51.9 (EE, 2.40); Phytate, 0.50, 36.4 (EE, 5.20); Energy, 75.1, 75.0 (EE, 0.74), basal and phytase diet, respectively. Phytase increased the digestibilities of phytate and phosphorus ($P<0.001$), but did not affect the AID of energy ($P>0.10$). Although phytase had a highly significant increase in phosphorus and phytate digestibility, it did not improve the AID of energy in growing pigs fed sorghum-soybean meal diets. Thus, these data indicate that phytates do not affect the digestibility of energy in growing pigs

Effect of phytase supplementation to sorghum-soybean meal diets on stomach and intestinal pH values, and apparent ileal digestibility of protein in pigs

Sánchez, E., Cervantes, M., Morales, A., Araiza, B.A. and Carrillo, G., Universidad Autónoma de Baja California, ICA, Mexicali, BC, México, 2100, Mexico; mcr_1102@yahoo.com

Phytates contained in several feed ingredients may interact with pancreatic proteases affecting their activities and reducing the digestibility of dietary protein, given optimal intestinal pH conditions occurs. An experiment was conducted to evaluate the effect of supplementing phytase to a sorghum-soybean meal diet for growing pigs on the AID of crude protein (CP), and pH of stomach and intestinal content. Twelve pigs (BW 22.3±1.98 kg) were fed one of two experimental diets during 14 days. Dietary treatments were: 1) basal sorghum-soybean meal diet, 2) basal diet plus phytase (500 FTU/kg diet). Both diets contained chromic oxide to determine apparent ileal digestibility of CP. All pigs were sacrificed the end of the experiment to collect contents from stomach and three segments of the small intestine (duodenum, jejunum, and ileum). pH was measured immediately after samples were collected. The results of pH and AID of CP values were: Stomach, 3.07, 3.12; Duodenum, 5.66, 5.63; Jejunum, 6.48, 6.50; Ileum, 6.54, 6.48; CP-AID, 67.6, 68.2%, basal and phytase-added diet, respectively. pH values in stomach and segments of the small intestine were not different among dietary treatments ($P>0.10$). In the small intestine, pH values were higher than the optimal values (below pH 4.5) needed for phytate and protein interactions to occur. Phytase supplementation did not affect the AID of CP ($P>0.10$). These data suggest that the activities of pancreatic proteases of growing pigs are not affected by the levels of phytates naturally occurring in feed ingredients. Thus, the data also suggest that phytates do not affect the digestibility of protein in pigs fed sorghum-soybean meal diets

Effect of a chelated source of copper on growth performance and tissue copper concentration in weaned piglets

Parker, D.S.[1], Gracia, M.I.[2], Harrell, R.J.[3], Richards, J.D.[3] and Stebbens, H.[1], [1]Novus Europe, Ave Marcel Thiry 200, 1200 Brussels, Belgium, [2]Imasde Agroalimentaria, Pozuelo de Alarcon, 28224 Madrid, Spain, [3]Novus International, 20 research Park Drive, 63303 St Charles, MO, USA; helen.stebbens@novusint.com

Copper (Cu) can be added to animal feed in an inorganic form, for example CuSO4, or in the form of a chelate. MINTREX® Cu is a chelate of Cu in which the hydroxy analogue of methionine (2-hydroxy-4-(methylthio)butanoic acid) is the ligand. In addition to Cu, the chelate is also a source of methionine activity in the diet. The objective of the study was to investigate the effect of the chelate on growth performance and Cu bioavailability in weaned piglets. 240 weaned piglets (26 days of age) were allocated to one of three dietary treatments with 8 replicates per treatment and 10 piglets per pen. The treatments were 6 mg/kg Cu from CuSO4 (Trt 1, Control), 170 mg/kg Cu from CuSO4 (Trt 2) and 170 mg/kg from MINTREX® Cu (Trt 3). Methionine levels were balanced across treatments to account for the methionine activity in the chelate. Piglets fed the chelated source grew 9% faster over the 42 day period than those fed the diets containing the inorganic Cu source with ADG values (g/d) of 346, 346 and 378 for Treatments 1-3 respectively (P=0.032). In the same period feed intake was higher for piglets fed the chelated Cu source when compared to piglets fed 170 mg/kg from CuSO4 (544 g/d and 510 g/d respectively (P=0.034). Piglets fed the chelated Cu source had higher liver Cu than pigs fed 170 mg/kg as CuSO4 (30.46 mg/kg vs. 18.37 mg/kg; P=0.06), and significantly higher liver Cu than pigs fed 6 mg/kg as CuSO4 (6.83 mg/kg; P<0.01). In conclusion piglets fed 170 mg/kg Cu from MINTREX® had a significantly higher ADG (+9%) and feed intake (+7%) when compared to piglets fed 170 mg/kg Cu from CuSO4. Cu levels in liver tissue indicate that chelated Cu was more bioavailable than the inorganic source of the trace mineral.

Bioavailability of zinc sources in piglets and broilers: a meta-analysis

Schlegel, P.[1], Sauvant, D.[2] and Jondreville, C.[3], [1]Agroscope Liebefeld-Posieux, Tioleyre 4, 1725 Posieux, Switzerland, [2]INRA-AgroParisTech, Rue Claude Bernard 16, 75231 Paris, France, [3]INRA-Nancy Université, Ave de la Forêt de Haye 2, 54505 Vandoeuvre-les-Nancy, France; patrick.schlegel@alp.admin.ch

Zinc (Zn) is essential for swine and poultry and native Zn contents in feedstuffs are too low to meet their requirement. Added Zn in organic form (OZn) is considered as more bioavailable than inorganic Zn (IZn). One broiler (28 exp., 175 obs.) and one piglet database (34 exp., 159 obs.) were used for a meta-analysis. Within dose-response effect from dietary Zn on dependent variables, the GLM procedure with native Zn, IZn and OZn as independent variables presented: In broilers, G:F, plasma Zn and bone Zn responded linearly (P<0.001) and bone Zn quadratically (P<0.001) to the three Zn sources. All variables presented a negative interaction between native Zn and added Zn as IZn or OZn (P<0.01). Coefficients of determinations for variables were above 0.90. The relative bioavailability (RBV) of OZn versus IZn were 98, 97 and 103 for G:F, plasma Zn and bone Zn, respectively. In piglets, absorbed Zn, liver Zn and bone Zn responded negatively to increasing native Zn (P<0.10). Absorbed Zn, plasma Zn, plasma ALP, liver Zn and bone Zn increased linearly (P<0.001) and plasma Zn, plasma ALP and liver Zn reduced quadratically (P<0.10) with added Zn. No interaction was calculated between DZNN and added Zn (P>0.10). Coefficients of determinations were above 0.89. The RBV of OZn versus IZn were 111, 99, 95, 88 and 100 for absorbed Zn, plasma Zn, plasma ALP, liver Zn and bone Zn, respectively. This meta-analysis concludes that native Zn was highly bioavailable to broilers, similar to added Zn and independent from source. In piglets however, native Zn bioavailability was reduced with increasing contents, most probably due to antagonism from diet components, such as phytate. Added Zn was not affected such a way. The amount of absorbed Zn from OZn was numerically higher than IZn, but its capacity to improve Zn status of piglets was equal to IZn.

Effect of Nupro™ on performance of weaned piglets

Taylor-Pickard, J.A.[1], Spring, P.[2], Mcardle, T.[1], Boyd, J.[3] and Wilson, S.[3], [1]Alltech, Ryhall Road, Stamford Lincs, PE91TZ, United Kingdom, [2]Swiss College of Agriculture, Laengasse 85, 3052 Zollikofen, Switzerland, [3]BOCM PAULS, Tucks Mill, Burston, Diss, Norfolk, IP22 5TJ, United Kingdom; jpickard@alltech.com

Nupro™ (Alltech Inc., Nicholasville KY, USA) is an extract derived from a specific strain of yeast containing hydrolyzed protein and high concentrations of nucleotides. The objective of the present trial was to evaluate the effect of Nupro™ on performance and health of weaned piglets. The trial was setup as a 2x2x3 factorial design with the factors diet (control and Nupro™), sex (boar and gilt) and weight band (light, medium, heavy). Diets were isocaloric and iso-nitrogenic and had the following specifications: First stage: 15.2 MJ DE, 21.5% CP, 1.70% lysine; second stage: 15.0 MJ DE, 21.0% CP, 1.60% lysine. Nupro was included at 2.0% and did replace 2.0% fish meal. A total of 210 piglets with an average weight of 8.27 kg were assigned by body weight and sex to 36 pens of 5 piglets each. The trial lasted 4 weeks and feed intake, body weight and FCR were determined weekly. Animal health and mortalities were recorded daily. The data were analyzed as a 2x2x3 factorial design by ANOVA and means were separated by the test of Tukey-Kramer. Overall performance and health was good: average daily gain: 424 g (SE 9.30 g); feed intake 556 g (SE 13.38 g); FCR 1.31 (SE 0.018). Feed intake of gilts was higher (575 vs. 536 g; $P<0.05$) and weight gain tended to be higher (435 vs. 413 g; $P=0.11$) than for boars. Weight band did not influence weight gain, but FCR was increasing from light to heavy piglets (1.24, 1.31, 1.38; $P<0.001$). Nupro™ showed a strong tendency to improve weight gain (436 vs. 412 g; $P= 0.08$) and FCR (1.29 vs. 1.33; $P=0.09$). Weight gain in week 4 was significantly improved by Nupro™ ($P<0.05$).

Optimal ileal digestible valine/lysine ratio for the performance of piglets

Millet, S.[1], De Boever, J.[1], Aluwé, M.[1], Segers, L.[2], Primot, Y.[3], De Paepe, M.[1] and De Brabander, D.[1], [1]Institute for Agricultural and Fisheries Research (ILVO), Animal Sciences Unit, Scheldeweg 68, 9090 Melle, Belgium, [2]Orffa Belgium N.V., Rijksweg 10G, 2880 Bornem, Belgium, [3]Ajinomoto Eurolysine S.A.S., 153, Rue de Courcelles, 75817 Paris Cedex 17, France; marijke.aluwe@ilvo.vlaanderen.be

The objective of this experiment was to evaluate the effect of increasing apparent ileal digestible (AID) valine/AID lysine ratios on the performance results of newly weaned piglets between 4 and 9 weeks of age.One hundred and eighty piglets (Piétrain boar x hybrid sow) were divided over 30 pens with 3 barrows and 3 sows. Each pen was randomly assigned to one out of 5 dietary treatments, leading to 6 replicates per treatment. One basal feed was formulated according to the ideal amino acid concept, with apparent ileal digestible lysine as the reference amino acid. Based on previous experiments, AID lysine content was slightly limiting performance (10.2 g/kg). Dietary AID methionine+cystine, threonine, tryptophan, isoleucine, valine, and leucine contents expressed as percent of AID lysine were 65, 72, 23, 54, 60 and 107%, respectively. Then free valine was added to obtain 5 AID valine to AID lysine ratios: 60, 65, 70, 75 and 80%. Performance results were subject to simple and polynomial regression analysis. The requirements were estimated using broken line models with either a linear or a quadratic ascending function. Dietary valine concentration had a clear impact on animal performance. Mainly feed intake and daily gain were affected. Requirement of AID valine expressed relative to AID lysine was 65% or 66% (for respectively average daily gain and average daily feed intake) using a broken line model with linear ascending function and 67% (average daily gain) or 68% (average daily feed intake) using a broken line model with quadratic ascending function. The results of the present experiment suggested even higher requirements in the first two weeks after weaning. Expressed as standardised ileal digestible VAL: LYS, the requirements range from 68% to 71%, depending on the model used.

The genetics of growth to maturity in commercial sheep

Pollott, G.E. and Galea, G., Royal Veterinary College, Veterinary Basic Sciences, Royal College Street, NW1 0TU London, United Kingdom; gpollott@rvc.ac.uk

Most growth models are used to investigate pre-pubertal growth in meat species. However, growth to maturity is a critical aspect of all breeding herds/flocks and mature weight itself plays a key role in both lifetime nutritional requirements and the level of greenhouse gas emissions. Recording of growth to maturity is rare in sheep flocks and so data of this sort are not commonly reported. A flock of 600 purebred ewes was recorded 3-times per year (mating, lambing, weaning) for liveweight over a 20 year period. A key feature of such ewes is that they are mated before reaching maturity and so their growth is affected by bearing and rearing their lambs. A range of growth models were fitted to the data and the parameters for each ewe for each model were estimated. The commonly used logistic, Gompertz and Brody equations were found to be unsuitable for these growth curves because they did not incorporate the flexibility to cope with the effects of pregnancy and lamb rearing on ewe liveweight. Two new growth models were used which fitted the data with a lower RMS than the three models mentioned above. These were the Hill model, commonly used in pharmacology, and a 3-stage linear model. The parameters of all curves, and mature weight, were analysed to estimate their heritabilities and genetic correlations. The correlation between growth as a lamb and growth during the post-pubertal period was found to be -0.33, suggesting that good growth as a lamb is not associated with good growth as a young ewe. Mature weight had no correlation with growth in either period. Critical parameters of the new models were found to be moderately heritable (0.35-0.45). Modelling of these results as selection objectives in selection programmes suggests ways in which a variety of outcomes may be met, including reducing mature weight whilst maintaining lamb growth, by selecting on the various parameters in the models.

A study of recursive relationships between performance and visual score traits in swine

Ibanez-Escriche, N.[1], López De Maturana, E.[2], Reixach, J.[3], Lleonart, N.[3] and Noguera, J.L.[1], [1]IRTA, Av. Rovira roure,191, 25198 Lleida, Spain, [2]INIA, Apdo. 8111, 28080 Madrid, Spain, [3]Selección Batallé, Av. Segadors, S/N, 17421 Riudarenes, Spain; noelia.ibanez@irta.es

The objective of this study was to analyze the recursive relationships of live weight nested within sex on lean depth, backfat thickness and muscle visual score in Pietrain breed. The musculature score was measured on a scale from 1 to 5, where 5 is the most desirable score. On-farm data consisted of records between 160 and 200 days from 25,282 fattening Pietrain belonging to Selección Batallé S.A. The analyses were carried out using Bayesian statistics via Gibbs Sampling, to solve linear-threshold models. Results from the recursive model did not show different genetic covariance components regarding sex. The posterior means of the heritabilities were high (0.4-0.5) for the four traits analyzed. However, in a previous single-trait analysis for muscle score, the posterior mean of the heritability was moderate (0.17). This fact can be explained for the high correlation obtained between the performance traits (weight, lean, backfat) and the muscle scores (0.6-0.9) that increased its heritability. The structural regression coefficients revealed a positive relationship between live weight and lean depth, backfat thickness and muscle score. The magnitude of these coefficients within trait was different among sex (P>90%). The effect of live weight on lean depth was greater in males than in females, whereas that effect on both backfat thickness and muscle score was greater in females. This means that the probability of having greater backfat or score for musculature is higher in females when live weight increases. These results would suggest that, although lean depth and muscle score are highly genetically correlated, the visual score of musculature could be related with a higher backfat deposition in the female muscles. However, more research would be needed to study deeply the relationship between these traits and confirm these outcomes

A reduced model variance component approach for mapping multiple QTL in F2 populations

Zimmer, D., Mayer, M. and Reinsch, N., Leibniz Institute for Farm Animal Biology, Genetics & Biometry, Wilhelm-Stahl-Allee 2, 18196 Dummerstorf, Germany; zimmer@fbn-dummerstorf.de

Variance component methods (VCM) have been proposed in the literature for mapping quantitative trait loci (QTL) in F_2 populations using relationship matrices of order equal to the number of individuals. This approach is computationally demanding, especially if additive genetic, dominance and epistatic effects are considered or the number of progeny is large. We propose a reduced model, which regards an average genotypic effect for each possible marker genotype. To set up customized reduced relationship matrices, conditional QTL genotype probabilities and elementary QTL relationship matrices (where QTL genotypes are assumed to be known) can also be used for interaction effects, as already has been shown for the main effects in the literature. The result is an efficient approach for computing the additive genetic, dominance and pairwise epistatic relationship matrices, where the dimensions of all relationship matrices are independent of sample size. A small simulation study indicates that the reduced VCM is considerably less computationally demanding, but competitive to the traditional approach in terms of the accuracy of the estimated QTL positions.

Zero inflated poisson, threshold and linear models for genetic analysis of mastitis in Norwegian Red cattle

Vazquez, A.I.[1], Heringstad, B.[2], Perez-Cabal, M.A.[3], Rodrigues-Motta, M.[1], Weigel, K.A.[1], Gianola, D.[1], Rosa, G.J.M.[1] and Chang, Y.M.[1], [1]University of Wisconsin-Madison, 1675 University Av., 53706, Madison, WI, USA, [2]Norwegian University of Life Sciences, P.O. Box 5003, 1432 Ås, Norway, [3]Universidad Politécnica de Madrid, Paseo De Juan Xxiii 11, 28040, Madrid, Spain; bjorg.heringstad@umb.no

Mastitis is typically coded as presence or absence of disease, and analyzed with either linear or threshold (probit) models. Presence may include cows with multiple episodes; the binary response implies a loss of information relative to counts of events. Zero Inflated Poisson (ZIP) models apply to counting variables with an excess of zeros, correcting an over-dispersion problem often present in Poisson model applications. Also, ordinal threshold (OT) models for ordered categorical responses are candidates. ZIP and OT models have rarely been used in genetic analysis of mastitis or evaluated versus traditionally used models. This study compared probit, ZIP, OT and linear models for either binary (linearB) or counts (linearC) responses. To evaluate the ability of the models to predict future data, 620,492 first-lactation Norwegian Red daughters of 3,064 sires were evaluated with a four-fold cross-validation, via mean squared error of prediction (MSEP) as end-point. Heritability (not defined for a ZIP model) were 0.03, 0.07, 0.08 and 0.03 for linearB, probit, OT, and linearC, respectively. Overall MSEP was 0.160, 0.161, 0.180, 0.157, and 0.163 for linearB, probit, OT, ZIP, and linearC models, indicating a superior prediction ability of the ZIP model. MSEP was also separately obtained for healthy and sick cows. The healthier cows were better predicted by ZIP model closely followed by probit. For diseased cows, MSEP was smallest for OT, followed by linearC model. ZIP model seems to be the best choice for this application, followed by the linearB model.

A simulation study of bias in variance component estimation due to ignoring environmental correlation between 'test-status' and a continuous trait in culled data

Arnason, T.[1], Albertsdottir, E.[1], Eriksson, S.[2], Fikse, W.F.[2] and Sigurdsson, A.[1], [1]Agric. Univ. of Iceland, Hvanneyri, IS-311 Borgarnes, Iceland, [2]Swedish Univ. of Agric. Sci., Dept. of Animal Breeding and Genetics, P.O. Box 7023, SE-750 07 Uppsala, Sweden; ihbcab@bahnhof.se

To attenuate bias in genetic evaluations due to culling, a binary trait 'test-status' can be defined with the value zero for culled animals and the value one for tested animals. A problem with this approach is that the environmental covariance between 'test-status' and the trait of interest is not estimable. A possible solution is to assume a zero environmental covariance. The aim of this study was to evaluate the consequences of this assumption on the estimates of genetic parameters by REML and Gibbs sampling, when the true environmental covariance deviates from zero. Data were simulated for two traits (one that culling was based on and a continuous trait) using the following true parameters, on the underlying scale: h^2=0.4; r_A=0.5; r_E=0.5, 0.0, or -0.5. The base population consisted of 25 males and 500 females. The first three generations were randomly selected, while mass selection on the continuous trait was applied to five subsequent generations where 25 sires and 500 dams produced 1500 offspring per generation. The culling frequency was 0.5 and 0.8 within each generation, and the phenotypically (on the underlying scale) worst animals were culled. Only non-culled animals had observations on the continuous trait. Each of ten replicates included 7500 records on 'test-status' and 9600 animals in the pedigree file. Results from bivariate analysis showed unbiased estimates of variance components and genetic parameters when true r_E=0.0. For r_E=0.5 variance components (15% bias) and especially $\hat{r}_A$ (50-80%) were underestimated, while heritability estimates were unbiased. For r_E=-0.5 heritability estimate of 'test-status' was unbiased, while genetic variance and heritability of the continuous trait together with $\hat{r}_A$ were overestimated (25-50%). The bias was larger for the higher culling frequency

Inclusion of externally computed genotype frequencies in the gene-flow method

Ytournel, F. and Simianer, H., Georg-August Universität Göttingen, Department of Animal Sciences - Animal Breeding and Genetics Group, Albrecht-Thaer-Weg 3, 37075 Göttingen, Germany; fytourn@gwdg.de

The evolution of the molecular and statistical methods provides the opportunity of a more precise knowledge of the positions of Quantitative Trait Loci and even the discovery of some causal mutations. This knowledge is now routinely used in breeding nuclei for some traits, but the outcomes on the whole population are not considered when selecting. The reference method to define a breeding program is the gene-flow method which has the so-called P-matrix as a central element. This matrix describes the transmission paths between the different cohorts, classically defined by the purpose of the group, the sex and age of the animals. We here subdivided these cohorts according to their genotypes and adapted the elements to the changes in the genotype frequencies occurring in the breeding nucleus. The new frequencies in the breeding nucleus were computed with a deterministic method for optimal selection which accounted for overlapping generations. We compared the results obtained (a) when keeping a constant P-matrix, (b) for various initial frequencies of the selected genes and (c) with selected and unselected breeding nuclei. The results showed a necessity to account for the changes in the frequencies in the genotypes in order not to underestimate the achieved Standard Discounted Expressions or gene proportions.

Kinship breeding in theory and practise

Nauta, W.J.[1], Baars, T.[2] and Cazemier, C.H.[1], [1]Louis Bolk Institute, Animal production, Hoofdstraat 24, NL=3972LA Amersfoort, Netherlands, [2]University of Kassel, Biodynamic Farming, Nordbahnhofstrasse 1a, D-37213 Witzenhausen, Germany; w.nauta@louisbolk.nl

Kinship breeding is a farm based breeding system and can be used as a basis of breeding within small populations like in organic farming. This system is used by a group of Dutch farmers that breed the native Dutch Friesian cow breed. There are 17 nucleus breeding farms in the Netherlands and France. These breeders breed their cows mainly with bulls that are bred out of the farm herd. In a total population of about 800 cows at the 15 Dutch farms, 56 breeding bulls from 35 different cow families were used for mating in 2007. With this number of bulls, the genetic variation is kept at a high level, which is important to avoid inbreeding and for the survival of the population of about 1,500 animals in the Netherlands, and about 4,000 in France, Germany, England and Ireland. In small populations the number of bulls used is the main key to keep the inbreeding trend at farm level below 1% per generation. On the other hand this system only makes sense, if the breeders create their unique subpopulation within the entire population. Like in hybrid-breeding in chicken and pig breeding systems, different genetic lines are available within the breeding population as a whole. To find out about how farmers have managed to use this system, data of pedigrees and production and fertility records are analyzed from the farmers and from the Dutch Herd Books. Primary results show that mild inbreeding is used at farms but did not have impact on production and longevity of cows. Further investigations will show us how farmers deal with the kinship breeding system. The results can help farmers in all animal sectors to set up a similar breeding system to meet the principles of region-based and integrity of production of animals.

Maternal lineage effects on dressage and show jumping performance of Hanoverian sport horses

König, S.[1], Schmidt, L.[1], Brügemann, K.[1], Von Borstel, U.U.[1], Schade, W.[2] and Christmann, L.[2], [1]University of Göttingen, Department of Animal Science, Albrecht-Thaer-Weg 3, 37075 Göttingen, Germany, [2]Hanoverian Breeding Association, Lindhooper Str. 92, 27283 Verden, Germany; skoenig2@gwdg.de

A total of 130,524 Hanoverian warmblood sport horses were traced back to 1,520 mare lines to assess the impact of mare lineage effects on dressage and show jumping performance. Analyzed traits related to own performance in show jumping (SJUMP) and dressage (DRE), and included for both disciplines indices from stallion performance tests, as well as sport horse competition results. Indices were analysed as separate traits depending on the length of performance test (70 d or 300 d). Competition results were analysed as weighted earnings (WE = sum of earnings in relation to the no. of placings) and as percentage of first and second ranks (PER). Variance components were estimated using REML for two multiple trait animal models. Fixed effects for stallion performance test indices were station test date and age of the stallion. Maternal lineage and additive genetic effects were modelled as random. Heritabilities ranged from 0.29 (SJUMP_70) to 0.36 (DRE_300). The relative proportion of the lineage variance component was 0.05 for DRE_300, 0.12 for DRE_70, 0.10 for SJUMP_300, and 0.09 for SJUMP_70. Statistical models for WE and PER included fixed effects of birth year, sex and competition class level, and the random mare lineage, permanent environment and additive genetic effects. Heritabilities ranged between 0.22 (SJUMP_WE) and 0.34 (DRE_PER), but lineage variances were close to zero. Hence, lineage effects were more important for results from stallion performance tests. However, since lineage effects are presumably based on mitochondrial DNA, such effects can only be transmitted on the female path of selection. Application of animal models for such research in dairy cows was hampered due to confounding effects between herd and cow lineage, but the widespread use of mare lines enabled reliable estimates in sport horses.

Analysis of (co)variance components in economic traits of Baluchi sheep with REML method

Hosseinpour Mashhadi, M.[1], Aminafshar, M.[2], Tabasi, N.[3], Bahari Kashani, R.[1] and Sobhanirad, S.[1], [1]Islamic Azad University Mashhad Branch, Animal Sci Dept, Mashhad, Iran, [2]Islamic Azad University Science and Research Branch, Tehran, Iran, [3]Bu Ali- Immunology Research Center, Mashhad, Iran; Mojtaba_h_m@yahoo.com

The objective of this investigation is to estimate the (co)variance components of direct and maternal additive genetic effect, maternal permanent environmental effect, genetic and phenotypic correlation for weight traits. Records of 5,913 Baluchi sheep which is native of eastern part of Iran were used. The (co)variance component of traits was estimated with REML method under Animal Model. Fixed factors affected on traits were age of dam, sex, type of birth, year and month of birth. Direct and maternal heritabilities (with standard error in parentheses) for Birth Weight (BW), Weaning Weight (WW), 6, 9 and 12 Month weights (6MW, 9MW, 12MW), Average Daily Gain PreWeaning (ADGPreW), Average Daily Gain PostWeaning (ADGPosW), and Fleece Weight (FW) were estimated to be 0.10 (0.024) and 0.083 (0.03), 0.068 (0.022) and 0.04 (0.021), 0.076 (0.026) and 0.033 (0.028), 0.14 (0.05) and 0.032 (0.025), 0.13 (0.04) and 0.032 (0.025), 0.068 (0.024) and 0.05 (0.025), 0.08 (0.028) and 0.027 (0.015), 0.08 (0.027) and 0.014 (0.013), respectively. Total heritability for these traits was estimated to be 0.19, 0.08, 0.16, 0.26, 0.24, 0.09, 0.08 and 0.11 respectively. Genetic and phenotypic correlations were estimated to be for BW and 9MW 0.56 and 0.31, BW and 12MW 0.59 and 0.31, WW and 9MW 0.92 and 0.5, WW and 12MW 0.94 and 0.61, 9MW and 12MW 0.95 and 0.82, ADGPreW and ADGPosW -0.07 and -0.17, respectively. The maternal heritability of BW and WW are similar to direct heritability and this reveals that the maternal effect can be an important component for weight traits prior to weaning. Total heritability is larger for post-weaning weight traits than for BW and WW. Selection of replacement lambs in the flock according to direct and maternal breeding values can be expected to bring about a response in the selected traits.

Estimation of direct and maternal genetic variances for calving ease in Croatian Holstein breed

Špehar, M.[1,2], Gorjanc, G.[2], Ivkić, Z.[1] and Bulić, V.[1], [1]Croatian Agricultural Agency, Department of cattle development, Ilica 101, 10000 Zagreb, Croatia (Hrvatska), [2]University of Ljubljana, Biotechnical Faculty, Department of Animal Science, Groblje 3, 1230 Domžale, Slovenia; mspehar@hpa.hr

The objective of this study was to estimate direct and maternal genetic parameters for calving ease in Croatian Holstein breed. Data for 80,562 first calvings were taken from the central database of the Croatian Agricultural Agency. Calving ease was scored on the scale from 1 to 4 (1 = no problem, 2 = slight problem, 3 = cow needed assistance, 4 = veterinary assistance). The total number of animals in pedigree was 160,371. Statistical model included the fixed effects of calving season, interaction between sex and age at calving in months, and interaction between region and year of calving. Herd and calving year interaction, service sire (direct genetic effect), and maternal genetic effect were included in the model as random effects. Variance components were estimated from Gaussian model using REML method as implemented in the VCE-6 program. The estimated variances (and corresponding ratios) were: 0.031±0.001 (0.147±0.003) for herd-year, 0.0031±0.0004 (0.015±0.002) for service sire, and 0.008±0.001 (0.038±0.006) for maternal genetic effect. Correlation between direct and maternal genetic effect was 0.674±0.095. Results provide genetic parameters needed for the application of genetic evaluation for calving ease in Croatian Holstein breed.

Estimation of genetic parameters for hip dysplasia in Czech Labrador retrievers

Vostry, L., Capkova, Z., Sebkova, N., Mach, K. and Majzlik, I., Czech Univ. of Life Sci, Prague, 165 21, Czech Republic; vostry@af.czu.cz

The genetic parameters, genetic trend and breeding value was estimated for development of the hip dysplasia (HD) grade in Labrador retrievers in the Czech Republic (n=3,145). Genetic parameters were estimated by linear model (LM) and threshold model (TM). The right and the left hip joints were evaluated separately as a repeated trait, using the FCI methodology. Several models were tested for the correct estimation of genetic parameters. All tested models comprised fixed effects of sex, assessor, year of birth, regression on age at evaluation and random direct genetic effect and the effect of dog's permanent environment. Models differed in the inclusion of these effects: random maternal effect and random effect of dam's permanent environment. Compared to TM, LM provided lower values of the coefficients of direct (0.28-0.29 vs. 0.35-0.39) and maternal heritability (0.01-0.02 vs. 0.07-0.10), repeatability (0.75-0.76 vs. 0.82-0.84) and of the correlation between direct and maternal effect (-0.56 - -0.20 vs. -0.70 - -0.28). A similar trend was observed in breeding value. In the tested models no statistical significance was proved for fixed regression on inbreeding coefficient and for the random effect of dam's permanent environment. In spite of similarity of LM to TM it is recommended to use the threshold model for the estimation of genetic parameters and subsequent breeding value for HD in Labrador retrievers kept in the Czech Republic.

Pedigree analysis and estimation of inbreeding effects on calving traits in an organised performance test for functional traits

Hinrichs, D., Almeroth, J. and Thaller, G., Institute of Animal Breeding and Husbandry, Christian-Albrechts-University, Olshausenstraße 40, D-24098 Kiel, Germany; dhinrichs@tierzucht.uni-kiel.de

The effect of inbreeding on calving traits (stillbirth and birth weight) was analysed in this study. Two models were used to estimate the effect of inbreeding. One model estimates the effect of inbreeding as a regression on the inbreeding coefficient of the mother, whereas the other model estimates the effect of inbreeding as a regression on the inbreeding coefficient of the calf. In addition, a pedigree analysis was done with respect to the effective number of founders and the ancestors with the biggest impact on two defined reference populations. The first contained all animals born between 1999 and 2003 (n=17,671) and the second includes all animals born between 2003 and 2008 (n=18,324). The pedigree file includes 70,938 animals. Phenotypic data for this study were collected on three large commercial milk farms. Data recording took place from February 1998 to December 2008 and 36,6610 calving events were recorded and 9.19% of the calves died within 48 hours after calving. All calves were weighted after birth and the average birth weight was 43.09 kg. The effective number of founders for the first reference population was 95 and decreased to 83 in the second reference population. In the first reference population 53.72% of the gene pool could be explained by ten ancestors with the highest marginal genetic contribution. In the second reference population the ten ancestors with the greatest impact explained 57.26% of the gene pool. The effect of inbreeding showed no significant effect on all traits of this study when the regression was done on the inbreeding coefficient of the dam. If we used the inbreeding coefficient of calf as regression variable the effect of inbreeding was not significant for birth weight. However, inbreeding showed a significant effect on stillbirth, i.e. the risk of stillbirth increases (0.26% per percent extra inbreeding) with increasing inbreeding coefficient of calf.

Genetic relationships between clinical mastitis and different somatic cell count traits in Austrian Fleckvieh cows

Koeck, A.[1], Heringstad, B.[2], Egger-Danner, C.[3], Fuerst, C.[3] and Fuerst-Waltl, B.[1], [1]University of Natural Resources and Applied Life Sciences, Department of Sustainable Agricultural Systems, Division of Livestock Sciences, Gregor-Mendel-Str. 33, 1180 Vienna, Austria, [2]Norwegian University of Life Sciences, Department of Animal and Aquacultural Sciences, P.O. Box 5003, N-1432 Ås, Norway, [3]ZuchtData EDV-Dienstleistungen GmbH, Dresdner Str. 89/19, 1200 Vienna, Austria; astrid.koeck@boku.ac.at

The objective was to explore genetic associations between clinical mastitis (CM) and different somatic cell count (SCC) traits in Austrian dual purpose Fleckvieh cows. Records on veterinary treatments of CM were available from the Austrian health monitoring project. For CM three intervals were considered: -10 to 50 d, 51 to 150 d and -10 to 150 d after calving. Within each interval, presence or absence of CM was scored as '1' or '0' based on whether or not the cow had recorded at least one veterinary treatment of CM. Lactation mean somatic cell score (LSCS) was the average of all testday SCS from 8 to 305 d after calving, and early lactation SCS (ESCS) was the average of the first 2 testdays. Subclinical mastitis (SM) was expressed as a binary trait based on prolonged elevated SCC (3 consecutive test-days with SCC>200,000). A total of 43,299 lactations from 26,748 cows were analyzed. Threshold liability models were applied for binary traits, while Gaussian models were used for ESCS and LSCS. A Bayesian approach using Gibbs sampling was used. Posterior means of heritability of liability to CM were 0.07 and 0.02 in the first and second interval, respectively, and 0.06 in the full period. Heritability estimates of ESCS, LSCS and liability to SM were higher (0.10 to 0.15). The posterior mean of the genetic correlation between CM in lactation period 1 and 2 was close to unity. Posterior means of genetic correlations between CM and SCC traits ranged from 0.64 to 0.77, whereas the genetic correlation between SM and LSCS was almost 1. As CM and SCC are describing different aspects of udder health, both traits should be considered for selection of bulls.

Microsatellite analysis of local and commercial chicken populations from Center Black Sea Region of Turkey

Mercan, L. and Okumus, A., Ondokuz Mayis University, Animal Science, Ondokuz Mayis University, Faculty of Agriculture, Department of Animal Science, 55139, SAMSUN, Turkey; lmercan@omu.edu.tr

The village/backyard chicken genotype samples selected from three towns and three villages from each Samsun, Sinop, Amasya, Ordu and Tokat districts from center Black Sea Region of Turkey (totally from 45 points) analyzed for genetic similarity, gene diversity ratios, allele frequencies between commercial chickens in population level by SSR (Simple Sequence Repeats-Microsatellites) markers with the aim of evaluating present potential of local chicken genetic resources in Center Black Sea Region of Turkey. Data were analyzed according to Unweighted Pair Group Method with Arithmetic Mean (UPGMA) and Principle Coordinates Analysis (PCoA) with the program NTSYSpc V.2.11 and were presented in graphics. Twenty eight microsatellite loci were used in the study. The mean Polymorphism Information Content (PIC) value of these loci was 0.86±0.06, mean allele number per locus for all populations was 12.96±4.97 and the mean allele number per population per locus was 2.33±0.19. In addition, populations' mean gene diversity ratio (h) was calculated as 0.675±0.040 and Chi-square analysis showed that there was no difference between the populations (P>0.05). It was determined that local chicken genotypes were rich genetic resources with their allelic diversity and the population obtained from the Kıran village of Kumru/Ordu (ORKK) did not share any alleles with commercial populations and also individual genotyping was needed for further studies to improve sensitivity and informativeness.

Session 12 Poster 15

The genetic resistance to gastrointestinal parasites in Bergamasca sheep breed: heritability and repeatability of faecal egg counts (FEC)

Cecchi, F.[1], Venditti, G.[2], Ciampolini, R.[1], D'avino, N.[2], Ciani, E.[3], Sebastiani, C.[2], Macchioni, F.[4], Filippini, F.[2], Calzoni, P.[2] and Biagetti, M.[2], [1]University of Pisa, Animal Production Department, Viale delle Piagge 2, 56124 Pisa, Italy, [2]IZS Umbria Marche, Via Gaetano Salvemini 1, 06126 Perugia, Italy, [3]University of Bari, General and Environmental Physiology Department, Via Amendola 165/a, 70126 Bari, Italy, [4]University of Pisa, Animal Pathology, Viale delle Piagge 2, 56124 Pisa, Italy; fcecchi@vet.unipi.it

Bergamasca is a sheep breed mainly reared for the production of meat. Native of the Clusone plateau and the Bergamo valleys, it is now mostly spread in Lombardy (Bergamo), but it is also commonly found in other regions of Northern and Central Italy particularly as a crossing breed. This study was carried out in a single Bergamasca herd from Matelica (MC) to estimate heritability and repeatability of EPG (Eggs Per Gram) and OPG (Oocyst Per Gram) count as a preliminary investigation on the genetic resistance to parasitic fauna. Individual fecal samples of 75 adult Bergamasca ewes and 39 lambs have been considered coming from 5 subsequent withdrawals (from May 2007 to February 2009). Fecal samples were analyzed in order to determine EPG or OPG using the FLOTAC technique. Besides Coccidia and gastro-intestinal Strongylids (including Nematodirus sp.), samples were also analyzed in order to evaluate the presence of Strongyloides sp, Trichuris spp and Dicrocoelium dendriticum. For the statistical analysis data regarding EPG and OPG were transformed to the natural logarithm according to the formula y = log (EPG/OPG+25) to correct for heterogeneity of variance and to produce approximately normally distributed data. A mixed linear model for repeated measures was used to test the effect of the season on the considered variables and to estimate the repeatability. The heritability coefficients of EPG and OPG using daughter-mother regression and repeatability resulted in null values for all considered parameters.

Session 12 Poster 16

Inducible Hsp-70.1 gene polymorphisms affect heat shock response of bovine mononuclear cells

Basiricò, L., Morera, P., Primi, V., Lacetera, N., Nardone, A. and Bernabucci, U., University of Tuscia, DIPA, via C. De Lellis s.n.c., 01100 Viterbo, Italy; bernab@unitus.it

Heat shock proteins (Hsp) are known to protect cells from several stressors. Nucleotide changes in the flanking regions [5'- and 3'-untranslated region(UTR)] of Hsp gene might affect inducibility, degree of expression or stability of Hsp70 mRNA. The present study aimed to investigate the association between inducible Hsp70.1 SNPs and heat shock (HS) response of peripheral blood mononuclear cells (PBMC) in dairy cows. Four-hundred fifty Italian Holstein cows were genotyped for two Hsp70.1 SNPs (g115 C/- and g348 G/T). Among those, 31 cows were selected to be representative of the following genotypes: CC/TG, C-/TG, --/TG, --/TT,C-/TT/, CC/GG. Blood samples were taken and PBMC were isolated by centrifugation on Ficoll-PaquePLUS. The PBMC were heated (43 °C in thermal bath) for 1 h (time 0) and then incubated at 39 °C in atmosphere of 5% CO_2 for 1, 2, 4, 8, 16 and 24 h. Cell viability was determined by XTT assay. Gene expression of Hsp70.1 was determined by real-time RT-PCRon 12 cows (4 per genotype) only for genotypes showing higher (--/TG), medium (C-/TT) and lower (CC/GG) viability. Expression of mRNA was normalized using RPS9 as housekeeping gene. Data were analyzed using the MIXED procedure by the Statistica-7 software package. Cell viability was higher (P<0.001) in --/TG genotype than other genotypes at time 0 and at 1, 2 e 4 h. Gene expression of Hsp70.1 showed higher (P<0.001) levels at 1 and 2 h in --/TG genotype compared with its counterparts. Exposure to HS affects differently cell viability and gene expression of Hsp70.1 in the three selected genotypes. These results indicate that the presence of SNPs (g115 C/- and g348 G/T)in the 5'-UTR region of inducible Hsp-70.1 ameliorates HS response and tolerance to heat of bovine PBMC. These mutation sites might be used as molecular genetic markers to assist selection for heat tolerance.The research was financially supported by the SELMOL-Project.

Use of some slaughter and quality traits to discriminate rabbit meat from different genotypes reared under organic and conventional systems

Paci, G., Preziuso, G., Bagliacca, M. and Cecchi, F., University of Pisa, Department of Animal Production, Viale Piagge, 2, 56124 Pisa, Italy; gpaci@vet.unipi.it

The aim of the trial was to identify some quality traits to differentiate rabbit meat from different genotypes reared under different housing systems: organic and conventional. 84 rabbits of local population (Group A) were housed in colony cages under organic system, according to an official organ of certification; 72 rabbits of the same population (Group B) and 72 hybrids (Group C) were housed in colony cages under conventional system. All rabbits were fed an organic diet ad libitum. At a weight of 2,400±100 g but at different ages (local population 102 days old; hybrids 90 days old) 30 animals of each group were slaughtered. Slaughter data, chilled carcass composition, reference carcass characteristics and meat quality traits were collected and statistically processed by stepwise discriminant analysis. Among investigated parameters live weight, drip loss, loin, muscle/bone, b* and L* colour traits were selected as predictors. The selected parameters allowed to differentiate the three groups and a total of 94%, 58% and 100% rabbits were correctly assigned to their original group. The selected parameters were analyzed by ANOVA and drip loss, loin, muscle/bone and L* showed the highest significant differences between groups. Drip loss was higher in Group A than in the other groups (37.6 g vs 30.8 g and 29.22 g, $P<0.05$). Loins of Groups A and B were higher than in Group C (22.8% and 21.5% vs 21.0%, $P<0.01$). Muscle/bone was better in Group C (4.66 vs 4.01 and 3.49, $P<0.01$). C and B Groups showed higher L* than Group A (61,89 and 59.15 vs 53.94, $P<0.01$). In conclusion the best accuracy of classification was in the classes characterized by local population reared under organic system and hybrid reared under conventional system. Some quality parameters could be considered efficient to discriminate rabbits from the different genotype and rearing systems and to improve the traceability process.

Long term economic consideration of progeny testing program in Iranian Holsteins

Joezy Shekalgorabi, S.[1], Shadparvar, A.A.[2], Vaez Torshizi, R.[1], Moradi Shahre Babak, M.[3] and Jorjani, H.[4], [1]Tarbiat Modares University, Pajouhesh Blvd., 1497713111, Tehran, Iran, [2]University of Guilan, Tehran-Rasht Road, Rasht, Iran, [3]University of Tehran, Agricultural Faculty, Karaj, Iran, [4]Swedish University of Agricultural Science, Interbull Center, 7023S-75007, Uppsala, Sweden; Joezy@modares.ac.ir

Cost-benefit ratio was demonstrated for young bulls sampling program in Holsteins of Iran. Returns from progeny testing program was divided into two groups, one from genetic improvement in different selection pathways and the other from culled bulls after progeny testing or at the end of reproductive period. The most important part of expenses was food and maintenance costs of young bulls during quarantine and waiting period. All costs and returns were discounted to their happening time. Interest rate was 8 and 6 percent for costs and returns, respectively. Gene flow theory was used for calculating cost-benefit ratio in a continuous selection program for 70 years. Cumulated cost-benefit ratio was 1.95 in year 70. This value was not the optimum value and could change by change in some management parameters like number of sampled young bulls or number of daughters per young bulls. Late onset of positive profit (year 23) was a result of large number of sex-age classes especially in dam pathways.

Determination of growth hormone receptor gene polymorphisms in East Anatolian Red cattle, South Anatolian Red cattle and Turkish Grey cattle

Akis, I., Oztabak, K.O. and Mengi, A., Istanbul University Faculty of Veterinary, Department of Biochemistry, Istanbul Universitesi Veteriner Fakültesi Biyokimya Anabilim Dalı Avcılar, 34320 İstanbul, Turkey; irazakis@gmail.com

The main objective of the genetic studies of livestock is the determination of polymorphisms in genes that have a role in metabolic processes affecting trait characteristics. The aim of this study is to develop a scientific basis for selection by determining genotype and allele distributions of growth hormone receptor gene polymorphisms. It is claimed that these polymorphisms have effects on both milk and meat quality and also on traits of native Turkish breeds East Anatolian Red cattle (EAR), South Anatolian Red cattle (SAR) and Turkish Grey cattle. DNA samples were isolated by using the standard ammonium acetate salt-out method. Target regions were amplified by using polymerase chain reaction and were digested by AluI, AccI, StuI, NsiI and Fnu4HI restriction enzymes. The allele and genotype frequencies were calculated by using PopGen32 software program. The + allele frequency of GHR/AluI polymorphism related with milk traits in the Turkish Grey cattle, and – allele frequencies of GHR/AluI polymorphism related with meat traits in EAR and SAR breeds were found to be low. The + allele frequency of GHR/StuI polymorphism affecting milk traits was found to be low in the EAR breed. The -/- genotype of GHR/NsiI polymorphism was also low in all of the three breeds. The + allele frequency of GHR/Fnu4HI polymorphism related with meat traits turned out to be low in SAR and EAR breeds. It was determined that EAR and SAR breeds were genetically closer to each other in terms of GHR gene, compared with Turkish Grey cattle. As a conclusion it can be claimed that native breeds have a disadvantage in terms of GHR gene polymorphisms, compared with high producing European breeds.

Single nucleotide polymorphisms in the porcine PCSK1 gene are associated with fat deposition and carcass traits in Italian Large White and Italian Duroc pigs

Russo, V.[1], Scotti, E.[1], Bertolini, F.[1], Buttazzoni, L.[2], Dall'olio, S.[1], Davoli, R.[1] and Fontanesi, L.[1], [1]University of Bologna, DIPROVAL, Sezione di Allevamenti Zootecnici, Via F.lli Rosselli 107, 42100, Reggio Emilia, Italy, [2]Associazione Nazionale Allevatori Suini (ANAS), Via Lazzaro Spallanzani 4/6, 00161, Roma, Italy; vincenzo.russo@unibo.it

The PCSK1 gene encodes the prohormone convertase 1/3 enzyme that is a key regulator of central and peripheral energy metabolism. Mutations in the human PCSK1 gene cause monogenic and common obesity. We sequenced about 4.7 kb of the PCSK1 gene (including 14 exons and intronic regions) in 21 pigs belonging to different breeds and identified 13 single nucleotide polymorphisms (SNPs), organized in 9 different haplotypes. Overall nucleotide diversity was 0.044% (±0.008). Tajima's D values indicated putative directional selection and balancing selection in Italian Duroc and Italian Large White breeds, respectively. This gene was assigned to porcine chromosome 2 by radiation hybrid mapping. Two SNPs were selected for association studies in four groups of performance tested heavy pigs: 1) 100 Italian Large White pigs selected according to extreme and divergent estimated breeding value (EBV) for back fat thickness (BFT; 50 with low and 50 with high EBV), 2) 100 Italian Duroc pigs selected according to extreme and divergent EBV for visible intermuscular fat (VIF; 50 with low and 50 with high EBV), 3) 270 Italian Large White and 4) 301 Italian Duroc pigs not selected for any phenotypic criteria and for which EBV for several traits were calculated. Allele frequencies differences for one SNP between the extreme tails of the first two groups of pigs were significant ($P<0.05$). In the 270 Italian Large White pigs, both SNPs showed association with average daily gain (ADG, $P<0.01$) and lean cuts (LC; $P<0.05$), whereas in the 301 Italian Duroc these SNPs were associated with ADG, LC ($P<0.001$), VIF ($P<0.001$ and $P<0.05$), BFT ($P<0.01$ and $P<0.1$), and feed:gain ratio ($P<0.01$).

Session 12 Poster 21

Bovine beta-casein allele A1: the source of beta-casomorphin 7 peptide, prevalence in Holstein bulls and its association with breeding value

Cieslinska, A.[1], Olenski, K.[1], Kostyra, E.[1], Szyda, J.[2] and Kaminski, S.[1], [1]Univ Warmia & Mazury, Oczapowski str 5, 10-718, Poland, [2]Univ Life Sciences, Kozuchowska str 7, 51-631 Wroclaw, Poland; stachel@uwm.edu.pl

Epidemiological studies suggest that beta-casein (CSN2) A1 protein variant might be one of the risk factors in the etiology of human diabetes and ischemic heart disease. It is hypothesized that during enzymatic digestion beta-casomorphin-7 (BTN7) peptide is released exclusively from CSN2 A1 variant and therefore it is thought to be harmful for human health. However, more research is needed to better understand the basis of this assumption: 1. to quantify BTN7 in digested milk, 2. to evaluate the frequency of the A1 allele in dairy cattle and, 3. to find out whether CSN2 is associated with milk production. In order to tackle these problems we determined the CSN2 genotype of 177 Holstein cows and 478 Holstein bulls by the use of the PCR-ACRS method. Eighteen cows in the same lactation period, but of different CSN2 genotypes were chosen to measure BTN7 content (by ELISA test) in the 30th, 100th and 200th day of the first lactation. Significant differences in BTN7 concentration between homozygotes and as well as between A1A2 and A2A2 cows were found. The frequency of the undesirable A1 allele was relatively high - 0.35. A linear regression model was used for testing the association between the polymorphism and breeding values for production traits. It was observed that the allele coding the A2 protein variant increases bulls' breeding values for milk protein yield and decreases breeding values for fat percentage. Although there are no clinical studies confirming the harmful effect of BTN7 on human health, it seems that selection for the A2 allele at CSN2 locus may be beneficial as it leads to increased protein yield, and at the same time, decreases the risk for human health.

Session 12 Poster 22

Relative quantification of p53 and VEGF-C gene expression in canine mammary tumors

Doosti, M.[1], Nassiri, M.R.[1,2], Movaseghi, A.[2,3], Tahmoorepur, M.[1,2], Ghovvati, S.[1], Kazemi, H.[1,3] and Solatani, M.[1], [1]Ferdowsi University of Mashhad, Animal Science Department, Azadi Sq., 91775-1163, Iran, [2]Ferdowsi University of Mashhad, Institute of Biotechnology, Azadi Sq., 9177948974, Iran, [3]Ferdowsi University of Mashhad, Department of Veterinary Pathology, Azadi Sq., 9177948975, Iran; Ghovvati@stu-mail.um.ac.ir

Breast cancer is one of the most common cancers through the world that ranked second with respect to its mortality rate. If malignant cancer detected and cured in its initial stages, lifespan in more than 90 percent of patients can be increased. Because of scientific and ethical limitations of conducting researches on human diseases, it is necessary to establish suitable animal models for these purposes. The dog is an appropriate model for studying breast cancer. The main goal of current study was to optimize SYBR Green based quantitative Real-time PCR (qRT-PCR) approach to measure relative expression levels of p53 and VEGF-C genes in normal and cancerous specimens. Eight normal mammary glands and 11 mammary gland tumors (including 7 benign and 4 malignant mammary tumors) were used in this study. Total RNA was extracted and p53, VEGF-C and B-Actin genes fragments were reverse transcribed. Real-time PCR assay were used for quantification of mRNA expression levels. Results were statistically analyzed using Student t-test. Findings showed that all of these three genes were expressed in both normal and cancerous samples but VEGF-C expression in the malignant mammary tumors was much higher than in the benign mammary tumors or normal mammary tissue ($P<0.001$). In contrast to the normal samples, 9%, 18% and 73% of malignant samples had shown higher, equal and lower of p53 gene expression, respectively.

Associations between Single Nucleotide Polymorphisms in GDF9 and BMP15 fecundity genes and litter size in two dairy sheep breeds of Greece

Kapeoldassi, K.[1], Kominakis, A.[2], Chadio, S.[1], Andreadou, M.[1] and Ikonomopoulos, I.[1], [1]Institute of Animal Anatomy and Physiology, Department of Animal Science, AUA, Iera Odos 75, 11855, Athens, Greece, [2]Institute of Animal Breeding & Husbandry, Iera Odos 75, 11855, Athens, Greece; shad@aua.gr

Mutation studies in different prolific sheep breeds have shown that the bone morphogenetic protein 15 (BMP15) and the growth differentiation factor 9 (GDF9) are major determinant of ovulation rate and litter size. Aim of the present study was to investigate whether Single Nucleotide Polymorphisms (SNPs) in these two fecundity genes are associated with litter size in the Chios and Karagouniko sheep of Greece. In total, 92 Chios and 96 Karagouniko ewes were genotyped in two different times (batches). Repeated records (n=189 and 259 for the Chios and Karagouniko breed, respectively) for litter size on the same animals were also obtained. Genetic analysis included examination of Hardy Weinberg equilibrium and allelic as well as genotypic differentiation between breeds and batches within breeds. Association between SNPs and litter size within breeds was assessed by application of a mixed model treating single genotypes as the fixed effect(s) and the ewe within breed as the random term while accounting for lactation number and month of lambing. Point mutations were detected for the G1, G4 and G8 regions in the GDF9 gene and B2 as well as B4 in the BMP15 gene. A clear allelic differentiation between the two breeds mainly for the GDF9 gene was observed. Significant allele substitution effects and dominance deviations for both genes were detected only in the Chios breed. Dominance deviations were -0.5±0.2, 0.65±0.22 and 0.42±0.15 lambs in the G1, G4 and G8 polymorphisms, respectively (GDF9 gene, Chios breed, pooled data). No significant effects were detected in the Karagouniko breed. Results were only partially confirmed within batches implying that any associations observed between SNPs and litter size should be interpreted with caution when sample or population stratification exists

---

Association of STAT5A and FGF2 gene mutations with conception rate in dairy cattle

Michailidou, S.[1], Oikonomou, G.[2], Michailidis, G.[1], Avdi, M.[1] and Banos, G.[2], [1]School of Agriculture, Aristotle University of Thessaloniki, 54124, Greece, [2]Faculty of Veterinary Medicine, Aristotle University of Thessaloniki, 54124, Greece; sofmicha@agro.auth.gr

Genetic selection for improved cow fertility is becoming increasingly important, since the latter has declined over the past decades and cannot be improved only by management. The objective of this study was to investigate the association of mutations of the signal transducer and activator of transcription 5A (STAT5A) and fibroblast growth factor 2 (FGF2) genes with fertility in cattle. These genes have been previously reported to be associated with *in vitro* fertilization and embryonic survival rate in cattle. A total of 120 1st lactation Holstein cows, raised on the same farm and under the same conditions, were divided into 2 performance groups (n=60 per group), the 1st consisting of animals that conceived and the 2nd that did not conceive with the 1st artificial insemination. Genotyping was performed using RFLP-PCR. In the 1st group, genotype CC of STAT5A had a frequency of 0.32, GC 0.45 and GG 0.23, while the frequency of allele C was 0.54 and of G 0.46. For FGF2, genotype AA was had a frequency of 0.15, AG 0.60 and GG 0.25, while the frequency of allele A was 0.45 and of G 0.55. In the 2nd group, CC had a frequency of 0.40, GC 0.42 and GG 0.18 (STAT5A), with frequencies of alleles C and A being 0.61 and 0.39, respectively. For FGF2, genotype AA had a frequency of 0.22, AG 0.50 and GG 0.28, while the frequency of allele A was 0.47 and of allele G 0.53. A Chi-square test was used to investigate differences in genotypic and allelic frequencies between the 2 groups. Differences were non-significantly greater than 0 (P>0.05) implying no association between STAT5A and FGF2 and fertility. Results suggest that the polymorphisms of these 2 genes cannot be used in gene-assisted selection to improve conception rate in cattle.

Session 12 Poster 25

Expression level of growth factor receptors GHR and IGF1R in liver and muscle during postnatal growth in pigs

Pierzchała, M., Urbański, P., Lisowski, P., Wyszyńska-Koko, J., Goluch, D. and Kamyczek, M., Institute of Genetics and Animal Breeding, Animal Immunogenetics, Jastrzębiec, str. Postępu 1, 05-552 Wólka Kosowska, Poland; m.pierzchala@ighz.pl

The GHR and IGF1R gene expression plays a crucial role for the growth rate in animals. However, these genes encode receptors essential for the transmission of signals in growth hormone signaling pathway for muscle cell growth in the embryo, but they are also important for the postnatal cell proliferation and differentiation. The action of GH on muscle cell size could be mediated by IGF-1, a potent muscle growth factor whose expression is induced by GH. Circulating IGF-1 is mostly derived from the liver and may act in an endocrine manner. In addition, GH-induced growth could also be mediated by local production of IGF-1 in target tissues, where IGF-1 may act in an autocrine/paracrine fashion. That participation of IGF-1 receptor signaling is required for the proper GH effects on skeletal muscle growth. The aim of present work was to study the mRNA transcription level of GHR and IGF1R genes in postnatal liver and muscle growth in 5 different pig breeds (Duroc, Polish Landrace, Pietrain, Pulawska and Polish Large White). The samples of liver and three muscles (m. longissimus dorsi, m. semimembranosus and m. gluteus medius) from six gilts (half-sibs) of each breed at six age stages (60, 90, 120, 150, 180 and 210 days) were collected. Between-breed effects and within-breed effects in relation to the muscles effects were studied. The results show significant variation in the level of IGF1R and GHR between different pig breeds. The highest difference was observed between polish indigenous pig breed - Pulawska (lowest level of transcription of GHR and IGF1R) and all others studied pig breeds, where the highest level of transcription was noticed in Polish Landrace and Pietrain. There were also noticed significant associations between age of pigs and level of transcripts: GHR in liver and IGF1R in muscles.

Session 13 Theatre 1

Experience from sheep in mountain environments

Kompan, D., University of Ljubljana, Biotechnical Faculty, Department of Animal Science, Groblje 3, 1230 Domzale, Slovenia; drago.kompan@bf.uni-lj.si

The most salient features in the composition of Slovenian lands are: the importance of forestry (around 60% of all land is forested), harsh relief (70% of all agricultural land is in less favoured areas) and a considerable share of meadows and pastures in total agricultural land (i.e., the share which ranks Slovenia the third in Europe). These relatively unfavourable production conditions diminish the competitiveness of Slovenian agriculture and limit the range of possible production orientations. Seasonal Alpine pasture has long tradition, from late Bronze Age period (17th to 10th centuries BC) and has not changed very much. More development on the sheep and goat alpine pasture was in 18th and in the beginning of 19th centuries, when hundreds of mountain pastures were alive. In that time the products from milk (cheese and salt curd) were produced for self consumption for farmers family in the winter time. In the past seasonal migration between valley and high pastures were traditional practice that has shaped much of the landscape in the Alps, as without it, most areas below 2,000 m would be forests. While today tourism and industry contribute much to Alpine economy, seasonal migration to high pastures is still practiced and the products are directly sold. The strategy of rural development in Slovenia is aimed at strengthening the multifunctional role of agriculture in Slovenia. On the other hand, preservation of typical cultural landscape is an important part of aesthetic and environmental identity of rural areas and it has a crucial importance in areas where agricultural activities are being abandoned. By nature-friendly and adequate technologies in mountains agriculture contributes to the maintenance of biodiversity and water and soil protection. This is possible with suitability of livestock, especially with sheep and goat production in mountain environments. However, high mountain pastures and traditional production systems are disappearing, and the existing production is oriented to selling products on the premises for relatively good price.

Management challenges of animal production on semiarid and Mediterranean rangelands

Ungar, E.D., Agricultural Research Organization - the Volcani Center, Department of Agronomy and Natural Resources, P.O.Box 6, Bet Dagan, 50250, Israel; eugene@volcani.agri.gov.il

The seasonally hot and dry conditions typical of many semiarid and Mediterranean regions create challenges for the management of animal production systems that are based on rangeland. A number of such challenges are highlighted using a simple model that explores some basic properties of grazing system dynamics and identifies key management decisions. The model comprises two rate functions which capture the essence of all grazing systems: grass growth and animal consumption. For simplicity, both rates are defined as functions of a single state variable – the herbage mass per unit area. Superimposition of the two functions yields a graphical model with emergent properties of directionality, and with points of equilibrium which may be stable or unstable. The graphical model represents a continuously growing pasture. In order to extend the model to conditions of highly seasonal forage production, the rates are integrated numerically using a dynamic form of the model. We then explore the implications of two key management decisions – animal density and grazing deferment at the start of the growing season – in terms of output per unit area and output per animal, using consumption as a surrogate. This is done firstly for green season dynamics and then for the entire annual cycle. We discuss the concept of carrying capacity and the implications for the producer who needs to make a living from a limited area of land.

Suckler cow production in nordic conditions: feeds, feeding and housing

Manninen, M., Finnish Food Safety Authority Evira, Plant Production and Feeds, Mustialankatu 3, 00790 Helsinki, Finland; merja.manninen@evira.fi

In Finland, suckler cow production is carried out in circumstances characterized by long winter periods and short grazing periods. All experiments with suckler cows in Finland during 1988-2005 were carried out at MTT Agrifood Research Finland, Tohmajärvi Research Station, Eastern Finland (62°20'N, 30°13'E), where winter conditions are arctic-continental. During the experimental years the mean temperature in January was -8.5 °C, and temperatures as low as -40 °C were observed. The snow depth could reach 100 cm. In Finland, the traditional winter housing system for suckler cows is insulated or uninsulated buildings, while there is a demand for less expensive housing systems. The experiments confirmed that production in cold conditions is possible in uninsulated or outdoor facilities when shelter against rain and wind, a dry resting place, adequate amounts of feed suitable for cold conditions and water are provided for the animals. The conventional feeds for suckler cows in Finland are grass silage, hay and straw, generally supplemented with a small amount of concentrate. Wilted grass silage with medium or high D value (640-690 g digestible organic matter in dry matter), offered as sole feed, proved suitable for spring calving adult suckler cows in cold conditions. Alternatively, grass silage can be replaced partly or totally by, e.g., treated straw, industrial by-products or whole-crop silage. Feeds provided at a lower feeding level, with inaccurate feeding, flat-rate feeding or feeding every third day were suitable winter feeding strategies for adult suckler cows. However, the cows must have an opportunity to replenish their live weights and body condition losses at pasture before the next winter. The new winter feeding strategies offer producers flexibility to plan the feeding based on the feeds available. Alternative feeds and feeding strategies should not have any detrimental effects on animal welfare in order to be acceptable to both farmers and consumers.

Maternal effects affect the genetic variation of meat production in reindeer

Mäki-Tanila, A.V.[1], Muuttoranta, K.[1] and Nieminen, M.[2], [1]MTT Agrifood Research Finland, Biotechnology and Food Research, P.O. Box 1, 31600 Jokioinen, Finland, [2]Finnish Game and Fisheries Research Institute, Reindeer Research Station, 99910 Kaamanen, Finland; Asko.Maki-Tanila@mtt.fi

Meat production is the main source of revenues in reindeer production. Most of the slaughtered animals are calves. Calves are born in the spring and they follow their dam over the first year until the following spring. Therefore the maternal effects influence strongly calf growth and survival and should be considered in planning the management and selection for meat production. The Reindeer Herders' Association has an experimental herd in Lapland. The herd is used for research by the Finnish Game and Fisheries Research Institute. The data are very unique for reindeer research because also the sires of calves are identified with DNA markers. Hence, it is possible to separate the maternal effects on the variation. There are weight records available for 1754 calves from the period 1987-2009, while the pedigree contains 3096 individuals. Considering the environmental effects, REML methodology was applied to estimate the variance components due to direct and maternal genetic (including dam's permanent environmental) effects on birth weight and growth. The small data set caused some limitations in dissecting the variation by the causal factors. Direct and maternal heritability estimates for birth weight were 0.30±0.11 and 0.19±0.11; for growth they were 0.35±0.13 and 0.22±0.12, respectively. Correlations between direct and maternal genetic effects for birth weight and growth were -0.18±0.08 and -0.75±0.19, respectively. Applying these results in reindeer breeding is challenging. Selection takes place in round-ups, without any knowledge of sires and as mass selection. The results can be used to find the guidelines for planning a breeding scheme for reindeer and support the importance of gathering the maternal information in practice.

Trends in longevity and reproductive indicators in Holstein dairy cows in Tunisia

Ben Salem, M.[1], Bouraoui, R.[2], Hammami, M.[3] and Marwa, H.[4], [1]INRA Tunisia, forage and animal production, rue Hédi Karray, 2049 Ariana, Tunisia, [2]ESA Mateur, animal production, mateur, 7030, Tunisia, [3]ESA Mateur, animal production, mateur, 7030, Turkmenistan, [4]ESA Mateur, animal produtcion, mateur, 7030, Tunisia; lahmar.mondher@iresa.agrinet.tn

The profitabilty of dairy herds in Tunisia has become a major concern to dairy producers. Among the major suggested underlying factors is reproduction and longivity of the cows. However, this needs to be confirmed. The objective of this work was then to investigate the trends in longivity and major reproductive indicators of dairy cows using individual cow data from a large herd over a 16-year period. Collected data included calving and insemination dates, lactation number and milk production and culling age and reasons. Data were first edited and major reproductive parameters and longevity were then calculated and trends investigated using regression procedures. Results showed that age average at first calving (31 months) was above the recommended optimum for Holstein cows. Respective averages of days open and calving interval were 134 and 411 days. Average service per conception was 1.7. Longevity was 65 months in the average and appeared to be largely associated with lactation number. Milk production per lactation increased from 4,300 kg in 1994 to 6,500 kg in 2008. Meanwhile, average calving interval increased by 7 days each year between 1993 and 2007, and was associated with increased days open. Average longevity decreased by 0.4 year during the same period. Reproductive failures accounted for 15% of the total culling rate. It was concluded that reduced reproductive performance and longevity of dairy cows in Tunisia is of concern and that various srategies are thus needed to attenuate further declines and to improve dairy herds' profitability. Reducing the age at first calving and targeting reasonable fisrt service and overall conception rates are two practices that can be used by farmers to achieve such an objective.

Robotic milking under arid environment

Halachmi, I., Van't Land, A., Ofir, G., Antler, A. and Maltz, E., ARO, Agri Engineering, Volcani, Israel; halachmi@volcani.agri.gov.il

Robotic milking could be the 'lifeboat' of the family farms. A robot (a) attracts the younger generation to the farm; (b) reduces the routine work, milking; (c) shifts workload to 'human timetable'; (d) frees time to management; (e) improves image - 'high tech' vs. 'low-tech' farming. (f) local tourism around a milking robot; (g) one family member can work outside while his spouse is in the farm. Robots were developed in North West Europe (forever green pastures, cubicles housing fitted to colder climate). In cold environment, concentrates (served in the robot) attract the cows to the robot. Under semi-arid condition, the cow cooling system has the potential to pull the cows towards the robot. However, the today situation generates a queue ('traffic jam') at the robot entrance. Therefore, the aim of this study was to compare two alternative cow traffics guided by the cow cooling systems, testing the hypothesis that relocating the cow cooling system, from the robot entrance to the robot exit may improve cow traffic towards the robot. The study was conducted in a single cowshed with two milking robots (Lely A3), each served 50-60 cows. One robot was equipped with electric ventilators and sprinklers (cow cooling system) at the robot entrance, and the otherwas equipped with cooling system at the robot exit. Both cooling systems ware identical (producer, model, measured air velocity, water flow, average droplet size). Measurements included: animal behavior (video camera), cow performance, natural wind direction and velocity, radiation, and air humidly. Significant differences were found in cow behaviour but no difference in production level. In hot days, the cow queue was shifted from the robot entrance to the forage feeding lane. Protein (%) 2.96 vs. 3.05; Fat(%) 3.4 vs. 3.3; milk yield (kg/day) 43.6 vs. 45.5; robot visits 3.8 vs. 3.7 at the control (cooling before the robot) and experiment (cooling after the robot) groups correspondingly. No more milkings and, consequently, no yield increments were observed.

Physiological and nutritional changes to maintain milk yield in late lactating dairy goats exposed to extreme heat stress conditions

Hamzaoui, S., Salama, A.A.K., Caja, G., Albanell, E., Flores, C. and Such, X., Grup de Recerca en Remugants (G2R), Universitat Autònoma de Barcelona, 08193 Bellaterra, Barcelona, Spain; ahmed.salama@uab.es

Murciano-Granadina dairy goats (n=8; 43.5±2.6 kg BW; 194±3 DIM) were assigned to 2 climatic treatments in a crossover design (35 d periods). Treatments were (temperature, ºC; humidity, %; THI, Thorn index): thermal neutral (TN, 15 to 20 °C and 35 to 45%; THI = 59 to 64), and heat stress (HS, 12-h day at 37 °C and 12-h night at 30.5 °C, 40%, THI = 85 and 77, respectively). Goats were fed 0.8 kg concentrate, 0.65 kg alfalfa pellets, and dehydrated fescue ad libitum. Concentrate was daily adjusted to maintain constant forage:concentrate ratio. Feed and water intake, and milk yield were recorded daily. Rectal temperature and respiration rate were recorded 3 times daily (8, 12 and 17 h). Milk and blood samples were collected weekly. Blood samples for acid-base balance indicators (d 25), as well as feces and urine for digestibility and N balance (d 31 to 35) were also collected. Rectal temperature (38.7 vs. 39.2 ºC), respiration (34 vs. 82/ min), water intake (5.5 vs. 11.1 L/d) and evaporation (1.1 vs. 3.3 L/d), were greater ($P<0.01$) in HS than in TN, while feed intake (2.0 vs. 1.6 kg DM/d) was lower ($P<0.01$). Blood NEFA (37 vs.12 mmol/L) and haptoglobin (0.134 vs. 0.105 ng/ml) were greater ($P<0.05$) in HS than TN only at d 7. Milk yield (1.23 L/d) did not vary, but milk of HS goats contained less ($P<0.05$) protein (3.36 vs. 3.84%) and casein (2.84 vs. 3.21%) than TN goats. Panting decreased blood CO_2 (22 vs. 26 mmol/L; $P<0.01$) in HS goats, but they maintained blood pH at a similar value to TN goats by lowering $HCO3^-$ (21 vs. 25 mmol/L; $P<0.01$) and increasing Cl^-(109 vs. 107; $P<0.05$) in blood. Digestibility of DM, OM, and ADF tended ($P<0.15$) to be greater in HS goats, which partially compensated for the reduction in intake. In conclusion, late lactating dairy goats were able to adapt to severe heat stress conditions maintaining milk yield but with reduced milk protein content.

Session 13 — Theatre 8

The effects of shade and betaine supplementation on tissue responses to heat stress in beef cattle

Digiacomo, K.[1], Leury, B.J.[1], Gaughan, J.B.[2], Loxton, I.[3] and Dunshea, F.R.[1], [1]The University of Melbourne, Department of Agriculture and Food Systems, Parkville, VIC 3010, Australia, [2]The University of Queensland, School of Animal Studies, Gatton, QLD 4343, Australia, [3]Beef Support Services Pty Ltd, P.O. Box 247, Yeppoon, QLD 4703, Australia; k.digiacomo@pgrad.unimelb.edu.au

Heat stress (HS) is an important issue for livestock production, being particularly costly for ruminants. Therefore, dietary and other methods to alleviate HS would result in increased productivity. Betaine (trimethylglycine) is an organic osmolyte and methyl donor that is commonly used as a nutritional supplement for monogastrics because of energy sparing effects. However, there are limited and conflicting data on the efficacy of betaine to reduce HS in ruminants. This study utilised 24 feed lot steers in a 2 x 2 factorial design to determine the effects of dietary betaine (0 or 40 g/head.day) and shade (no shade or shade) for 120 days. After 120 days the steers were slaughtered and tissue samples obtained from the neck muscle (M), omental adipose tissue (OM) and subcutaneous adipose tissue (SC). RNA was extracted and RT-PCR performed to analyse the gene expression of heat shock protein (HSP) 70, HSP90 and heat shock factor (HSF)-1normalised to β-actin. Tissue type significantly altered the response of all genes measured. HSP90 expression in muscle was significantly increased by shade ($P=0.006$). This effect held true for HSF1 and HSP90 across all tissues when analysed collectively. Betaine supplementation tended ($P<0.10$) to decrease HSP70 and HSF1 and increase HSP90. This is the first study to demonstrate the presence of HS related genes in adipose tissue of cattle. In conclusion, these data indicate that that HS related genes can be modified by both environmental and dietary factors.

Session 13 — Theatre 9

Effects of saline water on food and water intake, blood and urine electrolytes and biochemical and hematological parameters in male goats

Hadjigeorgiou, I., Kasomoulis, I., Gogas, A. and Zoidis, E., Faculty of Animal Science and Aquaculture, Agricultural University of Athens, Department of Nutrition Physiology and Feeding, 75 Iera Odos, 11855, Athens, Greece; ihadjig@aua.gr

The aim of this study was to investigate the saline water tolerance of male goats. A group of four castrated animals, aged 3-4 years, were used in a 5 weeks experiment, which was preceeded by a 2 weeks pre-trial period. Animals were offered alfalfa hay and concentrates at about maintenance level and allowed free access to water of five levels of salinity: 0, 0.5, 5, 10 and 20‰ NaCl. Food and water consumption were recorded daily, as it was for urine produced. Blood and urine samples were collected weekly during the trial, where several parameters were determined. Plasma concentrations of aldosterone, Na, K, glucose, creatinine, urea and proteins and hematological parameters were analyzed. Urine pH, specific weight and levels of Na, K and creatinine were also measured, as well as plasma and urine osmolality. Water intake increased until 10‰ NaCl (2.0 to 3.2 l/d) and decreased thereafter to reach 2.5 l/d. Food intake decreased (1.4 to 1.1 kg/d) and urine excretion increased (1.12 to 1.47 l/d) with increasing salinity. Increasing NaCl elevated concentrations of plasma Na (143 to 150 mmol/l), K (4.0 to 4.7 mmol/l), urea (26.5 to 47 mg/dl), proteins (6.3 to 8.2 g/dl), aldosterone (140 to 260 pg/ml) and osmolality (284 to 299 mosm/kg) whereas, glucose levels remained unaffected (~70 mg/dl). Urine parameters also increased: osmolality (317 to 1217 mosm/kg), specific weight (1018 to 1040), Na (55 to 377 mmol/l), K (144 to 329 mmol/l) and creatinine (69 to 116 mg/dl), whereas, pH (~8.5) was unaffected. Analysis of hematological parameters demonstrated increased levels of hematocrit (20.2 to 24.6%) and red blood cells (8.0 to 10.6 x106/μl) as indicators of mild dehydration while, all other parameters were not affected. The results indicate that goats can subsist on drinking saline water (up to 20‰ NaCl), for at least 2 weeks without harmful effects.

Dynamics of body fluids and hormonal profile of Egyptian Nubian goats under chronic heat stress conditions

Abou-Hashim, F.[1], Ashour, G.[1], Abdel Khalek, T.M.M.[2], Elshafie, M.H.[2], Abou-Ammou, F.F.[2] and Shafie, M.M.[1], [1]Faculty of Agriculture, Cairo University, Animal Production Department, 5 Gammaa St., Giza, 12613, Giza, Egypt, [2]Animal Production Research Institute, Agricultural Research Center, Sheep and Goats Department, 14 Nadi Elsayd, Dokki, Giza, Dokki, Giza, Egypt; fatmaabohashim@yahoo.com

Heat stress and usual outbreaks of severe hot weather waves during summer months are the major factors adversely affecting animal performance. The ability of animal to cope with new environmental temperatures depends on specific compensating mechanisms that activated via body's regulatory system. The impact of chronic heat stress on water balance, body fluids distribution, and hormonal profile was studied in 8 Egyptian-Nubian bucks. Bucks were kept continuously in thermal chambers and exposed to fixed ambient temperatures (40 °C as heat stressed group vs. 25 °C as control group) for 50 days. Heat stress of 40 °C significantly increased ($P<0.05$) total water intake, urine excretion volume, insensible water output and cortisol hormone. In addition, heat stress increased ($P>0.05$) total body water, plasma and blood volumes, extracellular and interstitial fluids and aldosterone hormone. Results suggest that cortisol may act to defend plasma volume depletion in heat stressed bucks. In conclusion, The Egyptian-Nubian goats are adapted to wide range of ambient temperatures, and could increase their water intake to regulate body temperature and dissipate more heat via skin and respiratory system without causing significant changes in their body fluids distribution.

Selenium yeast in the diet of dairy cows in hot climate

Pinheiro, M.G.[1], Oltramari, C.E.[2], Arcaro Júnior, I.[2], Arcaro, J.R.P.[2], Toledo, L.M.[2], Leme, P.R.[3], Manella, M.Q.[4], Pozzi, C.R.[2], Ambrósio, L.A.[2] and Freitas, E.C.[5], [1]APTA, DDD, Avenida Bandeirantes, 2419, 14030-670 Ribeirão Preto SP, Brazil, [2]APTA, IZ, Rua Heitor Penteado, 56, 13460-000 Nova Odessa SP, Brazil, [3]USP, FZEA, Avenida Duque de Caxias Norte, 225, 13635-900 Pirassununga SP, Brazil, [4]ALLTECH DO BRAZIL, Rua Curió, 312, 83705-560 Araucaria PR, Brazil, [5]CONNAN, Avenida Mário Pedro Vercellino, 877, 18550-000 Boituva SP, Brazil; mgpinheiro@apta.sp.gov.br

To obtain maximum performance of dairy cattle under stressing environment during the summer in the Southeast region of Brazil, cooling and nutritional adjustments are necessary. In this experiment, developed at the Instituto de Zootecnia, in Nova Odessa, State of São Paulo, 24 Holstein and Brown Swiss cows were distributed in a randomized blocks experimental design to study the effects of selenium yeast on the milk yield and quality The animals stayed in a free-stall and received the same diet with 0.277 mg/ kg DM of selenium, but from two distinct sources, selenium yeast (Sel-Plex 277 mg/kg DM) and sodium selenite (0.617 mg/kg DM). The addition of selenium yeast in the diet had no effect in the daily milk production (18 kg) and in the protein, lactose, solids not fat and total solids percentages (respectively, 3.06, 4.57, 8.58 and 12.69%), but increased ($P<0.01$) the fat percentage (3.96 and 4.15%) and decreased ($P<0.01$) somatic cells count (7.54 and 6.54 log2/ml), improving milk quality.

Session 14 Theatre 1

How to distinguish products made from a specific animal breed: stakes and ways of collective action in the European Union

Boutonnet, J.P., INRA SAD, UMR Innovation, 34060 Montpellier, France; boutonnet@supagro.inra.fr

In less favoured areas, local breeds have not be totally replaced by 'global' breeds, due to their good adaptation to specific harsh conditions: frequent droughts, necessity to move on long distances, from high mountain to lowlands conditions, etc. Nevertheless their productivity in terms of meat and/or milk is low, and does not allow livestock farmers to be competitive in global markets, unless they can extend on large land areas and manage large size flocks or herds. Small and medium livestock farmers and their communities have developed many alternative ways to survive. One of the most frequent is the elaboration of special products, of which particular characteristics are due to the region's soil-climate-animal breed complex and to the collective, regional know-how, adapted to these conditions. These traditional products are also well adapted to the wishes and demand of local populations that have developed some sort of 'consumers' knowledge' In the frame of trade liberalisation and globalisation, the distance between these producers and the potential consumers of these products is increasing. Thus 'consumers' knowledge' is weakening. It becomes more and more necessary to protect the use of the names of these products, under the threat of misuse or unfair competition. The communication presents some experiences of traditional products distinction, and explores different ways (Collective brands or Geographical Indications, among others) to achieve an efficient way to maintain these special products and keep the consumers' trust.

Session 14 Theatre 2

Local breeds as tool for a balanced regional development in the world of 'globalization': the case of Crete Island

Belibasaki, S. and Stefanakis, A., NAGREF-Veterinary Research Inst., P.O. Box 60272, 57001 Thessaloniki, Greece; belibasaki@vri.gr

The importance of the animal's diversity is increasingly recognized. Genetically diverse livestock populations provide a lot of options 'for future challenges' such as the emerging disease threats, nutritional requirements, societal needs, market demands, environmental changes, production of Protected Destination of Origin products, and better landscape use. Local breeds are very well adapted to their environment having the ability to use and influence their natural resources in a unique manner. The environment, continually interacting with the genome during the growth and development of an organism, creates behavioural responses that in social animals are transferred across generations. Scientists and managers often ignore the power of environmental adaptation and transfer breeds and production systems all over the world. On the other hand, farmers turn their interest on high yielding industrialized breeds, as they believe that the high production is the basic criterion for their survival in the world of globalization. These lead to the desolation of the Less Favorite Areas (LFA), immigration to capital cities, and rapid expansion of industrialized breeds and thus, bring the diversity of indigenous breeds in great concern. The case of Crete Island is a good example for the advantages of keeping local breeds to achieve a balanced development of marginal areas. The extensive and semi-extensive farming of more than 1,200,000 sheep and 400,000 goats of indigenous breeds, keeps people in LFA, develops the agro-tourism and produces PDO dairy products. All above are strongly related to the traditional culture and habits. In some industrialized breeds or their crosses with local ones, models of nutrition and energy balance showed that the available pastures can not cover daily milk yield more than 1000 g per day. Emphasis is, also, given to the comparative advantage of the production of traditional dairy PDO products, as by law, they have to be produced only by indigenous breeds.

Interactions between management of local breeds and valorization of products, examples of French cattle breeds

Lauvie, A. and Lambert-Derkimba, A., INRA SAD LRDE, Quartier Grossetti, 20250 Corte, France; anne.lauvie@corte.inra.fr

Valorization of local breeds is often presented as a good way to maintain or develop those breeds. This valorization can take various forms: quality food products, landscape conservation, association to specific farming systems, cultural heritage etc. We will focus in this communication on valorization through local food products. We will discuss how valorization processes re-question genetic management of local breeds and we will illustrate our discussion on French cattle breeds' examples. Involving a breed in a valorization project means involving new stakeholders (food processors, new types of farmers etc…), which means that: -New points of view on the breed and its genetic management are expressed and have to be taken into account. We will discuss the possible consequences on organization for genetic resources management -The exigencies due to the orientation towards a particular type of food product have to be taken into account in the genetic management (translation of specific characteristics in selection criteria). -The stakeholders have to deal with various projects or 'models' for the breeds and to choose a single genetic orientation. This phenomenon can lead to controversies or can delay the choice of a genetic orientation. We can conclude that if involving a local breed in a valorization project is a way to maintain and develop this breed, such a project is not easy to establish and to articulate with genetic management. We propose several aspects to consider before establishing such a project, in order to anticipate potential difficulties.

Session 14 Theatre 4

Marketing products based on breed in the Nordic region

Lund, B., Nordic Genetic Resource Center, NordGen, Sector Farm Animals, Box 115, N-1431 Aas, Norway; benedicte.lund@nordgen.org

NordGen farm animals (formally known as Nordic Gene Bank Farm Animals) presented in 2004 a report on the use of local breeds as a base for developing quality products. Since then we have observed both success stories and excamples of products which have not met the markets demand. The Global Plan of Action from 2007 states that each country is responsible for maintaining the genetic diversity of their own farm animal populations. 'Use them, or loose them', is one way of fulfilling that responsibility. The indigenous breeds have their specific traits, stories and traditions that could be utilized in developing and marketing attractive products with added value. For such product development to be successful many factors must be in place: • An idea of a product with a defined quality and history • Knowledge about a likely market demand and strategies for mapping and reaching the market • Realistic and adequate arrangement for practical issues such as logistics, with the right allies • Knowledge of available scientific knowledge as well as an overview of possibilities for political and financial stimuli and support Practical examples will be given. The particular breed and the product it self, as well as the tradition, the history and the landscape or production system may be part of the whole product. How the different stakeholders could contribute to such sustainable use will be discussed. NordGen's role on this issue in the Nordic region will be presented, as well as how the individual Nordic countries are stimulating for product development based on breed.

Added values of dairy cattle breeds

Cassandro, M., University of Padova, Department of Animal Science, viale della Università, 16 - Agripolis, Legnaro (PD), 35020, Italy; martino.cassandro@unipd.it

Aim of this study was to compare milk added value and methane emissions estimation from some dairy breeds. Several studies evidenced that milk from local dairy breeds is more suitable to be processed into cheese, so the development of payment systems that take into account the added value of milk for cheese production could support the conservation and valorisation of local animal genetic resources (AnGR). Cheese yield and link with history traditions and environment aspects might partially compensate their lower level of milk production. Breeders of Burlina, Reggiana, Rendena and Valdostana cattle breeds for ensuring their survival sell products to high-value in specialist markets realizing an added value per kg of milk of +44%, compared to Holstein Friesian cattle. Market oriented strategies on payment systems including added milk value could enhance profitability and interest in rearing and safeguarding of local AnGR, but this strategy is not available for all countries and areas. Therefore, new alternatives and strategies to enrich adding value of might include differences in greenhouse gasses emissions (GHG) between breeds. Indeed, e.g. local AnGR for their low-input and body weight are expected might reduce the environmental impact in term of methane emissions. Some results will be showed. Methane emissions were calculated by of dry matter intake estimation, utilizing production of milk and fat, and body weight estimations among breeds. A reduction of -13% of daily methane emissions/kg of metabolic body weight is expected for local breeds compared to cosmopolitan breeds. In conclusion, new phenotypes and traits are needed for define opportunity and economic incentives to preserve AnGR.

Strategies to add value to local breeds with particular reference to sheep and goats

Papachristoforou, C.[1], Koumas, A.[2] and Hadjipavlou, G.[2], [1]Cyprus University of Technology, Agricultural Science, Biotechnology and Food Science, P.O. Box 50329, 3603 Lemesos, Cyprus, [2]Agricultural Research Institute, Animal Production, P.O. Box 22016, 1516 Lefkosia, Cyprus; c.papachristoforou@cut.ac.cy

Local breeds (LBs) of farm animals are those found in only one country. The population of many LBs is either small or declining as a result of pressures by economic, political, social, environmental and other factors. LBs are unique genetic resources and therefore, valuable. However, the value assigned to LBs differs significantly between interest groups e.g. farmers, breeders, consumers, scientists, policy makers, local societies etc. Using several strategies to improve the relative value of LBs, will strengthen the prospects of their sustainable use. Marketing is perhaps the most powerful tool in this context. Finding or creating markets for meat, milk and wool products from local sheep and goat (S&G) breeds, contributes substantially to their sustainability. Marketing strategies include identification and promotion of traditional products linked to LBs, highlighting the quality and culture associated with these products, and facilitating access of LB farmers to commercial markets. Other important strategies relate to: promotion of local S&G breeds as the best suited ones for use in landscape management, silvopastoral systems and organic farming, by stressing the aspects of animal welfare and of benefits to human health from special characteristics of products; exploitation of unique LB characteristics in agrotourism (food, non-food products, souvenirs, recreation) and for educational purposes; sustainable use of LBs in breeding schemes, particularly those employing two- and three-way crossing; educating people about the value and importance of LBs; finally, research, can also serve as a very powerful tool to add value to LBs by creating new knowledge and providing the scientific evidence on their particular biological characteristics. The strategies for local S&G breeds will be illustrated by specific examples.

Session 14 Theatre 7

Are PDO projects adding value to local breeds: comparing two case studies, Nustrale and Cinta Senese pig breeds

Casabianca, F.[1], Maestrini, O.[1], Franci, O.[2] and Pugliese, C.[2], [1]INRA, SAD - LRDE, Quartier grossetti, F-20250 CORTE, France, Metropolitan, [2]University of Florence, Dept. Agricultural Biotechnology Section of Animal Science, Via delle Cascine 5, 50144 Firenze, Italy; fca@corte.inra.fr

Protected Designation of Origin is generally adding value to the product on the market. When based upon the mandatory use of local breeds, are PDO adding also value to these breeds and what kind of value? We compare two cases of previously endangered and now recovered local pig breeds: Nustrale in Corsica island (France) and Cinta Senese in Tuscany (Italy). Protection in Tuscany as Cinto Toscano is already obtained (for the pork meat) at national level (European registration still in progress), when the application for a PDO in Corsica as Prisuttu (dry cured ham of Corsica) is not yet completed. In both situations, breed census is growing and animals are reared in pasturelands and forests referring to traditional practices. 5 questions must be emphasized as main discussion: - Name to be protected. PDO name must avoid any confusion with the name of the breed. What decisions to be made by administrations, breed managers and PDO appliers? - Content of PDO specification. Code of practices regulates the use of the local breed within a livestock system. Is it valorizing rusticity and adaptation of the animals? - Environmental impacts. As herds are reared on the pasturelands, some degradation can occur. Is natural resource managed in a sustainable way? - Distribution of added value within the supply chain. Are lack of productivity and carcass adiposity compensated by raw material price? - Heritage value of such protection. Local breed brings a strong image to the product. Is PDO bringing to the farmers new attractiveness for the local breed? PDO projects mobilizing local breeds can add value for the breeders according the shape of the supply chain. Value is including symbolic and cultural elements able to attract new breeders and to help in recovering endangered local breeds.

Session 14 Theatre 8

Beef quality differentiation in the framework of Serrana de Teruel endangered breed conservation programme

Sanz, A.[1], Alberti, P.[1], Blasco, I.[1], Ripoll, G.[1], Alvarez-Rodriguez, J.[1], Bernues, A.[1], Olaizola, A.[2], Zaragoza, P.[2], Rodellar, C.[2], Sanz, A.[2], Martin-Burriel, I.[2], Picot, A.[3], Congost, S.[3], Abril, F.[3] and Vijil, E.[3], [1]CITA de Aragón, Animal Production, Avenida Montañana 930, 50059 Zaragoza, Spain, [2]Facultad de Veterinaria de Zaragoza, LAGENBIO, Miguel Servet 177, 50013 Zaragoza, Spain, [3]CTA de Aragón, DGA, Avenida Movera 580, 50194 Zaragoza, Spain; asanz@aragon.es

Serrana de Teruel (ST) is a dark or tabby-breed raised traditionally in South Aragon due to its great coping ability to harsh environments. Characterization of population structure, and morphological, zootecnic and genetic values of ST were conducted. Individuals showed medium to high homogeneity and harmony degree, being most of the animals straight profiled, eumetric and sublongilineus, although smaller in size than other close breeds. The thirty microsatellites analysed for biodiversity studies showed good diversity values despite its low effective population size (180 individuals in 2007). These studies provided basis for a conservation genetic programme and in 2007 ST was officially breed recognised. Germplasm banks were established, containing 6400 doses of semen from 7 males, and 40 embryos obtained from 10 males and 5 females. To assess their viability, ten embryos were transferred to receptor cows, being pregnant 4 of them. In order to guarantee the long term maintenance of ST, a prospective study of the meat value chain was carried out. A qualitative questionnaire was applied to all stages of the meat chain, from farmer to consumer, in the breed influence area. Concurrently, carcass and meat quality of ST calves was studied and several diversification alternatives for labelled calves market have been assessed (animals slaughtered at 470 and 700 kg live-weight, bulls and steers). Good performances and high quality products with no commercial constraints in the beef market were obtained. This study should provide the standard requirements for a labelled meat product that allows the farmer survival and assures ST breed conservation.

Valuation of autochthonous bovine genotypes: assessment of the meat quality of the Italian Podolian and the Greek Katerini cattle

Karatosidi, D.[1], Marsico, G.[1], Ligda, C.[2] and Tarricone, S.[1], [1]University of Bari, Animal Production, via Amendola 165/a, 70126 Bari, Italy, [2]National Agricultural Research Foundation, P.O. Box, 60 458, Thessaloniki, Greece; despinakaratosidi@yahoo.com

The aim of this study is to present the results of the comparison of the physico-chemical analysis of the characteristics of meat quality between the Greek Katerini cattle breed and the Italian Podolian cattle. The Katerini breed has a very low population less than 300 heads in 2 herds in Central Greece, while the Italian Podolian cattle comprises 24.000 heads in 600 herds. This study was initiated in order to define the unique characteristics of the breed and especially the meat quality characteristics that will describe and differentiate the meat of the Katerini cattle from the standard cattle meat. Although several researchs address rhe meat meat quality of the Italian Podolian cattle, there are no relevant references on the Katerini breed. The preliminary results presented in this paper refer to 4 Katerini and 4 Podolian calves. All animals are raised outdoors and they were slaughtered at 18 months of age. The first results of the physical analysis show that the meat of Katerini is less dark with better values of tenderness after cooking. The indexes of pH are normal and those of moisture content higher than the meat of Podolian cattle. Both protein and fat intramuscular percentage are low. As far as regards the results of meat fatty acid composition, the results show a higher percentage of SFA but a lower of MUFA.The concentration of ω3 and ω6 was found much more higher than meat of Podolian cattle. All the analysis were carried on Longissimus dorsi and aged 3 days at 4oC. The Italian market has already certified the podolian meat with a registered quality trade- mark ' 5R'. This experience could be useful in developing a certification procedure for a quality label of the Katerini breed meat, defining the production system, the region and the unique characteristics.

Towards (self) sustainability of local cattle breeds in Europe

Hiemstra, S.J., and EURECA Consortium, Wageningen University and Research Centre, Centre for Genetic Resources, the Netherlands (CGN), P.O. Box 65, 8200 AB Lelystad, Netherlands; sipkejoost.hiemstra@wur.nl

Local cattle breeds in Europe have important genetic, cultural, historical, socio-economic, and environmental values, but many are considered to be at risk of extinction. The general aim of the EU co-funded project EURECA (www.regionalcattlebreeds.eu) was to get a better understanding of factors affecting local cattle breeds' (self) sustainability in Europe. Data was collected from farmers and a wide range of stakeholders. Similarities and differences were identified by comparing 15 breed cases in 8 countries and cryopreservation programs in 4 countries. Additional data was collected in a wider survey among FAO National Coordinators in Europe. Across countries three aspects seem to be relevant for the current and future status of local breeds: 1) age of farmers, 2) cooperation among farmers, and 3) farmers' opinion on the appreciation of the local breed by the society. Herd books were rather common for the breeds surveyed, but more than 50% of the breeds lacked performance recording. For 25% of the breeds a specific breed-associated food product was reported. Breed-specific strategic opportunities and common factors for EU policy development were identified, using a quantified SWOT (strength, weakness, opportunity, threat) analysis. We concluded that both common policies and tailor-made breed specific strategies are needed to enhance breed (self) sustainability. Common policies should address generation transfer and positively influence cooperation among farmers, breed associations and society. SWOT analysis was also used to explore possibilities to add value to local breeds. Factors related to branded products were often mentioned as strength, while marketing of breed products was often considered as a weakness. While cultural values were perceived as strengths more than the environmental values, there seemed to be more opportunities attached to the environmental aspects.

Session 14 Theatre 11

Suitability of traditional breeds for organic and low input pig production systems

Leenhouwers, J.I. and Merks, J.W.M., IPG, Institute for Pig Genetics BV, P.O. Box 43, 6640AA Beuningen, Netherlands; jascha.leenhouwers@ipg.nl

Organic and low input pig farms in Europe generally use the same breeds and genetic lines as in conventional pig production. However, EU regulations on organic livestock farming limit the use of replacement breeding stock from conventional origin. Moreover, organic and low input pork producers are encouraged to use traditional breeds which may be more suitable for local conditions than conventional genetic lines. The objective of the present study is to evaluate the suitability of traditional versus conventional breeds for organic and low input pig production systems. Local breed performance data in different macro-climatic zones in Europe were collected as well as a literature meta analysis was performed, focussing on the suitability of traditional breeds with respect to different characteristics desired by the organic and low input sector. Results showed that traditional breeds generally have a lower growth rate, higher backfat thickness and poorer feed conversion ratio compared with conventional genetic lines kept under organic and low input conditions. Sow reproductive performance data show lower litter sizes and consequently lower number of piglets weaned per litter compared with conventional lines. These results indicate that the use of traditional breeds in commercial organic and low input pig production units is generally economically not feasible if the price/kg of pork is independent of the breed. However, in some countries local breeds have additional value for specific local products (e.g. Jamon Iberico in Spain and Mangalica ham in Hungary) which compensate the higher production costs. If conventional genetic lines are used for organic and low input production systems, the breeding goal must be adapted with focus on robustness traits such as piglet vitality, maternal abilities and sow longevity.

Session 14 Poster 12

Comparative studies on the aptitudes for meat production of the fattened lambs from local Romanian breeds

Ghita, E., Lazar, C., Pelmus, R. and Ropota, M., National Research Development Institute for Animal Biology and Nutrition, Animal Biology, 1, Calea Bucuresti, Balotesti, Ilfov, 077015, Romania; elena.ghita@ibna.ro

Worldwide, the fattened lamb meat (15-22 kg carcass) is a major goal of sheep production of most developed countries and of the large wool producers. Although the Romanian consumers prefer the nursing lamb meat, the lambs should be slaughtered at higher body weights, obtaining thus larger amounts of high quality lamb meat, which increases producers' income. The purpose of the paper is to analyse the aptitudes for meat production of the fattened lambs from local Romanian breeds. The research was done on 3 groups of 16 lambs each (n=48) from Karabash, Tsigai and Tsurcana sheep. The experiment started after the lambs were weaned (20-21 kg body weight) and ended when the lambs reached 37-39 kg body weight. During the fattening period the lambs had daily average weight gains of 0.258 kg Karabash, 0.191 kg Tsigai and 0.154 kg Tsurcana. The slaughtering yield, the commercial yield and the proportion of carcass parts were determined, as well as the meat to bone ratio, the gross chemical composition of the meat and the fatty acids content. The slaughtering yield was very significantly different ($P<0.001$) in the three breeds: 51.79% in Karabash, 46.82% in Tsigai and 42.81% in Tsurcana. The meat to bone ratio showed that Tsigai lambs had the highest proportion of meat, 2.63:1, followed by Karabash with 2.32:1 and by Tsurcana with 2.10:1. The fatty acids and cholesterol concentrations of the fat showed that the polyunsaturated fatty acids were in a much higher proportion in the meat from Tsigai lambs, and the cholesterol was in a much lower proportion in Karabash lambs. In conclusion, the studied local breeds produce lambs with poor aptitudes for meat production, which required their breeding by crossing with meat rams which will produce lamb carcasses competitive on the market.

Session 14 Poster 13

Changes in the Lithuanian Heavy Draught horse population after the introduction foreign genes

Sveistiene, R., Institute of Animal Science of LVA, Lithuanian Center for Farm Animal Genetic Resources Conservation, R.Zebenkos 12, LT-82317, Lithuania; ruta@lgi.lt

The Lithuanian Heavy Draught (LHD) breed began to emerge at the end of the 19th century. Native mares were crossbred with Brabant stallions, Percheron stallions, Ardennes stallions. Ardennes brought from Sweden had the biggest impact on the formation of the breed. Currently, LHD horses are experiencing a fast decline. The breed is recognised as protected. The genealogical analysis of LHD stallion progeny indicated that there might be found four stallions lines. The narrow genealogy and condition of some stallions does not satisfy the horse breeders. Therefore LHD Breeders' Association, in 1998 and 2000, brought Ardennes stallions to Lithuania from Sweden. The aim of our investigation was to evaluate the changes in the population after introduction foreign stallions chosen for the development of the breed in order to stop the disappearing of the genealogical structure of the LHD breed. The method of our study was to determine the genetic diversity within the breed, as well as genetic differences and genetic variation between LHD population structural units by the method of blood group and protein polymorphism investigation. The pedigree, exterior measurements and blood samples were collected from 439 horses. The influence of selected foreign stallions on LHD population was defined unlike. The offspring's of Flatentas stallion of Swedish Ardennes breed are of a desirable type and body conformation. The selected typical half-bred stallions will be included in the general programme for Lithuanian Heavy Draught horse breeding. The influence of selected stallion Flatentas on population genetic diversity was detected low and helps to improve some phenotypic traits of purebred LHD. The stallion Flatentas is characterized as most genetically and phenotypically suitable to extend genealogical structure of LHD population. But the proportion of foreign genes which are introduced in to the population can not be more than 10%.

Session 14 Poster 14

Effect on meat production of F1 crossbreds resulted from Alpine breed(♂) x Albanian local goat breeds(♀)

Kume, K. and Hajno, L., Centre for Agricultural Technology Transfer, Animal production, Rinas, Fushe Kruja, Kruje, Albania; kkume@icc-al.org

The local goat populations are the most important ones within genetic patrimony of Albania. These populations exhibit high level of diversity. Local breeds of goat are very well adapted to harsh local environmental conditions. They can survive and produce in restricted feeding and adverse living conditions. Referring to goat population size, they can be classified in two groups: (i)local breeds at risk of extinction farmed in small scale farms (ii)commercial local breeds. Within Albanian National Action Plan of conservation and sustainable use of Farm Animal Genetic Resources, two programs of genetic improvement of commercial local goat breeds have been compiled:(a)the massive selection program, (b)the program of cross breeding with exotic breeds. Crosses F1(Alpine breed ♂ x Local breed ♀)are the most frequently in Albania. Statistical data analysis of about 4080 kid`s live weights at the age of 0-6 months old, out of which 1560 local breed, 1424 F1, 1096 Alpine breed, showed that, the cross breeding improves the growth performance of local breed. ANOVA test shows a significant effect on kids` live weights of both genotype ($P<0.01$) and environment ($P<0.05$, 0.01). The averages kids live weight at birth (BW), at weaning (WW) and at the age of 6 months (6MW) of local breed, F1 and Alpine breed, were respectively 2.29a, 3.12b, 3.19b kg (BW), 10.73a, 13.21b, 12.68c kg (WW) and 21.04a, 26.13b, 24.84c kg (6MW). The differences between average kids live weight at six months (6MW) showed that the heterosis power is present ($P<0.01$). The Gompertz`s model of kids` growth of three genotype was estimated. ANOVA test for three Gompertz`s model parameters showed a significant effect of genotype ($P<0.001$), sex ($P<0.01$) and birth mode ($P<0.05$) The crossbreeding of Alpine goat breed with Albanian local goat breed is an opportunity to add value to local breeds. Amelioration of meat performance of local goat breed, increases up to 20 percent the annual income of farm.

Morphology characterization, health and dairy production evaluation in native Garfagnina goat

Corrias, F.[1], Salari, F.[2], Dal Prà, A.[1], Ragona, G.[1], Lombardo, A.[1], Mari, M.[1], Altomonte, I.[2], Colombani, G.[3], Pedri, P.[4], Scotti, B.[5], Brajon, G.[1] and Martini, M.[2], [1]I.Z.S. Regioni Lazio e Toscana, Via di Castelpulci, 50010, Firenze, Italy, [2]Università di Pisa, D.P.A., Viale delle Piagge 2, 56124, Pisa, Italy, [3]U.S.L. 2 Lucca, Via per S. Alessio, 55100, Monte San Quirico, Italy, [4]U.S.L. 1 Massa e Carrara, Via Don Minzoni, 54033,Carrara, Italy, [5]U.S.L. 12 Viareggio, Via Aurelia 335, 55041,Lido di Camaiore, Italy; mmartini@vet.unipi.it

The study was carried out on a goat population named Garfagnina from north-west Tuscany with the aim to evaluate breeding, morphology and health status in an estimated population of 2500 animals. Data were collected from 31 farms. Morphometric measurements revealed that the animals were homogeneous in terms of zoometric data while they shown a variability regarding coat. The milk gross composition was similar to that reported in literature. The lower average of SCC and SPC indicates good hygienic farm management and correct milking practices although milking is mainly manual. Milk coagulation was low, this suggests that Garfagnina goat milk could be used for direct consumption. The fatty acid composition confirm that C10:0, C14:0, C16:0, C18:0 and C18:1 account for more than 75% of total fatty acids in goat milk and that CLA is approximately 60% of the value in cow's and ewe's milk. A normal ecto-parasitism is present without sign of illness or scrape; a good fecal-score was recorded according to the low number of eimeria eggs/g found on some single fecal sample, strongylus spp. in contrast are always present on coprocultures of mass fecal sample. According to national rules, all the tested animals were Brucellosis free. Therefore in these 'rural breeding', all the animals were clinically healthy and serodiagnosis showed ($P<0.05$) 6.58% seroprevalence of CAEV, 5.26% of Paratuberculosis, 0.5% of Tularemia. The sanitary condition of the population reassures on the healthy status of the animals: sporadic positiveness seems to be correlated with wild ungulates in the same area.

Identification of established genetic variants associated with milk traits

Orford, M.[1], Tzamaloukas, O.[1], Papachristoforou, C.[1], Koumas, A.[2], Hadjipavlou, G.[2] and Miltiadou, D.[1], [1]Cyprus University of Technology, Department of Agricultural Sciences, Biotechnology and Food Science, P.O. Box 50329, 3603, Lemesos, Cyprus, [2]Agricultural Research Institute, P.O. Box 2016, 1516 Lefkosia, Cyprus; georgiah@arinet.ari.gov.cy

The present study investigated the genetic diversity of small ruminant breeds of Cyprus, regarding β-lactoglobulin (β-LG) genotypes and the existence of bovine single nucleotide polymorphisms (SNPs). In total, 366 animals belonging to the Chios and Cyprus fat-tailed sheep breeds, and to the Damascus and Machaeras goat breeds, were genotyped for the identification of the most common β-lactoglobulin variants (A and B), as well as for the presence of the growth hormone receptor (GHR) F279Y and the acylCoA:diacylglycerol acyltransferase 1 (DGAT1) K232A SNPs. With regard to β-lactoglobulin genotypes, two genetic variants (A and B) were identified for the sheep breeds and only one variant (A) for the goat breeds. The results showed that variant B of β-LG gene was not present in any caprine sample and therefore a fixed A-allelic genotype was suggested for both goat breeds. However, the present study showed that the Cyprus fat-tailed sheep was predominantly of the β-LG B type demonstrating significant differences in allelic frequencies ($P<0.001$) and genotypic distributions ($P<0.05$) compared to Chios sheep, revealing unusually high distribution of the BB genotype (38.5% of the total genotypes) as compared to Chios (4.5%) or other Mediterranean sheep breeds. With regard to bovine SNPs, the present study investigated the existence of GHR F279Y and the DGAT1 K232A SNPs, which have both been previously well documented in cattle as having strong effects on milk yield and composition. Although we were able to confirm the presence of both of these mutations in bovine sample controls by both allele specific PCR reactions and direct DNA sequencing, we were unable to detect them in all four major pure breeds of sheep and goats supporting the small ruminant dairy industry in Cyprus.

Power analysis of a population assignment test via SNP and STR markers with a view to breed authentication of sheep meat from native Southern Italy breeds

Cecchi, F.[1], Ciani, E.[2], Bramante, A.[1], Castellana, E.[2], D'Andrea, M.[3], Occidente, M.[4], Incoronato, C.[4], D'Angelo, F.[5] and Ciampolini, R.[1], [1]UNIPI, Animal Production Dept., Viale Piagge 2, 56124 Pisa, Italy, [2]UNIBA, General & Environm. Physiol. Dept., Via Amendola 165/a, 70126 Bari, Italy, [3]UNIMOL, Animal, Plant & Environm. Sciences Dept., Via F. de Sanctis, 86100 Campobasso, Italy, [4]ConSDABI, Loc. Piano Cappelle, 82100 Benevento, Italy, [5]UNIFG, Dept. of Prod. Sciences, Engin. & Economics for Agricult. Systems, Via Napoli 25, 71100 Foggia, Italy; elenaciani@biologia.uniba.it

Local sheep breeds from Southern Italy have suffered in the last decades a severe population size decline due to replacement with selected breeds. Presently, interest in preserving these resources has grown up and attempts are being made to encourage farmers to rear native sheep breeds. Authentication and verification of the breed origin of sheep meat may represent a valuable tool of commercial valorization. This study aims to compare the statistical power in assigning sheep samples to their true breed by using molecular information at SNP or STR loci. A total of 739 individuals, representative of two Italian sheep breeds (Sarda and Comisana) prevalent on the national lamb market and five local rare sheep breeds from Southern Italy (Bagnolese, Laticauda, Gentile di Puglia, Altamurana, Leccese) were typed at 19 STR and 104 SNP loci. A maximum likelihood-based assignment test was adopted to evaluate the proportion of correct breed allocation. STR markers performed better than SNP markers in all the seven breeds (with an average percentage of correct allocation equal to 99.6±0.5 and 85.9±15.5, respectively). Despite the low genetic differentiation among the considered breeds (overall F_{ST}=0.049), results suggest that there is enough room to optimize a molecular tool able to discriminate among the local sheep breeds from Southern Italy and the two national selected breeds.

Melanocortin 1 receptor (MC1R) gene polymorphisms in Modicana and Sardo-Modicana Italian cattle breeds

Guastella, A.M.[1], Sorbolini, S.[2], Zuccaro, A.[1], Pintus, E.[2], Macciotta, N.P.[2] and Marletta, D.[1], [1]University of Catania, DACPA Sez. di Scienze delle Produzioni Animali, via Valdisavoia. 5, 95123 Catania, Italy, [2]University of Sassary, Dipartimento di Scienze Zootecniche, via De Nicola, 9, 07100 Sassari, Italy; d.marletta@unict.it

Coat colour in mammals is a distinctive trait of selected breeds. It depends on the quantities and distribution of two pigment types, eumelanin (black/brown) and pheomelanin (red/yellow) and is primarily controlled by the Extension (E) locus that encodes for the Melanocortin 1 receptor (MC1R). In cattle four common alleles are present at this locus and most of the genetic variation is distributed between breeds. This particular genetic structure has been suited for the development of molecular protocols for genetic breed traceability. Modicana (MO) and Sardo-Modicana (SM) are, respectively, a Sicilian and a Sardinian endangered local cattle populations. The increasing market demand for typical products, has recently renewed the interest for cheese and meat produced by these breeds and in consequence for traceability protocols to guarantee the consumers and protect breeders. In this work PCR-RFLP, PCR-APLP methods and sequence analysis were performed to screen MO and SM cattle breeds for polymorphisms at the MC1R gene. Four alleles were observed in these breeds with different distribution. E+ and E1 allele were observed at good frequency in both cattle breeds (0.42 and 0.57 in MO; 0.52 and 0.47 in SM). The recessive allele e, that cause red coat colour, was identified with very low frequency (0.01). Finally a rare allele, characterized by a C667T transition was detected in both these breeds. In conclusion red coat in these two indigenous breed is genetically determined by the E+ and E1 alleles and is not associated with the homozygote genotype for the recesive e allele, as occurs in other red Italian and French breeds. These results indicate that MC1R polymorphism is not enough informative in MO and SM and cannot be an effective marker for the development of traceability protocols for their mono-breed products.

Donkey population in Montenegro and their exterior characteristics

Markovic, B. and Markovic, M., University of Montenegro, Biotechnical faculty, Department of Livestock science, Mihaila Lalica 1, 20000 Podgorica, Montenegro; bmarkovic@t-com.me

Donkey's population has been rearing during centuries in coastal and carst part of Montenegro, mainly in Ulcinj, Bar, Cetinje, Podgorica and parts of Danilovgrad municipality. Adaptability to the severe conditions, strengths, resistance and modest requirements in nutrition have favored donkey comparing to a horse in those regions. Together with intensification of agriculture and depopulation remote rural areas, economic importance of donkey is decreasing, thus its population has been drastically reduced. Based on the estimation, donkey's population in Montenegro count approximately 1500 to 2000 animals. Since there was no interest for research expressed, relevant scientific facts on donkey's population do not exist. The aim of this research is to determine basic exterior characteristics of the donkey's population in Montenegro. Based on the investigations conducted in the main rearing areas, donkeys of grey to brown color prevail in the total population, often with black rings around lower part of the legs, and occasionally a black dorsal stripe. Animals with dark to black pigmentation of hair covering are much lower presented, while animals with white pigmentation of hair covering are very rare. Measuring 65 adult animals (45 males and 20 females), average values for the main body traits were determined as follows: withers height 98,6 cm; chest depth 42,6 cm; body length 103,2; chest width 24,6 cm; chest circumference 117,9 and canon bone circumference 13,3 cm. Differences between male and female head were not significant ($P>0.05$).

Session 14 Poster 20

Exploring the risk factors to the heritage sheep breeds using multivariate analysis

Ligda, C.[1], Kotsaftiki, A.[2], Carson, A.[3] and Georgoudis, A.[2], [1]National Agricultural Research Foundation, P.O. Box 60 458, 57 001 Thessaloniki, Greece, [2]Aristotle University of Thessaloniki, Dept of Animal Production, 57 124 Thessaloniki, Greece, [3]The Sheep Trust, P.O. Box 373, York YO10 5YW, United Kingdom; chligda@otenet.gr

The aim of the present work was to assess the impact of the different threats to the heritage sheep breeds. The data were collected in the frame of the HERITAGESHEEP project funded under the EU Regulation 870/04 and referred to the social, political and environmental factors that influence the future of the sheep breeds. In total information from 45 sheep breeds from 5 European countries are represented in this data set. The collected data correspond to the breed societies' perceptions on the impact of the different threats to the breeds, ranking them from 5 to 1, according to their importance. After preliminary analysis, 18 variables corresponding to the 3 categories of the questionnaire as social, political and environmental were included in the principal components analysis. Two principal components were extracted in each category, explaining the 63% to 67% of the total variance of the category. In the group of the social threats, the first component is highly correlated with the ageing of the farmers and the ceasing of farming and the second with issues linked with the lack of skills and the urbanisation. The first component of the political factors is correlated with the impact of diversification and changing to non farming uses, while the second with the removal of headage payments and the application of environmental schemes. Finally, the two components of the environmental category were grouped as climatic and management impacts. In the next step, using hierarchical clustering two clusters of breeds were formed according to the six new variables. The results show significant differences between the participated countries on the threat factors of the sheep breeds.

Session 15 Theatre 1

Researches in progress in reproductive physiology to increase the efficiency of sheep production

Folch, J., Lahoz, B. and Alabart, J.L., CITA de Aragón, Av. Montañana 930, 50059 Zaragoza, Spain; jfolch@aragon.es

Age at first lambing, prolificacy and seasonal anoestrous, limit the profitability in the ovine meat production. In some breeds, single-genes mutations can be used to enhance prolificacy, but comparable mutation affecting seasonal sensibility have no been described yet. Genetic progress in polygenic selection for reproductive criteria is slow and complicate to apply in accelerated reproductive systems. Hopeful new strategies of selection use physiological parameters associated to productivity, such as the 'calm temperament' which is indirectly associated to ovulation rate, sexual behaviour and other physiological parameters. The presence of Progesterone in spring can also be an indicative parameter to detect non-seasonal animals, at least in Mediterranean breeds, where spontaneous oestrus and ovulation occur during non-seasonal season in some ewes. Breed and individual differences also exist in the facility to conceive by artificial insemination. The reasons of these differences are being studied at present. Research should be impelled to study parameters associated to the reproductive potential of the ewes, especially if they can be recorded in ewe-lamb at replacement. Male effect is a natural method to induce reproduction in anoestrous period. In Mediterranean breeds similar levels of fertility can be achieved all the year round by male effect, but the ewe response to the male stimulus depends on the nutrition level. More basic research should be impelled to clarify these interactions and more trials must be done to determine the strategies to be used in the different conditions in the field. Nutrition affects all the reproductive process. The effects of the supplementary feeding applied on the different steps of the reproductive process as well as the direct effect of some specific nutrients on reproductive performances are just now starting to be known. The application of these knowledges to sheep production will improve the efficiency in sheep reproduction, but a lot of research is still needed.

Session 15 Theatre 2

Photoperiod and socio-sexual relationships can be used to develop sustainable breeding techniques for goat reproduction

Delgadillo, J.A., Flores, J.A., Duarte, G., Vielma, J., Hernández, H. and Fernández, I.G., Universidad Autónoma Agraria Antonio Narro, Ciencias Médico Veterinarias, Periférico Raúl López Sánchez y Carretera a Santa Fe, 27054 Torreón, Coahuila, Mexico; joaldesa@yahoo.com

Reproductive seasonality observed in goat breeds causes a seasonality of milk and meat production. To extend the availability of these products all year round, some animals must bred during the anoestrous period. The 'male effect' is a biostimulation technique that induces and synchronizes the sexual activity in does during the anoestrous period. However, the major limitation of the 'male effect' is that the sexual response of does is weak or absent when it is performed during the mid-seasonal anoestrous probably because at this time males are also out-of-season breeding and display low sexual behaviour. The stimulation of the male sexual activity during the out-of-season breeding by exposure to a photoperiodic treatment improves the quality of the male signals (e.g. odor, sexual behaviour, vocalizations), increasing the proportion of females responding to the 'male effect'. The availability of sexually active males during the out-of-season breeding increases the opportunity to use the 'male effect' under different management systems conditions (i.e. intensive or extensive) resulting in a higher percentage of stimulated females, even under circumstances where using this biostimulation technique could be complicated. Combination of photoperiod and 'male effect' allows developing sustainable breeding techniques to induce the sexual activity of female goats during the seasonal anoestrous.

Effect of body condition on response of male goats to artificial long-days treatment

Flores, J.A., Lemiere, A., Secundino, S., Hernández, H., Duarte, G., Vielma, J. and Delgadillo, J.A., Universidad Autónoma Agraria Antonio Narro, CIRCA, Periférico Raúl López Sánchez, 27054, Mexico; flores_cabrera@hotmail.com

In well-nourished males in subtropical Mexico, the non-breeding season last from January to April. In these males subjected to 2.5 months of artificial long days, sexual activity is stimulated during the non-breeding season. Nutrition can influence their response to photoperiodic treatment. This study determined the sexual response of undernourished bucks to photoperiod treatment. Two groups of males (n=7) were exposed to 2.5 months of long days (16 h of light/day) starting in November 1. The undernourished bucks fed 0.5 maintenance requirements and had a body condition score (SBC) of 1.5±0.1, while the well-nourished bucks fed 1.5 maintenance requirements and had a BCS of 3.0±0.1. Scrotal circumference (SC) and sexual odour intensity were determined every 2 week from November 1st until May 30. Photoperiodic treatment stimulated an increase of SC and odour intensity during the non-breeding season in both groups of males. ANOVA revealed an effect of time in scrotal circumference ($P<0.001$) and odour intensity ($P<0.001$) and an interaction between group and time of experiment ($P<0.001$), in both variables. However, the response of undernourished males was delayed and lower than in well-nourished males. In well-nourished males, SC started to increase progressively from January 15 (25.0±0.2 cm) and peaked on March 30 (28.8±0.8 cm). Contrary, in undernourished males the SC started to increase until February 15 (23.0±0.5 cm) and peaked on April 30 (26.9±0.7 cm). In well-nourished male goats the odour intensity (score 0-4) increased from February 28 (0.86±0.1) and peaked on April 15 (2.0±0), while under-nourished males odor intensity increased from March 30 (0.3±0.1) and picked on April 30 (0.6±0.2). In conclusion, long day treatment stimulated sexual activity during the non-breeding season in nourished male goats; however, the response was delayed and lower than in well-nourished males.

Analysis of the interactions between productive, nutritive and reproductive parameters from an experimental flock of Latxa dairy sheep

Diez-Unquera, B., Beltrán De Heredia, I., Arranz, J., Amenabar, M.E., Mandaluniz, N., Ugarte, E. and Ruiz, R., NEIKER-Tecnalia, P.O. Box 46, E-01080 Vitoria-Gasteiz, Spain; bdiez@neiker.net

The efficiency of reproductive management is crucial for the economic profitability of livestock farming due to its direct impact on the associated costs and incomes, but also in relation to planning the productive schedule and labour requirements. Moreover, the distribution of lambings throughout the year exerts a significant effect upon milk yield and milk quality of the bulk tank in dairy sheep. In the case of the Latxa sheep, cervical artificial insemination (AI) with fresh semen is used for the purposes of the breeding programme since 1985. Average conception rates range between 41-56%, and due to the large variations observed between flocks and between years, farmers are really interested in improving fertility. However, the success of AI depends on a wide range of factors, most of which are difficult to be monitored under commercial conditions, such as the interactions with the feeding management and the nutritional status of the sheep. The experimental dairy flock managed in NEIKER-Tecnalia for the last 17 years has allowed recording an important amount of data, which analysis can provide interesting information for the local sector. The objective of this work was to assess the effect of the previous lambing and milk yields, body condition score (BCS) and flushing upon fertility and prolificity of Latxa sheep, as well as to parameterise the chances of reproductive success according to these factors. A database consisting of 1972 sheep inseminated between 1996 and 2008 was analysed. The BCS assessed 3 weeks prior to insemination, which was related to the age and milk yield during the previous season, as well as the evolution of BCS during the following 6 weeks exerted a significant effect on fertility and prolificity. Flushing positively affected BCS evolution and reproductive indexes. The equations for modelling the pattern of reproductive success at AI in Latxa are proposed.

Longer time to oestrus in ewes fed a high omega-3 diet

Gulliver, C.E.[1,2], Friend, M.A.[1,2], King, B.J.[1,2], Robertson, S.M.[1,2] and Clayton, E.H.[1], [1]EH Graham Centre for Agricultural Innovation, I & I NSW and Charles Sturt University, Wagga Wagga, NSW 2678, Australia, [2]Future Farm Industries CRC, The University of Western Australia, Crawley, WA 6009, Australia; cgulliver@csu.edu.au

Diets high in omega-3 (n-3) may affect ewe reproduction. Conserved forages, such as cereal silages, contain high concentrations of n-3 compared with cereal grains. The aim of the study was to determine the effect of feeding high n-3 silage on time to oestrus and ovulation rate (OR) in crossbred ewes. Thirty Merino x Border Leicester ewes were randomly allocated to one of two treatment groups according to their condition score (3.5±0.03) and liveweight (86.8±1.00 kg). Ewes were fed either a low n-3 diet (70% oat grain, 22% oat/pea silage, 8% cottonseed meal; n=15) or a high n-3 diet (100% oat/pea silage; n=15) together with a commercial vitamin/mineral premix. Ewes were housed in individual pens and fed at 1.2 x maintenance for 4 weeks prior to and 1 week post-mating. All ewes had their oestrous cycles synchronised with an injection of $PGF_{2\alpha}$ and an Eazibreed® Sheep CIDR. Oestrus was detected over a 5 day period from crayon marks by harnessed rams at the first natural oestrus after synchronisation. OR was measured over two oestrus cycles via transrectal ultrasound. Differences in time to oestrus and OR between groups were analysed using the Mixed model procedure and the proportion of ewes showing oestrus was analysed using proportional hazards regression analysis in the SAS program. Time to oestrus tended to be longer (4.4±0.89 vs 3.0±0.83 days; P=0.073) when ewes were fed a high n-3 diet compared with a low n-3 diet. OR was not significantly different when ewes were fed a diet high in n-3 (P=0.259). The longer time to oestrus in ewes fed the high n-3 diet may be related to decreased *in vivo* synthesis of inflammatory prostaglandins involved in the initiation of oestrus and ovulation. Further analyses will examine changes in $PGF_{2\alpha}$ synthesis over time and the relationship between $PGF_{2\alpha}$ and fatty acid profiles in plasma and red blood cells.

Effects of different photoperiod regimes on melatonin secretion pattern in seasonally anestrous Chios ewes

Abooie, F.[1], Sadeghipanah, H.[2], Zare Shahneh, A.[3], Mirhadi, A.[2] and Babaei, M.[2], [1]Islamic Azad University of Ghaemshahr, Department of Animal Science, Ghaemshahr, p. box: 163, Iran, [2]Animal Science Research Institute, Department of Animal Production and Management, Shahid Beheshti Street, Karaj, 3146618361, Iran, [3]University of Tehran, Department of Animal Science, Karaj, 3146618361, Iran; hassansadeghipanah@yahoo.com

In order to investigate the effect of two photoperiod regimes (short day & long day) on melatonin secretion pattern in seasonally anestrous Chios ewes, circadian variations of serum melatonin concentrations were determined during non-breeding season. In spring, eight 3-year-old non-pregnant ewes were allocated into two groups on the basis of photoperiodic patterns: 1- natural long day photoperiod group (LD); 2- artificial short day photoperiod group (SD). Artificial photoperiod program in SD group was started on April 12. On June 18, jugular veins of ewes were catheterized. From 12:00 June 21 to 12:00 June 22 (summer solstice), blood samples were consecutively collected at 2 h interval. On summer solstice, day lengths were 15 h 35 min (from 5:20 to 20:55) and 8 h (from 10:00 to 18:00), respectively in LD and SD groups. Serum melatonin concentrations were measured by ELIZA kit (IBL Company, Germany; RE54021). Statistical analysis carried out using SPSS software. The melatonin concentration in SD group showed an increase from 18 o'clock and reached its peak at 2 o'clock (117.31±12.59 pg/ml) and then was decreased. However, in the LD group, the melatonin concentration increased after 20 o'clock and reached its peak at 24 o'clock (99.95 46.28 pg/ml) and then was decreased. The total melatonin secretion throughout the 24 h period in the SD group (9964.08 pg/ml) was significantly (P=0.005) greater than LD group (6043.59 pg/ml). Totally, these results indicated that artificial short day photoperiod regime used current study can change melatonin secretion pattern (as a stimulant of hypothalamus-hypothesis-ovary axis) in seasonally anestrous Chios ewes.

Effect of artificial photoperiod on reproductive performance of seasonally anestrous Chios ewes

Sadeghipanah, H.[1], Abooie, F.[2], Zare Shahneh, A.[3], Aghashahi, A.L.[1], Asadzadeh, N.[1], Papi, N.[1], Mahdavi, A.[1], Mafakheri, S.[4] and Pahlevan Afshar, K.[5], [1]Animal Science Research Institute of Iran, Department of Animal Production and Management, Shahid Beheshti Street,Karaj, 3146618361, Iran, [2]Islamic Azad University of Ghaemshahr, Department of Animal Scienc, Ghaemshahr, p box: 163, Iran, [3]University of Tehran, Department of Animal Science, Karaj, 3146618361, Iran, [4]Kurdestan Research Center of Agricultural and Natural Resources, Department of Animal Science, Sanandaj, 6616936311, Iran, [5]Islamic Azad University of Abhar, Department of Animal Scienc, Abhar, 45615-1333, Iran; hassansadeghipanah@yahoo.com

In order to investigate the effect of artificial photoperiod on reproductive performance in seasonally anestrous ewes, 40 non-pregnant Chios ewes in spring were allocated into two groups on the basis of photoperiodic patterns: 1- natural long day photoperiod group (LD) 2- artificial short day photoperiod group (SD). On April 12, ewes In SD group were firstly exposed to 10 long days (15 h and 35 min lightness and 8 h and 25 min darkness; equal to local summer solstice) then in next 60 days, were gradually followed by exposure to short days (8L:16D) using light-sealed rooms; this treatment is referred to as LD→SD. Ten days after the 8L:16D program was fixed for the SD group, rams were introduced to both groups suddenly and simultaneously. Interval from ram introduction to first oestrus was shorter and estrus rate was greater in the SD group than in the LD group ($P<0.05$). Conception rate, parturition rate, prolificacy and fecundity in the SD group were nonsignificantly ($P>0.05$) better than in the LD group. The lamb crop at both lambing and weaning times, in SD group (respectively, 3.15 and 15.63 kg) was greater than the in LD group (respectively, 2.27 and 9.83 kg). Totally, these results indicated that artificial short day photoperiod used in this experiment, can improve reproductive performance in seasonally anestrous Chios ewes.

Characteristics of ram semen collected from Boujaâd sheep breed in Morocco under fresh conservation

Talbi, H.[1,2], El Amiri, B.[2], Derqaoui, L.[3], El Bennani, M.[3], Hilali, A.[1] and Druart, X.[4], [1]Faculty of Science and Technical Hassan 1er– Settat, Biologie, 577, Route de casa, Settat, 26 000, Morocco, [2]INRA, Regional Center for Agricultural Research, Settat, Productions Animales, 589, Settat, 26 000, Morocco, [3]Hassan II Agronomic and et Veterinary Institut Rabat, Reproduction, 6202, Madinate Al Irfan, Rabat, 10101, Morocco, [4]INRA, Reproductive Physiology and Behavior, Centre de recherche de Tours, Nouzilly, 37380, France; bouchraelamiri@hotmail.com

Artificial insemination (AI) is one of reproductive biotechnologies that has most contributed to the genetic improvement of different animal species. It is not a single act but a set of complementary techniques (semen production and storage, heat synchronization, insemination conditions). The technique of semen production and conservation is one of the main components of all these procedures. However, in Morocco the whole package of AI is not well known. In addition, the short life span of fresh semen is a major constraint on the use of AI in genetic improvement programs for sheep. Thus, the present study aims to investigate the Boujaâd breed ram fresh sperm conserved in skim milk or in Tris/egg yolk extender at different temperatures (4 °C, 37 °C or room temperature ranging from 20 to 25 °C) and different hours of storage (0, 2, 4, 8, 16, 24, 48, 72, 86, 120, 144 hours). For this, a flock of 9 rams (4 light and 5 heavy) were trained and collected every 15 days from March to May 2009. The main findings of the present experiment are: fresh storage of ram semen at 4 °C is superior to the storage at 37 °C and room temperature, Tris/egg yolk allows a better conservation than skim milk, the semen of light rams has the best ability to be conserved than the heavy rams semen. To conclude, when combining 4 °C temperature and Tris/egg yolk extender we obtain the best mobility. This study is on progress and the next steps will be the test of other temperatures during the breeding and non breeding season and the evaluation of fertilizing ability of fresh ram in these conservation conditions.

Seasonal variations in testicular size and semen quality of Boujaâd ram in Morocco

Derqaoui, L.[1], El Amiri, B.[2], Talbi, H.[3], Druart, X.[4], El Bennani, M.[1] and Hadrbach, M.[2], [1]IAV Hassan II, Rabat, 10101, Morocco, [2]INRA, Settat, 00000, Morocco, ^{3}H, Talbi, FST, 00000, Morocco, [4]INRA, PRC, Nouzilly, 37000, France; lderqaoui@yahoo.fr

Most of the native Moroccan sheep breeds namely Timahdite, Sardi, Beni Guil, Beni Hsen, and Boujaâd are seasonal breeders except the D'man. The objective of the current study is to evaluate the effect of season on testicular size and semen quality in Boujaâd rams. Nine mature rams kept in entire confinement were initially used from January through December after a six-month period of adaptation. Animals were fed wheat straw, barely, alfalfa and concentrate. Data was collected every other week. The major testicular parameters measured were scrotal circumference (SC) using a meter band, testicular length (TL) and diameter (TD) and epididymal tail diameter (ED) using a calliper. Semen was collected using an artificial vagina to evaluate volume (V) of ejaculate, sperm mass motility (MM), individual motility (IM) and concentration (C). Data was analyzed using SAS with season as theh main source of variation. The main results showed that season had a significant effect on testicular size and semen quality. In fact, testicular size was maximal (SC=31.33 cm; TL=9.51 cm, TD=6.22 cm and ED=3.01 cm) during the beginning of the normal breeding season in sheep in Morocco (May June) as compared the period of decreased sexual activity (November-January). Mass motility and individual sperm motility changed according to season. In fact, maximum MM (5) and IM (92.78%) were recorded respectively in June-July and April-May. IM was at its minimum (67.78%) in January. Sperm volume and concentration showed irregular patterns. In conclusion, testicular size and semen quality did change significantly according to season. These changes should be considered during mating to optimize the overall flock fertility.

Persistency of ram's sperm fatty acid profiles after removing fat sources from diet

Esmaeili, V.[1,2], Shahverdi, A.H.[2], Alizadeh, A.R.[1] and Towhidi, A.[3], [1]Islamic Azad University, Saveh branch, Animal Science Department, Saveh, 39187/366, Iran, [2]Royan Institute, Department of Embryology, Tehran, 19395/4644, Iran, [3]University Of Tehran, Animal Science Department, Faculty of Agric. Sci. Eng., Karaj, 315877871/4111, Iran; vahid.esmaeeli@yahoo.com

Researches have indicated that mammalian spermatozoa is characterized by a high proportion of polyunsaturated fatty acids (PUFA), especially the n-3 and n-6 series. But reliable data concerning dietary effect on FA profile in ram's sperm and persistency of ration FA in FA of ram's sperm has not been reported. Our aim was to determine the stability of FA in ram's sperm despite removing FA sources from diet. Nine Kalkohi rams were randomly assigned to 3 groups at Sep 2009. The treatments were diet supplemented (35 g/d/ram) by C16:0 (RP-10®), C18:2 (sunflower oil (SO)) and n-3 (fish oil (FO)) with a constant level of Vit E. 15 weeks after the start of the supplemented diet, rams were offered a basal diet without any FA sources for 35 days when FA was determined by gas chromatography in the sperm of each ram. Major FA in sperm consists of: C14:0, C16:0, C18:0, C18:1 cis, C18:2 cis and C22:6 (DHA). C14:0 ($P=0.8$) as well as C18:1 cis ($P=0.4$) percentage were similar among the treatments. C16:0 percentage decreased as FO was added to the diet (26.18, 27.03 vs. 20.35% of total FA in RP-10, SO and FO, respectively; $P<0.01$). C18:0 percentage was significantly decreased in RP-10 compared with other groups (10 vs. 14.3 and 15.5% of total sperm FA in RP-10, SO and FO, respectively; $P<0.01$). Interestingly, C22:6 percentage was highest in FO treatment (8.5, 8.9 vs. 13.6% of total sperm FA in RP-10, SO and FO, respectively; $P<0.01$) after 36 d. The different sperm FA profile among various groups suggests that dietary FA had significant impacts on sperm FA profile after 36 d. The effectiveness of FO was proved by maintaining different DHA percentages 36 d after removing the DHA source from diet. Physic-chemical changes in sperm characteristics may occur pursuant to sperm's FA shift.

Searching for genes of interest in sheep

Moreno, C.[1], Elsen, J.M.[1], Legarra, A.[1], Rupp, R.[1], Bouix, J.[1], Barillet, F.[1], Palhière, I.[1], Larroque, H.[1], Allain, D.[1], François, D.[1], Robert, C.[1], Tosser-Klopp, G.[2], Bodin, L.[1] and Mulsant, P.[2], [1]INRA, UR631 SAGA, BP 52627, F-31320 Castanet, France, [2]INRA, UMR444 INRA ENVT Génétique cellulaire, BP52627, F-31320 Castanet, France; Loys.Bodin@toulouse.inra.fr

In the past, several studies have been performed to look for genes affecting traits of interest using microsatellite markers in sheep. Even if numerous chromosomal regions have been detected, few genes and causal mutations have been identified. The recent availability of high density SNP chip in sheep is a technological revolution for genomic researches and breeding selection. Firstly, genomic research in small ruminants will be boosted towards the identification of causal mutations underlying large genetic effects on sustainability traits, or the identification of very closely linked genetic markers that allow selecting genes in ovine breeding programmes. Secondly, such new molecular tools allow considering a new type of selection: the genomic selection. The principle of this selection is to use marker effects (without any information about the underlying genes) estimated within a part of the population which is phenotyped and genotyped (called the training population) and applying these effects to the rest of the genotyped population. However, the genomic selection might not always be a profitable strategy in sheep.

Session 15 Poster 12

Artificial long days increase milk yield in local goats from subtropical Mexico milked twice daily

Flores, M.J.[1], Flores, J.A.[1], Elizundia, J.A.[2], Delgadillo, J.A.[1] and Hernández, H.[1], [1]Centro de Investigación en Reproducción Caprina-Universidad Autónoma Agraria Antonio Narro, Ciencias Medico Veterinarias, Periférico Raúl López Sánchez y Carretera a Santa Fé, 27054 Torreón, Coahuila, Mexico., Mexico, [2]Proffesional Private Practice, Angel Camino 161, Col. Valle del Nazas, 35070 Gómez Palacio, Durango México, Mexico; mflores_najera@hotmail.com

In ruminants, such as cow and sheep artificial long days exposure during natural decreasing daylength is a practice management tool to increase milk yield throughout lactation. In female goat native of template regions, exposition to long daily photoperiod increase 10% milk yield. However, in subtropical Mexican goats (which are generally milked once daily) is not know if the artificial long photoperiod modify the milk yield when goats are milked twice a day. The study was conducted out in the Laguna region in the state of Coahuila, Mexico (26 °N). One group of female goats milked twice a day was kept under natural decreasing daylength (SD2X; n=8), whereas the other group was submitted to artificial long days (LD2X; n=7: 16 h light: 8 h darkness). Milk yield and its components (fat, protein and lactose) were assessed in three phases through a 140-days period of lactation. Data from each phase were statistically compared between groups by using independent t test. During the phase I (day 0 to 28 of lactation suckling phase) mean daily milk yield did not differ ($P>0.05$) between goats from SD2X (2.5±0.3 kg) and LD2X (2.6±0.2 kg) groups. Nevertheless, in the phase II (day 29 to 84 of lactation early milking) milk yield was greater ($P<0.05$) in LD2X group (3.3±0.2 kg) than in SD2X group (2.8±0.2 kg), whereas during phase III (day 85 to 140 of lactation late milking) milk yield was not different ($P>0.05$) between the two groups. The milk composition was similar between the two groups during each phase. We concluded that in subtropical goats kidding during natural short days the exposition to artificial long days increase milk yield until 18% during early lactation when they were milked twice a day.

Sonorous emissions and sexual odor from male goats stimulates the ovarian response in anovulatory female goats

Vielma, J., Hernández, H., Ramírez, S., Flores, J.A., Duarte, G. and Delgadillo, J.A., Universidad Autónoma Agraria Antonio Narro, Centro de Investigación en Reproducción Caprina, Periférico Raúl López Sánchez SN, Torreón, Coahuila, 27054, Mexico; jesus_vielm@hotmail.com

The male effect is a multifactorial phenomenon. This study was carried out to determine if restricted contact between males and female goats induces ovarian response. A group of males (n=7) was put under artificial long days (16 h light/day) from November 1 to January 15 to stimulate his sexual activity during the natural period of sexual repose. Another group of males (n=7) was kept perceiving the natural variations of local photoperiod, with purpose that they were in sexual repose during the experiment. A male of each photoperiodic treatment was used. On June15, a group of females (n=10) was submitted in a restricted way to one sexually active male. In order to restrict the contact between the sexually active male and the females of the corresponding group, a solid barrier with wood of 2 m of height was interposed between male and females, this barrier allowed the auditory and olfactory communication but it prevents the visual and physical contact. Another group (n=10), located to 100 m of the previous one, was submitted to one male in sexual repose. The male in sexual repose was in full physical contact with female goats. In both groups, the respective male remained with the females during 18 d. The ovarian activity of the females was determined by transrectal ecography to 18 d of initiate the contact with males. The percentage of females with ovarian response was analyzed by chi square test. Proportion of goats that displayed ovarian activity was greater ($P<0.05$) in the group of females submitted to the restricted contact with sexually active male (4/10; 40%) that in the group of females submitted to full physical contact with male in sexual repose (0/10; 0%). We conclude that the sonorous emissions and the sexual odor from males, acting together, inducing an ovarian response weak in seasonally anovulatory female goats.

Session 15 Poster 14

In subtropical goats, a nursing/milking mixed management during the first 60 days postpartum delay the recovery of postpartum ovarian activity

Hernández, H., Ramírez, S., Flores, J.A., Duarte, G., Vielma, J. and Delgadillo, J.A., Universidad Autónoma Agraria Antonio Narro, CIRCA, Periférico Raúl López Sánchez S/N, Torreón Coahuila, 27054, Mexico; hernandezhoracio@hotmail.com

In subtropical regions some goat intensive producers applies during first 50-60 days postpartum a nursing/milking mixed management to the female goats with the aim to obtain a high weights of the kids at weaning and to reduce the weaning stress. However, limited information is available on how this management can affect the postpartum anoestrus length in these females. The objective of the present study was to determine if the nursing/suckling mixed management may reduce the postpartum anovulation period as compared with goats which kids were early weaned. After first week postpartum, 14 goats were assigned to a mixed management consistent in three 30-min nursing periods/day and milking twice daily (morning-afternoon; this was the mixed management group; MM) up to day 60 postpartum (end of the study). In the other 15 goats their kids were weaned at first week postpartum and then the mothers were submitted to milking twice daily (weaning group; W) until the end of the study. In all goats, ovarian activity was monitored after day 15 postpartum and from then on each two weeks by means of transrectal utrasonography. Proportion of goats that had ovulation was compared between groups by using chi square test. At 21 day postpartum none female goat from both groups had ovulation (0.0% in both groups; $P>0.05$). However, at day 50 postpartum a 100% of goats from W group had ovulated, compared with only 43% of goats from MM group ($P<0.05$). It was concluded that in subtropical goats under intensive conditions, the nursing/milking mixed management delay the recovery of postpartum ovarian activity compared with females goats which kids are early weaned.

Session 15 Poster 15

Effects of dietary fat on reproductive performance and blood metabolites of ewes

Alizadeh, A.R.[1], Azizi, F.[1], Karkodi, K.[1] and Ghoreishi, M.[2], [1]Islamic Azad University, Saveh branch, Animal Science Department, Saveh, 39187/366, Iran, [2]Jiroft Faculty of Agriculture,Shahid Bahonar University of Kerman, Animal Science Department, 7861887619, Iran; ali@ag.iut.ac.ir

Lipids have been used to increase energy density of the diet and FAs may have also direct positive effects on reproduction. It was hypothesis that FAs enhance progesterone concentration through increased cholesterol provision. To determine the effects of Megalac addition on ewes' reproductive parameters, a complete randomized design was used. Multiparous Kalkohi ewes (n=32; 3 years old) were randomly assigned to 2 groups which consumed isoenergetic and isonitrogenous control (C) or Megalac (M; 5% Megalac in diet DM) diet. All ewes were synchronized by prostaglandin injection in September. Ewes estrus was detected using teaser rams. Blood samples were collected from d 8, 10 and 12 of estrus cycle and ewes were bred via natural service and 3 estrus cycles were used for mating. Diets offered 4 wk prior to mating and up to 4 wk after mating. Pregnancy, lambing and twining rate were not significantly affected by fat inclusion in ewe diet ($P>0.05$). Similarly, Pregnancy length was unaltered by treatment ($P>0.05$). Cholesterol concentration significantly increased as Megalac added to the diet (52.7 vs. 49.7 mg/dl in M and C, respectively; $P<0.05$), whereas P4 concentration was similar between M and C group in d 8, 10 and 12 of estrus cycle (3.80 and 3.81 nmol/l for M and C, respectively; $P>0.05$). The similar reproduction parameters suggest that fatty acid content and profile of this level of Megalac may be insufficient for improving ewes' reproduction performance. The lack of change in P4 concentration alongside the increasing of blood cholesterol did not support cholesterol and P4 relation. Further studies are needed to determine impact of various fat sources and level on physiological responses of ewes.

Session 15 Poster 16

Induction and synchronization of estrus during anestrous season in North Moroccan goats using fluorogestone acetate vaginal sponges/eCG/cloprostenol or IMA-PRO2® protocols

Chentouf, M.[1], Molina, F.A.[2], Archa, B.[3] and Bister, J.L.[4], [1]INRA, Centre Régional de Tanger, 78, Bd Sidi Mohamed Ben Abdallah Tanger, 90010, Morocco, [2]IFAPA, Centro Hinojosa del Duque, Crtra del viso, Km 2, Cordoba, 14270, Spain, [3]ENA, Département de Production Animale, BP 40 Meknes, 50000, Morocco, [4]Université de Namur, Laboratoire de Physiologie Animale, 51, Rue de Bruxelles, Namur, B-5000, Belgium; mouad.chentouf@gmail.com

The efficiency of fluorogestone acetate (FGA) vaginal sponges/eCG/cloprostenol and IMA-PRO2® protocols for the induction and synchronization of estrus in North Moroccan goats was evaluated during seasonal anestrous. Twelve non cyclic does were randomly assigned to two treatments. Does in FGA group (n=6), were treated with intravaginal sponges containing 45 mg de FGA for 11 days followed by a single dose of 450 UI of eCG and 75 μg of cloprostenol 48 h prior to sponge removal. In IMA.PRO2 group (n=6), light and melatonin treated buck was used for the induction of the male effect, does received a single injection of 25 mg progesterone and 75 μg of cloprostenol 9 days after buck exposure. At the end of treatments, estrus was checked every 4 hours and blood samples were collected every 2 h during 24 h from heat detection for the determination of plasmatic level of LH. GLM Procedure of SAS was used to analyze the effect of theses protocols on the onset of estrus and LH peak. All does showed estrus and LH peak in FAG group and only 83% and 63% respectively in IMA.PRO2 group. In FGA group the onset of estrus and LH peak occurred sooner ($P<0.05$) in relation to the end of treatment (18.0±7.5 h and 30.0±5.5 h) compared to FGA group (32.4±7.4 h and 48.0±3.3 h). There were no differences ($P>0.05$) regarding the interval estrus - LH peak between FGA (12.0±5.6 h) and IMA.PRO2 (14.0±8.5 h) groups. In conclusion FGA protocol is more effective for estrus induction during non breeding season in North Moroccan goats than IMA.PRO2® but theses protocols allows the same level of synchronization of induced estrus and ovulation.

Seasonal variation of semen characteristics in Damascus buck goats

Pavlou, E., Michailidis, G. and Avdi, M., School of Agriculture, Department of Animal Production, Aristotle University of Thessaloniki, 54124, Greece; epavlou@agro.auth.gr

The aim of this study was to evaluate seasonal variations in ejaculate volume, sperm concentration, total number of spermatozoa per ejaculation, semen's mass and progressive motility, viability of spermatozoa and the percentage of abnormal spermatozoa in Damascus bucks. Semen characteristics were assessed in 10 mature bucks, raised on the same farm and under the same conditions, aged 8 months old and weight 43±3.8 kg at the beginning of the experiment. Semen collection and evaluation conducted once a week for a period of 12 months. The influence of seasonality was analyzed by Analysis of Variance, using General Linear Models. All bucks examined showed a similar pattern in all the reproductive parameters studied. Seasonal differences ($P<0.05$) were noted for all study parameters. Semen volume presented the highest value (1.02±0.018 ml) during the breeding season (autumn) and the lowest (0.7±0.017 ml) during the non-breeding season (spring). Mass and progressive motility as well as the viability of the semen also presented higher values (4.83±0.02 and 90±0.39%, respectively) during the breeding season and lower values (3.55±0.04 and 65.33±0.71%, respectively) during the non breeding season. In addition, semen concentration presented higher values during summer ((5.43±0.06)x109 spermatozoa/ml) and a decrease in values ((4.22±0.014)x109 spermatozoa/ml) during autumn. The total number of spermatozoa per ejaculation was also significantly higher during autumn ((5.08±0.07)x109) than in spring ((3.7±0.07)x109). The percentage of abnormal spermatozoa presented higher values outside of the natural breeding season (7.61±0.18%) and the lowest values during the breeding season (5.41±0.1%). The data presented in this study suggests that Damascus bucks semen quantity and quality exhibit a significant seasonal variation, with the best parameters obtained during the natural breeding season.

Involvement of Na+/H+ exchanger in intracellular pH regulation and motility in ram sperm cells

Guerra, L.[1], Ciani, E.[1], Silvestre, F.[2], Muzzachi, S.[1], Ferrara, M.T.[1], Castellana, E.[1], Guaricci, A.C.[2], Lacalandra, G.M.[2] and Casavola, V.[1], [1]University of Bari, General & Environmental Physiology Dept., Via Amendola 165/a, 70126 Bari, Italy, [2]University of Bari, Animal Production Dept., Str. Prov.le per Casamassima km 3, 70010, Italy; elenaciani@biologia.uniba.it

Alkalinization of the intracellular pH (pH_i) is critical to many aspects of sperm physiology. Many of these physiological processes involve a Na^+-dependent pH regulatory mechanism. Na^+/H^+ exchangers (NHEs) are transmembrane proteins mediating exchange of Na^+ and H^+ ions in various tissues. In mammalian sperm cells they seem to play a major role by determining the electroneutral transport of extracellular sodium for intracellular H^+. To date, ten NHE isoforms have been identified in mammals, of which three (NHE1, NHE5, sNHE) in sperm cells. We focused on NHE1 isoform in order to analyze the pH_i regulation and its role in sperm motility in the ovine species. Western blotting and immunofluorescence analysis revealed the presence of NHE1 using two different commercial antibodies. The protein localized mainly at the mid-piece and sub-equatorial head region. Functional characterization was performed using a pH_i-sensitive fluorescent dye (BCECF-AM) on sperm cell suspensions obtained after swim-up of fresh Altamurana ram semen. The rate of pH_i recovery from an NH_4Cl acid load resulted to be dependent on the extracellular $[Na^+]$. To confirm the involvement of NHE1 in the observed pH_i recovery we used two specific inhibitors (DMA and cariporide). Pre-incubation with both substances led to a marked reduction in the rate of pH_i recovery. Moreover, since it has been long accepted that pH_i is a vital regulator of spermatozoa motility, we tested, for the first time in the ovine species, the effect of NHE1 inhibition on sperm motility. Results from CASA showed a significant DMA and cariporide-dependent depression of sperm motility. It is therefore proposed that physiological function of NHE1 may be involved in modulation of pH_i with important consequences on sperm physiology.

Session 16 Theatre 1

Major causes affecting raw milk composition and its procession into curd in sheep and goats

Silanikove, N., Merin, U. and Leitner, G., Agr. Res. Org., Biology of lactation Lab., P.O. Box 6, 50 250 Bet Dagan, Israel; nsilanik@agri.huji.ac.il

In many countries, the vast majority of sheep and goats milk is processed into dairy products, particularly cheeses. Research in the last decade identified two major causes that affect negatively milk composition and its procession into curd; those are mastitis and milk produced in late lactation. Milk yield (MY) and milk gross composition are significantly affected by subclinical IMI in goats and sheep. In both species subclinical IMI was associated with increased plasmin (PL) activity in the infected glands. These changes were associated with a reduction in MY from the infected gland and with increased measures of casein (CN) degradation, casein degradation products (proteose peptones), decreased lactose, increased secretion of antimicrobial peptides (e.g., lactoferrin), acute phase proteins and increased activity of range of indigenous milk enzymes. These change, in turn are reflected by reduced curd yield and increased curd clotting time, indicating that the changes in milk composition negatively affect cheese yield and cheese quality. Though the effect of mastitis and late lactation is similar, the physiological basis is different: Changes in milk composition in mastitic animals relate to the immune response and those in late-lactating animals relates to metabolic pre-adaptation to involution. The impairment of milk quality during mastitis is greater in sheep the goats, whereas the impairment in milk quality in late-lactating animals is greater in goats comparing to sheep; the physiological basis is discussed. Most recent advances give hope that milk quality for curd production could be rapidly and even on-line be assessed and in the near future such a capability will be available at the farm tank level. These technology will allow more effective processing of raw milk into products that are sensitive to curd formation (cheeses) and products that are less influenced by such changes (e.g., pasteurized milk and yogurts)

Session 16 Theatre 2

Fatty acids in sheep milk related to diet and season

Elgersma, A., Wageningen University, Plant Sciences Group, P.O. Box 16, 6700 AA Wageningen, Netherlands; anjo.elgersma@wur.nl

Tank milk of 10 commercial dairy sheep farms in The Netherlands was sampled periodically in 2008/09. Management details and diet information were provided by the farmers. Fatty acid (FA) analyses were carried out and the FA profiles were related to the rations fed in the weeks preceding samplings. Seasonal variation could thus be related to actual changes in diet, and to lactation stage of herds with seasonal lambing. A literature survey revealed that the average concentration of mono-unsaturated FA (MUFA) found in sheep milk (23.1 g/100 g) was lower than in average Dutch cow milk (25.5 g/100 g), but that the poly-unsaturated FA (PUFA) concentration was higher (4.0 g/100 g versus 2.8 g/100 g). The concentrations measured in this experiment were on average higher than published values for sheep for both MUFA and PUFA (i.e., 24.0 g/100 g resp. 4.5 g/100 g), but showed large variation among farms and sampling dates. Besides, the measurement period varied among the various farms. PUFA concentrations ranged from 2.24 g/100 g in June on a farm with indoor feeding of grass silage plus concentrated to 6.11 g/100 g in July on a farm with unrestricted stocking on perennial ryegrass. CLA concentrations ranged from 0.39 g/100 g to 1.91 g/100 g. The concentration of CLA was correlated to that of total unsaturated FA, and grazing was positively associated with both. Also the omega-3 FA and MUFA concentrations were positively correlated and associated with grazing, albeit less strong. One farm with two breeds showed higher concentrations of saturated FA in 'Lacaune' sheep in early lactation than in milk from 'Zwartbles' animals in late lactation, but breed and lactation stage were confounded. Further studies would be needed to further unravel the relative effect of breed, lactation stage and diet on milk FA in sheep.

Effects of essential oils on milk production and composition and rumen microbiota in Chios dairy ewes

Giannenas, I., Chronis, E., Triantafillou, E., Giannakopoulos, C., Loukeri, S., Skoufos, J. and Kyriazakis, I., University of Thessaly, Veterinary Faculty, Trikalon 224, 43100 Karditsa, Greece; giannenas@vet.uth.gr

The effect of the addition of an essential oil (EO) mixture (Crina® Ruminants - containing a mixture of natural and nature-identical EO) on the performance of dairy ewes of the Chios breed was investigated. A total of 80 lactating ewes were allocated into 4 equal groups in a randomized block design with 4 replicates of 5 ewes housed in the same pen. The four groups were fed the same total mixed ration (TMR), a mix of corn silage, lucerne hay and wheat straw and concentrate based on cereals and oil cakes. Control was fed the basal diet without EO. The other groups were dietary supplemented with EO at levels of 50, 100 and 150 mg/kg of the concentrated feed, respectively. Individual milk yield was recorded daily and feed intake weekly on a pen basis during the first five months of lactation. Milk samples were analysed for chemical composition, somatic cell counts and urea. Rumen samples were analysed for pH, NH_3-N, protozoa, cellulolytic, hyper ammonia producing (HAP) and total viable bacteria. Results showed that inclusion of EO increased milk production per ewe; the effect being dose dependent (1.565; 1.681; 1.876; 2.119 l/d (SED±0.176) for the control, 50, 100 and 150 mg of EO/kg concentrate, respectively) and improved feed utilisation. Although the inclusion of EO did not affect milk composition, it lowered urea concentration and somatic cell counts in milk samples at the highest supplementation level compared to the control. Total viable and cellulolytic bacteria and protozoa were not influenced by EO supplementation; however HAP bacteria were reduced at the supplementation levels of 100 and 150 compared to control group. Rumen pH was not affected by EO supplementation, but NH3-N was reduced at the highest supplementation level compared to the control. In conclusion, EO supplementation may improve performance of the high yielding dairy Chios ewes; however underlying mechanisms leading to this, merit further investigation.

Effects of terpenes oral administration in blood plasma and milk concentration and some physicochemical characteristics of sheep milk

Poulopoulou, I.[1], Hadjigeorgiou, I.[1], Zoidis, E.[1], Avramidou, S.[2] and Masouras, T.[2], [1]Agricultural University Athens, Nutrition Physiology and Feeding, Faculty of Animal Science and Aquaculture, Iera Odos 75, 118 55 Athens, Greece, [2]Agricultural University Athens, Dairy Research, Faculty of Food Science and Technology, Iera Odos 75, 118 55 Athens, Greece; gpoulop@hotmail.com

Recently there has been an increasing consumer's demand for products of specific quality and hence for certification of the origins of the food they consume. Terpenes have been proposed as biomarkers of a grass based diet. In a 20 days experiment, 8 adult sheep were divided in two equal groups, representing control (C) and treatment (T) group. In the treatment group oral administration of 1 g of each terpene, α-pinene, limonene and b-caryophyllene, were applied. Milk production was recorded daily and blood plasma and milk samples were also collected. Blood plasma samples were extracted with organic solvents and the Solid Phase Micro-extraction Method using PDMS/CAR fiber was used for milk samples, before terpenes were identified on a GC-MS. Milk samples were also analyzed for fat content, protein, lactose, total solids, total solids without fat, ash and mineral contents. The results indicated terpenes not having an effect on milk production. Dosed terpenes were found only in plasma and milk samples of T group. Plasma contents of α-pinene and limonene reached up to 0.010 mg/l, while for b-caryophyllene was 0.003 mg/l, but varied greatly between days and animals. Terpenes concentration in milk samples reached up to 7 mg/l for α-pinene, 18 mg/l for limonene and 14 mg/l for b-caryophyllene, in a similar to plasma mode. Terpenes administration had effects on chemical characteristics of milk, since protein ($P<0.01$), lactose ($P<0.05$), ash ($P<0.01$) and total solids without fat ($P<0.001$) were lower in T group, while fat content and Ca, Mg, Na, K were not affected. It was concluded that terpenes can be integrated in certification schemes as biomarkers in animal products, but always used together with other indicators.

Role of endogenous proteolytic enzymes and of macrophages and neutrophils in cheesemaking ability of sheep milk

Albenzio, M., Caroprese, M., D'angelo, F., Ruggieri, D., Russo, D.E. and Sevi, A., University Of Foggia, PrIME, Via Napoli 25, 71100, Italy; m.albenzio@unifg.it

Sheep milk is destined totally for dairy product manufacture; the role of endogenous enzymes is critical during milk storage prior to cheesemaking because enzymes can cleave caseins and impair milk coagulating properties. Endogenous proteolytic milk enzymes are mainly represented by plasmin, cathepsin D, and elastase. A total of 27 bulk milk ewe samples were collected and analyzed for fat, total protein, and lactose, pH value, nitrogen fractions, casein nitrogen, milk renneting characteristics (clotting time, rate of clot formation, and clot firmness after 30 min), PL, PG, cathepsin D, and elastase content. Leukocyte differential count was performed by flow cytometry. Data were processed by ANOVA, using the GLM procedure of SAS system. Stage of lactation significantly affected fat, protein, casein, and whey protein percentage in milk; in general an increase was observed for these components with the advancement of lactation. SCC did not display significant differences throughout lactation, being always lower than 600,000 cells/ ml. Plasmin, plasminogen and plasminogen activator in ewe bulk milk were not significantly affected by stage of lactation whereas elastase content increased significantly during lactation, and cathepsin showed the highest content in mid lactation. The poorer clot firmness detected in early lactation milk could be an outcome of casein breakdown brought about by cathepsin D. Casein content was negatively correlated with clotting time and rate of clot formation ($r=-0.54$, $P<0.01$ and $r=-0.60$, $P<0.001$, respectively) and positively correlated with curd firmness ($P<0.001$). Macrophages had the highest levels at the beginning of lactation whereas PMNL that increased throughout lactation. Plasmin-plasminogen system does not vary significantly when SCC remains relatively low throughout lactation; cathepsin D can impair the coagulating behaviour of sheep milk; elastase content in ewe milk followed closely those found in PMNL.

Effect of probiotics supplementation of milk replacer on the quality of lamb meat

Santillo, A.[1], Marino, R.[1], Annicchiarico, G.[2], Russo, D.E.[1], Ruggieri, D.[1] and Albenzio, M.[1], [1]University of Foggia, PrIME, Via Napoli 25, 71100, Italy, [2]CRA, Via Napoli, 71100, Italy; a.santillo@unifg.it

Milk feeding is a major factor affecting nutritional characteristics of meat in unweaned lambs due to different fatty acid composition of ewe milk and milk substitute. Several beneficial effects are associated to the use of probiotics in animal feeding. This study was undertaken to assess the effect of milk replacer containing *Lactobacillus acidophilus* and a mix of *Bifidobacterium logum* and *Bifidobacterium lactis* on meat quality of lambs. Forty male Comisana lambs were divided into four experimental groups of ten each, grouped as maternal milk (MM), artificial milk (AM), artificial milk with *Lb. acidophilus* supplementation (AML), and artificial milk with a mix of *B. lactis* and *B. longum* supplementation together (AMB). Lambs were slaughtered at 42 d and meat was analysed for chemical composition, color and rheological properties and fatty acid. Data were processed by ANOVA using the GLM procedure of SAS system. The effect of milk source on principal composition of lambs meat evidenced that MM lambs had the highest fat content and the lowest protein content, whereas AM and AMB displayed lower fat and higher protein content. Moisture was affected by milk diet reporting the lowest and the highest value in AMB and AM respectively in accordance with hardness, cohesiveness, elasticity, gumminess, and chewiness measured in the same groups. Moreover moisture was negatively correlated with chewiness, fat content was positively correlated with cohesiveness and elasticity whereas protein was negatively correlated with the same parameters. Lamb feeding regimen influenced fatty acids levels in meat evidencing that maternal milk feeding lead to higher contents of myristic, palmitic, stearic, EPA, and DHA than artificial milk feeding. It is worth to note that trans vaccenic acid and linoleic acid were 2.8 and 1.5 times higher, respectively, in artificially reared lambs than in lambs fed maternal milk.

Effects of supplementation with linseed and olive cake on fatty acid composition and lipid oxidation of lamb meat

Conte, G.[1], Mele, M.[1], Pauselli, M.[2], Luciano, G.[3], Serra, A.[1], Morbidini, L.[2], Lanza, M.[3], Pennisi, P.[3] and Secchiari, P.[1], [1]Univ. of Pisa, DAGA, via San Michele degli Scalzi 2, 56100 Pisa, Italy, [2]Univ. of Perugia, DBA, Borgo XX Giugno 74, 06121 Perugia, Italy, [3]Univ. of Catania, DSAAPA, Via Valdisavoia 5, 95123 Catania, Italy; gconte@agr.unipi.it

The incorporation of linseed in the diet induce an increase of linolenic acid (LNA), n-3 PUFA and conjugated linoleic acid (CLA) content in the meat fat. However, feeding animals with linseed increases the susceptibility of the meat fat to oxidation. Concomitant increases in dietary antioxidant are therefore necessary to prevent flavour deterioration due to lipid oxidation. The use of olive cake as a source of antioxidant substance may be an alternative to Vitamin E supplementation. The experiment was conducted on 32 Appenninica lambs, assigned to 4 experimental groups of 8 animals each: control (C), linseed (L), olive cake (O) and olive cake + linseed (OL). FA composition and 2-thiobarbituric acid reactive substances (TBARS) analysis were performed on longissimus dorsi muscle samples. The presence of linseed in the diet increased significantly PUFA n-3 (+56%) and trans FA (+50%) content in meat fat and decreased the n-6/n-3 ratio lower than 4. The linseed supplementation led also to an increase of CLA content which reached the maximum level with L diet (+58%). Although olive cake is rich in oleic acid, the content of oleic acid in intramuscular fat of lambs fed O diet did not significantly differ to that in meat fat from lambs fed C diet. At time 0 the amount of TBARs in lamb meat was not different across diets. After 4 days of storage at 4 °C the amount of TBARs significantly increased only in meat of lambs fed C and L diet. The presence of olive cake in the diet seemed to exert a protective effect against lipid oxidation with minimal changes in FA composition of intramuscular fat in comparison to meat from lamb fed L diet. The use of olive cake in lamb feeding may be interesting in the Mediterranean areas where the olive oil industry is widespread.

Effect of feeding linseed to lambs on growth parameters, adipose tissue metabolism and meat fatty acid composition

Arana, A.[1], Mendizabal, J.A.[1], Insausti, K.[1], Maeztu, F.[2], Eguinoa, P.[2], Sarries, V.[1], Beriain, M.J.[1], Soret, B.[1] and Purroy, A.[1], [1]Universidad Pública de Navarra, Animal Production, campus de Arrosadia s.n., 31006 Pamplona, Spain, [2]Instituto Técnico y de Gestión Ganadero, Research, Avda. Serapio Huici, 31620 Villava, Spain; aarana@unavarra.es

The aim of this work was to study the effect of feeding two levels of linseed (5 and 10%) on lambs during fattening. Growth parameters, amount of fat, size and number of adipocytes, activity of the lipogenic enzymes G3-PDH, FAS, G6-PDH and ICDH and fatty acid composition of subcutaneous and intramuscular fat were studied in Navarra breed lambs slaughtered at 26 kg LW. Thirty six male lambs were assigned to three groups: control (fed on a barley and soya concentrate), L-5 and L-10, which received the same concentrate feed but including a 5% or a 10% of linseed, respectively. Lambs were studied from 15 to 26 kg live weight. Results showed that there were no significant differences between the three lamb groups on growth, carcass, Longissimus dorsi area and lipid content, adipocyte size and number and adipose enzyme activities. Fatty acid composition of intramuscular fat was significantly different between the control and the two groups that were fed linseed. Subcutaneous and intramuscular fat from groups L-5 and L-10 showed higher levels of linolenic acid and n3, together with a lower n6/n3 ratio ($P<0.01$). The intramuscular fat of the L-10 group showed higher n3 and linolenic acid content than L-5 lambs ($P=0.07$) but no significant differences were found in the subcutaneous fat. Therefore, it could be concluded that feeding 5 or 10% linseed feed to lambs had no effect on growth parameters and on adipose tissue deposition but increased the content in unsaturated (linolenic and n3) fatty acids in the intramuscular fat.

Session 16 Theatre 9

Fatty acid profile of meat, liver and heart from lambs fed oil components

Borys, B.[1], Siminska, E.[2] and Bernacka, H.[2], [1]National Research Institute of Animal Production, Experimental Station Koluda Wielka, Parkowa str. 1, 88-160 Janikowo, Poland, [2]University of Technology and Life Sciences, Mazowiecka str. 28, 85-844 Bydgoszcz, Poland; bronislaw.borys@onet.eu

The effects of fattening lambs with sunflower cake (SC) and linseeds (L) and adding vitamin E on the fatty acid profile of fat from m. longissimus lumborum (LLF), liver (LF) and heart (HF) were studied. 18 rams were fattened in 3 groups to 35 kg b.w. with a cereal and rapeseed meal-based mixture (control group C), a mixture with SC and L (SCL; 23.5 and 5%, respectively) and a mixture with vitamin E (SCL+E). Fatty acid composition was analysed by gas chromatography (AOAC 905.02). Results were analysed by two-way analysis of variance. LF contained more SFA than LLF and HF (56.7 vs. 48.4 and 45.9%), with higher C17:0 and C18:0 and lower C14:0 and C16:0 compared to LLF fat. SFA content of HF was in between. MUFA was most abundant in LLF, followed by LF and HF (42.6, 33.9 and 29.9%, respectively). This corresponded to 38.7, 29.2 and 26.3% of the main acid (C18:1), respectively. HF had a higher proportion of PUFA (25.3%) compared to LF (11.3) and LLF (9.0%). A similar pattern was found for C18:2 (19.6 vs. 9.1 and 7.0%), with C18:3 n3 content of 0.82 vs. 0.73 and 0.64%. Best health parameters were observed for HF, with similar values for LF and LLF (PUFA:SFA of 0.57 vs. 0.21 and 0.19, and DFA:OFA of 3.96 vs. 3.40 and 2.58). CLA was most abundant in the liver (41.2 mg/100 g), followed by heart (6.2) and muscle (4.5 mg/100 g). Feeding oil components had no significant effect on total SFA in the organs, with a decrease in C16:0 and C17:0 and an increase in C18:0. Feeding oil components decreased MUFA (SCL 33.3, C 39.8%) and increased PUFA (16.8 vs. 12.1%). The organs of SCL lambs had less CLA than C lambs (16.3 vs. 19.4 mg/100 g). In general, SC and L feeding was beneficial for health parameters of fat from the organs (PUFA:SFA of 0.35 for SCL and 0.27 for C; DFA:OFA of 3.51 vs. 2.92), and the addition of vitamin E to the mixture enhanced this beneficial effect.

Session 16 Theatre 10

Meat quality and lipid composition of four sheep breeds reared by the traditional Greek sheep farming

Bizelis, I.[1], Koutsouli, P.[1], Sinanoglou, V.[2], Symeon, G.[1] and Mantis, F.[2], [1]Agricultural University of Athens, Animal Breeding & Husbandry, 75, Iera Odos, GR 118 55, Athens, Greece, [2]Technological Educational Institution of Athens, Food Technology, Egaleo, GR 12210, Greece; jmpiz@aua.gr

Under the traditional Greek sheep farming, lambs are slaughtered in very young age because consumers are in fond of young lamb meat and sheep milk enjoys high prices. Lamb carcasses are marketed without any labelling referring to breed, specific age at slaughter or diet. The aim of the present study was the evaluation of meat quality and lipid composition of four sheep breeds reared in Greece under traditional farming practises. Indigenous Greek breeds, as Kallarytiki, Karagouniki and Chios, and Lacaune breed were used. Ten male lambs, randomly selected from each breed, were slaughtered at the age of 40±5 days in regional abattoirs. 24 hours after slaughter the weight of the cold carcasses was recorded as well as carcass length, thigh perimeter, pH24 of the thigh and pH24 of Longissimus dorsi. Colour (L, a*, b*), water holding capacity, cook loss, shear values, intramuscular fat content and chemical fat of kidney fat were also measured. The extracted pure fat was then submitted to further analysis in order to evaluate lipid and fatty acid composition. The data were analyzed using GLM procedures. Meat quality was significantly affected by the breed in terms of pH24 of the thigh and Longissimus dorsi, colour (L, a*, b*), water holding capacity, cook loss, shear values, intramuscular fat content and chemical fat of kidney fat. Also, significant differences were found at lipids profile and ω-6/ω-3, PUFA and PUFA/SFA ratio as well as in atherogenic and thrombogenic indices. All muscle samples, had significant proportion of phospholipids, predominated of phosphatidylcholine. Although ruminant meats normally have a low PUFA/SFA ratio, the examined muscles contained a range of ω-6 and ω-3 PUFA of potential significance in human nutrition. Intramuscular fat had a favourable ω-6/ω-3 PUFA ratio

Session 16 Theatre 11

Effects of the Texel Muscling QTL (TM-QTL) on lamb tenderness

Lambe, N.R.[1], Macfarlane, J.M.[1], Richardson, R.I.[2], Nevison, I.[3], Haresign, W.[4], Matika, O.[5] and Bunger, L.[1], [1]SAC, Kings Buildings, Edinburgh, United Kingdom, [2]University of Bristol, Division of Farm Animal Science, Langford, Bristol, United Kingdom, [3]BioSS, Kings Buildings, Edinburgh, United Kingdom, [4]IBERS, University of Wales Aberystwyth, Aberystwyth, United Kingdom, [5]The Roslin Institute and R(D)SVS, University of Edinburgh, Midlothian, United Kingdom; Nicola.Lambe@sac.ac.uk

A QTL identified on Chr 18 in Texel sheep (TM-QTL) increases loin muscling. This study assesses TM-QTL effects on Texel lamb tenderness. Grass-fed lambs (male and female), from 7 TM-QTL carrier sires, were ultrasound scanned then slaughtered at 20w. After aging for ~1w, shear force tenderness was measured in a loin (ShF_loin) and leg (ShF_leg) muscle. Paternally-inherited TM-QTL genotypes classified lambs as non-carriers (n=85), carriers (n=118) or unknown (n=6). An untrained taste panel also assessed loin texture (TP_loin) on 20 non-carriers and 20 carriers. Mixed models were fitted to validate direct effects of TM-QTL on loin muscle depth (MLL_D) and weight (MLL_wt), and to estimate indirect effects on ShF_loin, ShF_leg and TP_loin. TP_loin was adjusted for sample order and assessor. Other traits were adjusted for carcass weight and other relevant factors. TM-QTL carriers had significantly (P<0.001) higher MLL_D (+5%) and MLL_wt (+6%) than non-carriers, but effects of TM-QTL on ShF_loin, ShF_leg and TP_loin were non-significant (all P>0.3). Genotyping of dams further improved the classification into 40, 53, 17, 34 and 65 as non-carriers, heterozygotes inheriting the TM-QTL allele from sire and dam, homozygous TM-QTL and unknown, respectively. The re-classification revealed a polar-overdominant imprinting mode of inheritance, with only lambs inheriting TM-QTL alleles from their sire, and not their dam, showing increased MLL_D (+6%) and MLL_wt (+4%). However, genotype effects on tenderness remained statistically non-significant (P>0.1). These results imply that TM-QTL in Texel sheep increases loin muscling without evidence of it affecting tenderness.

Session 16 Theatre 12

The quality of a new sheep meat product: effect of salting and ageing process

Teixeira, A., Rodrigues, M.J., Pereira, E. and Rodrigues, S., CIMO, Escola Superior Agrária - Instituto Politécnico de Bragança, Campus Sta Apolónia Aptd 1172, 5301-855, Portugal; teixeira@ipb.pt

With the main objective to find a strategy to give added value to meat from culled animals a study was conducted to create a new product promoting meat with very low commercial price and sufficiently versatile. Thirteen culled ewes from Churra Galega Bragançana breed, between 2 and 9 years old were used. According to the definition of processing method the effects of ageing, salting and drying on physical characteristics of meat such as color and water activity were assessed in subscapularis and semimembranosus muscles. Animals were slaughtered in the Bragança commercial abattoir, with an average carcass weight of 20 kg. After slaughter carcasses were refrigerated at a temperature of 4 °C and 6 and 7 carcasses aged for 2 and 8 days, respectively. Carcasses were dived into quarters and deboned. The boneless meat was then submitted to a salting process for 72 hours followed by more 72 hours air-dried at °C. Meat characteristics such as pH, color, water activity, water-holding capacity, texture and determination of heminic pigments were also evaluated in the longissimus thoracis et lumborum muscle. The meat pH, water-holding capacity and pigments were not affected by meat ageing process which influenced meat color. All parameters values decreased with increasing days of aging and Chroma (C*) values were higher (32.6 vs 22.8) than Hue (H*) values (52.9 vs 54.3), giving a brighter color meat. But the most affected meat quality parameter by aging process was texture making the meat tender, 6.2 and 9.0 kgf for 8 or 2 days ageing, respectively. During the salting process, there was a decrease in brightness (L*), red (a*) and yellow (b*) indexes. So, H* value increased from 25.6 to 56.6 and C* value dropped from 171.8 to 30.6 producing a dark meat as result of salting and dry-air. During salting as well as during air-drying the values of meat water activity (aW) dropped from 0.86 to 0.79 and from 0.79 to 0.76 which was extremely important for final product conservation.

Hygiene of bulk-tank parlors and milk quality of dairy sheep farms in Thrace

Tzatzimakis, G.[1], Alexopoulos, A.[1], Bezirtzogloub, E.[1], Sinapis, S.[2] and Abas, Z.[1], [1]Democritus University of Thrace, Agricultural Development, 123 Padazidou Str., GR-68200, Greece, [2]Aristotle University of Thesaloniki, School of Agriculture, Department of Animal Production, GR 54124, Greece; abas@agro.duth.gr

The aim of this study was the monitoring of milk quality produced in 21 dairy ewe's farms from the regions of Xanthi and Evros, in the north-eastern Greece. Milk samples were collected every 15 days throughout the dairy period (March-June 2008). For the study a questionnaire was used. The questionnaire was conducted by personal interview of the owners of the farm. With the questionnaire we select information about flock characteristics, health status, handling practices etc. From each farm, air was also sampled for microbiological analysis. Milk samples were examined for their chemical components like fat, protein, lactose, non-fat dry matter (NFDM) and somatic cells count (SCC). At the same time, the same milk samples were examined for Total Bacterial Counts (TBC), coliform count, Staphylococcus aureus, environmental streptococcal count and preliminary incubation count (PIC). The possible correlation among different bacterial species and their interaction with SCC and chemical components of milk was also considered. It was examined whether farm management practices could influence the hygiene and the quality of milk. Our results show that as an average TBCs were 5.48 log cfu/ml, SCC: 6.05 log cells/ml, coliforms: 4.49 log cfu/ml, Staphylococcus aureus: 3.94 log cfu/ml, environmental streptococcal counts: 4,95 log cfu/ml and PIC: 5.7 log cfu/ml. The mean fat, protein, lactose and NFDM were 6.17%, 5.28%, 4.73% and 10.95% respectively. The study revealed significant positive correlation between TBC and PIC (0.825), while SCC was marginal positive correlated with protein and NFDM. No statistically significant correlations observed among SCC with any of the bacterial species. Herd size and farm management practices had considerable influence on SCC and bacterial species, except of coliform count.

Session 16 Poster 14

Effect of magnesium oxide and maturation on meat quality of culled ewes

Ribeiro, E.L.A., Constantino, C., Bridi, A.M., Castro, F.A.B., Koritiaki, N.A., Tarsitano, M.A. and Mizubuti, I.Y., Universidade Estadual de Londrina, Zootecnia, CCA, 86051-990, Londrina/PR, Brazil; elar@uel.br

The objective of this study was to evaluate the influence of magnesium supplementation to culled ewes in feedlot and maturation period on meat characteristics. Eighteen Santa Ines ewes were used, with average age of 6 years. Three different levels (0.0; 0.1 and 0.2%) of magnesium oxide were added to the concentrate ration. The roughage/concentrate ratio was 60/40. Sorghum silage was used as roughage. Diets (9.3% CP and 61.5% TDN) were supplied ad libitum. Ewes were randomly assigned to one of the three levels of magnesium. After 42 days of confinement ewes were slaughtered. Following a 24 hour chilling period, Longissimus lumborum muscles were collected. Muscles from each animal were submitted to each of the three periods of maturation (0, 4 and 8 days) following a Split-plot design. For maturation, samples were vacuum sealed and stored at temperature of 5 °C. Level of magnesium and its interaction with maturation period did not affect ($P>0.05$) the studied traits. Averages for body weight at slaughter and cold carcass were 55.8 and 23.3 kg, respectively. Sensorial characteristics (tenderness, juiciness, overall acceptability and odor), evaluated by a trained panel, differed ($P<0.05$) only in the intensity of odor, which decreased linearly with the maturation. The average value for tenderness was 6.3 (1 = very though to 7 = very tender). When tenderness was evaluated by the Warner-Bratzler shear force, values (kgf) decreased linearly ($\hat{Y} = 3.45 - 0.055x$, $P<0.05$) with maturation. Other traits (pH, color, myofibril fragmentation index) were positively affected by maturation. On the other hand, mesophilic and psycophilic microorganisms increased linearly ($P<0.05$) with maturation. However, the range of averages for all traits were within values considered normal to good quality sheep meat. It can be concluded that supplementation with magnesium did not affect meat quality of ewes, and maturation up to eight days can improve its quality.

Session 16 Poster 15

In vitro evaluation of terpene's effects on aspects of animal physiology: cytotoxicity and rumen degradation

Poulopoulou, I.[1], Pitulis, N.[2], Mountzouris, K.C.[1], Hadjigeorgiou, I.[1] and Xylouri, E.[2], [1]Agricultural University Athens, Nutrition Physiology and Feeding, Faculty of Animal Science and Aquaculture, Iera Odos 75, 118 55 Athens, Greece, [2]Agricultural University Athens, Anatomy and Physiology of Farm Animals, Faculty of Animal Science and Aquaculture, Iera Odos 75, 118 55 Athens, Greece; gpoulop@hotmail.com

Plants beyond nutrients provide herbivores with plant secondary metabolites (PSM), which, at certain levels can lead to toxicity. The aim of this study was to evaluate the potential cytotoxic effects of three terpenes, α-pinene, limonene and β-caryophyllene using *in vitro* techniques and to assess the detoxification capacity of rumen for these substances. Rabbit kidney cell line RK13, grown in DMEM containing 10% FBS, was used as a substrate model. A reduction assay of methylthiazolyldiphenyl-tetrazolium bromide (MTT) was used to assess cell metabolic activity. Cells were treated for 24 hours with each terpene, as well as their combination in a range of concentrations (100 to 3.1 μg/ml) run in triplicate. Limonene and β-caryophyllene arrested totally cell proliferation at 100 μg/ml, while the combination of three produced total arrest at 50 μg/ml. A-pinene produced 50% reduction in cell proliferation at the concentration of 100 μg/ml. For rumen degradation assessment, the 3 terpenes were incubated in sheep and goat rumen fluid at a concentration of 100 μg/ml and degradation rate was measured at 0, 2, 4, 8, 20 and 24 hours. Degradation was assessed also in supernatant and microbial cell fractions after centrifugation. All three terpenes were degraded by the three rumen liquor forms, but to different extend. Degradation in sheep or goat rumen fluid reached for: a) α-pinene 80% and 60% respectively, b) limonene 50% and 45% respectively and c) b-caryophyllenne 55% for both species. The results indicate terpenes might have a toxic effect on animal cells, but their incubation in rumen can reduce quantities available for absorption.

Session 16 Poster 16

Welfare-friendly management practices on farm and slaughter weight effects on meat quality of Negra Serrana kids

Alcalde, M.J.[1], Horcada, A.[1], Alvarez, R.[1] and Pérez-Almero, J.L.[2], [1]Univ Seville, Ctra Utrera Km1, 41013 Seville, Spain, [2]IFAPA, Las Torres, 41200 Alcala del Rio (Sevilla), Spain; aldea@us.es

An evaluation system in which has into account – in 13 sections and 120 items –, for instance, vigilance, buildings, environment and resources, food, healthiness and behaviour, handling of adult animals and kids, records and freedom of movement was put into practice on 15 extensive conditions farms belongings to Negra Serrana goat breed in the South of Spain. According with 0.14. In a multivariante data analysis, only L* was affected by weight (p±0.22 and b* was 11.70±0.46, a* was 4.50±0,42, L* was 49.31±0.02, WHC was 17.10±0.29 kg) the meat quality (pH, Water Holding Capacity (WHC) and CIE L*, a* and b* measured in Longissimus lumborum) at 3 days post-slaughter were evaluated in 60 kids. All the kids were reared with their dams until their slaughter. The average pH was 5.61±0.11 and 16.46±h the punctuation obtained, two farms were chosen: one with higher welfare-friendly practices and other with lower welfare-friendly ones. In that conditions and with two slaughter weight groups (average ones were 10.05<0.05), obviously darker meat corresponded with higher weight. On the other hand, the welfare level influenced the WHC (P<0.01) and pH (P<0.05), kids meat bellowing to the farm with high welfare level was higher percentage of expelled juice and lower pH. None of the interactions were significantly different. Finally WHC and L* in a discriminant analysis were able to classified the 76.7% of the kids into their four groups.

Effects of welfare conditions on post-mortem evolution of kid muscle proteins

Soriano, J.D.[1], Suárez, M.D.[1], Martínez, T.F.[1], Perez-Almero, J.L.[2] and Alcalde, M.J.[3], [1]Universidad de Almería, Biología Aplicada, La Cañada, 04120 Almería, Spain, [2]IFAPA-Las Torres, Alcalá del Río, 41200 Sevilla, Spain, [3]Universidad de Sevilla, Departamento de Ciencias Agroforestales, Carretera Utrera Km 1, 41013 Sevilla, Spain; jsg131@alboran.ual.es

This work was aimed to assess the influence of on-farm management conditions and transport time on the electrophoretic pattern of Longísimus dorsi toraci muscle proteins during post-mortem storage. Sixty-four kids (Blanca Celtibérica breed) (10.05±0.15 kg), reared in extensive systems in two farms in Huéscar (Granada), dessigned as high-welfare (HW) and low-welfare (LW). This classification was based on a 120 items survey related to welfare topics. Animals were transported to the slaughterhouse under two transport conditions: short transport (one hour) and long transport (6 hours). The kids were slaughtered immediately, and carcasses were kept for 24 hours at 4 °C. After, different portions of Longisimus dorsi toraci muscle were vacuum packed and stored at 4 °C in a cold room. Muscle samples were taken at 3, 8, and 21 days after slaughtering, frozen in liquid nitrogen, and stored at -80 °C until the extraction of muscle proteins, and further electrophoretic separation. Densitometry of gels was performed for quantifying the relative contribution of each individual fraction to the total optical density of each lane. The relative optical density shown significant influence of management (high and low welfare) on muscle proteins. The transport time significantly influenced the relative contribution of 43 and 48 KDa bands of those kids kept in high welfare farms; however, these fractions were not affected in animals reared in low welfare farms. Storage time also exerted certain effects on the relative contribution of some of the muscle protein fractions. These findings suggest the possibilities of electrophoretic separations of muscle proteins as a feasible procedure aimed to assess the influence of previous welfare conditions on kid meat properties during post-mortem storage.

Session 16 Poster 18

Influence of management and transportation time on the post-mortem changes in protein content of kid muscle

Soriano, J.D.[1], Suárez, M.D.[1], Martínez, T.F.[1], Perez-Almero, J.L.[2] and Alcalde, M.J.[3], [1]Universidad de Almería, Biología Aplicada, La Cañada, 04120 Almería, Spain, [2]IFAPA-Las Torres, Alcalá del Río, 41200 Sevilla, Spain, [3]Universidad de Sevilla, Ciencias agroforestales, Carretera Utrera Km 1, 41013 Sevilla, Spain; jsg131@alboran.ual.es

The quality of kid meat is influenced by factors related to handling, transport, and post-mortem storage; these factors determine, with others, the intensity of post-mortem proteolysis in muscle cells. In order to determine the effects of the type of handling and transport time prior to slaughtering on post-mortem proteolysis in goats muscle, Sixty-four kids (Blanca Celtibérica breed) with an average slaughter weigh of 10.05±0.15 kg were were reared in extensive systems in two farms in Huéscar (Granada), dessigned as high-welfare (HW) and low-welfare (LW). This classification was based on a 120-item survey related to welfare topics. Within each farm, the selected animals were transported to the slaughterhouse under two transport conditions: short transport (one hour) and long transport (6 hours). The animals were slaughtered immediately on arrival, their carcasses were kept for 24 hours in cold chamber, and then a portion of the Longisimus dorsi toraci were packed at vacuum and stored at 4 °C. Muscle samples were taken at 3, 8, and 21 days after slaughtering, frozen in liquid nitrogen and stored at -80 °C until being used for the determination of different parameters (amounts of free amino acids and sarcoplasmic and myofibrillar proteins in muscle extract)s. Animals kept in high welfare farms showed higher content of myofibrillar protein and low sarcoplasmic protein content than those reared under lower welfare conditions, these differences are more pronounced at 21 days of storage, which may be indicative of increased protein hydrolysis in these animals. No effect of the transport time was observed on these parameters. These findings suggest the influence of management conditions previously to slaughter on post-mortem proteolysis in kid muscle during cold storage.

Influence of part-time grazing management on lipid fractions (fatty acids and triglycerides) of sheep´s milk

Valdivielso, I.[1], Barrón, L.J.R.[1], Amores, G.[1], Virto, M.[1], Arranz, J.[2], Beltrán De Heredia, I.[2], Ruiz, R.[2], Ruiz De Gordoa, J.C.[1], Nájera, A.I.[1], Albisu, M.[1], Pérez-Elortondo, F.J.[1], Mandaluniz, N.[2] and De Renobales, M.[1], [1]Universidad del País Vasco, Paseo de la Universidad, 7, E-01006 Vitoria-Gasteiz, Spain, [2]NEIKER-Tecnalia, P.O.Box 46, E-01080 Vitoria-Gasteiz, Spain; izaskunvz@hotmail.com

Part-time grazing is a traditional flock management used in the Basque Country (Northern Spain) in which pasture feeding is supplemented indoor with forage (alfalfa and pasture hay) and concentrate to meet milk production requirements. This study evaluated the effect of part-time grazing management on the content of fatty acids and triglycerides in the milk. The experiment was conducted during 4 weeks from late April until mid-May. Sheep were separated into 4 homogeneous groups of 12 sheep each, and randomly assigned to 3 different alfalfa hay supplements: 300 g/day (G1), 600 g/day (G2), and 900 g/day (G3). G1-G3 animals were allowed to graze outdoors for 4 hours. The control group (G0) received 600 g alfalfa hay/day and was not allowed to graze outdoors. All animals received 500 g concentrate/day at milking. Milk samples (evening and morning milking combined) were taken once a week. Fatty acids were analyzed by GC-FID and triglycerides by HPLC-ELSD. Part-time grazing significantly ($P<0.05$) increased the amount of unsaturated fatty acids, particularly the amount of c9t11 CLA isomer and that of trans-vaccenic acid. Relative percentages of triglycerides of partition number (PN) 40 and 34 were also significantly affected by grazing. When the 3 different amounts of alfalfa hay were compared (G1, G2 and G3), milk fat from G1 had the highest level ($P<0.05$) of unsaturated fatty acids, c9t11 CLA isomer and trans-vaccenic acid. The higher amount of alfalfa hay, the higher relative percentage of PN 48 triglycerides and the lower relative percentage of PN 38 triglycerides was observed. This study was financed by UPV/EHU Cátedra UNESCO 02/05 and INIA TRA2006-00100-C02.

Breed effect on the meat colour and visible spectrum of Spanish suckling kids

Alcalde, M.J.[1], Ripoll, G.[2], Sañudo, C.[3], Horcada, A.[1], Texeira, A.[4] and Panea, B.[2], [1]Univ Seville, Ctra Utrera Km1, 41013 Seville, Spain, [2]CITA, Av Montañana 930, 50059 Zaragoza, Spain, [3]Univ Zaragoza, Miguel Servet, 177, 50013 Zaragoza, Spain, [4]CIMO Inst.PolitecnicoBragança, Apdo1172, 5301-855 Bragança, Portugal; aldea@us.es

The effect of the breed (Blanca Celtibérica (BC), Moncaína (MO), Negra Serrana-Castiza (NE), Blanca Andaluza (BA), Pirenaica (PI), Malagueña (MA) and Murciano-Granadina (MU)) on the colour (CIE L* a* b* h* and C*) and visible reflectance (between 400 and 700 nm wavelength region) were evaluated in 105 male kids ranging 4.2±0.12 kg cold carcass weight. All the kids were reared with their dams until slaughter. The variables were measured at 24 hours post-slaughter in Longissimus lumborum after 1 h blooming with a spectrophotometer CM-700d. The average values of colour parameters were L* 48.40±0.51, a* 3.09±0.19, b* 9.11±0.34, C* 9.94±0.30 h* 68.90±1.63, for L*,a*, b*, C* and h*, respectively. There were found significant differences ($p<0.001$) among breeds on all the variables analysed. BC had the palest and most light-coloured meat whereas BC had the most yellowish and vivid colour and the highest hue angle. Whilst PI had the darker and reder meat, and the lowest chroma, hue angle and b* values. Being b* and C* the variables that can explained a higher percentage of the variation of the results. A discriminant analysis was not able to classified carcasses into their breeds, since only the 58.8.% of animals were accurately classified correctly into their breeds, the inclusion of the spectra variables did not improve significantly the % of classification (62.7%).

Selection and evaluation of autochthonous lactic acid bacteria for the manufacture of a traditional goat cheese of Garfagnana

Nuvoloni, R., Fratini, F., Faedda, L., Pedonese, F., Turchi, B., Ebani, V.V. and Cerri, D., University of Pisa, Animal Pathology, Prophilaxe and Food Hygiene, viale delle Piagge 2, 56100 Pisa, Italy; bturchi83@gmail.com

The aim of this study was to identify and select the lactic acid bacteria (LAB), isolated from raw milk and ripened goat cheese manufactured in Garfagnana (Italy), for the development of an autochthonous starter. The suitability of this starter was tested in a preliminary cheese-making trial. In the first phase goats' raw milk and cheese samples, at different ripening times (2, 7, 14, 21 and 45 days), were collected from three farmers, representative of the area of production. Fifty-five strains of mesophilic LAB isolated from milk and cheese were screened for salt tolerance, acidifying, proteolytic and aminopeptidase activities. Five different mixed cultures of selected Lactococcus lactis subsp. lactis (4 strains) and Lactobacillus paracasei (4 strains), chosen on the basis of their technological attributes, were tested for the ability to grow and acidify in association. Finally the autochthonous starter was prepared using the best association composed by 1 strain of Lactococcus lactis subsp. lactis and 2 strains of Lactobacillus paracasei subsp. paracasei. In the second phase the cheese-making trial was performed at one farm by comparing three batches of experimental cheese, manufactured using the autochthonous starter, with control cheeses, produced using a commercial starter. The evolution of pH and lactic acid microflora was monitored in both types of cheese. The addition of the autochthonous starter resulted in a good acidification during the entire ripening time. The organoleptic evaluation on the experimental and control cheeses after 45 days of ripening, carried out by an expert staff of the cheese factories, detected the best characteristics in cheeses made with the autochthonous starter. Particularly, texture, flavour and taste of the experimental cheese were described as agreeable and typical.

Study on the possibility of making cheese from all milk proteins and mozzarella-type cheese from sheep milk

Pakulski, T. and Pakulska, E., National Research Institute of Animal Production, Experimental Station Koluda Wielka, Parkowa str. 1, 88-160 Janikowo, Poland; etpakulscy@poczta.onet.pl

The aim of this study was to determine the possibility of making cheese from all milk proteins (AMP) and mozzarella-type cheese (Moz), with semi-hard maturing cheese (SHM) as the control product. AMP cheese was made as follows: milk with $CaCl_2$ (2/3 of the dose) was pasteurized at 90-92 °C and cooled, adding the rest of $CaCl_2$ and cheese cultures, and treated after 30 min with rennet; after milk curdled, cheese was moulded, pressed and salted. In the Moz cheese making process, rennet and cheese cultures were added to milk, a block of cheese was moulded from curdled milk and cooled for 24 h, after which the cheese was cut, scalded with hot water (85 °C) and hot cheese mass was 'plastified'; cheese was moulded, cooled, salted and packed. SHM cheese was obtained from rennet-curdled milk with cheese cultures; moulded cheese was pressed and salted. AMP and SHM cheeses were matured for 4-6 weeks at 10-12 °C and 80% humidity. Cheeses were made from the milk of the prolific-dairy line of Koluda sheep in two successive years, and 4 batches of each cheese were produced per year. Milk contained an average of 17.26 solids, 5.48 protein and 6.66% fat. The yield of cheeses was similar and averaged 21.2%. AMP, Moz and SHM cheeses contained 52.4, 55.3 and 51.6 solids; 23.3, 22.3 and 21.7 protein, and 21.1, 21.7 and 21.3% fat, respectively. No differences were found between the cheeses in the content of fatty acids (CLA, SFA, UFA, MUFA, PUFA,Ω-6 and Ω-3) and minerals (Ca, Cu, Fe, Mg, Na and P). Slightly higher milk protein retention (by 2.6%) was obtained for AMP compared to Moz and SHM cheeses. The results obtained show that new cheese types can be made from sheep milk.

Effect of production method on composition of cheese from Merino sheep milk

Pakulski, T. and Pakulska, E., National Research Institute of Animal Production, Experimental Station Koluda Wielka, Parkowa str. 1, 88-160 Janikowo, Poland; etpakulscy@poczta.onet.pl

The effect of different production technologies on yield and composition of sheep milk cheeses was compared. Three types of cheese were made: I) semi-hard maturing cheese from scalded mass; II) rennet white non-maturing cheese; III) curd cheese from all milk proteins. Cheeses were made as follows: I) pasteurized milk was acidified with cheese cultures and rennet was added. Blocks of cheese were moulded from cheese mass and pressed for 2-2.5 h, after which cheese mass was cut and treated for 1-1.5 min with hot (75 °C) brine. Cheeses were moulded again, pressed for 20 h and left in ripening room (10-12 °C, humidity up to 80%) for 4-6 weeks; II) milk was curdled with rennet, cheese was moulded from curdled cheese mass, pressed for 24 h and salted to obtain ready-to-eat product; III) cheese was obtained by the acid-rennet method as a result of long-term (<20 h) curdling of milk with rennet at reduced temperature (25-30 °C); cheese mass was transferred into moulds and pressed for several hours. No salt was added at any stage of the technological process when cheese III was made. Each cheese was produced 4 times from Merino milk containing an average of 20.63 solids, 6.88 protein and 8.29% fat. The yield of cheese mass was the highest (P<0.01) for cheese III (43.51), intermediate for cheese II (32.86) and the lowest for cheese I (24.51%). Cheese III was characterized by the highest (P<0.01) retention of solids (III – 77.2, II – 66.2, I – 63.4) and milk fat (III – 88.7, II – 70.4, I – 61.9%). Cheese I contained more (P<0.01) of all components than the other two cheeses. Mean content of solids, ash, protein and fat was 53.03, 3.92, 24.86 and 21.12 in cheese I, 41.24, 2.72, 16.62 and 17.61 in cheese II, and 36.14, 1.09, 14.29 and 16.49% respectively in cheese III. The protein/fat ratio was 1.19 (I), 0.95 (II) and 0.87 (III). Production method had an effect on cheese yield and composition.

Aromatic profile of 'canestrato pugliese' PDO cheese made from different sheep breeds of South Italy

Annicchiarico, G., Claps, S., Bruno, A., Pizzillo, M., Caputo, A.R., Di Napoli, M.A. and Fedele, V., CRA, CRA-ZOE, Unità di ricerca per la Zootecnia Estensiva, Via Appia, Bella Scalo, 85054, Italy; giovanni.annicchiarico@entecra.it

Canestrato pugliese is one of the most important and well known type of hard ripened sheep cheese for the Puglia region (Southern Italy) economy. Importance has been confirmed by the Protected Designation of Origin in 1995. In spite of the popularity, there are very few studies on the characteristics of cheese made from different ewe's breed milk. The focus of the work is to verify the relation between aromatic profile and breed and territory to promote the socio-economical development of area and safeguard the Animal Genetic Resource (AnGR). In this study the volatile profile and sensory characteristics of Canestrato pugliese cheese made from milk of three different breeds (Gentile di Puglia, Altamurana and Comisana) were compared. The experiment was carried out in winter and all animals were fed on native pasture, supplied with commercial concentrate two time a day. Three cheese-makings were carried out for three consecutive days. VOC's and sensory properties, of ripened cheese 12 months, was performed on nine samples and in duplicate. VOC's were extracted by 'purge and trap' system coupled to a gas chromatograph and detected with a MS detector. Data were analysed by GLM and the means were compared by LSD test. Sensory data were normalised before submitted to ANOVA repeated measure procedures. The results showed that breed affected the aromatic profile and sensory properties. Comisana's cheese showed, with the exclusion of ketones, aldeydes and sesquiterpenes, a higher content of alcohols, esters, acids, aromatic hydrocarbures and terpenes than other breeds. The same patterns was observed for the sensory properties. Results indicate that VOC's compounds and sensory properties vary according to the sheep breed.

Chemical composition and quality of lamb meat as related to fat content of *m. longissimus lumborum*

Borys, B.[1], Grzeskowiak, E.[2] and Borys, A.[2], [1]National Research Institute of Animal Production, Experimental Station Koluda Wielka, Parkowa str. 1, 88-160 Janikowo, Poland, [2]Meat and Fat Research Institute, Jubilerska str. 4, 04-190 Warszawa, Poland; bronislaw.borys@onet.eu

Differences in lamb meat quality were studied in relation to intramuscular fat (IMF) content. The meat was obtained from 90 rams of prolific Koluda sheep and their crosses with Ile de France. Lambs were fattened to 32-37 kg b.w. Complete mixtures were fed ad libitum with hay or forage. m. longissimus lumborum (LL) was analysed for water, protein and fat content, pH_{24} and electrical conductivity (EC_{24}), water holding capacity, drip loss, weight loss, tenderness after cooking (WB), marbling and colour score, L*, a* and b* colour values, and sensory score of grilled loin chops. The material was graded according to fat content of LL into three classes; L – low, <1.50% (14 pcs.), O – optimal, 1.51-2.50% (44 pcs.) and H – high, >2.51% (32 pcs.). Significant differences were estimated between the classes for the above meat quality traits. The increase in LL fat was paralleled by a significant decrease in water and protein content (L - 76.5 and 21.3%; O - 75.9 and 21.0%; H - 75.3 and 20.6%, respectively). No differences were found in pH_{24}, with a tendency towards increased EC_{24} and water holding capacity in higher fatness classes. O meat was more tender than L and H (by 8.5%). With similar L*, b* and a* colour values, significantly poorer colour scores were obtained for H meat (3.58 vs. 4.10 pts. for L and O). Marbling score increased with the increase in fatness from 1.55 to 1.77 pts. O meat had 12.7% lower natural losses than L and H meat, with similar cooking loss. The sensory scores of meat were not differentiated by fat content. Overall, the increase in IMF content adversely affected the basic chemical composition of the meat, but had no effect on sensory scores. The effect on physico-chemical traits varied, with the most favourable traits being found for meat with a 1.5-2.5% fat content, which is considered optimal in terms of lamb meat quality.

Session 17 Theatre 1

Ban of castration: product quality matters

Støier, S., DMRI/Technological Institute, Meat Quality, Maglegaardsvej 2, 4000 Roskilde, Denmark; sst@teknologisk.dk

Stopping castration of pigs is increasingly interesting, when animal welfare is considered. However, it is important to address the quality matters related to production of entire males. A major problem is the risk of boar taint. Meat from some entire males develops an unpleasant flavour - boar taint - that is generally not accepted by consumers. Boar taint is perceived through a combination of odour, flavour and taste in pork and pork products during cooking and eating. Boar taint can largely be attributed to three compounds; androstenone, skatole and to a lesser extent indole. Castration reduces the content of androstenone and skatole in the carcass. Even though the three compounds cannot explain all variation in boar taint determined by a sensory panel or consumers, they are regarded as the main compounds. Furthermore the carcass composition of entire males differs from gilts and castrates. The forepart is heavier and the ham is lighter. The higher meat content results in changes in lean/fat distribution, and the fat quality deviates; the fat in entire males is more unsaturated resulting in more soft fat, which for some products like bellies can cause quality problems. Another aspect is the rind quality. General entire males are more aggressive compared to gilts and castrates resulting in increased skin damages due to fighting. Given the meat industry will begin to produce entire males in large scale, the consequences for product quality have to be addressed. A method for sorting out the tainted carcasses is needed. Furthermore, the possibilities of keeping the amount of tainted carcasses at a low level have to be investigated. It is of the outmost importance that the rejected carcasses and meat can realize reasonable prices. Therefore, possibilities have to be examined of using meat from rejected carcasses for products in which the flavour would not have a negative influence on product acceptability. Procedures for handling the more aggressive male pigs have to be developed and implemented.

Consumers' attitudes and preferences towards pig castration: the trade-off between animal welfare and hedonism

Gil, J.M. and Kallas, Z., CREDA-UPC-IRTA, Esteve terradas, 8, 08060-Castelldefels (Barcelona), Spain; chema.gil@upc.edu

This paper seeks to analyze consumers' attitudes and preferences toward pig welfare and tries to assess the trade-off they make between the different attributes of the fresh pork meat. The empirical analysis is applied on consumers-level data collected through a survey to a sample of Spanish and British consumers during last year. The survey included three steps: 1) an initial questionnaire to elicit consumers' attitudes towards animal welfare issues (including pig castration) and sensorial quality of the meat; 2) a hedonic sensorial analysis (three meat pieces with different boar taint); and 3) a brief questionnaire to test for changes after tasting the products. The methodological approach is based on a combination of alternative tools to tackle with attitudes, preferences and willingness-to-pay issues. Cluster analysis is used to segment consumers based on results from sensory analysis. For each cluster, attitudes and preferences are analyzed through Likert scales and the Analytical Hierarchy Process (AHP). Willingness to pay for animal welfare and sensorial quality attributes are assessed also through Contingent Valuation. Results indicate that although significant differences exist between countries: 1) animal welfare issues are marginally relevant when buying pork; 2) castration is not an issue; 3) hedonism (taste and odor attributes) is preferred over animal welfare considerations as consumers' are willing to pay a higher price for a good sensory product that for animal welfare considerations.

Androstenone sensitivity of European consumers: the Spanish, French and English case

Blanch, M.[1], Panella-Riera, N.[1], Chevillon, P.[2], Gonzàlez, J.[1], Gil, M.[1], Gispert, M.[1], Font I Furnols, M.[1] and Oliver, M.A.[1], [1]IRTA-Monells, Finca Camps i Armet, 17121 Monells, Spain, [2]IFIP, La Motte au Vicomte, 35651 Le Rheu, France; marta.blanch@irta.cat

Androstenone (AND) is one of the main compounds responsible for the boar taint and its perception is genetically determined. The aim of this work was to evaluate consumers sensitivity to AND in three countries: Spain (ES, n=293), France (FR, n=138) and United Kingdom (UK, n=147). Consumers were checked for AND sensitivity by smelling crystals of pure substance, they were asked about the odour on a nine-point intensity scale (0=imperceptible, 8=extremely strong) and preference (I like/Neutral, I don't like). Consumers were classified as 'Insensitive' (0), 'Low sensitive' (1-3), 'Middle sensitive' (4-5) and 'High sensitive' (6-8). The Chi-square and Fisher's exact tests were used for data analysis. Forty-seven per cent of consumers were classified as 'Middle-High sensitive' to androstenone. Women were more sensitive than men in ES and UK (ES: 59.1% vs 46.8%, $P<0.05$; UK: 58.9% vs 36.5%, $P<0.01$), whereas sensitivity to AND was found to increase by age only in ES (from 40.0% in 18-25 years old to 64.4% in >60 years old, $P<0.05$). In all the countries, the percentage of consumers that dislike the AND odour was higher in 'Middle-High sensitive' consumers than in 'Low sensitive' ones (ES: 71.2% vs 31.8%, $P<0.001$; FR: 70.9% vs 54.2%, $P<0.05$; UK: 64.3 vs 36.8%, $P<0.05$). The percentage of potential consumers that may reject tainted meat due to androstenone was between 30.6 and 35.4% in UK, 37.9 and 40.3% in ES, and 40.6 and 49.3% in FR (those that did not like the smell and were 'Middle-High sensitive' or 'Low-Middle-High sensitive', respectively).

Consumer's opinion on alternatives for unanaesthetized piglet castration

Van Beirendonck, S.[1], Driessen, B.[1,2] and Geers, R.[2], [1]K.H.Kempen, Kleinhoefstraat 4, B-2440 Geel, Belgium, [2]K.U.Leuven, Laboratory for Quality Care in Animal Production, Bijzondere weg 12, B-3360 Lovenjoel, Belgium; sanne.van.beirendonck@khk.be

Unanaesthetized piglet castration is still a routine management practice in most countries today. Because of growing concern on animal welfare however, it is seriously questioned. A search for alternatives is in progress but besides animal welfare, the opinion of the consumer should also be considered. With this in mind 1018 people (aged 16-80 years) were interrogated in Flanders (the Dutch speaking part of Belgium), equally spread over the different regions and different social classes. Questions were about their awareness of piglet castration in general and several alternatives for unanaesthetized piglet castration like breeding intact males, immunocastration, and castration under anaesthesia combined with analgesia for the pain afterwards. The results show that 54.13% of the population still isn't aware of unanaesthetized piglet castration. After being given information about the problem and the alternatives at present, 54.97% preferred castration under anaesthesia, 26.45% was willing to accept both anaesthetized castration and immunocastration, 7.08% favoured immunocastration, 5.6% chose both anaesthetized and unanaesthetized castration, 2.75% preferred unanaesthetized castration, 1.67% accepted unanaesthetized, anaesthetized castration and immunocastration, 0.69% wanted to stop castration (with the risk of boar taint), 0.39% favoured both no castration and unanaesthetized castration, and 0.39% stated they wanted no castration and that they would stop eating pork. However, when it comes to the additional costs of these alternatives, the majority of the respondents is not willing to pay extra for their pork.

Removing the taint

Backus, G.B.C., Wageningen UR, LEI, P.O. Box 29703, 2502 LS Den Haag, Netherlands; ge.backus@wur.nl

The Dutch research programme 'Removing the taint' is an integrated approach that must ultimately lead to the production and marketing of entire male pigs. This approach consists of breeding and management measures, combined with a feed strategy and risk management on the slaughter line. The aim of the consumer study is to gain insight into consumer acceptance towards boar taint and the potential impact of this acceptance and attitude on the purchasing behaviour of the consumer. The cut chosen was loin, directly behind the ribs. Consumers will be subjected to the following research approach: pre-test in which attitudes will be measured towards castration; sensory test in which consumers rate three types of meat, a kitchen session in which boar tainted meat is cooked, a post-test in which the attitude towards castration is measured again. The aim of the breeding part of the project is to reach genetic solutions, while maintaining efficiency in reproduction and fattening. The breeding project focuses on developing and selecting low boar taint crossbreeds. Data analysis concerns charting differences between and within lines. The risk management project focuses on optimising a risk-based detection method 'human nose'. The 'economic evaluation' project looks for cost effective combinations of possible measures. The farm level part of the study uses behaviour observations to investigate whether behavioural problems occur among boars, on two commercial farms and a research station. The data to be collected is managed decentrally. Crucial features are the unique farm number for the individual farm, the tattoo number of the individual pig and the number assigned to the carcass by the slaughterhouse with the associated slaughter date. Recording this data makes it possible to determine consumer acceptance in relation to farms and crossbred products.

Expected effects on carcass and pork quality when surgical castration is omitted: results of a meta-analysis study

Pauly, C., Ampuero, S. and Bee, G., Agroscope Liebefeld Posieux, Posieux, 1725, Switzerland; giuseppe.bee@alp.admin.ch

Alternatives to the common castration (C) practice of piglets are surgical castration under anaesthesia and rearing entire males (EM) or immunocastrates (IC). It is well established that boar taint hinders the breakthrough of these alternatives. Less is known how avoiding surgical castration would affect carcass and meat quality traits. The objective of this meta-analysis was to estimate the impact of lack of castration or immunocastration on these traits. In order to build the database 26 published and 2 unpublished studies containing results of carcass characteristics and meat quality from EM, C, IC and female (F) pigs were used. In all publications 1 group of EM was present which was used as the control group in the statistical analysis. The dataset included results from 2683 EM, 3427 C, 96 IC and 3736 F and 9 traits: lean meat-%, intramuscular fat-%, initial and ultimate pH, L*-value, drip loss-%, shear force and sensory tenderness. From the published treatment means and the pooled standard deviations, the empirical effect sizes of each study were computed as the difference between treatment means of C, IC, F and the EM means, divided by the pooled standard deviation. Data were analysed as multiple-treatment studies, which accounts for the correlation of the effect sizes as introduced by the common control group. The most marked effect of castration method and gender was found in lean meat and intramuscular fat percentage. Compared to EM, carcass leanness was estimated to be 2.69, 1.77 and 0.42% greater and intramuscular fat level 0.60, 0.30 and 0.25% lower than in C, IC and F, respectively. Although significant effect sizes were found for all meat quality traits, only the difference in shear force between IC and EM was of relevant magnitude (-0.33 kg). Contrarily, tenderness and juiciness assessed by sensory analysis are not expected to differ between EM and C, IC and F. This meta-analysis revealed that the implementation of IC and EM production should not be hindered by meat quality concerns.

Quantitative genetic opportunities to ban castration

Merks, J.W.M., Bergsma, R., Bloemhof, S., Roelofs-Prins, D.T. and Knol, E.F., IPG, Institute for Pig Genetics BV, P.O. Box 43, 6640 AA Beuningen, Netherlands; Jan.Merks@ipg.nl

Boar taint, an unpleasant odour in pork of non-castrated male pigs, is mainly caused by higher levels of three compounds; androstenone, skatole and indole. The overall goal of this study was to investigate the genetic opportunities to reduce boar taint compounds. Fat samples and data on production traits were collected from 1,539 purebred entire males of a Duroc line for genetic parameters and correlation with production traits. In addition the fat samples were collected from 1,034 purebred entire males of 3 dam lines for genetic parameters and correlations with reproduction traits. The fat samples were analysed by HPLC for androstenone, skatole and indole. The average levels of androstenone, skatole, and indole in the boar line were 1.71±1.42 µg/g, 75±80 ng/g, and 48±54 ng/g, respectively. The average values in the dam lines were similar for androstenone but higher for skatole (up to 300 ng/g). The heritability estimates were 0.64±0.08, 0.36±0.07, and 0.26±0.06 for androstenone, skatole and indole, respectively in the boar line and similar values in the 3 dam lines. Genetic correlations among the boar taint compounds were very high, particularly between indole and skatole (0.83±0.06). Tthe preliminary results for correlations of boar taint compounds with daily gain and carcass traits were very low and all close to zero (e.g. -0.14 to 0.21). The preliminary correlations of boar taint components with reproduction traits were for most traits close to zero but not between androstenone and age at first mating (-0.24±0.24) and weaning to oestrus interval (-0.44±0.31). These results indicate that selection against the boar taint compounds is the most effective way of reducing the boar taint problem and should not have any significant adverse effect on daily gain, carcass traits and reproduction traits if selection against boar taint is done in a balanced way.

Impact of the non castration of male pigs on growth performance and behaviour- comparison with barrows and gilts

Quiniou, N., Courboulay, V., Salaün, Y. and Chevillon, P., IFIP, BP35104, 35651 LeRheu, France; nathalie.quiniou@ifip.asso.fr

Growth performance, carcass quality, behaviour and general condition of crossbred (Pietrain x Large White) x (Large White x Landrace) boars, barrows and gilts fed ad libitum and group-housed both during the post-weaning (8-9 pigs/pen) and the fattening (6 pigs/pen) periods. Pigs were all slaughtered on the same day. Between 28 and 63 d of age (post-weaning period), growth performance was not significantly influenced by the gender. Between 63 and 152 days of age (fattening period), daily feed intake of boars was lower than that of barrows (2.41 and 2.70 kg/d, respectively). Their average daily gain was similar not significantly different (1032 and 1069 g/d, respectively), even if barrows presented a higher growth rate at the beginning of the fattening period. Consequently, castration was associated with an increase of feed conversion ratio (2.62 vs. 2.26) and carcass fatness. Gilts' feed conversion ratio (2.48) was intermediate between those of boars and barrows. According to simulations performed with the InraPorc software over the 25-116 kg body weight range, the digestible lysine requirement per unit of net energy intake was on average 0.1 g/MJ higher for boars than for gilts and barrows. Barrows were less active than gilts and boars and had more leg problems at the end of the fattening period (lameness, bursitis). Boars presented higher lesion scores (wounds/scratches) during the first six weeks of fattening and more social behaviour. Gilts were more interested by pen features than by other pigs. Economic context needs to be considered in order to decide if dietary nutritional values have to be adapted on the basis of boars' requirements or if specific diets have to be provided to boars and gilts. More knowledge is required to characterise the boars' behaviour and general condition under restricted feeding conditions or when slaughtering occurs in different conditions (older pigs, many departures to the slaughter house).

Impact of single-sex and mixed-sex group housing of boars physically castrated or vaccinated against GnRF (Improvac®) on feeding duration and shoulder lesions

Schmidt, T.[1], Grodzycki, M.[2], Paulick, M.[1], Rau, F.[1] and Von Borell, E.[1], [1]Martin-Luther-University, Theodor-Lieser Str. 11, 06120 Halle, Germany, [2]Pfizer, Animal Health, Berlin, Germany; eberhard.vonborell@landw.uni-halle.de

As alternative to painful physical castration of male piglets a vaccine against GnRF (Improvac) is licensed in Europe. Vaccinated pigs behave like entire males up to the second vaccination and may exhibit aggressive behavior towards each other. We analysed feeding duration (FD, min per day) and indirectly measured agonistic interactions by shoulder lesion scores (LS) of physically castrated pigs (PC) and pigs vaccinated with Improvac (IC) to examine if penning with females would affect these parameters in males. PC and IC males (Pi[LRxLW], n=149, 4 replicates, 11-28wks of age, vaccinations: 12& 23wks) were penned in single-sex (ss) groups of 10 (T1&T2) or mixed (ms) with females (5:5, T3&T4). A single-space feeder (IVOG®) recorded the FD of each pig. LS were scored weekly ranging from scratches to severe wounds (scores1-6). Linear models with repeated measures design were used for analysis (fixed effects: treatment, replicate [+observer for LS data], JMP®, SAS Institute). T1 showed a greater FD (ls mean: 54.8) than T2 (38.3, $P<0.001$) and T4 (40.4, $P<0.001$) in wk20 (T3: 46.7). T2 showed greater LS (2.02) in wk19-23 than T1 (1.63, $P<0.01$). LS for ms-groups were intermediate (T3:1.83, T4:1.72). There was no treatment effect in either FD in wk25 or LS in wk24-28 after effective castration of IC males. Competition for feeder access could have reduced FD in IC groups before the second vaccination. Higher LS in ss-IC males in wk19-23 indicate more agonistic interactions. In contrast, LS of ms-IC groups were not greater than in PC treatments suggesting less aggression compared to ss-IC groups. However, mean LS were only marginally >2, indicating almost no severe injuries. This suggests that the impact of increased aggression in the IC males does not seem to constitute a major welfare problem in either ss or ms-groups.

Influence of housing condition on behaviour and sexual development in male pigs

Prunier, A., Brillouet, A., Tallet, C. and Bonneau, M., INRA, SENAH, 35590 Saint-Gilles, France; armelle.prunier@rennes.inra.fr

Rearing entire pigs may lead to meat quality and welfare problems due to sexual development. One way to reduce these problems could be to enrich the housing environment as it may influence behaviour and sexual development. Males were castrated at 5-6 days of age or left entire. From 84 days of age they were reared in groups of 10 either in a barren (1 m^2/animal, slatted floor) or enriched (2.5 m^2/animal, straw bedding and outdoor run) housing (n=20 pigs/experimental group) and fed ad libitum. Mounting behaviour was observed 3 times a month from 3 to 5 months of age by continuous sampling for 1 hour. Skin lesions (fresh lesions > 2 cm) were counted on both sides of the pigs, one day before observations. Blood samples were collected at 90, 120, 153 days of age for testosterone measurement. Animals were slaughtered at 121±9 kg live weight and 161±1 days of age (mean±SD). Testes were freed from surrounding tissues and weighed. The percentage of males with ≥ 1 mounting behaviour was higher in entire than in castrated males at all ages but was never influenced by housing. In entire males, testosterone increased with age but was not influenced by housing. Analysis of covariance showed that testosterone measured at 5 months of age was negatively related with live weight at that age. Absolute or relative testis weight was not influenced by housing. Relative testis weight was not related to testosterone measured at 3 months of age but increased with testosterone measured at 4 and 5 months of age. Number of skin lesions was not influenced by castration but was higher in pigs raised on slatted floor at all ages. At 4 and 5 months of age, the number of lesions increased with live weight in both types of males. Within entire males, the number of skin lesions increased also with testosterone at 5 months of age but not before. Overall, housing had no influence on sexual development but enrichment decreased skin lesions. Some links were shown between testosterone and testis development or skin lesions.

Alternatives to castration in pigs: an E-learning tutorial

Blanch, M.[1], Cook, J.[2], Panella-Riera, N.[1], Oliver, M.A.[1] and Thomas, C.[3], [1]IRTA, Finca Camps i Armet, 17121 Monells (Girona), Spain, [2]elearning solutions, Cotswold Road, Bristol BS3 4NT, United Kingdom, [3]EAAP, Via G Tomassetti, 0161 Rome, Italy; cledwyn.thomas@googlemail.com

e-Learning or web-based learning methods provide a flexible and innovative means of informing and training those who either do not have access to face to face to learning or are busy during normal teaching times. A web-based interactive tutorial was developed for an EU contract with the Health and Consumers Directorate-General (SANCO) entitled 'Study on the improved methods for animal-friendly production, in particular on ALternatives to the CAStration of pigs and on alternatives to the DEhorning of cattle' (Alcasde http://www.alcasde.eu). This material aims to give an overview of pig castration in Europe, the reasons for it and alternative approaches. It is intended for veterinarians, meat technologists, pig sector (producers, slaughterhouses, meat industries, etc.), and consumer organisations. It has been designed to engage the user actively through extensive use of interactive questions, interactive maps, and pop ups. The tutorial enables students to be aware of the extent of pig castration practices across the EU, know the main reasons why pigs are castrated and understand the background science of the effects of castration. They will be able to describe the process of pig castration and its consequences, be aware of alternatives to pig castration, their advantages and disadvantages and also the attitudes of producers' and consumers' towards alternatives to pig castration. The course is broken down into sections on pig castration in the EU, consequences of raising entire male pigs, methods of pig castration, alternatives to castration and finally attitudes of producers and consumers. The material concludes with an interactive summary of the main points.

Effect of dietary chicory on boar taint

Zammerini, D.[1], Wood, J.D.[1], Whittington, F.M.[1], Nute, G.R.[1], Hughes, S.I.[1] and Hazzledine, M.[2], [1]University of Bristol, Division of Farm Animal Science, University of Bristol, School of Clinical Veterinary Science, Lower Langford, Bristol, BS40 5DU, United Kingdom, [2]Premier Nutrition Products Ltd, Brereton Business Park, The Levels, Rugeley, Staffordshire, WS15 1RD, United Kingdom; dz5058@bristol.ac.uk

A total of 360 entire male pigs were used to evaluate the effects of a short feeding period with inclusion of chicory (*Chicorium intybus* L.), a source of inulin, before slaughter on skatole and androstenone levels in backfat. The pigs had been divided into 4 groups fed different levels of chicory: 0, 30, 60 and 90 g/kg diet. For each group 30 entire pigs were sampled at 3 different times: a first time (called week 0) to measure the base level of skatole and androstenone in all the pigs, then the supplement of chicory was introduced and the pigs were sampled after 1 and 2 weeks on the test diet. All 360 backfat samples from the neck region were assessed for skatole concentration; androstenone was measured in 110 pigs (all 90 g/kg pigs). Samples of backfat were presented to a 10 member taste panel after cooking for 'sniff' tests to determine if reducing skatole had also reduced boar taint. Chicory fed at the level of 90 g/kg for 2 weeks was successful in reducing skatole to a level well below the 'threshold' for this compound (0.2 µg/g) with only 1 pig with a skatole value over the threshold. In the 90 g/kg group there was a downward trend in skatole by 1 week and 0.55 of pigs had levels between 0 and 0.05 µg/g, typical of levels in castrated males. The other levels of chicory (30 and 60 g/kg) were not effective. the concentration of androstenone increased slightly in the pigs fed 90 g/kg chicory after 2 weeks. The 90 g/kg chicory group had values as high as in the other treatments. The results show that the inclusion of dried chicory in the diet for 2 weeks was effective in reducing skatole concentrations. however no improvement in the odour scores occurred, probably because androstenone remained high.

Castration under analgesia or local anesthesia: impact on pain and labour demand

Courboulay, V.[1], Hémonic, A.[1], Gadonna, M.[1] and Prunier, A.[2,3], [1]IFIP - Institut du Porc, BP 35104, 35651 Le Rheu cedex, France, [2]INRA, UMR1079 SENAH, 35590 Saint-Gilles, France, [3]Agrocampus Ouest, UMR1079 SENAH, 35000 Rennes, France; valerie.courboulay@ifip.asso.fr

Different solutions were investigated in order to reduce pain associated with piglet castration. Four treatments were compared in a first experiment: sham castration (S), castration without analgesia or anesthesia (V), castration with local anesthesia (1 ml lidocaïne 2% / testis, L) and castration with anti-inflammatory treatment (0.75 ml ketoprofene 1% / piglet, K). All treatments were distributed within litter. Behaviour, vocalisations and the duration of the surgery were recorded during castration. Behaviour was also recorded during the following hour every 2 minutes using scan sampling on 24 piglets per treatment (D0). Observations were repeated for 1 h starting 24 h after castration (D1). Blood samples were taken 30 minutes after castration on other piglets submitted to the same protocol in order to measure plasma cortisol (18 pigs/group). Piglets were weighed on D-1, D3 and at weaning. Extra labour due to the use of anaesthetic prior to castration was investigated in a second trial according to the same protocol with two different operators. All measurements indicated that pain at castration was the same for K and V piglets and reduced by an injection of lidocaïne (P<10-4). Nevertheless, cortisol levels were similar in L and V piglets but higher than in K and S piglets (196, 177, 128 and 67 ng/ml, respectively, for V, L, K and S groups, P<10-4). After castration on D0, K piglets tended to behave like S ones. Exploring and standing were more frequent whereas huddling up, isolation and desynchronization at suckling were less frequent in K and S groups than in V and L ones even though differences were not always significant. Tail wagging tended to differ between treatments at D1 (1.6%, 3.5%, 0.1% and 8.6% respectively, for K, L, S and V groups, NS). Local anaesthesia prior to castration increased labour demand by 39 to 52% but this duration could be reduced by training.

Session 17 Poster 14

Effect of housing system and slaughter strategy on physiological measurements, skin lesions and androstenone and skatol levels in entire male pigs

Fabrega, E.[1], Soler, J.[1], Tibau, J.[1], Hortós, M.[1], Puigvert, X.[2] and Dalmau, A.[1], [1]IRTA, Spain, [2]Universitat de Girona, Spain; emma.fabrega@irta.cat

Entire male production has been considered an alternative to surgical castration in piglets. The objective of this study was to evaluate the effect of different entire male housing systems (HS) and slaughter strategy (SS) on physiological and behavioural responses and welfare. One hundred and twenty (Large WhitexLandrace) xDuroc piglets were used. Pigs were allocated in 12 pens, 4 pens of males in visual contact with males (MM), 4 of males in visual contact with females (MF) and 4 of females in visual contact with males (FM). Two pens per housing system were slaughtered by split marketing (SM) in three days and the other 2 pens were slaughtered penwise (PW) when individual or pen mean body weight was around 120 kg, respectively. Saliva samples of all males were collected at around 90 days of age and the day before the first slaughter (at a mean age of 156 days). A third sample was taken for the pigs remaining in the pen after SM. Each pen was video recorded every week for three hours from 90 days of age until slaughter. Skin lesions according to the Welfare Quality® protocol were recorded every three weeks (scale 0-2) and on line according to MLC scale (1-3). No significant effect of HS on cortisol levels was observed, although MF presented higher absolute values. For both HS, the second sample was significantly higher than the first ($P<0.05$). HS or SS did not significantly affect the skin lesions evaluated on farm or on line, although MM pigs showed a tendency to present less skin lesions compared to MF ($P<0.1$). A 24% of the samples in each HS and SS presented levels of androstenone higher than 2 µg/g and 42% presented skatole concentrations between 0.05-01 µg/g. However, no effect of HS or SS on mean androstenone or skatole concentrations was observed. In the present study, the influence of housing system or slaughter strategy on cortisol, skin lesions or androstenone and skatol levels was not notorious.

Session 17 Poster 15

Relationship between the hepatic SUL2A1 protein expression and backfat androstenone level in pigs of three breeds

Panella-Riera, N.[1], Mackay, J.[2], Chevillon, P.[3], Oliver, M.A.[1], Bonneau, M.[4] and Doran, O.[2], [1]IRTA-Monells, Finca Camps i Armet, 17121, Spain, [2]University of the West of England, School of Life Sciences, Coldharbour lane, BS161QY, United Kingdom, [3]IFIP, Institute du porc, Le Rheu, 3565, France, [4]INRA, St. Gilles, 35590, France; nuria.panella@irta.es

Excessive accumulation of androstenone in pig adipose tissue contributes to the phenomenon of boar taint, an off-odour of some cooked pork. One of the reasons for high androstenone accumulation is a low rate of androstenone metabolism in pig liver. The main enzymes involved in the hepatic androstenone metabolism are 3β-hydroxysteroid dehydrogenase (3β-HSD) and hydroxysteroid sulfotransferase (SULT2A1). It has been previously suggested that the mechanisms regulating androstenone metabolism are breed specific but the nature of the breed-specificity is not well understood and the previous study mainly focused on 3β-HSD. The present study investigated the relationship between the hepatic SULT2A1 protein expression and subcutaneous adipose tissue androstenone level in three genetically diverse breeds: Pietrain (P), Duroc x Pietrain (D) and Large White x Pietrain (LW). SULT2A1 protein expression was analysed in isolated cytosol by Western blotting. Androstenone content was determined by high resolution gas chromatography. The highest androstenone level was observed in (LW) crosses followed by (P) with the lowest level in (D). Expression of the hepatic SULT2A1 protein followed the opposite pattern with the highest level in the (D) pigs followed by (P) and the lowest values in (LW). It has been suggested that breed-specific expression of SULT2A1, an enzyme catalysing conjugative stage of hepatic androstenone metabolism, is one of the factors determining breed-differences in androstenone accumulation.

Androstenone sensitivity in Spain: differences between urban and rural consumers

Panella-Riera, N., Blanch, M., Gonzàlez, J., Gil, M., Tibau, J., Gispert, M., Font I Furnols, M. and Oliver, M.A., IRTA-Monells, Finca Camps i Armet, 17121, Spain; nuria.panella@irta.es

Spanish consumers' sensitivity to androstenone (AND) - one of the main compounds responsible for the boar taint - was assessed in urban and rural consumers by smelling the crystals of pure substance. It is well known that the sensitivity to this substance is genetically determined. A total of 489 consumers (198 urban and 291 rural) were asked about their capability to smell AND (Can you smell it? No: 'insensitive'; Yes: 'sensitive') and the intensity (imperceptible, low, medium, high) and preference (I like, neutral, I don't like) of the smell. The SAS Proc Freq procedure with the Chi-square and Fisher's exact tests was used for data analysis. In general, about the thirty-three percent (32.7%) of the total population were 'insensitive' (anosmic) to the androstenone. Women were more sensitive than men (72.4% vs 61.9%, P=0.014) and the sensitivity also increased by age which was especially seen in rural consumers (from 55.4% in the 18-25 years old to 72.5% in > 60 years old, P=0.0108). Fifty-nine percent of the consumers had lived or were living in a rural environment, while the 41% were living in a city. Thus, differences were also seen when comparing rural and urban consumers. Results showed that rural consumers were more sensitive than urban ones (rural: 75.6% vs urban: 55.1%, P<0.0001). These differences were bigger in sensitive men (urban: 43%; rural: 72.5%) than in sensitive women (urban 64.3%; rural: 79%). Results showed that 16.5% of the rural consumers and 6.6% of the urban ones liked the AND smell, and 14.4% in rural and 12.6% in urban found it a neutral smell. The potential risk of consumers that may reject tainted meat due to androstenone ('I don't like the smell') was around 35% in urban and 44% in rural consumers.

(Missing) link between different boar taint detection methods

Aluwé, M.[1], Bekaert, K.[1], Tuyttens, F.[1], De Smet, S.[2], De Brabander, D.L.[1] and Millet, S.[1], [1]Institute for Agricultural and Fisheries Research (ILVO), Animal Sciences Unit, Scheldeweg 68, 9090 Melle, Belgium, [2]Ghent University, Faculty of Bioscience Engineering, Department of Animal Production, Proefhoevestraat 10, 9090 Melle, Belgium; Marijke.Aluwe@ilvo.vlaanderen.be

Research on the management of boar taint is hampered by the lack of a gold standard for measuring boar taint. In our studies, boar taint level (n=465) was tested using several detection methods: a hot iron method, a standardised and a home consumer panel (meat), an expert panel (fat, meat), and laboratory analysis of skatole, androstenone and indole (fat, serum). Armpit odour was also investigated for a number of boars after noticing a strong off-odour in the armpit area from some boars with strong boar taint according to the hot iron method. Pearson correlations among the values for the different boar taint detection methods were determined to evaluate the link between detection methods. Boar taint prevalence varied between 4 and 37%, according to the used detection method. The hot iron method was correlated moderately well with the other detection methods (r=0.13-0.42). The evaluator was androstenone-sensitive and familiar with this evaluation. The expert panel was also moderately well correlated with the other methods. Although these experts were intensively trained, this effort did not result in higher correlations with the other detection methods compared to the hot iron method. Correlation with the armpit method was 0.38 for the hot iron method and between 0.17 and 0.32 for the expert panel. The armpit method was not significantly correlated with the consumer panels. The armpit method might be a fast and easy alternative for the hot iron method as there is no need to heat the fat or meat and it is experienced as an easy matrix for evaluation compared to fat or meat. A more extensive study is needed to explore its possibilities. The link between the different detection methods is only moderate, so combining information from different methods may give more reliable results.

The effect of Meloxyvet® administration on the behaviour of piglets after castration

Van Beirendonck, S.[1], Driessen, B.[1,2] and Geers, R.[2], [1]K.H.Kempen, Kleinhoefstraat 4, B-2440 Geel, Belgium, [2]K.U.Leuven, Laboratory for Quality Care in Animal Production, Bijzondere weg 12, B-3360 Lovenjoel, Belgium; sanne.van.beirendonck@khk.be

Unanaesthetized piglet castration is a routine management practice that is much under discussion these days because of animal welfare concerns. In a search for alternatives the possible effect of an analgesic administered at the time of castration (Meloxyvet®, p.o., 0.2 mg/kg; 1.5 mg/ml) was investigated. Piglets were castrated within the first week of life, and were divided into two groups. In the first (control) group, piglets were castrated without anaesthesia or analgesia, like the common practice today. The second group of piglets was castrated right after they received Meloxyvet®. Piglets were housed in identical pens in the same room. Behaviour was scored based on an ethogram described in literature on the day of castration, and four subsequent days (except for Saturday and Sunday). All piglets were observed 10 minutes before noon and 10 minutes in the afternoon. Overall, no significant differences in behaviour between the different treatment groups were observed. The different observation periods were analyzed separately, because the analgesic needs some time before it starts working properly. Meloxyvet® was only administered at the time of castration, which explains why no differences in (pain related) behaviour between the treatment groups were observed the first 10 minutes after castration. However, several hours later, when the analgesic should already have started to kick in, also no differences in (pain related) behaviour were expressed. This may suggest that the administration of meloxicam (p.o., 0.2 mg/kg, 1.5 mg/ml) does not relieve the pain caused by castration, or that any effect on the behaviour of piglets was perceived, or that the timing and dose of application was not appropriate.

Welfare assessment of males and females on pig farms

Courboulay, V.[1], Temple, D.[2], Dalmau, A.[2], Fabrega, E.[2], Velarde, A.[2] and Chevillon, P.[1], [1]IFIP - Institut du Porc, BP 35104, 35651 Le Rheu cedex, France, [2]IRTA, Animal Welfare Unit, Finca Camps i Armet, s/n, 17121 Monells, Girona, Spain; valerie.courboulay@ifip.asso.fr

Female and male pig welfare was assessed on three Spanish and six French farms. In Spain, both genders were raised separately in groups of 4 to 16 whereas in France males were either mixed with females and/or castrated males or kept separately. Pen size varied from 6 to 30 pigs. Social, investigative and sexual behaviour, human/animal relationship, lesion scores, lameness and pressure injuries were recorded according to the methods described in Welfare Quality ® during one visit when the age of the animals ranged from 128 to 137-day old (Spain) and 136 to 150-day old (France). All mounting behaviour (M) and attempts to mount (T) were also recorded continuously during 5 minutes at the beginning of the visit. The frequency of mounting behaviour was low, but a high variability was found among farms. Five attempts to mount and two mountings behaviour were recorded within the 565 females observed and 95 and 21 respectively for the 561 entire males. A limited number of males carried out the major part of attempts to mount. In general, males were less reactive to the presence of the observer. They were also less active than the females and once disturbed, they returned more rapidly than females to a resting posture. Lesion scores were more important when pigs were mixed but there was no difference between genders: 63.3% and 68.6% of the females and the males, respectively, were not injured. When raised in single sex pens, male pigs tended to have more lesions than the females ($P<0.1$). There was no difference in lameness between genders. Raising males up to this age did not result in more problems than the ones usually met by the farmers with other pigs (castrated males). Further observations are needed for older animals.

Boar taint levels and performance data in Pietrain sired crossbred males in Germany

Frieden, L.[1], Mörlein, D.[2], Meier-Dinkel, L.[2], Boeker, P.[1] and Tholen, E.[1], [1]Institute of Animal Science, University of Bonn, Department of Animal Genetics, Endenicher Allee 15, 53115 Bonn, Germany, [2]Institute of Animal Science, Georg-August-University of Göttingen, Albrecht-Thaer-Weg 3, 37075 Göttingen, Germany; lfri@itw.uni-bonn.de

For animal welfare reasons, fattening of boars as an alternative of surgical castration of male piglets is currently discussed in Germany. In a project, funded by the Federal Ministry of Food, Agriculture and Consumer Protection (BMELV) and Federal Institute of Agriculture and Food (BLE), a total of 1000 commercial crossbred boars are under investigation. The objectives of the study are a) to analyse the performance and frequency of carcasses having boar taint of Pietrain×F1 boars, b) the elucidation of detection methods of boar taint and c) the evaluation of genetic foundation of boar taint and its relationship to maternal and paternal fertility, in order to reduce boar taint problem within Pietrain sired crossbreds. Entire male progenies of artificial insemination boars reflecting different Pietrain sire lines across Germany were tested on station. Pigs were allocated to two different slaughter weights (85 kg and 95 kg) and were housed in single pens or in groups of twelve pigs. Back fat samples were taken at the 6th/7th rib and were analysed for skatole, indole (HPLC-FD) and 5α-Androst-16-en-3-one (androstenone, GC-MS). Samples of pork loin chops were evaluated by a trained sensory panel. Preliminary results show that boars have in comparison to sows and castrates a) a significant ($P<0.05$) better feed conversion (boars: 1:2.15, castrates: 1:2.40, sows: 1:2.29), b) no significant differences in daily gain, c) higher lean meat percentage (boars: 62.4%, castrates: 58.6% and sows: 61.8%). The frequency of carcasses which exceed androstenone (500 ng/gr fat) or skatole (250 ng/gr fat) cut-off levels were 13.9% for androstenone, 9.7% for skatole, and 4.2% for androstenone and skatole. Additionally, results of sensory panel analysis and electronic detection methods will be presented.

Recent trends in mastitis and fertility indicators in the United States and reasons for change

Norman, H.D., Animal Improvement Programs Laboratory, Agricultural Research Service, US Department of Agriculture, 10300 Baltimore Ave., Bldg. 005, Room 306, BARC-West, Beltsville, MD 20705-2350, USA; duane.norman@ars.usda.gov

Milk quality and reproductive performance of US dairy cattle have changed in recent years compared to expectations from earlier trends. Although comprehensive data for milk quality are difficult to obtain for most of the US dairy industry, somatic cell counts show that milk quality is improving at an extremely favorable rate. Mean somatic cell count for herds enrolled in Dairy Herd Improvement testing declined from 322,000 to 233,000 cells/ml between 2001 and 2009. Increased herd size accounted for part of the improvement because milk quality historically has been better in larger herds. Only a small fraction of the improved milk quality is the result of genetics. For US cows bred in 2006, Holstein means for fertility indicators were 26.9 months for age at first calving, 86 days to first service (DFS), 31% for first-service conception rate (CR), 30% for all-service CR, 47% for first-service nonreturn rate at 70 days (NR70), 44% for all-service NR70, 2.5 services per lactation, 38.2 days between consecutive services, and 422 days for calving interval (CI). Jersey means were 25.6 months for age at first calving, 84 days for DFS, 39% for first-service CR, 35% for all-service CR, 53% for NR70, 48% for all-service NR70, 2.3 services per lactation, 35.5 days between consecutive services, and 410 days for CI. Use of timed artificial insemination after synchronized estrus likely has reduced DFS, CR, and CI and increased services per lactation. Use of sexed semen, which became commercially available in the United States in 2005, increased to 18% for heifer services and 0.4% for cow services by 2008. Although 90% of calves from sexed semen had the desired sex, CR was only 70% as high as with conventional semen for heifers and 83% as high for cows. Since 2002, phenotypic performance for CR and CI and genetic merit for mastitis resistance and fertility have stopped their historical declines and begun to improve.

Economic and animal aspects of Mastitis and Fertility

Coffey, M.P.[1], Wall, E.[1], Pritchard, T.[1] and Amer, P.[2], [1]SAC, Sustainable Livestock Systems, Sir Stephen Watson Building, EH26 0PH, United Kingdom, [2]Abacus Bio Limited, P.O. Box 5585, Dunedin 9058, New Zealand; mike.coffey@sac.ac.uk

Mastitis is an expensive disease of dairy cattle. Until recently, the UK relied on somatic cell count (SCC) as an indicator of udder health and for the construction of udder health sub-indices in its national profit index £PLI. SCC records are easily available in very large numbers making it an ideal phenotype for genetic evaluations. The correlation between SCC and mastitis is reasonable (~0.74) but not unity implying there is even more benefit from using direct mastitis records. The UK has begun recording mastitis events in its national recording programs which will eventually lead to a greater rate of improvement in udder health and associated economic and welfare benefit. This increase in benefit has already been seen when fertility records were made available from recording schemes and a national fertility index was constructed. Tentative trends suggest a leveling out in the rate of decline in fertility. Again, a broader breeding goal utilizing direct measures of economically important traits has economic and welfare benefits – farmers and cows benefit simultaneously.

Economics of mastitis

Huijps, K.[1,2] and Hogeveen, H.[2], [1]CRV, Wassenaarweg 20, 6800 AL Arnhem, Netherlands, [2]Faculty Veterinary Medicine, Farm Animal Health, Yalelaan 7, 3584 CL Utrecht, Netherlands; kirsten.huijps@crv4all.com

Mastitis is one of the most costly diseases in the dairy sector and improvement of the udder health situation plays an important role in achieving an efficient and economically sound business. Goal of this study was to calculate farm specific costs of mastitis and benefits of management measures to be used in decision support. The total costs of mastitis varied between €65 and €180 per average cow present at the farm per year. The farm specific costs of mastitis were calculated for 78 farmers, after they gave their estimation of these costs. It appeared that the majority of the farmers (72%) underestimated their costs of mastitis. The estimated costs ranged from €17 to €200. Many different management measures to improve the udder health status are known, but insight in the costs and efficacies of these measures was missing. This study analysed 18 management measures. To assess the effects of the measures, an extensive literature search was carried out. Next, expert sessions were organised to get estimations for missing values in literature. The effects of the measures varied for different udder health situations. The large variation in the effects found, indicate that there exist a substantial uncertainty and variation in the estimated effects of the measures. When the costs were included, keep cows standing after milking, clean milking clusters after milking a clinical case, use a separate cloth for every cow, and wear milkers' gloves proved to be most cost-effective. Because the measures with the highest effect were not always most cost-efficient, it is important to take cost-efficiency of measures also into account. With the results from this study it is possible to give insight in the farm specific costs of mastitis and the benefits of different management measures. It remains a challenge to recommend the economically most optimal management measure to improve the udder health situation. Resources such as money and labour are limited, and thus choices have to be made.

Cow-specific treatment of clinical mastitis: an economic approach

Steeneveld, W.[1], Barkema, H.[2], Van Werven, T.[1] and Hogeveen, H.[1,3], [1]Utrecht University, Utrecht, Netherlands, [2]University of Calgary, Calgary, Canada, [3]Wageningen University, Wageningen, Netherlands; w.steeneveld@uu.nl

Under Dutch circumstances, clinical mastitis (CM) cases in dairy cows are treated the same, usually with a standard intramammary antibiotic treatment. Different antibiotic treatments are, however, available for CM, differing in antimicrobial compound, route of application, duration and costs. Cow factors (parity, stage of lactation and somatic cell count history) influence the probability of cure, and therefore influence the cost-benefit of treatment. In this study, it is determined if cow-specific treatment of CM is economically worthwhile. With a simulation model, 20,000 CM cases were simulated. For each simulated CM case, the consequences of using different antimicrobial treatment regimes (standard intramammary, extended intramammary, combination intramammary+systemic, and combination extended intramammary+systemic) were simulated simultaneously. Finally, total costs of the treatment regimes were compared. All input for the model was based on literature information and authors knowledge. Cure for each individual cow depended on the choice of treatment, the causal pathogen and the cow factors parity, stage of lactation and somatic cell count. Total costs for each case were dependent on treatment costs (including treatment costs for new CM cases), milk production losses and costs for culling. Mean total costs for the 4 treatments were €181, €197, €201 and €215, respectively. Mean probabilities of cure for the 4 treatments were 0.53, 0.67, 0.67 and 0.72, respectively. For all different CM cases, the standard intramammary treatment had the lowest total costs. The benefits of lower costs for milk production losses and culling for cases treated with the intensive treatments, did not outweigh the higher treatment costs. In conclusion, although effectiveness of different antimicrobial treatments does differ between cows, differentiation of treatments for different CM cases does not give economic benefits.

Economic aspects of dairy fertility in the USA

De Vries, A., University of Florida, Dept. of Animal Sciences, Bldg. 499, Shealy Drive, Gainesville, FL 32608, USA; devries@ufl.edu

Dairy cattle fertility in the USA decreased until approximately the year 2000 after which improvements are observed. These improvements are likely the results of an increased emphasis on genetic selection for health and fertility since the 1990s, and better reproductive management. Reproductive management receives much attention in extension programs, but economic aspects need to be carefully compared. Poor estrus detection is perceived to be a problem and failure to get pregnant is a major reason for culling. Timed artificial insemination (timed AI), which uses estrus synchronization, and the use of natural service bulls are two breeding programs that are widely used. The 2007 NAHMS Dairy survey reported that 58% of all surveyed dairy farms used timed-AI programs to manage reproduction in heifers, cows, or either. NAHMS also reported that natural service bulls were used for the first insemination by 33% of the surveyed dairy farms for heifers and 22% of the surveyed dairy farms for cows. Many dairy farms use a mixture of timed-AI and natural service bulls. Reproductive summary statistics from DairyMetrics (www.drms.org) for early 2010 show an average of 95 days to first insemination, 158 days open, 14.1 month calving interval, 43% estrus detection rate, 44% first insemination conception rate, and 16% pregnancy rate in 14,378 herds. Pregnancy rate is the key reproductive measure in the USA and is calculated as the fraction open eligible cows for insemination that conceived in a 21-day period. A 1-percentage point increase in average pregnancy rate is valued at approximately $25/cow per year. The value is greater in herds with lower reproductive performance. Most economic analyses show that timed-AI is more profitable than AI based on observed estrus. Well managed natural service bull programs can be cost competitive with AI programs. The use of drugs in timed-AI programs could become a consumer issue while natural service bulls may pose a safety risk. In conclusion, opportunity remains to improve dairy cattle reproductive management and associated profitability in the USA.

Session 18 Theatre 6

The economic importance of fertility in beef cattle breeding

Åby, B.A.[1], Vangen, O.[1], Sehested, E.[2] and Aass, L.[1], [1]Norwegian University of LIfe Sciences, Department of Animal and Aquacultural Sciences, P.O. Box 5003, 1432 Ås, Norway, [2]GENO Breeding and A.I. Association, P.O. Box 5003, 1432 Ås, Norway; bente.aby@umb.no

Many studies on the economic importance of functional traits have been carried out in dairy cattle, and more emphasis are put on them in breeding goals for many dairy breeds. However, regarding beef cattle, the research and inclusion of functional traits in breeding goals is limited. Published bio-economical models for beef cattle usually includes a limited amount of functional traits, thus the knowledge about the real importance of these traits is limited. It is to be expected that fertility traits account for a large proportion of the functional traits as income from slaughter normally represent the only source of income. The aim of this study was to estimate the economic importance of functional traits with empasis on fertility traits. A bio-economic model was developed simulating the lifetime production of one suckler cow of an intensive breed kept under semi-intensive production conditions (intensive indoor feeding of fattening animals, extensive cow-suckler cow enterprice). Data for base situation was obtained from the Norwegian beef cattle recording scheme. A typical management system according to Norwegian practice was assumed: indoor winter feeding (roughage + concentrates) and summer pasture. Economic values were estimated as the increase in profit from a 0.1% increase in the mean of the trait considered. The model considers six production and eight functional traits (six fertility traits). Preliminary results suggest that fertility traits were more than twice as important as production traits. Relative economic values for fertility traits were: stillbirth 64.0%, age at first calving 15.5%, calving interval 10.5%, calving difficulty 5.5%, twinning frequency 5% and birth weight 0%. Results suggest that fertility traits are economically the most important traits and should, to a larger extent than at present, be included in breeding goals for beef cattle.

Session 18 Theatre 7

Health monitoring system in Austrian Dual Purpose Fleckvieh cattle: incidences and prevalences

Schwarzenbacher, H.[1], Obritzhauser, W.[2], Fürst-Waltl, B.[3], Köck, A.[3] and Egger-Danner, C.[1], [1]ZuchtData, Dresdner Straße 89/19, 1200, Austria, [2]Österreichische Tierärztekammer, Biberstraße 22, 1010, Austria, [3]University of Natural Resources and Applied Life Sciences Vienna, Gregor Mendel Straße 33, 1180, Austria; Schwarzenbacher@Zuchtdata.at

In 2006 a project to establish a health monitoring system for cattle based on veterinarian diagnoses in Austria has been started. The data is primarily used for breeding value estimation and for management purposes. Besides that the system provides for the first time reliable information in Austrian cattle populations on frequencies of diseases that are important with regard to economic and animal welfare aspects. Only validated records from farms with at least 0.1 first diagnoses per cow and year were taken into account. Data from the year 2008 on length of observational episodes and frequency of first diagnoses in 29 diseases from 79,444 Fleckvieh animals on 3,183 farms was included in the analysis. Diseases were grouped in the complexes calf diseases, digestive tract, metabolism, fertility, udder health, feet and legs and miscellaneous. The most frequent disease complex in Austrian Fleckvieh is fertility with incidences and prevalences of 17.04% and 22.47%, respectively. Within the fertility complex ovarian cysts and stillbirth show the highest incidence with 7.37 and 5.18%. After fertility, udder health is with an incidence and prevalence of 13.21 and 17.82% the disease complex with highest frequencies. Surprisingly, acute mastitis is more frequent than chronic mastitis with incidences of 8.21% and 5.88%, respectively. Metabolic diseases are surprisingly rare with frequencies of 3.12% (incidence) and 3.23% (prevalence). This might be due to a lower rate of interventions by veterinarians in metabolic diseases compared to udder and fertility diseases, leading to underestimation of the incidence. The results from Austrian Fleckvieh cattle show the importance to improve health by selection on health traits and management measures.

Session 18 Theatre 8

Economic status in dairy herds ranked by a reproductive performance indicator that accounts for the voluntary waiting period

Emanuelson, U.[1], Löf, E.[1,2] and Gustafsson, H.[1,2], [1]Swedish University of Agricultural Sciences, Clinical Sciences, P.O. Box 7054, SE-75007 Uppsala, Sweden, [2]Swedish Dairy Association, P.O. Box 210, SE-10124 Stockholm, Sweden; ulf.emanuelson@kv.slu.se

A good reproductive performance is a central driving force in dairy herds to achieve a profitable production. Monitoring the performance is therefore an important component of herd management and several reproductive performance indicators are in use. In a previous study we compared how well a number of indicators could discriminate between herds with good or bad reproductive management efficiency and between herds with good or bad biological reproductive status. The results showed that a reproductive performance indicator, that accounts for the voluntary waiting period (VWP) of a herd, i.e. the percentage cows pregnant after the herd VWP plus 30 days (PPVWP+30), was to be preferred when both the purpose of the monitoring, ease of use, and preparedness for differences in management and future changes in management was considered. The aim of the current study was to evaluate how PPVWP+30 relates to the technical and economic status of a herd. A dynamic, stochastic simulation model, SimHerd, was used to generate data representing herds with different reproductive status. A total of 18 different scenarios were simulated by altering VWP, reproductive management efficiency and biological reproductive status of the herd. Each scenario was simulated over 10 years with 50 replications. The technical results showed a wide range of the observed values of PPVWP+30. The average PPVWP+30 for herds with good reproductive management efficiency and good biological reproductive status was thus 0.38 while it was 0.10 for herds with poor reproductive management efficiency and poor biological reproductive status. The average calving intervals for these two scenarios were 378 and 415 days, respectively. The overall correlation between the herd net return and PPVWP+30 and average calving interval was 0.75 and 0.66, respectively.

Session 18 Theatre 9

Correlated selection responses for fertility after selection for protein yield or mastitis in a selection experiment with Norwegian Red cows

Heringstad, B.[1,2] and Larsgard, A.G.[2], [1]Department of Animal and Aquacultural Sciences, Norwegian University of Life Sciences, P.O. Box 5003, N-1432 Ås, Norway, [2]Geno Breeding and A. I. Association, Ås, N-1432, Norway; bjorg.heringstad@umb.no

Correlated selection responses in female fertility were found in a selection experiment with two groups of Norwegian Red cows selected for high protein yield (HPY) and low mastitis frequency (LCM), respectively. Genetic trends were calculated for non-return rate within 56 days (NR56) for heifers, first lactation cows, and 2nd and 3rd lactation cows, calving interval between 1st and 2nd calving (CI), and interval from calving to first insemination (CFI) for first lactation cows and for 2nd and 3rd lactation cows. Genetic trends for the two selection groups were calculated as mean EBV per cow generation, for cow generations 0 to 6. Permutation tests, where cows were assigned randomly to 2 groups, were used to test whether the mean EBV in the two selection groups were significantly different. A total of 5,001 cows from the selection experiment had fertility EBVs, of which 2,806 were HPY and 2,195 were LCM cows. Permutations tests showed significant genetic differences between LCM and HPY for all fertility traits except CFI for 2nd and 3rd lactation cows. Observed differences were with few exceptions far outside the range from the permutation test, implying that observed differences are unlikely to have occurred by chance. LCM cows were in general genetically better for fertility than HPY cows, with higher NR56 in heifers and cows, shorter CI and shorter CFI in 1st lactation. The genetic differences after 6 cow generations were 2.5%-units NR56 in heifers, 2%-units NR56 in cows, and 4 days CI. No difference was found for CFI in 2nd and 3rd lactation.

Session 18 Poster 10

Investigation of SCC in the milk of Holstein cows in Greece

Pilafidis, O.[1], Tsiokos, D.[2] and Georgoudis, A.[2], [1]Greek Ministry for Rural Development and Food, Genetic Improvement Centre of Drama, KGBZ - DRAMAS, DRAMA - 661 00, Greece, [2]Aristotle University of Thessaloniki, Fac. of Agriculture - Dept of Animal Production, Greece, University Campus, Thessaloniki - 541 24, Greece; andgeorg@agro.auth.gr

Mastitis is still one of the most important reasons for culling in the Greek dairy sector affecting a large part of the dairy production chain. Mastitis affects the health of milk-producing animals, has consequences for the profitability of dairy farms and the animal welfare. Moreover, mastitis negatively influences the milk quality having consequences for the dairy processing industry. This study reflects the current knowledge of mastitis and mastitis control in Greece and also gives suggestions for further control measures. The study included the results of the analysis of qualitative characteristics of milk samples of individual cows from a recorded Holstein population of 4.613 cows from 34 dairy herds. The total number of records was 28.136 covering the period from January 2007 to March 2008. The average 305 days production was 8.760 kg milk and the content of fat, protein and lactose in % was 4.10, 3.52 and 4.91 respectively. The average SCC was 485.000/ml. So far, the quality of the milk delivered to the dairy factories was satisfactory. However, the results show that the dairy herds have a significant proportion of animals with subclinical mastitis (20%) so that farmers suffer severe economic losses primarily due to reduced production and by culling of animals. Specifically, producers loose 550 kg. of milk on average in each lactation, which quantity is equivalent to a loss of about 230 to 250 € per cow per year. This can be avoided by systematic monitoring of individual animal samples each month along with the establishment of a commonly accepted system of payment for the milk producers. Furthermore due to ongoing scientific and practical efforts, the control of mastitis in Greece, taking into consideration the increasing complexity of the farming systems, is improving either by prevention or by adequate measures (e.g. therapy).

Session 18 Poster 11

Economic losses resulting from subclinical mastitis in dairy cows

Jemeljanovs, A., Cerina, S. and Konosonoka, I.H., Latvia University of Agriculture, Research Institute of Biotechnology and Veterinary Medicine Sigra, Instituta street No 1, LV 2150, Sigulda, Latvia; sigra@lis.lv

The study aims to ascertain the number of losses due to subclinical mastitis in dairy cows in Latvia. It has been seen that the most frequent cause of subclinical mastitis is *Staphylococcus aureus* in dairy cows. The disease leads to major economic losses to farms with a high density of cows. The farms with a number of cows exceeding 500 have approximately 30-40% of cattle with mastitis. The disease in farms has caused significant losses, most of which consists of a reduction in productivity (45-50%), treatment costs (20-25%), elimination of cow (30-35%). On average 10.5% of all eliminated cows is being removed from the herd due to udder diseases. It is estimated that a decrease in productivity due to mastitis in cows reaches the loss of approx. 113.00 Euro a year. This amount of economic loss consists of reduced milk production -68%, improper milk elimination -11%, expenditures for veterinary services and medicinal preparations -8%, increased labour input -1.9%, and herd restoration related with the number of liquidated diseased cows -11.1%. It should be noted that 22% of diseased cows in the farm may have been confronted with a repeated mastitis during the lactating period in addition to economic losses. Approx 1.65 kg milk per cow has been lost due to mastitis which may reach 503 kg during the whole period of lactation. To minimize economic losses resulted from subclinical mastitis it is necessary to comply with animal welfare requirements, to ensure the proper functioning of the milking equipment, the proper preparation of milking the cows, balanced feed and the use of vaccines for the treatment of mastitis according to the scheme. A vaccine for mastitis prophylaxis and treatment *S. aureus* has been developed and experimentally tested. The number of diseased cows has been reduced by 65% after the first vaccination.

Factors influencing the results of resynchronisation programs with and without progesterone supplementation

Tsousis, G.[1], Forro, A.[2], Sharifi, A.R.[3] and Bollwein, H.[2], [1]Aristotle University of Thessaloniki, School of Veterinary Medicine, Clinic of Farm Animals, St. Voutyra 11 str., 54627, Thessaloniki, Greece, [2]University of Veterinary Medicine Hanover, Clinic for Cattle, Bischofsholer Damm 15, 30173, Hanover, Germany, [3]University of Göttingen, Institute of Animal Breeding and Genetics, Albrecht-Thaer-Weg 3, 37075, Göttingen, Germany; tsousis@vet.auth.gr

The aim of this study was to examine the effect of a progesterone (P4) supplementation during the first seven days of resynchronisation with the OvSynch protocol on pregnancy rates (PR). Initially, German Holstein Friesian cows were randomly synchronized at Day 60 after parturition with two different OvSynch protocols (with the 2nd GnRH injection either at 48 or 60 h after the PGF2α injection). At Day 4 p.i. half of the animals were supplemented with progesterone for 14 days. On Day 33 p.i. a pregnancy diagnosis was conducted with the use of an ultrasound device. Non pregnant animals were then resynchronised with the OvSynch protocol and in a random 50% P4 was supplemented (P4+) during the first seven days of the protocol. A generalized linear model with the SAS Proc GLIMMIX was used to analyse the data of 133 animals. Variables showing $P<0.1$ remained in the model. The previous synchronisation protocol and the previous P4 supplementation had no effect on the PR after ReSynch. Additionally there was no overall difference between the P4+ and P4- animals (37.3% vs. 45.5%, respectively). An interaction of the P4 supplementation with the Body Condition Score of the animals at the day of the ReSynch was significant. Specifically, P4- animals showed better PR in BCS until 2.75, whereas more P4+ cows became pregnant in BCS > 3.25. In conclusion, the synchronisation protocols used did not seem to influence the PR after resynchronisation of dairy cows. In this study, a progesterone supplementation during the ReSynch interacted with the body condition of the animals. This finding should be taken into account when such a protocol is applied.

Effects of Vitex Agnus-Castus on serum progesterone concentrations in dairy cows

Farhoodi, M., Khorshid, M. and Ataee, O., Faculty of Veterinary Medicine, Islamic Azad University-Karaj Branch, Clinical Sciences, Moazen Blvd, Islamic Azad University-Karaj Branch, Karaj, Iran; farhoudi@kiau.ac.ir

Vitex agnus-castus is a native shrub of the Mediterranean region. It is dopaminergic and has been used as a remedy for low progesterone concentrations, corpus luteum deficiency, etc. in women, for more than 2,500 years. In this study, Holstein dairy cows were divided into control (n=7) and treatment (n=7) groups. After 21 days a prescription of 50 ml hydroalcoholic extract of the plant in treatment group was orally assigned, both groups were synchronized by two intra muscular injections of PGF2α in 11 days apart. Prescription of herbal extract in the treatment group continued for 23 days, till the end of synchronized cycle. During this oestrus cycle blood samples were collected and progesterone concentrations of separated serums were measured by RIA, moreover, ovaries and uterus of cows were examined by ultrasonography. Results showed that 45 days prescription of the extract in the treatment group increases the average serum progesterone concentration versus the control group (2.88±0.63 ng/ml, and 2.19±0.57 ng/ml, respectively) with about 32% during the study cycle, but it is not statistically significant ($P>0.05$). In attention to worries about probable health risks of using hormones in reproductive management in herds for meat and milk consumers and according to an increasing effect of Vitex on progesterone concentration which in this study has been revealed, perhaps it can be a safe alternative remedy in the reproduction management of dairy cows.

Session 18 Poster 14

Antagonistic effects of *Lactococcus lactis* PTCC 1403 on bovine mastitis pathogens: *in vitro* study

Farhoudi, M., Mottaghian, P., Mousakhani, F. and Dini, P., Faculty of Veterinary Medicine, Islamic Azad University-Karaj Branch, Clinical Sciences, Moazen Blvd, Islamic Azad University-Karaj Branch, Karaj, Iran; farhoudi@kiau.ac.ir

Bovine mastitis is the most common cause of antibiotic administration in dairy herds. Due to the importance of decreasing antibiotic residues in milk for public health and also high expenses and to some extent inefficiency of antibiotic treatments, it is necessary to investigate alternative methods for prevention and treatment of mastitis. Nowadays administration of probiotics is considered as an alternative method for prevention and treatment of infections. The aim of this study was to assess the *in vitro* antagonistic effects of *Lactococcus lactis* PTCC 1403 as a probiotic on bovine mastitis pathogens. The antagonistic activity of *L. lactis* PTCC 1403 was evaluated by using agar spot test against contagious and environmental mastitis pathogen strains isolated from mastitis milk specimens referred to the laboratory. Antibiotic susceptibilities of pathogenic strains to commonly used antibiotics were also investigated using disc diffusion method. *L. lactis* PTCC 1403 produced antagonistic effects against all the pathogenic strains including both gram negative and gram positive pathogens except Clostridium perfringens. The greatest inhibitory effects were against Pseudomonas aeroginosa and Arcanobacterium pyogenes with an average inhibition zone of 15.2 mm and 14.8 mm respectively. It was also shown that environmental mastitis pathogens were more susceptible to inhibitory effects of *L. lactis* PTCC 1403 than contagious mastitis pathogens. Therefore, we concluded that *in vitro* growth inhibition of mastitis pathogens by probiotic potentials of *L. lactis* PTCC 1403 may contribute to the development of *in vivo* approaches for treatment and prevention of bovine mastitis.

Session 19 Theatre 1

Horse behavioural genetics: a review

Benhajali, H., Université Rennes 1, UMR 6552 Ethologie Animale et Humaine, 263 Avenue du Général Leclerc, 35042, France; benhajali@hotmail.com

The existence of a genetic component has been shown in a variety of behavioural traits like fear, aggression and emotionality. Selection experiments for specific behaviours have been successfully accomplished in a variety of species such as foxes, mink and chickens and high heritabilities were estimated for some traits. In some countries like New Zeeland and USA, behavioural reaction to human handling has been included in selection of beef cattle. In horses, as animals are often dispersed in a variety of environmental conditions, it is still difficult to estimate genetic parameters for behavioural traits. However, Behavioural similarities in animals issued from the same sire or belonging to the same breed are observed in foals and adult horses both in freely expressed behaviours (play, distance to the dam) as well as in reactions to experimental situations (emotional or learning tests). Stereotypic behaviours seem to appear more frequently in some breeds and / or families. Individual variations in foals behaviour can be observed at early ages and seem to predict some adult tendencies. It seems therefore that including some behavioural traits in selection is possible in horses especially since the first estimated heritabilities are promising. A chronological review of the history of research in behavioural genetics is presented with a particular interest for horses.

Evaluation of mental traits at young horse performance test

Jönsson, L.[1], Viklund, Å.[1], Höög, Y.[2], Rundgren, M.[2] and Philipsson, J.[1], [1]Swedish University of Agricultural Sciences, Animal Breeding and Genetics, Box 7023, 75007 Uppsala, Sweden, [2]Swedish University of Agricultural Sciences, Animal Nutrition and Management, Box 7024, 75007 Uppsala, Sweden; asa.viklund@hgen.slu.se

In a questionnaire to breeders, trainers, riders and riding schools in Sweden, 41% of the respondents meant that temperament in horses was inadequately considered in the genetic evaluation. Riding Horse Quality Test (RHQT) is a performance test for 4-year-old Swedish Warmblood horses (SWB). Besides scores for health, conformation, gaits and jumping, a score for temperament (1-10) is given by each judging team. The objective of this study was to investigate the possibility to assess defined mental traits at an existing performance test, and if those traits were correlated to the scores for temperament. Mental traits were observed for 60 horses during RHQT by documenting degree of expression on a linear scale for the traits calm, lively, shy from behind, spooky, tense, prone to freeze, flighty, attentive towards the environment and towards the handler. Calm was negatively correlated to lively and tense, whereas flighty was positively correlated to tense. The trait attentive towards the handler was positively correlated to temperament scores from all judges (0.46 to 0.71). Mental traits negatively correlated to temperament scores were tense, spooky and prone to freeze. Genetic analyses of traits evaluated at RHQT showed that temperament scores given by judges for jumping and gaits resulted in heritabilities of 0.17 and 0.41 respectively. High genetic correlations (0.76-0.97) between scores for gaits or jumping and respective temperament indicated that the temperament scores largely reflect the talent of the horse. The study indicates that it might be possible to evaluate specific mental traits during a performance test and that the studied mental traits would be complementary to the temperament score given by the regular judges. Further studies may focus on 3-year-old horses as they are less handled and more prone to react to the environment.

Should applied temperament tests in horses be conducted with or without a human rider and/or handler?

König Von Borstel, U., Euent, S., Graf, P. and Gauly, M., University of Göttingen, DNTW, A.-Thaer-Weg 3, 37075 Göttingen, Germany; koenigvb@gwdg.de

Temperament tests in horses increasingly receive attention both in research and practical horse breeding; However, the question whether or not riders or handlers should be part of the test is still subject of a controversial debate. Therefore, the aim of the present study was to compare horses' heart rate (HR) and behaviour in the same temperament test when being ridden (R), led (L), and released free (F). Behavioural measurements included scores and linear measurements for reactivity (Rs), activity (A), time to calm down (T) and emotionality (E), recorded during the approach (1) and/or during confrontation with the stimulus (2). 65 horses were each confronted 3 times (1xR, 1xL, 1xF in balanced order) with 3 novel and/or sudden stimuli. Mixed model analysis indicated that L resulted in lowest ($P<0.05$ throughout) reactions as measured by A1, A2, E1, E2 and Rs2, while R produced highest (A1, T2, HR) or medium (E1, E2, Rs2) reactions. F resulted either in highest (A2, E1, E2, Rs2) or in lowest (A1, T2, HR) reactions. The repeatability across tests for HR (0.57) was higher than for any behavioural measurement: The latter ranged from values below 0.10 (A1, A2, T2) to values between 0.30 and 0.45 (E1, E2, Rs2). In contrast, correlations between pairs of test-types were highest for Rs2 (R&L: 0.66), followed by T2 (L&F: 0.54), but lower for HR (R&L:0.34; L&F:0.38; R&F:0.15 ($P=0.09$)). Overall, the results show that a rider or handler influences, but not completely masks, the horses' intrinsic behaviour in a temperament test. If a combination of observed variables is chosen with care, a valid assessment of a horse's temperament may be possible both when leading or riding a horse, as well as when observing the horse when let loose. However, when a temperament test aims to supply horse breeders or riders with results applicable to practice, it is recommended to use a testing situation that resembles the practical circumstances most closely, i.e. testing riding horses under a rider.

Practical assessment of reactivity and associations to rideability and performance traits

Rothmann, J.[1,2], Søndergaard, E.[3], Christensen, O.F.[2] and Ladewig, J.[1], [1]University of Copenhagen, Faculty of Life Sciences, Grønnegaardsvej 8, 1870 Frederiksberg, Denmark, [2]University of Aarhus, Faculty of Agricultural Sciences, P.O.Box 50, 8830 Tjele, Denmark, [3]AgroTech, Udkærsvej 15, Skejby, 8200 Århus N, Denmark; jrn@dsr.life.ku.dk

Reactivity in horses has an impact on the welfare of horses and on human injuries. Behaviour tests have been developed to assess reactivity but several of the tests are difficult to apply in traditional breeding programs. The purpose of this study was to investigate the possibility of measuring reactivity using a behaviour score based on the level and degree of arousal and aversive behaviour in a practical situation. Further, the purpose was to investigate how reactivity associates with rideability and performance traits. A total of 322 Warmblood horses were scored during the field tests, and a questionnaire was filled in by the owners. No significant effects of location, trainer and length of training were found on the behaviour score, but location and trainer had an effect on rideability and performance traits. Therefore, the associations between reactivity and performance traits were investigated both by computing partial correlations adjusting for location and trainer (Pearson, rp) and Spearman correlations (rs). A low correlation was found between ratings from owners and reactivity (rs=0.15, P=0.02), indicating that horses considered nervous by their owner also were scored as highly reactive. Likewise, a low correlation were found between rideability and reactivity (rp=-0.16 P=0.02) and between reactivity and free jumping (rp=-0.14, P=0.03). On the contrary, no association between reactivity and the performance traits in dressage was found. The conclusion is that, to some extent, it is possible to measure reactivity in a practical situation despite the different locations and background of the horses. The results also suggest that highly reactive horses received lower grade in both rideability and free jumping. However, further research is needed to support these conclusions.

Implementation of temperament tests in performance tests on station and field

Graf, P., Koenig Von Borstel, U. and Gauly, M., Livestock Production Group, Department of Animal Sciences, Albrecht-Thaer-Weg 3, 37075 Goettingen, Germany; koenigvb@gwdg.de

As more and more breeding associations realize the importance of behaviour traits a move towards a more objective assessment of temperament in performance tests can be seen. So the aim of this study was to evaluate a temperament test for implementation into performance tests in practice. A total of 254 horses were tested at 10 different performance tests in Germany. 14 professional test riders rode the horses. Based on the literature and on our previous studies, an indoor course of three stimuli (1: blue ball lying on the ground; 2: grayish-brown floor mat; 3: blue ball rolling towards the horse) was designed. A trained observer took live behavioural observations and the test riders gave a grade for each horse. Statistics were calculated using the procedure Univariate in SAS Version 9.2. Horses' scores in the present test were lower and had higher standard deviations, than corresponding scores from the conventional personality evaluation. Mean reactivity scores (Rs) were highest with stimulus 1 (8.1±1.7). Rs scores for stimulus 2 resembled a normal distribution most closely (skewness = 0.18, kurtosis = -0.28), though not statistically confirmed (Kolmogorov-Smirnov: P<0.01), while scores for stimuli 1 and 3 were skewed towards the high scores (1: skewness = -1.54, kurtosis = 3.40; 3: skewness = -1.27, kurtosis: 1.92). Correlations of scores between stimuli ranged between 0.49 (Stimulus 2 & 3, P<0.001) and 0.62 (Stimulus 1 & 2) with the correlation between stimulus 1 and 3 (0.53, P<0.001) taking the rank in the middle. The values suggest, that similar features are tested by the stimuli. The strong agreement between test-rider and observer scores indicates a high inter-observer reliability, and proves the robustness of the observation scale. The present study showed that an implementation of temperament tests into horse performance tests is possible. An improvement of the current assessment of horses' personality traits may be possible with a temperament test.

Analysis of factors affecting gestation losses of mares

Langlois, B., Blouin, C. and Chaffaux, S., INRA, Animal genetics, Bige-INRA-CRJ, 78 352 Jouy-en-Josas, France; bertrand.langlois@jouy.inra.fr

The files for ultrasound diagnosis of gestating mares belonging to the French equine herd recorded for three consecutive years were joined with the files for foal birth of these same mares, allowing the statistical analysis of factors of pregnancy loss. For 28 872 positive diagnoses of gestation, 2 898 losses were recorded, that is a global rate of gestation interruption of 9.12%. The etiology of these interruptions is mainly extrinsic: the year and month of insemination as well as region. The intrinsic causes that are implicated are breed of the father (heavy breeds lose fewer pregnancies than pure-bred breeds), age of the mother (losses are lower in mares of 7 to 10 years of age) and status (mares with foals have fewer pregnancy losses than mares not having foaled the previous year), as well as fetuses with consanguinity (when this increases, the pregnancy losses increase as well). However, the additive effect is extremely low; it corresponds to a heritability below 5% and few effects of environment, common to breeding mares, were identified. This therefore, gives little hope of being able to select against the 'gestation loss' trait.

Revitalization of Medimurje horse: mitochondrial DNA evidence

Cubric-Curik, V.[1], Frkonja, A.[1], Kostelic, A.[1], Bokor, A.[2], Ferencakovic, M.[1], Tariba, B.[1], Druml, T.[3] and Curik, I.[1], [1]Faculty of Agriculture University of Zagreb, Department of Animal Science, Svetosimunska 25, 10000 Zagreb, Croatia (Hrvatska), [2]University of Kaposvár, Faculty of Animal Sciences, Department of Production and Breeding of Ruminants and Horse, Guba S. u. 40., Kaposvár, 7400, Hungary, [3]BOKU, Gregor-Mendel-Strasse 33, A-19 Vienna, Austria; vcubric@agr.hr

Medimurje horse is autochthonous breed named after northern region of Croatia that is separated by two rivers Mura and Drava. The history of this draft horse is linked to the Austro-Hungarian Empire where it was spread over the region that today is Austria, Croatia, Hungary and Slovenia. So, the breed is also recognized under names Muraközi lo (in Hungarian language), Murinsulaner (in German language) or Medimurski konj (in Croatian language). The origin of the breed is linked to the Noric horse, Arab horse and cold blooded horses from Belgian, Croatia and Germany. Today, there is only 38 horses registered as purebred and the breed is critically endangered in Croatia, while found in traces in Hungary and Slovenia. As a part of revitalization plan of Medimurje horse, we have performed mtDNA analysis. Our goal was to establish maternal lines for future pedigree and mating policy as well as to compare maternal origin of Medimurje horse with other neighboring horse breeds. Thus, we took samples from local population of Posavina horse (7), Croatian coldblood horse (8), Medimurje horse (37) and from Hungary Muraközi lo (23), Noric horse from Austria (38), Traditional Arab horse from Bosnia (10) and Bosnian mountain horse (5). In the analyses we also included published sequences referred to the 610-bp fragment of mtDNA (see Aberle et al. 2007). Only for Croatian and Hungarian samples related to Medimurje horse, we found 20 different mtDNA haplotypes arising from 32 variable sites. Shared haplotypes and 'mutational vicinity' of large number of haplotypes indicated closeness of Croatian and Hungarian populations of Medimurje (Muraközi) horse.

Session 19 Theatre 8

Genetics of brachygnathism in Peruvian Paso Horse

Mantovani, R.[1], Giosmin, L.[1], Sturaro, E.[1], Cecchinato, A.[1] and Wong Ponce, L.R.[2], [1]University of Padua, Department of Animal Science, Agripolis, Viale Universita, 16, 35020 - Legnaro (PD), Italy, [2]ANCPCPP, Registro Genealógico, Asociacion Nacional de Criadores y Proprietarios de Caballo Peruano de Paso, Bellavista, 456 - Miraflores, Lima 18, Peru; roberto.mantovani@unipd.it

The aim of this study was to assess the prevalence of brachygnathism and to investigate a possible genetic variation of such a trait in the Peruvian Paso Horse. A technical report form accounting for six degrees of a 'mouth closing' scale was design to collect field data from Peruvian Paso studs. An overall of 495 horses, 169 stallions and 326 mares were evaluated in 34 studs near Lima, grouped in 9 different districts that were used as main environmental effect. Data editing and preliminary analysis lead to use only four of the six different degrees, considering grade 0 as absence, 1 as moderate, 2 as mild and 3, 4 and 5 as severe defects. The incidence of moderate defect resulted 32.1% (26.6% on males, M and 35.0% on females, F), while mild or severe defect showed respectively an incidence of 13.3 (8.9% M and 15.6% F) and 5.7% (6.5% M and 5.2% F). Three different analyses were conducted: a linear animal model (LAM1) with brachygnathism classified into four categories (previously described), a linear animal model (LAM2) with brachygnathism classified into two categories (not affected, grade 0 and 1 or affected, grade 2-5) and a sire threshold model (TSM) with brachygnathism classified into two categories. All models included the systematic effects of sex and district, the LAM1 and LAM2 accounted for the additive genetic effect of animals (1,314 animals in the pedigree) while the TSM accounted for the additive genetic effect of sires. Heritability estimates of brachygnathism ranged from 0.09 to 0.20 with LAM1 and from 0.19 to 0.20 with LAM2, respectively. The estimation on the liability scale with TSM resulted 0.22. Results suggest that additive genetic variation of brachygnathism seems large enough in Peruvian Paso Horse to be exploited in specific breeding programs.

Session 19 Theatre 9

Analyses of performance traits in the Norwegian Warmblood

Furre, S., Heringstad, B. and Vangen, O., Norwegian University of Life Sciences, Department of Animal and Aquacultural Sciences, P. Box 5003, NO-1432 Ås, Norway; siri.furre@umb.no

Young horse performance testing is highly appreciated as a tool in breeding organisations for riding-horses throughout Europe for predicting the young horses' value as a performance- and/or breeding animal. The aim of this study is to present the first genetic analyses of traits recorded in the Norwegian Warmblood population, NWB, including heritability estimates. In Norway the performance test, the Young Horse Quality Test (YHQT), has been organised each year since 1981. The data analysed comprised 584 mares, geldings and stallions that had completed the YHQT from 1981 to 2006. Traits tested were health, temperament, conformation and performance traits such as gaits and jumping ability. The traits in this study were analysed for effects of year, sex, country of birth and starting group (four groups each test day where the first 25 percent of the horses, accordig to the start list, was in start group one, the next 25 percent was in start group two and so on). Year had significant effect on all traits, and showed an increase in the overall phenotypic quality of the horses. Stallions received higher scores for conformation and jumping than mares and geldings, and higher scores ($P<0.05$) for gaits than mares. Imported horses received significantly higher scores for conformation, gaits and jumping ability, both free-jumping and jumping under rider, than the Norwegian born horses. Starting group has signifikant effect on the performance scores. Horses starting in groups 3 and 4 received significantly better scores than horses starting in groups 1 and 2 ($P<0.05$). Heritability estimates were low to medium (0.06-0.18) for health and temperament, and medium (0.26-0.33) for conformation, walk, trot and canter under rider as well as the overall score for gaits under rider. Results obtained in this study suggest that data from the Norwegian Warmblood population is of sufficient quality to allow further statistical analyses and breeding value estimations.

Session 19 Theatre 10

Genetic trends for performance of Swedish Warmblood horses

Viklund, Å., Näsholm, A., Strandberg, E. and Philipsson, J., Swedish University of Agricultural Sciences, Animal Breeding and Genetics, Box 7023, 75007 Uppsala, Sweden; asa.viklund@hgen.slu.se

Genetic trends for performance in dressage and show jumping of Swedish Warmblood (SWB) horses were studied. Breeding objective for SWB is to produce internationally competitive horses in both disciplines. Data included competition results for 43,337 horses since 1961, results from 19,307 4-year-old horses tested at Riding Horse Quality Test (RHQT) since 1973, results from 10,911 horses tested at young horse test for 3-year-olds (YHT) since 1999, and additionally 160,000 horses in the pedigrees. For genetic evaluation a BLUP animal model has been used since 1986, initially based on RHQT data only. In 2006, a BLUP index system using YHT and RHQT data together with competition data was introduced. Genetic progress in both disciplines increased substantially in the middle of 1980s due to stronger selection of stallions based on a stallion performance test introduced in 1977 and testing of young horses in RHQT. Show jumping reached the highest rate from the beginning of 1990s as a result of importation and strong selection among tested stallions. Trends of breeding stallions and brood mares were analyzed separately and compared with geldings as a reference population. The selection differential for show jumping stallions was for a 10-year period as high as 1.2 genetic standard deviations but has leveled off as the genetic level of the population has increased. For dressage stallions the selection differential was kept rather constant at 0.6 genetic standard deviations. For brood mares the average genetic level was similar to that of the reference population for both disciplines. However, the introduction of YHT increased the proportion of well tested mares selected for breeding. In the future emphasis should be put on more effective selection of superior stallions, both from other populations, but foremost from the SWB population, which has reached high genetic levels in a few decades time, and to encourage breeders to use only the better tested mares for breeding.

Session 19 Theatre 11

Breeding evaluation of the linear conformation traits analyzed in Spanish Purebred Horses (PRE)

Sanchez, M.J.[1], Gómez, M.D.[1], Azor, P.J.[1,2], Molina, A.[1] and Valera, M.[3], [1]University of Cordoba, Genetics, C.U. Rabanales. Ed. Gregor Mendel Pl Baja, 14071 Cordoba, Spain, [2]ANCCE, Cortijo del Cuarto. Bellavista, 41014 Seville, Spain, [3]University of Seville, Agroforestal sciences, E.U.I.T.A. Ctra. Utrera km1, 41013 Seville, Spain; pottokamdg@gmail.com

A linear assessment methodology was developed for the morphological data collection in the Spanish Purebred Horse. The final design included 31 linear traits. In the last two years, a total of 2153 registered records from 1515 horses were collected. Genetic parameters were estimated following a REML methodology (animal model) for he 31 collected traits. Sex (2 classes), region (where the stud of birth is located, 44), event*judge (33) and age (8) were included as fixed effects in the model. The animal was also included as random effect All the ancestors of the recorded horses were added to the pedigree file for the genetic evaluation, including at least four generations for each analyzed horse, giving a total figure of 8,793 horses. Heretability values ranged between 0.185±0.063 and 0.395±0.047 for the head and neck traits (8 traits), 0.080±0.048 and 0.352±0.051 for the body traits (5), 0.108±0.057 and 0.369±0.050 for the forelimb traits (6), 0.039±0.007 and 0.285±0.062 for the hindlimb traits (10) and 0.247±0.041 and 0.266±0.046 for the general traits (2). Linear traits related to riding and dressage performance were selected and combined in multicharacter indeces for the evaluation of the morphological aptitude of the animals for riding and dressage ability. Dressage index consisted of traits related to head, neck, body, forelimb and hindlimb. Riding index consisted of traits related to harmony, breed type and functional chracteristics.

Breeding value indexes in the selection of Hungarian Sporthorses

Posta, J., Mihók, S. and Komlósi, I., University of Debrecen, Institute of Animal Science, Böszörményi str. 138., Debrecen, H-4032, Hungary; postaj@agr.unideb.hu

Mare performance tests for the Hungarian Sporthorse population were evaluated. Data from the period of 1993-2009 were used, covering scores of 618 3-year-old and 310 4-year-old mares, 109 of them were tested at both ages. Ten conformational, three free jumping performance and four movement analyses traits were scored on the tests. Breeding value estimation was based on BLUP animal model. Test year, age and owner were included in the model as fixed effects. Variance components were estimated with VCE-6 software package. Heritabilities ranged from 0.32 (frame) to 0.50 (saddle region) for conformation traits, from 0.39 (jumping style) to 0.49 (jumping ability) for free jumping traits and from 0.20 (walk) to 0.47 (canter) for movement analysis traits. Breeding value indexes were constructed for each trait group. Conformation index was computed based on the weighted scores of the breeding values of conformational traits. The conformational score scales were used as weightings. Free jumping and movement indexes contain the proper breeding values with equal weights. A total index was also constructed using conformation index, two times the free jumping index and two times the movement index. Each breeding values and breeding value indexes were presented with the mean 100 and standard deviation of 20 for the easier understanding.

Usability of detailed information on movement characteristics of mares and foals for breeding purposes in the German Warmblood horse

Becker, A.-C.[1], Stock, K.F.[1,2] and Distl, O.[1], [1]University of Veterinary Medicine Hannover, Institute for Animal Breeding and Genetics, Buenteweg 17p, 30559 Hannover, Germany, [2]Vereinigte Informationssysteme Tierhaltung w.V., Heideweg 1, 27283 Verden / Aller, Germany; Kathrin.Stock@vit.de

Correctness and quality of gaits are routinely evaluated in mares on the occasion of studbook inspection and are already used for genetic evaluation. However, certain characteristics of individual motion pattern may be already visible at younger age. Detailed judgements of movement and conformation of 2,631 foals and 2,945 mares provided the basis of multivariate genetic analyses which were performed in linear animal models with REML. All 5,576 horses were presented to the official evaluation commission of the Oldenburg Society in 2009. Mean evaluation age was 2.3 months in the foals and 9.8 years in the mares. Linear scales were used for quality of walk and trot, and conformation of legs and back. Binary coding was used for motion characteristics like indications of imbalance, irregular tail posture, irregular tail tone and irregular motion pattern in hind legs. Judgement was equally performed in foals and mares in addition to the standard evaluation. In both age groups, deviations from regular motion pattern occurred very rarely (maximum prevalence of 2%). Slight signs of imbalance were seen more often, with abnormal tail tone or posture being recognized in 5.1% of foals and 3.9% of mares. After transformation to the underlying liability scale for the binary traits, heritability estimates were mostly moderate, ranging between 0.16 and 0.50 (standard error 0.06-0.15). Additive genetic correlations between analogous traits in foals and mares were highly positive (> 0.95), implying possible combined use of foal and mare data. According to our results, breeding progress in the Warmblood horse may benefit from extension of currently used evaluation schemes and more extensive use of information on motion pattern in foals and mares.

Genetic evaluation of dressage performances in Belgian sport horses

Peeters, K.[1], Ducro, B.J.[1] and Janssens, S.[2], [1]WUR, Animal Breeding and Genomics Centre, P.O. Box 338, 6700 AH Wageningen, Netherlands, [2]K.U. Leuven, Department Biosystems, Kasteelpark Arenberg 30, 3001 Leuven, Belgium; katrijn.peeters@wur.nl

The objective of this study was to find a method on how to genetically evaluate dressage performances. The Belgian Equestrian Federation and the Rural Riders Association, provided 100,303 and 173,917 repeated measurements on 7,620 and 12,248 horses, respectively. Within these datasets, two problems pose. First, dressage performances get recorded on different levels, varying from starters' level until Grand Prix level. Therefore, level-transformations were performed in order to rate the dressage performances by their true value; implicating that the competition results were upgraded according to the level on which the animal competed. Second, dressage performances are highly influenced by the quality of the rider. Therefore, a random rider-effect was added to the model. The effect of these measures on the heritability (h^2) and repeatability (r) and on the selection index was investigated. In general, the applied level-transformations resulted in an increase of h^2 and r, indicating that they uncover a stronger genetic predisposition for 'dressage talent'. In addition the level-transformations have a substantial influence on the ranking of the estimated breeding values (EBVs). The inclusion of a random rider-effect resulted in a decrease of h^2 and r. Confounding of the rider-effect with the genetics of the horse resulted in an overestimation of h^2 and r when the rider-effect was ignored and an underestimation of h^2 and r when the rider-effect was included. Yet, no substantial influence of the inclusion of a random rider-effect was found on the ranking of the EBVs. In practice the studbook should perform a level-transformation if it fits within the scope of its breeding goal. Additionally, confounding of the rider-effect with the genetics of the horse is a problem. Yet, since the inclusion of a random rider-effect did not substantially influence the selection index, one could discuss its importance in the model.

Genetic evaluation of sport horses for performance in Eventing competitions in Great Britain

Stewart, I.D.[1], Brotherstone, S.[1] and Woolliams, J.A.[2], [1]Institute of Evolutionary Biology, University of Edinburgh, Kings Buildings, West Mains Road, Edinburgh EH9 3JT, United Kingdom, [2]The Roslin Institute, Royal (Dick) School of Veterinary Studies, University of Edinburgh, Roslin, Midlothian, EH25 9PS, United Kingdom; I.D.Stewart@sms.ed.ac.uk

In Great Britain, Eventing competitions have four ability levels for all three disciplines. A genetic evaluation requires 12*12 genetic and environmental (co)variance matrices. The objective of this work was to investigate different methods for production of these large positive definite (co)variance matrices. Competition results (penalty points) were obtained from British Eventing and converted to normal scores. Mixed effects models and REML in AsReml (Gilmour et al., 2006) were used to estimate variance components. Horse gender, age and competition class were considered as fixed effects, and sire, horse and rider as random effects. Two methods were investigated. 1) A series (66) of bivariate models, for every discipline-grade combination and subsequent use of a Cholesky decomposition to produce data for a 12-trait multivariate analysis. 2) The 'Impute' function in AsReml was used to conduct a 12-trait analysis. This splits the full model into 3 submodels and oscillates between submodels in each iteration. In (2), grade was modelled using an antedependence model of order 1 and correlations were assumed constant across grades within disciplines. The likelihood of the Cholesky decomposition (co)variance matrices was significantly greater than the Impute method. Heritabilities from the Cholesky decomposition were highest for show jumping (16.2% in advanced to 8.9% in prenovice). Within discipline genetic correlations were high; dressage (0.984-0.720), show jumping (0.982-0.783) and cross-country (0.988-0.328). Between discipline genetic correlations were low – moderate. The Cholesky decomposition method is most appropriate for the current analysis. A 12*12 covariance matrix for the production of breeding values has been produced.

A preliminary analysis of conservation programme for cold blooded horses in Poland

Polak, G.M., National Research Institute of Animal Production, Department of Animal Genetic Resources - National Focal Point, ul. Wspolna, 30, 00-930 WARSAW, Poland; Grazyna.Polak@minrol.gov.pl

The cold blooded horses were bred in Poland from the middle of XIX century. In the first 50 years of XX century six different populations of cold blooded horses were present within our territory: sokolski, sztumski, oszmanski, garwolinski, lowicki and kopczyk podlaski. These local varieties were created on the basis of autochthonous mares crossed with stallions imported from France (Breton, Ardennes), as well as Belgium and Germany. After the II World War the changes in the agricultural economy resulted in substantial decrease of population of local cold blooded horses. In 2008, an introduction of the animal genetic recourses conservation programme for cold blooded horses of sokolski (lightest type) and sztumski (heaviest type) type has created opportunity for restitution of these two native breeds. The present study is a first attempt to analyze situation in these breeds. Currently there are about 320,000 horses kept in Poland, with only about 40,000 of cold blooded horses entered into the stud books. Out of those 40,000, only about of 1000 represent sokolski, sztumski type and qualify to participate in the conservation programme. The general criteria for participation in the conservation programme include typical morphological conformation and desired, well defined pedigree. In the first year of implementation, 320 sokolski and 220 sztumski mares were qualified to join the programme. Initial group of mares represented respectively 11 and 9 regions of the country. In the second year, already 86% of mares of sokolski, and 65% of mares of sztumski breed, entering the programme, represented three most typical regions of origin of each breed. In 2008, the average body measurements (height at withers and metacarpal circumstance) were as follows: 156.5 cm and 24.9 cm for sokolski, and 157.1 cm and 25.2 for sztumski mares. In 2009, both parameters indicated enhancement of morphological differences between these two breeds.

Genetic parameters for body measurements in Menorca Horses: preliminary results for the development of a linear assessment methodology

Gómez, M.D.[1,2], Sanchez, M.J.[2], Sole, M.[2], Valera, M.[3] and Molina, A.[2], [1]A.C.P.Caballos De Raza Menorquina, C, Bijuters 17, 07760 Ciutadella De Menorca, Spain, [2]University Of Cordoba, Genetics, C.U. Rabanales. Ed. Gregor Mendel Pj. Baja, 14071 Cordoba, Spain, [3]University Of Seville, Agro-Forestal Science, E.U.I.T.A. Ctra. Utrera Km1, 41013 Seville, Spain; pottokamdg@gmail.com

The Menorca Horse is an endangered breed located in the Balearic Island. Its breeding programme is being developed by the breeders Association since 2007, in order to select the reproducers and to maintain the genetic variability. They are selected by functional (Classic Dressage and a special type of Dressage from Menorca called Menorca Dressage) and conformational objectives. In order to develop a linear assessment methodology for the selection of conformation traits, genetic and statistical analysis of 46 body measurements were made. A total of 179 horses (85 males and 94 females), aged between 3 and 15 years-old, were included in the study. All of them were highly related by their pedigree. Genetic parameters were estimated following a REML methodology (animal model). Sex (2 classes), stud (18) and age (8) were included as fixed effects in the model. The animal was also included as random effect. Heritability values ranged between 0.27±0.215 and 0.88±0.15 for the body lengths (9 traits), 0.17±0.011 and 0.64±0.233 for the body widths (4), 0.03±0.011 and 0.20±0.077 for the angles (8), 0.28±0.168 and 0.80±0.006 for the perimeters (6), 0.08±0.020 and 0.71±0.186 for the head and neck traits (8), 0.06±0.011 and 0.74±0.188 for the forelimb traits (3), 0.06±0.011 and 0.28±0.232 for the back and loins traits (2), and 0.03±0.011 and 0.69±0.231 for the hindlimb traits (6). Genetic correlations between traits were also estimated. The 62.72% of them were higher than 0.75, and 29.84% of them ranged between 0.25 and 0.50.

Repeatabilities of jumping parameters of KWPN stallions on the second step of selection

Lewczuk, D., Institute of Genetics and Animal Breeding Jastrzebiec, Polish Academy of Science, ul.Postepu 1, 05-552-Wolka Kosowska, Poland; d.lewczuk@ighz.pl

Data were collected during the second step of the KWPN stallion selection. In total 53 horses and 353 jumps were filmed from the audience place – press sector, situated perpendicular to the final jump in the line. The linear measurements were achieved by treating the current height of the obstacle as a scale (100%), the temporal measurements were achieved by calculations of frames of the film (25 frames per second). Linear and temporal variables were calculated from the selected frames. The repeatabilities were calculated using procedure Mixed from the SAS program calculated from the model with the random effect of the horse and fixed effects of the obstacle (different heights of verticals and oxers) and successive number of the jump. The repeatabilities of temporal parameters were 0.27 for landing time and 0.42 for taking off time. The repeatabilities for linear measurements of taking off and landing was 0.46 and 0.53 respectively. The lowest values were achieved for lifting of hind legs (0.4). Higher values were calculated for front legs and reached 0.5. The elevation of bascule points were above 0.5 and the position of the head 0.4. The effect of the obstacle was statistically significant for all parameters, that was not observed for the effect of the successive number of the jump.

The methods of horses performance tests in the Czech Republic

Navratil, J., Jezkova, P. and Stadnik, L., Czech University of Life Sciences Prague, Kamycka 129, Prague 6 - Suchdol, CZ - 165 21, Czech Republic; stadnik@af.czu.cz

The aim of this abstract is to outline the development of performance tests (PT) in the Czech Rep., their history and the current methods of implementation at the various breeds. Utility of horse depends mainly from their performance. It already aware of our ancestors. The first description PT comes from antiquity and makes it Kikkuliš for chittite king. In 1954 was create czech testing method, the basis of testing method of Testing institute at Westercelle. From 1956 to 2000 were PT adjusted by cz norm ČSN 466310 - Breeding horses. The PT are obligate for Old Kladruby horses and stallions of Shagya-arabien, cold-blooded breeds and Hucul. Warm-blooded stallions can compensate PT by tests of sporting. Hafling and Pony stallions can participace in PT, but are not compulsory. For mares except Old Kladruby Horse mares PT are optional. The results of PT mares of Czech warm-blooded horse at reference period are biased by low number of daughters of individual stallions (1033 mare of 252 stallions; x = 4,1 mare/stallion for all watching period and 2,06 - 2,98 mare/stallion in the separate years). Between the stallions, who had a higher number of daughters (10 and higher) was dominated stallion Ballast s.v., the worst were Minerál s.v. and Przedswit XVI-64. In evaluating the PT mares are not used a Ten-scale and the resulting marks are occur in a very narrow margin (from 6,1 to 9). The narrow margin of marks awarded to rated mares show close range of their rated fathers, which greatly hampers their objective evaluation.

Assessment of genetic variability and relationships among the main founder-lines of the Spanish Menorca horse breed using molecular coancestry information

Azor, P.J.[1], Solé, M.[1], Valera, M.[2] and Gómez, M.D.[1], [1]University of Cordoba, Dpt. Genetics, C.U. Rabanales. Ed. Mendel, pl. baja., 14071 Córdoba, Spain, [2]University of Seville, Dpt. Agro-forestry Science, Ctra. Utrera, km 1, 41013 Seville, Spain; pottokamdg@gmail.com

The Menorca Horse is an Spanish endangered horse breed, mainly located in the Balearic Island. Although there are some important populations in other European countries, as France, Italy, Holland... It is characterized by its black coat colour and its relationship with the popular activities in the Island and its environment. A study of the actual genetic variability has been assessed in order to quantify the genetic diversity of the population within its breeding and conservation programme. The genotypes of 753 registered horses were analyzed for 16 microsatellite markers in the Central Laboratory of Veterinary (Algete), belonging to the Spanish Ministry of Environment and Rural and Marine Affairs. The average number of alleles per locus was 7.69, and the average expected and observed heterocigosities were 0.710 and 0.723, respectively. The average molecular coancestry value (fij) for the whole population was 0.291 and the mean self-coancestry value was 0.638. Genetic relationships among three groups of horses by their founder-lines (same line via father and mother) were also estimated using molecular coancestry information. A total of 134 horses were included in the analysis. All of them belongs to the main founder-lines registered in this breed: Mudaino (M), Torretrençada (T) and Son Quart (S). For these three groups, the average molecular coancestry was 0.312 and the mean self-coancestry was 0.661. Mean kinship distance between groups was 0.322. The between groups molecular coancestry values (fij) varied from 0.263 (for the S-T groups pair) to 0.288 (for the M-T pair). Molecular coancestry values within groups varied from 0.318 (for S group) to 0.399 (for the T group). F-statistics, FIS, FST, and FIT for the groups analysed were, respectively -0.026, 0.038 and 0.013.

Analysis of genetic structure of 'Maremmano tradizionale' horse by means of microsatellite markers

Matassino, D.[1], Blasi, M.[2], Costanza, M.T.[3], Incoronato, C.[1], Negrini, R.[4], Occidente, M.[1], Pane, F.[1], Paoletti, F.[3], Pasquariello, P.[1] and Ciani, F.[1], [1]ConSDABI-CC (Mediterranean biodiversity) National Focal Point italiano FAO AnGR, Località Piano Cappelle, 82100 Benevento, Italy, [2]LGS, Via Bergamo, 26100 Cremona, Italy, [3]ARSIAL, Via R. Lanciani 38, 00162 Roma, Italy, [4]Università Cattolica del S. Cuore, Via Emilia Parmense 84, 29122 Piacenza, Italy; consdabi@consdabi.org

The present contribute is included in a wide safeguard program of 'Maremmano Tradizionale' horse ancient autochthonous genetic type (AAGT) that traces back to a mesomorph horse broadly represented by nascent civilizations of Near East and Northern Africa since 2nd millennium B.C.. This investigation was aimed at giving a first contribution to the knowledge of genetic structure of this AAGT through the preliminary typification of 134 subjects (97♀♀ and 37♂♂) at 11 microsatellite loci and the comparison of these data with those of other horse genetic types (GTs) {Maremmano [N=210 (132 ♀♀ e 78 ♂♂)]; Bardigiano [N=114 (39 ♀♀ and 75 ♂♂)]; Murgese [N=50 (18 ♀♀ and 32♂♂)]} typified at Laboratory of Genetics and Services (Cremona Italy). The significance of the differences was tested by Student's 't' test. In the limits of the observation field, the preliminary results evidenced that 'Maremmano Tradizionale' shows a satisfactory genetic variability: Na =9.09±1.92, c.v.% = 21; Ne = 4.99±1.64, c.v.% =33; Ho=0.735±0.086, c.v.% =12; He=0.779±0.073, c.v.% =9; this variability is statistically higher in female subjects. If compared with the other three GTs, 'Maremmano tradizionale' showed a higher variability estimated through Na, Ne, Ho and He. A pairwise comparison between the 4 genetic types highlighted that 'Maremmano tradizionale AAGT – Maremmano GT' pair has the lowest Fst value (0.0022), whereas 'Murgese-Bardigiano' the highest value (0.1100); all the six possible comparisons were significant ($P<0.05$). In conclusion, the two populations of Maremmano would derive from the same ancestor.

The impact of the breeding policy on the genetic structure of the Spanish Sport Horse

Bartolomé, E.[1], Cervantes, I.[2], Gutiérrez, J.P.[2] and Valera, M.[1], [1]Department of Agro-forestry Sciences. University of Seville., Ctra. Utrera, 1, 41013, Spain, [2]Department of Animal Production. UCM, Avda. Puerta del Hierro, 28040, Spain; ebartolome@us.es

The development of the Spanish equine industry has motivated the creation of a new breed, the Spanish-Sport horse (SSH), a composite breed with genetic influence from other Spanish and foreign breeds. At the beginning, this Studbook included any individual participating in sports competitions and not registered in any other Studbook. The SSH Studbook was closed in 2004 but is still possible to include individuals of other Spanish breeds as parents of a SSH. Moreover, it is possible to include Sport Horses registered in foreign Studbooks (SSHex) recognized by theWorld Breeding Federation for Sport Horses. Animals registered in this Studbook (from their foundation to December 2009), were used in this study. The total available records were 34017. The reference populations finally included the registered individuals after de Studbook was closed (4289 individuals). Since 2005, the 26.5% of the stallions were Spanish Purebred (SPB), 16.6% were SSH, 9.7% were Arab horses (A) and 47.2% other breeds, regarding females 43.0% were SSH, 12.9% SSHex and the rest were other breeds. The contributions of the founders belonging to the paternal breeds showed that in origin the Thoroughbred was the highest contributor (18%) followed by the SPB (15%). The inbreeding coefficient and average relatedness were 0.66% and 0.16%, respectively. The mean of equivalent complete generations was 3.8. The impact of a new regulation of the Studbook in which only individuals SSH could act as parents was considered in order to attain the possible change in the genetic structure and the effective size. The realized effective size using SSH individuals was 225.8, but if we removed SSH individuals with an external parent from the dataset, the Ne decreased to 51.4. Even though SPB stallions were preferred, aggregated founder contribution from breeds reflects that Thoroughbred genetic background was rather indirectly selected.

Session 19 Poster 23

Pedigree analysis of the German Paint Horse

Fuerst-Waltl, B., Mitsching, A. and Baumung, R., University of Natural Resources and Applied Life Sciences Vienna, Department of Sustainable Agricultural Systems, Gregor Mendel-Str. 33, A-1180 Vienna, Austria; birgit.fuerst-waltl@boku.ac.at

The genetic variability of the German Paint Horse population was analysed by pedigree analysis. Data were provided by the Paint Horse Club Germany e.V. (PHCG). The total pedigree file included 14,313 horses. The reference population was defined as Paint Horses born in Germany in the years 2000 to 2009 with both parents known. These 1,661 animals could be traced back to 4,216 founders. The effective number of founders was 561, the effective number of ancestors 207 and the effective number of founder genomes 139. The chestnut coloured Quarter Horse stallion 'Doc Bar' born in 1956 was found to be the most important ancestor and had a marginal contribution of 2.9%, followed by 'Cherokee Blanca' a black Tobiano Paint Horse born in 1993 with a marginal contribution of 2.8% and 'Cats Coco Dancer' a chesnut Tobiano Paint Horse born in 1991 with a marginal contribution of 2.3%. The 5 most important ancestors explain about 11% of the total genetic variability in the reference population, 50% is explained by 124 most important ancestors. The average complete generation equivalent for the reference population ranged from 4.1 to 4.7 depending on year of birth. Thus, results should be carefully interpreted due to partly incomplete pedigrees as no full pedigree information is provided for imported breeding stock. Next steps will include further analyses based on supplemented pedigrees.

Session 19 Poster 24

Pedigree analysis of the German Icelandic Horse

Baumung, R., Geng, M. and Fuerst-Waltl, B., Univ. of Natural Resources and Applied Life Sciences, Gregor Mendel-Str. 33, 1180 Vienna, Austria; roswitha.baumung@boku.ac.at

The genetic variability of the German Icelandic Horse population was analysed with pedigree analysis. Data were provided by World Fengur. The total pedigree file included 325,044 horses. The reference population was defined as Icelandic Horses born in Germany in the years 2000 to 2009 with both parents known. The average complete generation equivalent of this reference population increased from 6.9 for birth cohort 2000 to 7.7 for birth cohort 2009. These 10,839 animals could be traced back to 5,692 founders. The effective number of founders was 143, the effective number of ancestors 44 and the effective number of founder genomes 22. The stallion 'Hrafn' born in 1968 was found to be the most important ancestor and had a marginal contribution of 7.6%, followed by 'Sörli' born in 1964 with a marginal contribution of 7.2% and 'Sörli' born in 1916 with a marginal contribution of 6.1%. The 10 most important ancestors were all born in Iceland, 8 out of them were stallions, 7 were black. They explain about 40% of the total genetic variability in the reference population, 50% is explained by the 19 most important ancestors.

Session 19 Poster 25

Coat greying process in Old Kladruber horse

Hofmanova, B., Majzlik, I., Vostrý, L. and Mach, K., Czech University of Life Sciences, Dept.of Animal Science and Ethology, Kamýcká 129, 16521Prague 6, Czech Republic; majzlik@af.czu.cz

The aim of this study was to check and explain the dynamics and its parameter in greying process of gene resource of Old Kladruber Horse that is a typical grey breed. The data collection was carried out in four main studs by measuring (using Spectrophotometer Minolta 2500D on four body parts) and inspecting 376 horses repeatedly during four consecutive years with total of 702 records. In statistical analyses the GLM procedures of SAS package were used and REML VC5 respectively. The influence of line, sex, stud and year of the evaluation on speed of greying were analysed. Our results confirm all knowledge on greying known in other grey horse breeds (Sölkner et al.2004). The significant impact on greying has been shown within the age of the horse followed by sex (higher level of greying in mares). The four different body parts measured for the greying showed no significant diferences. The heritability of greying process characterized by L* reached 0.52 (L* - the parameter of black - white colour axis in range 0 - 100, i.e. higher L* value means more 'white' grey horse). The changes of greying are described using four 'growth functions', out of which the best fit was obtained by logistic function (R^2= 0.96). This project was supported by grant MSM 604 607 09 01

Session 20 Theatre 1

Milk and its evolution

Oftedal, O.T., Smithsonian Environmental Research Center, 647 Contees Wharf Rd, Edgewater MD 21037, USA; oftedalo@si.edu

Milk originated as a skin secretion providing moisture to eggs in Carboniferous/Permian synapsids. Generic antimicrobial and secretory constituents, such as lysozyme and secretory calcium-binding phosphoproteins (SCPP), were subsequently modified for nutritional purposes. Lysozyme evolved into α-lactalbumin, key to synthesis of lactose and lactose-based oligosaccharides. Ancestral SCPP evolved into proto-casein. Lipids produced by apocrine secretion evolved into milk fat globules with specific membrane proteins. Each milk constituent likely had an earlier function in skin secretions; the genetic and secretory machinery of ancestral mammary glands probably resembled extant apocrine glands. The origin of milk may be nearly as old (300 million years) as the appearance of fully terrestrial vertebrates (amniotes). Early synapsids laid parchment-shelled eggs intolerant of desiccation and dependent on skin secretions. By the Triassic, mammaliaforms (mammalian ancestors) were endothermic (requiring fluid to replace incubatory water losses), very small in size (making large eggs impossible) and had limited tooth replacement (indicating delayed onset of feeding and reliance on milk). A nutritionally complete milk replaced yolk nutrients and supported immature hatchlings, as in living monotremes. A prolonged lactation with large changes in milk composition appears ancestral to mammals. In monotremes and marsupials sequential lactation stages involve secretion of different milk constituents. The evolution of placental structures in eutherian mammals allowed prepartum nutrient transfer to embryos, birth of precocial young and contraction of lactation. Eutherian mammals with altricial young (e.g., bears) exhibit large changes in milk composition over lactation, but those with precocial young (e.g., equids, ruminants) produce milks that change little, but vary greatly among taxa. Milk fat ranges from <0.5% in rhinos to 61% in seals. The convergent evolution of low fat milks in hominids, ruminants and equids may have facilitated development of domesticated dairy species.

Session 20 Theatre 2

Colostrum immunoglobulins profile is determined by placenta structure

Argüello, A.[1], Castro, N.[1], Morales-Delanuez, A.[1], Moreno-Indias, I.[1], Sánchez-Macías, D.[1], Torres, A.[1], Ruiz-Díaz, M.D.[1], Hernández-Castellano, L.E.[1] and Capote, J.[2], [1]Universidad de Las Palmas de Gran Canaria, Animal Science, Transmontaña s/n, 35413 Las Palmas, Spain, [2]Instituto Canario de Investigaciones Agrarias, Animal Production, Aptdo. 60, La Laguna, Tenerife, Spain; aarguello@dpat.ulpgc.es

Colostrum is the first mammary gland secretion after partum in placental mammals. It is possible to observe two different groups of animals in reference to colostrum composition. The first group is characterized for an IgG rich colostrum and non transfer of immunoglobulins during the prepartum period. These animals shown wide placentas with low contact between mother and fetus classified as epitheliochorial (sow, mare) or synepitheliochorial (cow, ewe, goat). In this group, offspring needs the immunoglobulins (mainly IgG) from colostrum to avoid the failure of passive immune transfer that increases the possibilities to die during the first two months of life. The second group of animals is characterized for colostrums rich in IgA and presence of immunoglobulin transfer from mother to fetus through placenta during pregnancy. These animals display thin placentas with high contact between dam and fetus classified as endotheliocorial (dog, cat) or hemocorial (human, monkey). The offspring in that group does not depend of colostrum feed for survival but this colostrum fed improves the intestinal lumen immunity troughs secretory IgA. It is important to remark the different origins of IgG or secretory IgA in both groups of animals, while IgG is transferred using an exclusive receptor from blood to colostrum, secretory IgA is produced for B cells around lactocytes. In conclusion, the question that remains none answered is, does the placenta structure determine the colostrum composition or vice versa?.

An odorant-binding protein in bovine colostrum

Fukuda, K.[1], Senda, A.[1], Ishii, T.[2], Morita, M.[3], Terabayashi, T.[4] and Urashima, T.[1], [1]Obihiro University of Agriculture and Veterinary Medicine, Animal and Food Hygiene, Nishi 2-11, Inada-cho, Obihiro, Hokkaido, 080-8555, Japan, [2]Obihiro University of Agriculture and Veterinary Medicine, Pathobiological Science, Nishi 2-11, Inada-cho, Obihiro, Hokkaido, 080-8555, Japan, [3]Toko Pharmaceutical Industries Co., Ltd., Tokyo Laboratories, 199-14 Shikahama, Adachi-ku, Tokyo, 123-0864, Japan, [4]Kitasato University, Chemistry, 1-15-1 Kitasato, Sagamihara, Kanagawa, 228-8555, Japan; fuku@obihiro.ac.jp

Odorant-binding proteins (OBPs) are water-soluble proteins with molecular masses of around 19 kDa. Biological function of OBPs is supposed to be a transporter of small hydrophobic molecules such as odorants. We have obtained evidence that a putative odorant-binding protein (designated bcOBP) occurs in bovine colostrum. To investigate the expression pattern of bcOBP, hybridoma cell line that produced monoclonal antibody against bcOBP was established. The cDNA of bcOBP was synthesized by overlapping PCR. Using pET28a, N-terminal His-tagged recombinant bcOBP (rbcOBP) was expressed in *E. coli* BL21 Star(DE3)pLysS and purified by Ni+-Sepharose affinity chromatography. Three female BALB/c mice were immunized by 100 μg of rbcOBP. Splenocytes were collected from the immunized mice and fused with myeloma cells. Theoretical pI and molecular mass of bcOBP were calculated to be 4.57 and 19604.18, respectively. The highest sequence similarity (83%) was observed with a potential pheromone transporter, Allergen Bos d 2. An OBP in bovine nasal mucosa showed relatively low sequence similarity (52%). The rbcOBP yielded 5 mg/L in purified form. It exhibited binding affinity to a ligand, 1-aminoanthracene (K_d = 0.35 μM), indicating to be functionally active. The immuno response against rbcOBP was robust, giving about 2.0 of OD at 492 nm. Expression pattern of bcOBP in milk is under investigation by Western blotting. Its biological function is unclear, but pheromone transport could be considered.

Evolution of milk oligosaccharides and lactose; a hypothesis

Urashima, T.[1] and Messer, M.[2], [1]Obihiro University of Agriculture & Veterinary Medicine, Food hygiene, Nishi2sen 11banchi, Inada cho, Obihiro, Hokkaido, 080-8555, Japan, [2]The University of Sydney, Molecular and Microbial Biosciences, Sydney, NSW, 2006, Australia; urashima@obihiro.ac.jp

Mammalian milk or colostrum contains 0 to 10% of carbohydrate, of which free lactose usually constitutes more than 80%. Lactose is synthesized within lactating mammary glands from UDP-galactose and glucose by a transgalactosylation catalyzed by a complex of β4galactosyltransferase and α-lactalbumin (α-LA). α-LA is believed to have evolved from c-type lysozyme. Mammalian milk or colostrum usually contains a variety of oligosaccharides in addition to free lactose. Each oligosaccharide has a lactose unit at its reducing end; this unit acts as a precurser that is essential for its biosynthesis. It is generally believed that milk oligosaccharides act as prebiotics as well as being receptor analogues that act as anti-infection factors. We propose the following hypothesis. The proto-lacteal secretions of the primitive mammary glands of the common ancestor of mammals contained fat and protein including lysozyme, but no lactose or oligosaccharides because of the absence of α-LA. When α-LA first appeared as a result of its evolution from lysozyme, its content within the lactating mammary glands was low and lactose was therefore synthesized at a slow rate. Because of the presence of glycosyltransferases, almost all of the nascent lactose was utilized for the biosynthesis of oligosaccharides. The predominant saccharides in the proto-lacteal secretions or primitive milk produced by this common ancestor were therefore oligosaccharides rather than free lactose. Susequent to this initial period, the oligosaccharides began to serve as anti-infection factors. They were then recruited as a significant energy source for the neonates, which was achieved by an increase in the synthesis of α-LA. This produced a concomitant increase in the concentration of lactose in the milk and lactose therefore became an important energy source for most eutherians while oligosaccharides continued to serve mainly as anti-microbial agents.

Placental-Udder axis: role on colostrum production and lactation development

Argüello, A.[1], Castro, N.[1], Morales-Delanuez, A.[1], Moreno-Indias, I.[1], Sánchez-Macías, D.[1], Krupij, A.[1], Ruiz-Díaz, M.D.[1], Hernández-Castellano, L.E.[1] and Capote, J.[2], [1]Universidad de Las Palmas de Gran Canaria, Animal Science, Transmontaña s/n, 35413 Las Palmas, Spain, [2]Instituto Canario de Investigaciones Agrarias, Animal Production, Aptdo. 60, La Laguna, Tenerife, Spain; ncastro@dpat.ulpgc.es

Placental growth is essential for udder development and colostrum production. In fact, placenta and udder draw an endocrine axis, where mammogenesis needs the products secreted by placenta for a complete development. One of the main products secreted by the placenta is placental lactogen (PL). PL (bovine) is a 200-amino acid long glycoprotein hormone that exhibits both lactogenic and somatogenic properties. The apparent molecular masses of purified native PL molecules exceed 23041 Da. At least six isoelectric variants (pI: 4.85-6.3) of PL were described in cotyledonary extracts and three different PL isoforms (pI: 4.85-5.25) were found in fetal sera. The PL mRNA is transcribed in trophoblastic cells after day 30 of pregnancy in bovine but not before day 25 during the peri-implantation period. Most of the binuclear cells express PL proteins from day 60 of gestation. The PL proteins are stored in membrane-bound secretory granules in binuclear cells occupying up to 50% of the cytoplasm. The mechanism controlling the synthesis and secretion of PL in ruminant species has not yet been elucidated. In ruminants, the biological functions are not fully understood yet. In the latter species, PL molecules are supposed to have multiple biological effects related to luteotrophic activity, mammogenesis, lactogenesis, and fetal growth although these predictions have not always been supported by empirical observations. It is well known that lactogenic hormones are required for full lobulo-alveolar growth in the mammary gland, and ovarian steroids are needed for ductal growth, but the relationship between PL and colostrum production still is unclear.

Productive performance of primiparous Baladi cows and their F1 crossbreds with French Abondance and Tarentaise

Hafez, Y.M., Animal Production Department, Faculty of Agriculture, Cairo University, El Gamaa street, 12613, Giza, Egypt, Egypt; yasseinhafez@yahoo.com

A total of twenty four primiparous lactating cows (8 Baladi (B), 8 Baladi x Abondance (B x A) and 8 Baladi x Tarentaise (B x T)) were used to compare their potential in milk production. Production data (milk yield, days in milk) were recorded. Fortnightly milk and blood samples were collected starting two weeks postpartum till drying off to analyze the gross chemical composition of milk and quantify IGF-1 and leptin in blood. Baladi cows were the lowest ($P<0.0001$) in milk yield 727.7 kg in 156.3 days in milk. No significant differences were detected in milk yield between the two examined crossbred cows (1322.3 and 1656.6 kg milk in 191.5 and 236.0 days in milk for B x A and B X T, respectively). Data of the gross chemical composition of milk (fat, protein, lactose, total solids and solids not fat percentages) showed non significant differences among the examined genotypes. Slight increase in milk protein, lactose, total solids and solids not fat were noticed in B x A crossbred compared to the other genotypes. Values of milk urea nitrogen (mg/dl) were in the normal range without any significant differences among the studied genotypes. Values of IGF-1 (ng/ml) were lower ($P<0.05$) in B cows (143.3) than B x A (195.8 ng/ml) and B x T (239.2 ng/ml) crossbreds. The values of leptin (ng/ml) were higher in B x A (14.7) than B x T (13.4) and B (12.27). Upgrading Baladi cows with either French Abondance or French Tarentaise improved significantly milk yield with longer days in milk without any noticeable change in the studied milk constituents. Also, the measured blood parameters may add some explanations to the superiority of Baladi x Abondance, Baladi x Tarentaise over Baladi cows in milk production. More research is needed to investigate the impact of the upgrading of Baladi cows with European genotypes concerning reproductive performance.

Session 21 Theatre 1

Estimation of NDF degradation parameters in practice

Weisbjerg, M.R., Aarhus University, Faucly of Agricultural Sciences, P.O. Box 50, 8830 Tjele, Denmark; martin.weisbjerg@agrsci.dk

The importance of fibre in ruminant nutrition has got increased recognition. Fibre is important for physical structure to maintain a healthy rumen, for rumen fill and thereby feed intake, and for the energy supply. NDF has worldwide become the method for assessment of fibre concentration in ruminant feeds, although the method in not unproblematic and first recently has been approved as AOAC method. High producing dairy cows depend on a high feed intake. A high feed intake based on high forage proportion, to avoid digestive disorders like acidosis caused by overload of concentrate, requires forages with highly digestible organic matter (OM). Neutral detergent solubles (NDS = OM – NDF) have a true digestibility close to 100%, and therefore NDF concentration and digestibility almost fully explain variations in OM digestibility. This makes NDF degradation parameters the most important feed characteristic in forages, and the effect of a percent unit increase in NDF digestibility in a dairy cow ration has been estimated to be 0.25 kg milk/day. In modern feed/ration evaluation models NDF is evaluated based on potential degradability (dNDF) or indegradability (iNDF) and rate of degradation (k_d), and assessment of these important parameters on forages in practice is of great importance. In practice, NIRs is the method of choice as it is cheap and fast. However, NIRs require a reference method for calibration, and the existence of a correlation between the parameter in question and the reflectance pattern. Until now NIRs calibration on iNDF is promising and is used in practice, whereas calibration of k_d seems problematic. Reference methods for NDF degradation parameters have mainly been in situ, *in vitro*, and *in vitro* gas production methods, which all are very resource demanding and require access to rumen fistulated animals. In the NorFor system a method is used in practice for estimating k_d based on concentration of ash and NDF in combination with OM digestibility. This opens for estimation of k_d based on conventional feed analysis.

Session 21 Theatre 2

Wheat for broiler chickens is not a standard raw material

Gutiérrez Del Álamo, A.[1], Pérez De Ayala, P.[1], Verstegen, M.W.A.[2], Den Hartog, L.A.[2] and Villamide, M.J.[3], [1]Nutreco Poultry and Rabbit Research Centre, 45950 Casarrubios del Monte, Toledo, Spain, [2]Wageningen University, Animal Nutrition Group, P.O. Box 338, 6700 AH Wageningen, Netherlands, [3]UPM, E.T.S.I. Agrónomos, Departamento de Producción Animal, 28040 Madrid, Spain; a.gutierrez@nutreco.com

In Europe, there are some raw materials that due to their cost-quality properties (in terms of energy) are considered as standards in feed formulation. Inside the group of cereals, wheat is the main ingredient contributing up to 650 g/kg of the diet for finishing broilers. Although considered a moderately uniform raw material, evidence has shown that the nutritive value of wheat for broilers vary considerably (between 8.49 and 15.9 MJ/kg dry matter) and it is affected by several intrinsic and extrinsic factors. This variability is of importance to the feed industry whose primary objective is to optimize the efficiency of poultry production. Research has focused on the variability in nitrogen-corrected apparent metabolizable energy (AMEn) among wheat grains with special emphasis on the carbohydrate fraction. The two main reasons for that are firstly, that wheat is essentially used as an energy source due to its high starch (ST) content and secondly, that the non-starch polysaccharides fraction (NSP) of wheat negatively affects its nutritive value even when present at low quantities. Feed manufacturers have tried to counterbalance the variability in the nutritive value of wheat grains by measuring the ST content in incoming grains and also by adding NSP-degrading enzymes to the feed. Is this enough? The answer is clearly no. The influence of ST in wheat AMEn depends on its content and digestibility whereas the influence of the NSP fraction in wheat AMEn depends mostly on the soluble NSP fraction. The NSP-degrading enzymes do not always work as expected, and different broiler performance is observed. The factors associated with variation in the nutritive value between wheat samples are still not well understood and their study may be the answer to the real problems observed at the field.

Effect of frequency of corn silage chemical analysis and ration adjustment on milk production and profitability

Valergakis, G.E.[1], Souglis, E.[2], Zanakis, G.[3], Arsenos, G.[1] and Banos, G.[1], [1]Faculty of Veterinary Medicine, Aristotle University of Thessaloniki, Department of Animal Production, BOX 393, 54124 Thessaloniki, Greece, [2]American Farm School, BOX 23, 55102 Thessaloniki, Greece, [3]Pioneer Hi-Bred Hellas SA, Agr. School, 57001 Thermi, Greece; geval@vet.auth.gr

The aim of this study was to evaluate the effect of frequency of corn silage (CS) chemical analysis and ration adjustment on predicted milk production (PMP) and income over feed cost (IOFC). During a 3-year period, CS samples from a single dairy farm were weekly collected and analyzed (186 samples, 16 silos). Using a total mixed ration (TMR) for high producing dairy cows (45.5 kg of milk/day), weekly analysis results were compared (analysis of variance) over the 3-year period to the following scenarios: Book values (BV) + no ration adjustment (C-1), BV + daily adjustment of the amount of TMR based on intake (C-2), BV + weekly analysis of CS for dry matter (DM) and adjustment of the amount of CS (C-3), analysis on silo opening (ASO) + no ration adjustment (O-1), ASO + daily adjustment of the amount of TMR based on intake (O-2), ASO + weekly analysis of CS for DM and adjustment of the amount of CS (O-3), monthly analysis (MA) + no ration adjustment (M-1), MA + daily adjustment of the amount of TMR based on intake (M-2) and MA + weekly analysis of CS for DM and adjustment of the amount of CS (M-3). There was a statistically significant effect of scenario on PMP and IOCF ($P<0.001$). Scenarios not involving CS analysis or ration adjustment were consistently statistically inferior to others. Scenarios involving daily TMR adjustment were not statistically different from those based on CS adjustment. However, the latter were usually more profitable. Daily adjustment of the amount of TMR offered is a labour intensive and fairly inaccurate task; ration adjustment based on weekly analysis of CS for DM proved a worthy alternative. M-1 was more profitable than O-1 (8.5 cents/cow/day, $P<0.05$); there were no statistically significant differences between O-3 and M-3.

Associative effects of different feed combinations assessed by using a gas production system

Tagliapietra, F., Guadagnin, M., Cattani, M., Schiavon, S. and Bailoni, L., University of Padova, Department of Animal Science, Viale della università, Legnaro, PD, 35020, Italy; franco.tagliapietra@unipd.it

Associative effects due to different combinations of feeds were studied using an automatic batch culture gas production (GP) system. Two roughages (Silybum marianum, SM; Chrysanthemum coronarium, CC) and 3 by-products (tomato peels, TP; citrus pulp, CP; and apple pomace, AP) were incubated alone (0.500±0.001 g) and in combination 50:50 (1 roughages and 1 by-product), in 3 replications. The 5 feeds and the 6 mixtures were incubated in 280 ml bottles for 120 h with 75 ml of buffered rumen fluid collected with a probe from dry cows. The changes of headspace pressure were converted in volume terms. Venting occurred when the internal pressure reached 3.4 kPa. For the 6 mixtures the expected values of GP (absence of associative effects) were computed as the mean of the GP values achieved for the corresponding feeds incubated alone. The differences between measured and expected GP values of the 6 mixtures were computed for 6, 12, 24 and 48 h of incubation. Within mixture and incubation time these differences were subjected to ANOVA, using replication as source of variation, to test if they differed from 0. Although in all the combinations these differences of GP were always positive, no significant effects were observed at various times for: SM with TP, SM with CP and CC with TP. Associative effects were observed for: SM with AP (at 6 h: +10 ml gas corresponding to +40%, $P<0.01$; at 12 h: +10 ml gas corresponding to +18%; $P<0.01$), CC with AP (at 6 h: +7 ml gas corresponding to +26%, $P<0.05$; at 12 h: +7 ml gas corresponding to +12%; $P<0.10$), CC with CP (at 12 h: +7 ml gas corresponding to +12%, $P<0.10$; at 24 h: +7 ml gas corresponding to +8%; $P<0.10$). No significant associative effects were observed after 24 h, or later. Combination of poorly fermentable roughages with rapidly fermentable by-products (CP and AP) increased the rate of GP mainly during the first 12 h of incubation, the effects were less accentuated later.

Session 21 Theatre 5

The acetyl bromide lignin method to quantify lignin: contaminants in the acid detergent lignin residue

Ramos, M.H.[1], Fukushima, R.S.[2], Kerley, M.S.[1], Porter, J.H.[1] and Kallenbach, R.[1], [1]University of Missouri, Animal Science, 920 E Campus Dr, 65211 Columbia, MO, USA, [2]University of Sao Paulo, Veterinary and Animal Science College, Av. Duque de Caxias Norte, 225, 13630-900 Pirassununga, SP, Brazil; rsfukush@usp.br

In a previous abstract we reported lignin data in 45 grass samples utilizing the acetyl bromide lignin (ABL) method, the traditional acid detergent lignin (ADL) and the ABL procedure utilizing acid detergent fiber (ADF) instead of cell wall (CW). Because the results of ABL method run on ADF (ABLadf) were considerably lower than ADL and ABL we hypothesized that the acid detergent solution was dissolving lignin during ADF preparation. However, this explained only partially the low values of ABLadf. On the other hand, the gravimetric ADL could be inflated because of contaminants, such as ash, protein and cutin. To verify this hypothesis, contaminants in ADL were evaluated and compared with ABLadf. ADL ash was measured in furnace oven, nitrogen in a Leco analyzer and cutin by difference through oxidation of lignin with sodium chlorite. Ash, protein and cutin contents in ADL ranged from 108.1 to 562.6 g kg-1; 36.8 to 130.9 g kg-1 and 86.7 to 707.3 g kg-1, respectively. Nitrogen was also measured in the ADL sodium chlorite residue (cutin) which indicated similar values to ADL nitrogen content. To prevent any duplicity, nitrogen data was considered only once. Overall, ADL contaminants amounted from 322.4 to 854.5 g kg-1. When forage ADL (which ranged from 21.2 to 70.0 g kg-1 DM) were corrected for those contaminants, data varied from 3.7 to 37.6 g kg-1 which were similar to ABLadf numbers (2.4 to 33.7 g kg-1). Data strongly supports the hypothesis of ADL contamination with non-lignin compounds. ABL has no such interference because it is a spectrophotometric reading of lignin phenolic ring and could be an alternative to quantify lignin. In conclusion, this unbearable contamination (in several cases it was over 80%) suggests reevaluation of ADL for use in laboratory analysis.

Session 21 Theatre 6

Sources of variation of quality traits of herd bulk milk used for Trentingrana cheese production

Cologna, N., Tiezzi, F., De Marchi, M., Penasa, M., Cecchinato, A. and Bittante, G., University of Padova, Department of Animal Science, Viale dell Università, 35020 - Legnaro, Italy; nicola.cologna@unipd.it

Trentingrana (TG) is a Protected Designation of Origin hard-cheese manufactured in Trento province, a mountain area of north-eastern Italian Alps. Lasting years, a loss of quality has been observed in TG production, mainly expressed as increasing incidence of discarded wheels and worsening of sensory aspects. Possible factors affecting TG quality may be found at farm, dairy factory, and ripening store level. The first step was to characterize milk collected by dairies and destined to TG production. Therefore, objective of this study was to investigate the sources of variation of herd bulk milk used for manufacturing TG cheese. A total of 93,725 individual herd bulk milks were collected twice per month by 15 dairies from 2002 to 2008, and analysed for several quality traits: fat, protein, somatic cell score (SCS), log10 of total microbial count (LBC), log10 of clostridial count (LCC), and urea. Also, milk yield per milking (MY) was available for each herd. Data were analysed with a linear model that accounted for fixed effects of dairy (D), year (Y), month (M), herd nested within D, and first order interactions between main effects. Fat, protein, SCS, LBC, LCC, and urea averaged 3.88±0.32%, 3.39±0.20%, 3.96±1.10 units, 4.86±0.57 ufc/ml, 2.03±0.11 ufc/ml, and 0.03±0.01 mg/ml, respectively. The most important sources of variation were: herd for MY, M for fat, protein, SCS and LBC, and Y for LCC and urea. The effect of M exhibited a clear circannual variation with the best quality of milk expressed in winter and the worst in summer. Almost all quality traits improved across the studied period, and seem to be not correlated with the decrease in cheese quality. Further investigations are needed to account also for coagulations properties of herd bulk milks as well as herd features such as breeds, and feeding and management strategies.

Changes on raw bovine milk composition across the season

Roca-Fernández, A.I.[1]*, Nijkamp, R.*[2]*, Van Valenberg, H.J.F.*[3]*, González-Rodríguez, A.*[1] *and Elgersma, A.*[2]*,* [1]*Agrarian Research Centre of Mabegondo, P.O. Box 10, 15080, Spain,* [2]*Wageningen University, P.O. Box 16, 6700 AA, Netherlands,* [3]*Wageningen University, P.O. Box 8129, 6700 EV, Netherlands; anairf@ciam.es*

Large fluctuations in the raw bovine milk of many countries have been observed in the past decades because of changes in the feeding regimes, breeding strategies and cattle management. In spite of the fact that dairy manufactures try to keep the levels of the main milk components constant all year around, largest seasonal variations in its values are found and mostly are of dietary origin. The aim of this study was to investigate the variations on the Dutch bovine raw milk across the season. Weekly in 2005, bulk milk samples collected from 17 dairy plants situated in the Netherlands were pooled together and routinely analyzed for milk composition (fat, protein, lactose, casein, cell count, urea and freezing point). On the main milk components, lactose showed the smallest (0.04 g 100 g^{-1}) and fat the highest (0.19 g 100 g^{-1}) seasonal variations, with protein (0.08 g 100 g^{-1}) in between. This is in line with the general assumption that fat is the most sensitive component of milk to dietary changes and lactose is the least sensitive, with protein in between. Fat, protein and casein showed a significant ($P<0.001$) minimum value during the summer in July (4.10, 3.39 and 2.64 g 100 g^{-1}, respectively) and a maximum value during the winter in January and December (4.57, 3.56 and 2.78 g 100 g^{-1}, respectively). Cell count and urea had a significant ($P<0.001$) minimum value during the winter in November and December (167 x 10^3 cells ml^{-1} and 22 mg 100 g^{-1}, respectively) and a maximum value during the summer in August (217 x 10^3 cells ml^{-1} and 26 mg 100 g^{-1}, respectively). Lactose showed a significant ($P<0.001$) minimum value during the autumn in October (4.46 g 100 g^{-1}) and a maximum value during the spring in May (4.55 g 100 g^{-1}). Freezing point had a significant ($P<0.012$) minimum value during the summer in July (-0.517 °C) and a maximum value during the winter in February (-0.521 °C).

Are the chemical compositions of Almond hull of several Almond varieties the same?

Alizadeh, A.R.[1]*, Jafari, S.*[1] *and Imani, A.*[2]*,* [1]*Islamic Azad University, Saveh branch, Animal Science Department, Saveh, 39187/366, Iran,* [2]*Seed and plant Improvement Institute (SPII), Horticulture Research Department, Karaj, 31375-764, Iran; ali@ag.iut.ac.ir*

Because of changing economic considerations, by-products are receiving increased attention from nutritionists. Almond hulls are obtained by drying that portion of Almond fruit that surrounds the hard shell. This hull consists 35 percent of harvesting Almond, but there are few researches in these fields. This experiment was conducted to investigate different hull chemical composition of different varieties. Rabbi (R), Mamaii (M), Shahrud15 (SH 15) and Shokufe (SH) are common Almond varieties produced in certain regions of Iran. Hull compositions were analyzed by AOAC methods. Data were analyzed using the GLM procedure of SAS. Range of hull dry matter (DM) in varieties was 96-92%. Crude protein (CP) content was up to 2-3% DM. The NDF and ADF content of hull significantly affected by Almond varieties and SH15 and SH hah highest fiber content (32.6, 32.4, 29.4 and 28.05% NDF; 25.2, 25.1, 19.8 and 18.8% of DM ADF in SH15, SH, M and R, respectively; $P<0.01$). The ash content was most affected by varieties and SH15 had twofold ash content compared with others. Ether extract (EE) concentration in all Almond hull was negligible and up to 1 percent of DM. Interestingly, NFC concentration had significantly differences between varieties and NFC extent of R variety was the highest (60.1, 58.8, 58.2 and 50.4% of DM in R, M, SH and SH15, respectively; $P<0.01$).Results demonstrated that several Almond hulls can be introduced as an economical and rich feedstuff for livestock. Fiber, ash and NFC content of this by-product are important parameters in varieties. Also, the comparison between Californian results and ours, suggest that CP (2-3% VS. 5-6% DM) and EE (>1% VS. 3-4% DM) of several Almond hull are different.

Session 21 Poster 9

Wheat grain in dairy diets: an economical choice overlooked

Amiri, F., Amanlou, H., Khanaki, H. and Nikkhah, A., Zanjan University, Animal Sciences, Faculty of Agriculture, 313-45195, Iran; anikkha@yahoo.com

The objective was to determine effects of wheat grain (WG) level and particle size (PS) on metabolism and production of Holstein cows. Eight midlactation cows [176±8 days in milk; 554±13 kg body weight (BW); mean ± SE] were used in a 4 × 4 replicated Latin square design study with 4 21-d periods. Each period had 14 days of adaptation and 7 days of data collection. Treatments included feeding either 20% or 10% WG in either fine or coarse particles. Alfalfa hay-based total mixed rations with forage to concentrate ratio of 47.5:52.5 were offered individually 3 times daily at 0900, 1600 and 2300 h. Tail veins blood was sampled at 0 and 2 h relative to feeding. Data were analyzed as a linear Mixed Model with fixed effects of WG, PS and the interaction, and random effects of period, cow and residuals. WG at 20% instead of 10% of diet DM tended to increase blood BHBA (0.64 vs. 0.54 mmol/L, P=0.06), decreased blood total proteins (8.28 vs. 8.46, P=0.03), and tended to decrease albumin (3.72 vs. 3.83, P=0.06) concentrations, with no PS and interaction effects. Dry matter intake increased (19.9 vs. 19.4 kg/d, P<0.01) when WG replaced half of dietary barley grain i.e., 10% dietary WG. Milk NEL yield and milk NEL/NEL intake were similar. Milk yield and contents of solids as well as urine pH and BW were unaffected. Fecal pH tended to be higher (6.8 vs. 6.7, P=0.10) with diets containing 20% vs. 10% WG. Cows on coarse WG gained 0.06 whereas cows on fine WG tended to loose 0.1 unit BCS (P=0.10). The proportion of NEL intake deposited in tissues was higher with coarse than with fine WG particles (18 vs. 16%) at 10% but not at 20% dietary WG. Total tract apparent DM, but not NDF, digestibility was greater for coarse than for fine particles (70 vs. 65, P<0.01). Ultrasound examination revealed that external back fat thickness was greater when 20% WG was fed in fine rather than coarse particles (23 vs. 20.4 mm, P<0.05). Results suggest feasible economical and nutritional use of WG in dairy rations at up to 20% of diet DM.

Session 21 Poster 10

Comparative study of herbage nutritional quality between mountainous grasslands in north western Greece

Mountousis, I., Roukos, C. and Papanikolaou, K., Aristotle University, Faculty of Agriculture, Thessaloniki, 54006, Greece; roukxris@gmail.com

Introduction The objective of this study was to compare the variations in herbage production and chemical composition and dry matter digestibility, over a grazing period in two Greek mountainous grasslands in north-western Greece. Materials and Methods The study was conducted in two mountainous grasslands, on Mt Varnoudas, NW Greece (grassland 1) and in Epirus region W-NW Greece (grassland 2) from May to October of the year 2008, which elevated 1,250 m a.s.l. The geological substrate of grassland 1 consists of granite and micas shcists, while that of grassland 2 consists of limestone. Monthly average air temperatures were 9.0 °C and 11.1 °C for grasslands 1 and 2, respectively, while total annual precipitations were 409.8 mm and 1,027.60 mm. In order to study the effects of the geo-climatic conditions and the harvest month on herbage production, chemical composition and IVDMD, samples were accomplished by cutting herbage biomass at 2 cm above soil surface. Samples were analysed for NDF, ADF, lignin and CP contents and IVDMD. Data was analyzed statistically using ANOVA and correlation. Results Herbage production, chemical components and IVDMD were affected (P<0.001) by the harvest month in both grasslands, as well as by the geo-climatic conditions (except ADF and lignin). Mean herbage production in grassland 1, was significantly lower (619.83±80.646 kg DM/Ha) than in grassland 2 (1760.07±92.97 kg DM/Ha). NDF, ADF and lignin contents increased as the experimental period processed, showing similar fluctuation in both grasslands. Mean CP content of grassland 1 was lower (7.90% DM) than of grassland 2 (12.20% DM). Mean IVDMD was also lower in grassland 1 (50.17% DM) than in grassland 2 (58.51% DM). Conclusions Additional protein sources should be supplied in order to meet the maintenance requirements of ruminants in grassland 1, especially after mid-summer. Better management practices are needed, in order to improve herbage quality.

Trace element content in Swiss cereal grains

Schlegel, P., Bracher, A. and Hess, H.D., Agroscope Liebefeld-Posieux, Tioleyre 4, 1725 Posieux, Switzerland; patrick.schlegel@alp.admin.ch

The trace elements copper (Cu), iron (Fe), manganese (Mn), zinc (Zn) and selenium (Se) are essential for living organisms. The knowledge of their contents and variability in feed ingredients is necessary for precise feed formulation. The aim of this study was to determine the content of Cu, Fe, Mn, Zn and Se in cereals commonly used in animal nutrition. One hundred nineteen cereal samples were taken during 2007 from 20 cereal collecting centers which are distributed within 8 defined geographical regions of Switzerland. Cereals were wheat (n=26), barley (n=27), oat (n=19), triticale (n=25) and maize (n=22). Samples were analyzed by ICP-EOS for Cu, Fe, Mn and Zn contents and by GF-AAS for Se. The average content and standard deviation of Cu, Fe, Mn, Zn and Se respectively are expressed in mg per kg dry matter. Wheat: 3.4±0.6; 39.7±11.2; 31.8±6.5; 25.7±3.9; 0.031±0.034. Barley: 4.9±0.5; 34.8±5.7; 14.1±1.4; 25.2±3.1; 0.023±0.016 Oat: 3.8±0.4; 53.2±24.9; 41.7±10.7; 26.9±4.1; 0.014±0.011. Triticale: 5.3±0.7; 33.4±11.5; 30.4±7.9; 32.1±3.2; 0.014±0.012. Maize: < 2.5; 18.4±1.1; 4.8±1.0; 17.9±1.0; < 0.012. Maize presented the lowest trace element contents. In comparison with published data from neighboring countries, the present values are low for Cu, Fe and Se. The Cu content in barley, the Mn content in maize and the Zn content in wheat and triticale were influenced (P<0.05) by the geographical region. To conclude, the present indicates that Swiss cereals are low in Cu and Se, mainly. Further research will be needed to gain more knowledge on the geographical influence from a region or a country on trace mineral contents in cereal grains.

Nutritive value of sainfoin harvested at two phenological stages as fresh and preserved as wrapped silage bales

Theodoridou, K., Aufrere, J., Andueza, D. and Baumont, R., INRA UR 1213 Herbivores, INRA, Centre de Clermont-Fd / Theix, 63122 Saint-Genès-Champanelle, France; katerina.theodoridou@clermont.inra.fr

Sainfoin is a temperate legume plant containing condensed tannins (CT) which are polyphenols able to bind proteins and protect them from degradation in the rumen. There are few references on the nutritive value of this plant, and particularly when preserved as silage. The objective of this study was to compare the nutritive value of sainfoin (*Onobrychis viciifolia* cv. Perly) preserved as fresh or as wrapped silage bales (WS). A sainfoin culture was harvested at the end of flowering (first vegetation cycle) and at the beginning of flowering (five weeks regrowth) and compared with the corresponding WS. Six castrated male Texel sheep (12 months old, 60±3 kg live weight) fitted with rumen cannula were used for digestibility and nitrogen (N) balance measurements. The sheep offered daily 60 gDM of forage per kg W0.75. Data underwent analysis of variance to test the growth stage (GS) and the effect of preservation. GS was considered as a repeated variable. The biological activity (BA) of sainfoin CT, determined by the Radial Diffusion Assay, was higher for fresh sainfoin compared to WS and decreased with the GS for fresh (1.58 to 1.02 g eq. tannic acid) and for WS (0.72 to 0.69 g eq. tannic acid). Organic matter digestibility decreased (P<0.001) with the phenological stage. It was significantly higher for fresh forage (0.607±0.008) than for WS (0.540±0.008) only at the end of flowering. N digestibility was lower (P<0.001) at the end (0.574±0.009) than at the beginning of flowering (0.628±0.009) and significantly higher for fresh (0.633±0.009) than for WS (0.574±0.009). Body N retained (g/d) was not affected significantly by the preservation method but was higher (P<0.001) at the beginning of flowering compared to the end of flowering. Although the preservation of sainfoin as WS decreased the BA of CT, its nutritive value was not modified at the beginning of flowering compared to the corresponding fresh forage.

The effects of different compounds of some essential oils on *in vitro* gas production and estimated parameters

Kilic, U.[1], Boga, M.[2], Gorgulu, M.[3] and Sahan, Z.[3], [1]Ondokuz Mayis University, Department of Animal Science, Faculty of Agriculture, 55139, Turkey, [2]Nigde University, Animal Science, Bor Vocational School, 51000, Turkey, [3]Cukurova University, Department of Animal Science, Faculty of Agriculture, 01130, Turkey; unalk@omu.edu.tr

The aim of this study was to determine the effect of essential oils of Oregano; ORE (Origanum vulgare), black seed; BSD (Nigella sativa), Laurel; LAU (Laurus nobilis), cummin; CUM (Cumminum cyminum), garlic; GAR (Allium sativum), anise; ANI (Pinpinella anisum) and cinnamon; CIN (Cinnamomum verum) on *in vitro* gas production and gas production kinetics of barley, wheat straw and soybean meal. *In vitro* gas productions were determined in *in vitro* gas production technique by supplying rumen liquor from three infertile Holstein cows. The study was carried out in a completely randomised design in 7 (essential oil) x 4 (dose) factorial arrangement. Barley, soybean meal and wheat straw were incubated for 3, 6, 9, 12, 24, 48, 72 and 96 h. 0, 50, 100, 150 ppm doses were tested for all essential oils. The findings of the present study indicated that essential oils, doses and essential oils x doses interaction were significant ($P<0.01$). Each essential oil, each incubation time and each feedstuff responded differently to dose treatment. The addition of CUM resulted in the highest *in vitro* gas production values in barley, soybean meal and wheat straw, whereas the addition of ORE-150 caused the lowest values. Doses of BDS or CUM additions did not affect *in vitro* gas production pattern of all three feedstuffs ($P>0.05$). Highest OMD, ME and NE_L values were determined when barley, soybean meal and wheat straw were provided with CUM ($P<0.05$). Dose treatment of essential oils had no effect on incubation pH of all three feedstuffs ($P>0.05$). In conclusion, CUM could be used to improve nutrient digestibility and energy content and ORE could be used to control degradation of highly degradable starch and protein sources.

The acetyl bromide lignin method to quantify lignin: fibrous preparation utilization

Ramos, M.H.[1], Fukushima, R.S.[2], Kerley, M.S.[1], Porter, J.H.[1] and Kallenbach, R.[1], [1]University of Missouri, Animal Science, 920 E. Campus Dr, 65211 Columbia, MO, USA, [2]University of Sao Paulo, Veterinary and Animal Science College, Av. Duque de Caxias Norte, 225, 13630-900 Pirassununga, SP, Brazil; rsfukush@usp.br

The spectrophotometric acetyl bromide lignin (ABL) method quantifies lignin by dissolving it in the acetyl bromide reagent and reading absorbance at 280 nm. To avoid interference of non-lignin substances, cell wall (CW) preparation is employed. This work quantified lignin in 45 grass samples utilizing ABL and the traditional acid detergent lignin (ADL) method which employs acid detergent fiber (ADF). Because the time required to isolate CW is long as compared to isolation of ADF, we also run the ABL procedure utilizing ADF instead of CW. Both 280 nm readings and 240-320 nm scannings were performed on a UV spectrophotometer. ABL results were similar to ADL and within the range reported in the literature. However, when the ABL procedure was run on ADF instead of CW, results were considerably lower. One possible explanation could be that acetyl bromide was unable to completely dissolve lignin content of ADF. However, 72% sulfuric acid treatment of this residual ADF showed no remaining lignin. Another reason could be that the acid detergent solution (ADS) employed to prepare ADF was dissolving and removing some lignin. In fact, spectrophotometric scanning of ADS wash of several forage CW showed peak that resembled lignin. CW preparation contains lignin but not non-lignin phenolics (tannins, flavonoids, etc.). Extracted lignins from 8 grass species (purity around 95%) were treated with ADS and 72% sulfuric acid which dissolved 34% and 8.2% of lignin, respectively. That means almost half (42%) of grass lignin was solubilized and carried away by these two ADF solutions. When the ADS wash of purified lignins were scanned, 280 nm peaks were detected, strikingly similar to ADS wash of CW. In conclusion, because the ADS dissolves lignin, the resulting ADF should be reevaluated for use in the ADL method.

Morphological evaluation of forage degradation *in vitro*

Van De Vyver, W.F.J. and Cruywagen, C.W., Stellenbosch University, Department of Animal Sciences, Privatebag X1, Matieland, 7602, South Africa; wvdv@sun.ac.za

Exogenous fibrolytic enzymes (EFE) as additives in ruminant feeds have been researched worldwide. Promising effects on DMI, digestibility and production in especially dairy cows have been observed. However, poor or negative effects are also reported and the need arises for clarity on the mode-of-action of EFE. Three forages, treated with EFE, were evaluated *in vitro* and at microscopic level, in an attempt to determine the effect on tissue degradation. Weeping love grass, Kikuyu and Lucerne were used as substrates incubated in rumen fluid inoculated media, *in vitro*. Substrates were either treated with an EFE cocktail or distilled water (Control) prior to incubation. The gas production profile was calculated with an automated system based on the Reading pressure technique. A non-linear model was used to determine the kinetic coefficients. NDF and DM degradability was determined *in vitro* using the ANKOM Daisy incubation system. The main focus was however a qualitative assessment of the degradation of the plant tissue at morphological level over a 24 h period. The section to slide technique was used to mount plant tissues on microscope slides for incubation in rumen fluid media. Images were acquired using the Olympus Cell R system coupled to a MT 20 illumination apparatus. Degradation of cell wall components were quantified using image analysis software of the same system. Results showed a significant increase in total gas production due to EFE for the Kikuyu and Lucerne substrates only ($P<0.05$). *In vitro* true digestibility was significantly higher for EFE treated Kikuyu at 6 h of incubation ($P<0.05$). Clear morphological differences were observed for all tissue types over the incubation period, but differences due to EFE treatment could not be significantly quantified. It was concluded that image analysis can be useful to quantify changes in cell wall over an incubation period but that the addition of exogenous enzymes could not be quantified by this system.

Effects of electron beam and gamma irradiation on free gossypol, ruminal degradation and *in vitro* protein digestibility of cottonseed meal

Taghinejad-Roudbaneh, M.[1], Azizi, S.[2] and Shawrang, P.[3], [1]Islamic Azad University - Tabriz Branch, Tabriz, P.O. Box 51589, Iran, [2]Urmia University, Urmia, P.O. Box 57155-1177, Iran, [3]Nuclear Science and Technology Research Institute, Karaj, P.O. Box 31485-498, Iran; taghinejad_mehdi@yahoo.com

Proteins of cottonseed meal (CSM), which is used as a protein supplement for ruminants, are extensively degraded in the rumen. On other hand, free form of gossypol that is present in CSM is toxic for animal. Radiation processing has been recognized as a reliable and safe method for improving the nutritional value and inactivation or removal of certain antinutritional factors in foods/feeds. The major aim of the present study was to ascertain the impact of gamma (γ) and electron beam (EB) irradiation at doses of 15, 30 and 45 kGy on free gossypol contents, ruminal Crude Protein (CP) degradibility and *in vitro* CP digestibility of CSM. Nylon bags of untreated or irradiated CSM were suspended in the rumen of three ruminally fistulated Gezel rams for up to 48 h and resulting data were fitted to nonlinear degradation model to calculate degradation parameters of DM and CP. Results in this study showed that EB irradiation had similar effect to γ-irradiation and they decreased ($P<0.01$) free gossypol content of irradiated CSM. Gamma and EB irradiation at doses equal/more than 30 kGy were decreased effective degradibility of CP, compared to untreated sample ($P<0.05$). On the contrary digestibility of ruminally undegraded CP of irradiated CSM was slightly improved ($P<0.05$). The results propose that irradiation processing of CSM can decrease its gossypol content and ruminal protein degradability without negative effects on its intestinal CP digestibility. In conclusion, irradiation processing of CSM could be successfully utilized to improve its nutritional quality for ruminants, hence application and commercialization of irradiation processing technology is more worthwhile. Further study is needed to elucidate the economical benefits of this process, also technical feasibility for setting up an industrial process.

Large scale genomic evaluation in dairy cattle

Liu, Z., vit w.V., Genetic evaluation and biometrics, Heideweg 1, 27283 Verden, Germany; zengting.liu@vit.de

Conventional phenotypic genetic evaluation in dairy cattle has been very successful for dairy cattle breeding in last three decades. With massive SNP genotype data available for an increasingly large number of animals, significantly higher and more accurate genomic pre-selection can be expected. In major dairy breeding countries, such as Germany, national genomic reference population which comprised more than 5,500 genotyped bulls with daughters, was recently enlarged by adding 12,000 genotyped bulls from three European partner countries. As a consequence, more than 17,000 genotyped bulls with daughters were available for genomic evaluation for German Holsteins. However, the huge size of the new reference population posed a technical challenge for genomic evaluation, including SNP effect estimation, calculating reliabilities of direct genomic values, and combining genomic with conventional phenotypic information. The objective of this study was to investigate the feasibility of our developed methods for SNP effect estimation and reliability calculation as well as combination of genomic with conventional EBVs for the large-scale genomic data. A total of 45,181 SNP markers were selected for a marker BLUP model to estimate SNP effects. Due to the use of a special computing algorithm, iteration on data with special residual update, computing requirement increased only linearly with the number of genotyped and phenotyped bulls. Because the reliability of direct genomic value (DGV), sum of all SNP effect estimates, was obtained by inverting a genomic relationship matrix, it became infeasible to calculate the reliabilities for the new reference population with many more genotyped bulls. Therefore, an approximation method was developed to calculate the reliabilities. Estimates of DGV were combined with conventional EBV using a BLUP approach, which worked efficiently for the larger genomic data as well. Via a validation study, it was verified that the enlargement of the reference population led to higher genomic predictive ability and less biased prediction.

Modelling aspects of association studies of quantitative traits in farmed livestock

Pollott, G.E., Royal Veterinary College, Veterinary Basic Sciences, Royal College Street, NW1 0TU, United Kingdom; gpollott@rvc.ac.uk

Much of the methodology for association studies has been worked out on Mendelian or complex traits in human populations. Searching for polymorphisms associated with quantitative traits in livestock populations requires very different methods. Our group has published such studies on both calf death and several growth traits in Holstein heifers using a candidate gene/SNP approach with up to 20 SNPs. The use of genome-wide SNP chips in this population presents a number of challenges which need resolution. When SNP numbers are small then the use of usual mixed animal model methodology is practical but when considering 54,000 markers then this approach is computable but alternatives may be more desirable. This study compared a number of alternatives for estimating genome-wide SNP effects. Aulechenko *et al.* (2004) suggest fitting an appropriate mixed animal model, excluding the SNP of interest, and storing the residuals for each animal. These residuals can then be analysed using simple methods in programs such as GenAble or PLINK to look for significant SNPs with a simpler model. The use of mixed model methodology in genetic studies has been based on the simultaneous fitting of all effects in the model. The approach of Aulechenko *et al.* ignores this principal but it can be used as part of a preliminary screening process. Once significant SNPs have been found these can be reanalysed with the full mode mixed animal model, including the significant SNP. This paper presents results from analysing milk yield in 500 cows using the methods outlined above. Small differences were found between the Aulechenko method and the mixed model approach. Many significant SNPs were identified and investigated in more depth. The mixed animal model approach is also useful for identifying epistasis involving larger numbers of loci. Fitting higher order interactions between SNPs is possible and models tested using this approach have highlighted the possibilities.

Genome partitioning of genetic variance for production traits in Holstein cattle

Pimentel, E.C.G., Simianer, H. and König, S., Georg-August University Göttingen, Department of Animal Sciences, Albrecht-Thaer-Weg 3, 37075 Göttingen, Germany; epiment@gwdg.de

Genetic variation of complex traits is usually assumed to follow an infinitesimal mode of gene action, with many genes contributing with a very small proportion of variance and distributed across the whole genome. In conventional genetic evaluations, usually no assumption is made about this distribution. The objective of this work was to estimate the contribution of each autosomic chromosome to the genetic variance of milk yield, fat and protein percentage and somatic cell score in Holstein cattle. The data set comprised 2294 Holstein-Friesian bulls genotyped for 39,557 SNPs markers. Marker-derived relationship coefficients for each pair of bulls were computed, using either marker genotypes observed on the whole genome or on subsets of markers relative to each chromosome. For each trait and each assumed genetic covariance structure, variance components were estimated. The contribution of each chromosome to genetic variation was inferred by contrasting a full model relating the trait to the whole genome with a reduced model relating the trait to the respective subset of the genome. For the traits considered in this study, almost all the chromosomes were shown to contribute to genetic variation. The amount of variance proportion attributed to each chromosome was found to be associated with its physical length. These results suggest an agreement with the hypotheses that the studied traits are influenced by a large number of genes distributed throughout the whole genome. Nevertheless, traits that are influenced by genes with very large effects are expected to show a larger proportion of the genetic variance associated with the chromosomes where these genes are. For instance, the proportion of genetic variance for fat percentage associated with chromosome 14 was larger than would be predicted from chromosome size alone. One potential application of the knowledge about the distribution of genetic variance of a trait across the genome may be in setting priors for genomic evaluations.

The role of covariance between genome segments for genetic variation and selection

Simianer, H. and Pimentel, E.C.G., Georg-August-University Goettingen, Animal Breeding and Genetics Group, Albrecht-Thaer-Weg 3, 37075 Göttingen, Germany; hsimian@gwdg.de

The basic idea of genomic selection is that the sum of estimated SNP effects across the genome yields an estimate of the breeding value. Hence, sums of estimated SNP effects across single chromosomes can be interpreted as estimated chromosomal breeding values (cBVs), which in total add up to the total genomic breeding value. We estimated 29 autosomal cBVs from a data set comprising of 2,307 Holstein-Friesian bulls with 39,557 autosomal SNPs usable after filtering. The trait considered was somatic cell score. The total additive genomic variance is the sum of the variances of the cBVs plus twice the sum of all covariances. In the analysed data set, the covariances between cBVs explained about 50 per cent of the total genomic variance. When selecting a proportion p of animals with the highest genomic breeding values, the total genomic variance was reduced in almost perfect agreement with the expected variance of a truncated normal. Since the sum of the variances of the cBVs remained constant with selection, this reduction is exclusively due to changes in the covariance part, which decreases rapidly and turns negative with $p \sim 0.75$. With one round of simulated reproduction of the selected group, the deviation of the covariance from zero was halved, which is in full agreement with Bulmer's theory of genetic variance under selection. It is also shown, that a dissection of the genome in smaller segments of equal length even increases the impact of the covariances between segments, accounting for 62 (with 10 cM segments) or 81 (with 1 cM segments) per cent of the total genomic variance, respectively. Understanding the role of genomic covariances in the genetic make up of populations under selection using genomic tools will lead to a better understanding of how genetic improvement works on the genomic level and thus may help to design more efficient breeding strategies.

How much genetic variation is explained by dense SNP chips?

Woolliams, J.A.[1], Vanraden, P.M.[2] and Daetwyler, H.D.[1,3], [1]University of Edinburgh, The Roslin Institute & R(D)SVS, Roslin, Midlothian, EH25 9PS, United Kingdom, [2]USDA-ARS, Animal Improvement Programs Laboratory, Beltsville, MD 20705-2350, USA, [3]Department of Primary Industries Victoria, Biosciences Research Division, 1 Park Drive, Bundoora 3083, Victoria, Australia; john.woolliams@roslin.ed.ac.uk

The fraction (q^2) additive genetic variance (V_A) explained by SNP chips is a topic of high interest: (i) it determines the accuracy of genomic selection (GS) and so is relevant to all breeding schemes using GS; (ii) understanding its origin helps to set targets for technology development for other species with less advanced genomics; and (iii) it places into perspective q^2 derived from GWAS – related to 'missing heritability'. The theory is developed from the expression for squared accuracy (r^2) of genomic evaluation as the ratio of (i) the product of the heritability h^2 and the number of records per locus affecting the trait, and (ii) the same term plus 1. This can be generalised to embody the number of independent segments, a property of the species genome rather than the number of loci affecting a trait, and $q^2<1$. Then, for a series of evaluations for a trait, the reciprocal of r^2 is linearly related to the reciprocal of the number of records used. In the regression, the intercept is the reciprocal of q^2, whilst the slope will depend on the species genome, the trait, and why the variation remains untagged. In testing this with USDA cattle data, it was estimated that for the Illumina Bovine SNP50 BeadChip q^2=0.8, leading to a maximum accuracy ~0.9. q^2 is an expectation and an attribute of the chip. For particular traits the ultimate r^2 using a chip may be more or less than the q^2, for example if a single SNP is included in the chip and explains all V_A, then r^2 will ultimately be 1. However for most traits used with a chip this will not be the case, and the theory provides a prediction of the adequacy of the chip and a realistic assessment of the impact of increasing the number of records on achieved accuracy.

Localizing quantitative trait loci with the modified version of the Bayesian Information Criterion: a need for a multiple testing adjustment

Bogdan, M., Wroclaw University of Technology, Wybrzeze Wyspianskiego 27, 51-511 Wroclaw, Poland; Malgorzata.Bogdan@pwr.wroc.pl

The first step in locating QTLs usually relies on identifying associations between marker genotypes and the value of the trait in question. Since quantitative traits may be influenced by several important genes, the relationship between trait values and marker genotypes is often modeled using multiple regression. The most difficult part in fitting a multiple regression model lies in the identification of important predictors. This could, in principle, be addressed by employing one of many criteria for model selection. However, there is a lot of evidence that in this case popular model selection criteria, like the Akaike Information Criterion (AIC) or the Bayesian Information Criterion (BIC), have a strong tendency to overestimate the number of QTLs. In our earlier articles we explained this phenomenon by relating it to the well known problem of multiple testing. To address this issue we developed the modified version of BIC (mBIC), which enables the incorporation of the prior knowledge on a number of regressors and prevents overestimation. In the subsequent articles the criterion was further extended to work with traits, which do not have a normal distribution. In this talk we will present mBIC and discuss its relationship to the popular Bonferroni correction for multiple testing. We will also present a new version of mBIC, related to the Benjamini and Hochberg procedure, which controls the False Discovery Rate. We will illustrate the performance of modified versions of BIC with extensive simulations in the context of QTL mapping in experimental populations and genome-wide association studies, as well as with real data analyzes. Apart from other things, our simulations clearly demonstrate the advantage of using the regression model over methods based on individual marker tests.

Are there genes regulating environmental variation? A case study using yield in maize

Yang, Y.[1], *Schön, C.*[2] *and Sorensen, D.*[1], [1]*Aarhus University, Genetics and Biotechnology, Blichers Allé 20, PB 50, 8830 Tjele, Denmark,* [2]*Technische Universität München, Center of Life and Food Sciences, Emil-Ramann-Str. 4, 85350 Freising, Germany; sorensen@humo.dk*

A possible genetic control of environmental variation is relevant in studies of maintenance of variation and in artificial selection. In animal and plant breeding, it opens for the possibility of selecting for robustness to environmental change and for uniformity. In this paper, first, the current evidence for the presence of genetic effects operating on environmental variance is reviewed and we cast light on various factors that can lead to erroneous inferences from statistical analyses. Secondly, new compelling evidence is presented using yield in maize as experimental material. The data, consisting of replicates of the same genotype, are particularly well suited to investigate the problem. The variance between observations within a genotype reflects environmental variation, and a different variance across genotypes indicates that a genetic component operates at the level of the environmental variance. Results of three statistical analyses are presented and contrasted: a non-parametric approach based on resampling methods, a robust approach based on asymptotic approximations, and a Markov chain Monte Carlo implementation of a fully Bayesian parametric model. The first two methods build on relatively few assumptions and yield strong support for the existence of genes acting on environmental variance. The fully parametric Bayesian model provides for richer inferences but at the cost of introducing a larger set of assumptions. Results from the Bayesian analysis indicate that the largest source of environmental variance heterogeneity is due to genotype by location interaction effects. However, analyses carried out on subsets of the data belonging to a common environment favour models of genetically structured environmental variance heterogeneity.

First results from a divergent selection experiment for environmental variability of birth weight in *Mus musculus*

Nieto, B.[1], *Salgado, C.*[1], *Cervantes, I.*[1], *Pérez-Cabal, M.A.*[1], *Ibáñez-Escriche, N.*[2] *and Gutiérrez, J.P.*[1], [1]*University Complutense of Madrid, Animal Production, Avda Puerta de Hierro s/n, E-28040 Madrid, Spain,* [2]*IRTA, Lleida, E-25198 Lleida, Spain; gutgar@vet.ucm.es*

The control of the environmental variability by genetic selection is considered nowadays an important challenge. Some analytical models have been developed to deal with this concept, which assume the existence of a pool of genes controlling the mean value of the trait and another pool of genes controlling its environmental variability. These models have been shown to work properly under simulated data but there are very few selection experiments showing their usefulness in real data. Here we present the first results obtained from a divergent selection experiment for environmental variability of birth weight in mus musculus. The design of the experiment implies that the males are evaluated by the birth weight of their offspring when they are mated to 4 females belonging to an inbred line. The inbred females ensure that all the genetic variability in the data comes from the differences among males. Data were analyzed with the model proposed by SanCristobal *et al.* (1998) and the selection was carried out within line based on the breeding value concerning environmental variability. A reference selection response is calculated simulating the selection of 6 males out of 30 each generation in a classical way. Then, simulated annealing is used to carry out a weighted selection procedure by ensuring the maximum response given that the mean coancestry of the new solution does not exceed that from the solution of the reference selection. The evolution of the phenotypic variances of the birth weights in the nucleus as well as in the evaluation litters showed very few or null differences between the high and low lines. However genetic trends computed as the mean breeding values of selected males seem to show clear differences. More generations might be needed before reconsidering modifications in the experimental design.

Session 22 — Theatre 9

Prediction of 305 days milk production using artificial neural network in Iranian dairy cattle

Abbassi Daloii, T., Tahmoorespur, M., Nassiri, M.R., Abbassi Daloii, M., Hosseini, K. and Zabetian, M., Ferdowsi University of Mashhad, Animal Science, Azadi Sq., 91775-1163, Iran; tabbassid@gmail.com

More recently, ability of Artificial Neural Network (ANN) in prediction of the production has led to increase in its applications in many fields of agriculture. Several researchers have shown that ANN has powerful pattern classification and pattern recognition capabilities. It has been applied for prediction of individual milk, fat and protein yields and prediction of some disease in dairy cattle. In the last decade, many studies have shown that ANN is relatively successful for prediction of production traits. The aim of this study was to evaluate the accuracy of prediction of 305 days milk production in dairy cattle by bpANN model (back propagation ANN). The designed network had 3 layers including input, hidden and output with 11, 50 and 1 neuron(s), respectively. From all of 100 records of data 75%, 12.5% and 12.5% was assigned to training, verification and test category, respectively. Test-day records were taken at 28-day intervals. In this study 5 test-day period records were used (from TD1 to TD5) and each test-day record contained information about previous test-day records. Accuracy of prediction varied from TD1= 0.77 to TD5= 0.85 and was the highest when 5 test-day records were used. The results indicated that illustrated ANN can readily be adapted with any new data that added to the network and solve complex problems. Also, the results demonstrated our ANN model has the ability to predict 305 days milk yields with test-day records. In addition using more test-day records can increase the ability of bpANN for prediction of 305 days milk production in dairy cattle.

Session 22 — Poster 10

Analysis the probability of pregnancy after the first insemination in Iranian Holstein cow using a logistic statistical approach

Bahri Binabaj, F.[1], Farhangfar, H.[2], Shamshirgaran, Y.[1] and Taheri, A.[1], [1]Ferdowsi University of Mashhad, Animal Science, Faculty of Agriculture, Ferdowsi University of Mashhad, Mashhad, Iran, 91775-1163, Iran, [2]The University of Birjand, Animal Science, Faculty of Agriculture, The University of Birjand, Birjan, Iran, 97175/331, Iran; yas.shamshirgaran@gmail.com

To evaluate the effects of some environmental factors on the probability of being pregnant after just one insemination in Iranian Holstein cows, a total of 38,074 records obtained from 10,726 cows at different parities, which were inseminated during 1985-2009 were utilized. All records were collected from a very large dairy herd which comprised two units. A generalized statistical linear model (logistic regression) was applied as the statistical model. In the model, fixed effects of year period and month of insemination, technician, parity and herd unit were included. The statistical model was run using GLIMMIX procedure of SAS software. Odds ratio estimates for year period 1 (1985 to 1990) in comparison to periods 2 (1991 to 2000) and 3 (2001 to 2008) were 1.106 and 1.133 respectively, which means a chance of being pregnant after the first insemination in year period 1 in comparison to periods 2 and 3 was respectively 10.6% and 13.3% higher. Odds ratio estimates for a herd unit shows that a chance of being pregnant after the first insemination in unit 1 is 15.4% higher than in unit 2.

Comparison of fixed and random regression test-day models for genetic evaluation of milk yield trait of Iranian Holstein cows

Shamshirgaran, Y.[1], Aslaminejhad, A.A.[1], Tahmoorespour, M.[1], Farhangfar, H.[2], Bahri Binabaj, F.[1] and Taheri, A.[1], [1]Ferdowsi University of Mashhad, Animal Science, Faculty of Agriculture, Ferdowsi University of Mashhad, Mashhad, Iran, 91775-1163, Iran, [2]Birjand University, Animal Science, Faculty of Agriculture, Birjand University, Birjand, Iran, 97175/331, Iran; yas.shamshirgaran@gmail.com

In this research, Fixed Regression Test-Day Model (FRM) and Random Regression Test-Day Model (RRM) were studied for genetic evaluation of milk yield of Iranian Holsteins dairy cattle. Breeding values and genetic parameters of milk yield from two models were compared. A total of 164,391 monthly test day milk records obtained from 19,217 first lactation Holstein cows (three times milking a day) calving between 1991 to 2008 were analyzed. The records were distributed among 179 herds and were used to estimate genetic parameters and to predict breeding values. The contemporary groups of herd- year- month- of production were fitted as fixed effects in the models. Also linear and quadratic forms of age at calving and Holstein gene percentage were fitted as covariates. The random effects of the models were additive genetic and permanent environmental effects. In the random regression model, orthogonal Legendre polynomial up to order 4 (cubic) were implemented to take account of genetic and environmental aspects of milk production over the course of lactation. The heritability estimated from the FRM was found to be 0.15. From the RRM, the average estimated heritability of monthly test day milk production was higher for the second half of the lactation than that of the first half of the lactation period. The lowest and the highest heritabilities were found for the first (0.102) and the sixth (0.235) month of lactation, respectively. Genetic, permanent environmental and phenotypic correlations decreased as the interval between consecutive test days increased. Breeding value of animals predicted from FRM and RRM were also compared. The results showed similar ranking of animals based on their breeding values from both models.

Environmental and genetics effects influencing the ponderal development up to weaning time of Nellore animals raised in the Southeast Brazil

Vieira, D.H.[1], Medeiros, L.F.D.[2], Souza, J.C.D.[2], Oliveira, J.P.[2], Pinto, J.B.[2], Pedrosa, I.A.[2], Surge, C.A.[3], Silveira, J.P.F.[3] and Vieira Junior, L.C.[3], [1]Centro de Criação de Animais de Laboratório/FIOCRUZ, Rio de Janeiro, 21040-360/Rio de Janeiro-RJ, Brazil, [2]IZ/UFRRJ, Seropédica, 23851-970/Seropédica-RJ, Brazil, [3]FMVZ/UNESP, Post-Graduation student in Animal Science, Botucatu, 18610-130/Botucatu-SP, Brazil; joaopaulo_franco@ig.com.br

The growth records from the Farmer Association of Nellore Catlle of Rio de Janeiro (Nelorio), correspondent to a 3275 Nellore breed calves, born between 1999 and 2007, raised on pasture in five herds in Rio de Janeiro State, were collected. These records enabled study the environmental effects and an estimation of the heritability (h^2) for weaning weight (205 days of age). The matematical model included the fixed effects of sex, year, season and herd, the covariate age of the cow and the random effect of the bull. The variance components used to estimate the heritability, were obtained by Derivative Free Restricted Maximum Likelihood, using MTDFREML program. The fixed effects of sex, year and herd were significant ($P<0.05$), whereas the season of birth and the age of the cow at calving were not significant ($P<0.05$). The least squares mean for the weaning weight was 165.42 + 1.75 kg (VC = 15.25%). Males were an average, 13.50 kg heavier than females. Differences among herds were evident, and some showed high mean values in relation to the overall mean value. The heritability for weaning weight showed moderate to high magnitude ($h^2 = 0.45 + 0.05$). The diferences due to sex (superiority of males) suggest that different feeding for males and females may be used in order to decrease the necessary time to reach the suitable weight for slaughter (males) and to get on earlier time for coupling (female).

Pre- and early postnatal dietary interactions on development, metabolic and endocrine function from birth to adulthood in sheep

Nielsen, M.O.[1], Kongsted, A.H.[1], Hellgren, L.I.[2], Tygesen, M.P.[1], Johnsen, L.[1] and Husted, S.M.[1], [1]Faculty of Life Sciences, University of Copenhagen, Grønnegårdsvej 7, 1870 Frederiksberg C, Denmark, [2]Technical University of Denmark, Søltofts Plads, building 221, 2800 Kgs. Lyngby, Denmark; mon@life.ku.dk

The objective of the study was to assess 1) the impact of late gestation undernutrition (LG-UN) on metabolic and endocrine function in sheep, and 2) the interactions with postnatal diet on phenotypical expression of foetal programming from birth to adulthood. Twenty twin-pregnant ewes were fed a NORM (~requirements for energy and protein) or LOW (50% of NORM) diet the last 6wks of gestation (term=147d). From 3d-6mo *post-partum* (around puberty), twin lambs were assigned to each their feeding: CONV (hay) or HCHF (High-Fat-High-Carbohydrate: dairy cream+popped maize). Male lambs were slaughtered at 6mo. Female off-spring were raised on pasture/silage from 6mo-2yrs (young adulthood) and then slaughtered. Post-natal growth until age 6mo was determined exclusively by the postnatal diet. LG-UN resulted in lower adult body size with most tissues/organs proportionately smaller, but thyroid and adrenals were increased. HCHF lambs at 6mo had massive adipose and hepatic fat infiltration. Functional hepatocyte mass and kidney weight was reduced. By age 2yrs, postnatal diet effects on body proportions had disappeared, except for higher abdominal-to-renal fat ratio in HCHF animals. However, in adults LG-UN effects became consistently manifested as altered glucose-insulin homeostasis, thyroid hormone levels, lipid deposition and fatty acid profiles in hepatic (structural) lipids. In conclusion, LG-UN effects become manifested in early adulthood, with implications for key endocrine systems involved in regulation of growth and metabolic function. This can be involved in earlier termination of growth and smaller adult body size in LG-UN individuals. In adolescent animals, the actual postnatal diet is the main determinant of development. Effects of the extreme HCHF diet could thus be reversed by subsequent dietary correction.

Foetal programming in pigs: phenotypic consequences, the role of genetic variation and epigenetic mechanisms

Wimmers, K., Nuchchanart, W., Murani, E. and Ponsuksili, S., Leibniz Institute for Farm Animal Biology (FBN), Wilhelm-Stahl-Allee 2, 18196 Dummerstorf, Germany; wimmers@fbn-dummerstorf.de

Foetal programming (FP) encompasses persistent effects on the ontogenesis in response to environmental signals during intrauterine life. In farm animal intrauterine growth retardation (IUGR) and low birth weight are associated with reduced performance. In humans programming of foetal development is associated with an increased risk of metabolic diseases. Pigs share many similarities in physiology and genome with humans and therefore provide a good model to study the phenomenon of FP. The molecular mechanisms by which the mismatch of pre- and postnatal environmental conditions leads to adverse consequences remain unclear. There is growing evidence that epigenetic changes, in particular altered DNA methylation, in regulatory and growth-related genes play a significant role in FP. DNA methylation patterns are largely established in utero and alteration through extrinsic signals may cause persistent changes in gene expression, however intrinsic genetic effects may modulate the response to exogenous signals. In order to identify pathways and genes sensitive to FP, to evidence and quantify the role of DNA-methylation and genetic variation, German Landrace and Pietrain sows were fed gestation diets supplemented with metabolites and co-factors of the methionine pathways. Offspring were sampled and monitored at 35, 63, and 91 pre- (dpc) and 150 days post-partum (dpp). Whole genome methylation, pre- and postnatal growth and hepatic gene expression were altered due to age, diet, breed, and their interaction. Gene expression of intermediate metabolism, cell growth, proliferation, differentiation, and communication was modulated in both breeds, however, partially in different direction and magnitude and at different ages. This study complements our second experiment using different levels of protein supply during gestation and elucidates the role of genetic and epigenetic variation as the molecular basis of genotype-environment interactions.

Maternal undernutrition programs leptin expression in offspring

Chadio, S., Stavropoulou, M., Kotsampasi, B., Charismiadou, M., Papadomichelakis, G. and Menegatos, I., Agricultural University of Athens, Anatomy and Physiology of Domestic Animals, 75 Iera odos, 11855 Athens, Greece; shad@aua.gr

The objective of the present study was to investigate the effects of maternal undernutrition imposed during different periods of gestation on leptin expression in offspring at two different times postnatal in order to examine if adipose tissue can be reprogrammed as a consequence of maternal nutrient restriction. Pregnant ewes were fed to 100% throughout pregnancy (Control) or to 50% from 0-30 (R1) or from 31-100 day of gestation (R2). Lambs were born naturally and fed to appetite throughout the study period. At 4 and 10 months of age a sample of tail subcutaneous adipose tissue was obtained from all lambs and leptin mRNA expression was detected by semiquantitive RT-PCR. The activity of a number of lipogenic enzymes was also measured by spectophotometric methods. Data were analyzed by ANOVA for repeated measures with treatment and age as fixed effects and their interactions, followed by Duncan's test. A sex specific effect for leptin expression was observed, with male lambs born to mothers from both R1 and R2 groups exhibited a greater mRNA abundance for leptin at 4 months of age compared to Control, while female lambs of the same groups showed a decreased leptin expression at the age of 10 months ($P<0.05$). When data were analyzed in terms of age within groups a significant reduction ($P<0.001$) with age was detected for leptin mRNA in both sexes and in all three groups examined. No differences were detected for enzyme activities at 10 months of age. In conclusion, maternal nutrition during pregnancy seems to program postnatal leptin expression in offspring with an increase in 4 months old male lambs, not more apparent after lambs entered puberty (10 months). However, to which extent this early enhanced expression reflects altered adipocyte function that could lead to differences in fat metabolism which may become obvious later in life needs further elucidation.

Developmental programming of health and fertility in farm animals

Karamitri, A., Gardner, D.S. and Sinclair, K.D., University of Nottingham, Sutton Bonington Campus, LE12 5RD, United Kingdom; angeliki.karamitri@nottingham.ac.uk

The early-life developmental environment is a vulnerable period of the lifespan during which adverse environmental factors have the potential to disturb the processes of cell proliferation and differentiation or to alter patterns of epigenetic remodelling. It is well established that improved understanding of the physiological and molecular processes involved in the developmental period is central to the evolution of clinical approaches to protect the offspring from cardiovascular or metabolic syndrome related diseases. This is known as 'Developmental Origins of Health and Disease (DOHaD)' hypothesis and extends to programming of fecundity and fertility in mammals. Numerous animal studies have been conducted to evaluate the validity of DOHaD hypothesis and to provide important information on the underlying physiological mechanisms providing evidence of a causal relationship between early-life exposures and risk factors in adult life. Farm animals provide an excellent model as their pre- and postnatal development and physiology approximates that of humans. In the agricultural industry, knowledge of the factors that have impact on live weight, milk production and reproduction is important in terms of agricultural economy. The general concept is that malnourishment restricted either to in utero development or during infancy can predispose to late-onset non-communicable diseases in adult life and may reduce fecundity and fertility. The effects of gross nutrient or protein deficiencies in maternal diet during pregnancy are well documented, although less is known about the effects of specific nutrients or the timing or the actual mechanisms responsible for nutrient programming. Here we present experimental studies that have investigated reductions in specific dietary inputs during critical periods of early development, and review the available evidence regarding developmental programming of reproduction and fertility.

High level iodine supplementation of the pregnant ewe alters serum IgG level and gene expression in the small intestine of the newborn lamb

Boland, T.M.[1], Kenny, D.A.[2], Hogan, D.[1], Browne, J.A.[1] and Magee, D.A.[1], [1]University College Dublin, Belfield, Dublin 4, 0000, Ireland, [2]Teagasc, Grange, Dunsany, Co. Meath, Ireland; tommy.boland@ucd.ie

High iodine intake by the pregnant ewe reduces colostral immunoglobulin G (IgG) absorption by the newborn lamb. Such a failure of passive transfer can increase the risk of disease and mortality. A major site of IgG absorption in the newborn lamb is the ileum and a number of genes including Fc fragment of IgG receptor transporter Alpha (FCGRT) and polymeric immunoglobulin receptor (pIGR) have been associated with post natal IgG absorption. The objective of this study was to determine the impact of level of pre partum iodine supplementation on postnatal expression of selected genes in ileal tissue of the lamb. Twin bearing ewes were offered either zero (n=10; C-group) supplementary iodine or 27 mg per ewe per day of supplementary iodine (calcium iodate;n=10; I-group) for the final four weeks of pregnancy. At 1 h *post partum* one of each twin pair was euthanized and ileal tissue harvested and stored for subsequent gene expression analysis. The remaining sibling was blood sampled at 24 h post partum for serum IgG concentration. A panel of 10 selected genes were analysed using real time quantitative reverse transcription polymerase chain reaction (qRT-PCR). Progeny of I-group had lower serum IgG at 24 post partum ($P<0.001$). There was also a tendency for reduced serum free tri-iodothyronine concentrations at 1 h post partum in this group ($P=0.07$). Lambs born to I-group ewes had higher mRNA expression of β2-microglobulin (B2M) and a lower expression of upstream stimulator factor 2 (USF2), interleukin-4 (Il-4) and thyroid hormone receptor β (THRβ) ($P<0.05$). These findings suggest that the reduction in the postnatal IgG absorptive capacity in the I-group is the result of in utero pre-programming for down regulation in the mRNA expression of genes associated with IgG absorption potentially mediated through the induction of altered thyroid hormone metabolism.

---

900 MHz electromagnetic radiation induces immunosuppression in young chicken

Gabr, A.[1], Chronopoulou, R.[1], Kominakis, A.[1], Samaras, T.[2], Politis, I.[1] and Deligeorgis, S.[1], [1]Institute Animal Breeding & Husbandry, Animal Science, Iera Odos 75, 11855, Athens, Greece, [2]RadioCommunication Laboratory, Department of Physics, Aristotle University of Thessaloniki, 54124 Thessaloniki, Greece; amrgabr2003@yahoo.com

The objective of the present study was to investigate whether exposure of chicken eggs on 900 MHz electromagnetic radiation during early embryonic development affects immune response ability of young chicken. In total, 240 eggs (in 2 batches) of a commercial chicken layer line (RedPro) were randomly assigned in two incubators: one serving as the control (no radiation) and the other as the exposure unit. The exposure protocol included continuous 900 MHz electromagnetic radiation during the first five days of embryonic development. The average (± SD) specific absorption rate in the exposed eggs was determined both experimentally and theoretically and it was calculated as high as 0.13±0.02 mW/kg. Temperature rise and variation due to radiation were not significant. Blood samples were collected at 1, 6 or 10 days-old chicken after hatching. The following parameters were determined: total membrane-bound urokinase plasminogen activator as well as superoxide anion production on phagocytes and nitric oxide on monocytes. Data were subjected to multivariate analysis of variance (MANOVA) in which treatment group, age at sampling and batch were used as the class variables. Prior to analysis, all parameters were logarithmically transformed to meet MANOVA assumptions. Data analysis showed an overall statistically significant effect (Wilks' Lamda=0.4034, F-value(DF)=9.47(32), $P<0.001$) of radiation on all parameters studied with lower values for the exposed eggs. Age ($P=0.931$) and batch ($P=0.858$) had no statistically significant effect on the parameters studied. In conclusion, 900 MHz electromagnetic radiation during early embryonic development induced immunosuppression in young chicken.

Effect of in ovo injection of royal jelly on hatchability outcomes in broiler chickens

Moghaddam, A.A.[1], Karimi, I.[1], Borji, M.[2], Bahadori, S.[2] and Komazani, D.[2], [1]Razi University, Department of Biochemistry, Physiology and Pharmacology, Faculty of Veterinary Medicine, Razi University, 67154-8-5414, Kermanshah, Iran, [2]Agriculture and Natural Sources Research Center, Animal Science, Central Province, Arak, Iran; asgharmoghaddam2000@yahoo.co.uk

This study was conducted to investigate the effects of in ovo administration of Royal jelly (RJ) in broiler. On Day 7 of incubation, 150 eggs were divided into five groups: (i) NC: Control eggs received no injection, Air-sac injected eggs received (ii) APS: 0.5 ml of saline, (iii) ARJ: 0.5 ml of pure RJ, Yolk-injected eggs received (iv) YPS: 0.5 ml of saline, and (v) YRJ: 0.5 ml of pure RJ. In ovo injection of RJ into both air sac (33.3%) and yolk (60%) decreased hatchability compared with NC (90%) group ($P<0.05$). The hatchability of ARJ group (33.3%) decreased ($P<0.05$) compared with all groups (90, 70, 66.6 and 60% for NC, APS, YPS and YRJ groups, respectively). Chick body weight relative to its initial egg weight increased in both ARJ (72.9%) and YRJ (73%) groups compared with NC (70%) and/or APS (69.9%) groups ($P<0.05$). Heart (0.41 and 0.39 g for ARJ and YRJ, respectively), liver (0.83 and 0.80 g for ARJ and YRJ, respectively) and testis (0.02 and 0.02 g for ARJ and YRJ, respectively) weights were higher in RJ-injected groups compared with NC (0.36, 0.74 and 0.013 g for heart, liver and testis, respectively) and PS-treated (0.34, 0.71and 0.013 g for heart, liver and testis, respectively) groups ($P<0.05$). The heart and liver weights in ARJ group were higher ($P<0.05$) than those in APS group, while these parameters in ARJ and YRJ groups were similar to YPS group (0.37 and 0.77 g for heart and liver, respectively). However, in YPS group significant increases in weights of heart and liver were observed in comparison to ARJ group ($P<0.05$). The weight of testis was higher in ARJ than in APS group. The results have shown that in ovo injection of RJ despite of decreasing of hatchability rate, would improve weights of hatchlings and their liver, heart, and testis weights.

Lamb meat quality is not affected by the level of maternal nutrition between day 30 and day 80 of pregnancy in sheep

Sen, U.[1], Sirin, E.[2], Ensoy, U.[2], Ulutas, Z.[2] and Kuran, M.[1], [1]Ondokuz Mayis University, Department of Animal Science, Kurupelit, 55139 Samsun, Turkey, [2]Gaziosmanpasa University, Department of Animal Science, Tasliciftlik, 60250 Tokat, Turkey; ugursen_55@hotmail.com

It has been previously shown that maternal nutrition between day 30 and day 80 of pregnancy alter birth weight and postnatal muscle growth and development resulting in altered muscle fiber types in semitendinosus (ST) muscle in sheep. This study, therefore, aimed to investigate whether such a nutrition management during pregnancy in sheep has any effect on lamb meat quality. Mature Karayaka ewes were allocated randomly into three treatment groups and were fed as follows: daily requirement for maintenance (control group, C; n=16) or 0.5×maintenance (undernutrition, UN; n=29) or 1.75×maintenance (overnutrition, ON; n=17). The diets were offered for 50 days between days 30 and 80 of pregnancy. Lambs born were subjected to a fattening period for 60 days following weaning on day 90 and slaughtered on day 150. Samples from muscles longissimus dorsi (LD) and ST were collected for meat quality parameters (pH and colour at 1 and 24 hours post mortem, shear force, dripping loss and cooking loss, and moisture, protein, intramuscular fat, and ash contents). There were no significant differences between treatment groups in terms of meat quality parameters investigated. These results indicate that although the levels of nutrition during pregnancy in sheep alter lamb birth weight and skeletal muscle growth and development, lamb meat quality is not affected. Supported by TUBITAK (TBAG-U/148), Turkey.

Alterations of foetal hepatic gene expression in response to gestational dietary protein level in pigs

Wimmers, K., Oster, M., Metges, C.C., Murani, E. and Ponsuksili, S., Leibniz Institute for Farm Animal Biology (FBN), Wilhelm-Stahl-Allee 2, 18196 Dummerstorf, Germany; wimmers@fbn-dummerstorf.de

In animal breeding intrauterine growth retardation (IUGR) has a negative impact on postnatal performance and health. In various rat models a relationship between the amount of maternal protein intake and birth weight, metabolic health and body composition of the offspring was shown ('thrifty phenotype'). There are clues that diet-dependent modifications of metabolism persist until adulthood. This leads to the hypothesis that the offspring's transcriptome is persistently regulated depending on the maternal diet. We aim to identify molecular pathways and candidate genes with relevance regarding the foetal initiation of postnatal growth and development. Therefore nulliparous German Landrace gilts were fed an isocaloric diet (~15.4 ME MJ/kg DM) containing protein levels of 30% (high protein – HP), 6% (low protein – LP) or 12% (control protein – CP) during their whole pregnancy. Offspring's liver tissue was collected at 95th day post conceptium (dpc), 1st, 28th, and 185th day post natum (dpn) (n=48 per stage) and used to hybridize onto genome-wide GeneChip® Porcine Genome Array (Affymetrix). At 1st dpn the highest number of regulated genes was found between HP vs. CP (550 genes up-, 351 down-regulated at $P<0.05$; $q \leq 0.30$). Due to the maternal HP diet genes belonging to the OXPHOS and the ubiquinone biosynthesis pathways were down-regulated. Furthermore, genes of the lipid metabolism (7DCHR, GPX4) were down-regulated as well. Moreover, levels of AMPK-α2 and PGC-1α were up-regulated. These findings point to energy sensitive pathways and their downstream transcription factors that are important in the adaptive response to the prenatal nutritional supply and may be associated with IUGR and subsequently also postnatal impaired performance.

Influence of body condition score at lambing on milk yield and quality and growth of suckling lambs in Ojinegra sheep

Ripoll-Bosch, R., Álvarez-Rodríguez, J., Blasco, I. and Joy, M., CITA, Av. Montañana 930, 50059 Zaragoza, Spain; raimonripoll@gmail.com

The aim of this study was to assess the effect of body condition score (BCS) at lambing on mother-offspring performances. After lambing, twenty two Ojinegra ewes (48.0±6.3 kg) were assigned according to their BCS to H (2.75-3) or L (2.5-1.75) treatment groups. Ewes were fed concentrate (1 kg/ewe/d; 13.5% crude protein and 14.3% crude fibre) and straw ad libitum. Milk production was estimated weekly by the oxitocin and machine milking technique (4 h interval). Blood samples were collected to analyse plasma metabolites (non-esterified fatty acids (NEFA) and β-hydroxybutyrate (BHB)). Live-weight (LW) was recorded weekly and BCS fortnightly. BCS at lambing did not affect any parameter studied. Week of lactation had a significant effect on milk crude protein content ($P<0.01$) and protein yield ($P<0.05$). Milk production tended to be greater in the first week than that in subsequent weeks (937 vs. 789 g/day, $P<0.10$). The mean milk fat content was 5.63%. Peripheral NEFA was higher at 1st week post-partum than that at weeks 2-3 (0.76 vs. 0.44 mmol/l; $P<0.05$). The lowest NEFA levels were observed at weeks 4 to 6 post-partum (0.06 mmol/l, $P<0.05$). BHB remained steady throughout lactation (0.48 mmol/l). Lamb birth weights were similar in both treatment groups (3.6±0.5 kg). Average daily gain tended to be lower in H lambs (H=128±47 vs. L=168±46 g/d; $P<0.10$), but dam BCS did not affect lamb LW at weaning (9.7±1.8 kg, $P>0.05$). Ewe LW and BCS were affected by the interaction between BCS category and week of lactation ($P<0.01$). The H group lost body condition from the 1st to the 3rd week of lactation (2.81 vs. 2.56, $P<0.05$) but remained constant afterwards (2.44; $P>0.10$). L group gain body condition during the same period (2.17 vs. 2.56, $P<0.05$) and it was maintained afterwards (2.58; $P>0.10$). Ewe LW followed the same trend as BCS. In conclusion, milk yield and quality and growth of suckling lambs were not affected by BCS at lambing.

Effect of royal jelly injection in eggs on embryonic growth, hatchability and gonadotropin levels of chicks

Moghaddam, A.A.[1], Karimi, I.[1], Borji, M.[2], Bahadori, S.[2] and Komazani, D.[2], [1]Razi University, Department of Clinical Sciences, Faculty of Veterinary Medicine, Razi University, 67154-8-5414, Kermanshah, Iran, [2]Agriculture and Natural Sources Research Center, Animal Science, Central Province, Arak, Iran; asgharmoghaddam2000@yahoo.co.uk

This study was conducted to evaluate the effects of royal jelly (RJ) injection into fertile eggs of layer breed. On Day 7 of incubation, 270 fertile eggs were allocated into 9 experimental groups: (i) NC: Control eggs received no injection, Air-sac injected eggs received (ii) APSA: 0.5 ml of saline+antibiotic, (iii) ARJ: 0.5 ml of pure RJ, (iv) ARJF: 0.5 ml of filtered RJ, (v) ARJA: 0.5 ml of RJ+antibiotic, Yolk-sac injected eggs received (vi) YPSA: 0.5 ml of saline+antibiotic, (vii) YRJ: 0.5 ml of pure RJ, (viii) YRJF: 0.5 ml of filtered RJ, and (ix) YRJA: 0.5 ml of RJ+antibiotic. The hatchability was similar in control (80%) and APSA (70%) groups and it was higher ($P<0.05$) in these groups than treatment groups (46.7, 43.3, 43.3, 66.7, 66.7, 46.7, 66.3% for ARJ, ARJF, ARJA, YPSA, YRJ, YRJF, YRJA groups, respectively). The body weights of chickens from the air sac injected groups were higher than those from other groups ($P<0.05$). The body weight relative to its initial egg weight increased and corrected body weight of chicken from the injected eggs with RJ and RJ + antibiotic was higher when compared to other groups ($P<0.05$). The internal organ weights were higher in RJ injected into air sac than those in other groups ($P<0.05$). The secretion rate of LH (1.5, 1.4, 2.8, 1.3, 2.9, 1.4, 3.4, 1.4, 3.3 mIU/ml for each corresponding treatment groups respectively) and FSH (0.6, 0.6, 1.1, 0.6, 1.2, 0.6, 1.4, 0.6, 1.3 mIU/ml) was different ($P<0.05$) between treatment groups. The results have shown that injection of RJ alone or with antibiotic improved embryo developments and secretion rate of LH and FSH hormones.

Aquaculture and food production: challenges in light of global change

Rosenthal, H., University Kiel, Schifferstrasse 48, 21629 Neu Wulmstorf, Germany; haro.train@t-online.de

Aquaculture has gained substantial importance in aquatic food production, exceeding global capture fisheries (a brief overview on global and regional production trends will be provided as well as the major production systems in operation today). Although conventional aquaculture has been practiced for millennia in several parts of the world (primarily in Asia), modern aquaculture has become a mature industry that has reached a critical mass in several production sectors, serving global markets, achieving standard quality products in several branches of the industry and interacting efficiently with international networks in marketing, quality control and investment planning. With the globalisation of markets and under global climate change, aquaculture will - in the future - exposed to new challenges to (a) meet the market needs in a world of changing consumer habits and regionally changing demand patterns, (b) produce health food, (c) support local and coastal communities in areas where employment will otherwise dwindle away, (d) serve other industries as new resource supply for added value products, (e) support dwindling natural populations important in fisheries through hatchery supply for stocking, (f) participate in maintaining productivity in natural waters in light of shifting environmental conditions (through selective breeding), and (g) serve as a tool to maintain highly endangered species and re-establish valuable species that are at the brink of extinction. These challenges for aquaculture in serving a wider range of stakeholders will be addressed by presenting case histories for each of the above listed target groups. From assessing these case examples, major research needs for European and overseas organisations will be defined and discussed. An attempt will be made to rank these research priorities while recognizing regional differences. Finally, the various scenarios predicting the developmental trend of aquaculture in major continents (Europe, Asia, North America) will be presented and critically discussed.

Genomic tools and approaches for the population management of Mediterranean aquacultured fish

Magoulas, A. and Sarropoulou, E., Hellenic Centre for Marine Research, Institute of Marine Biology and Genetics, P.O. Box 2214, GR 710 03 Heraklion, Greece; magoulas@her.hcmr.gr

Applications of large-scale genomics have revolutionized selective breeding programs of livestock, with the bovine research community having the lead. Genomic approaches are in the process of unraveling the genetic basis of various phenotypic traits within a system-biology scope. Such progress has been made possible due to advances in technology and in the production of genomic tools, ranging from whole genome and transcriptome sequencing to functional analysis, proteomics, metabolomics and data management and analysis. Fish model species have recently brought about comparative tools and precious insights in fish genome research. However, the potential of new technologies and especially new generation sequencing and proteomics has not yet been fully applied in the Mediterranean fish species. Here, for the two most commercially important species, Sparus aurata and Dicentrarchus labrax, the integrative use of aptly developed genomic resources will be presented, including linkage maps, BAC-end and transcriptome sequencing. This has led to low density but targeted QTL scans giving insights into the genetic basis of quantitative traits, such as growth, sex determination and disease resistance. In addition, we present first data of expression profiling by cDNA sequencing and by deep sequencing, which gave insights into the function of the immune system and disease resistance mechanisms in fish. The production of whole genome and transcriptome data, as well as the production of targeted SNP-chips seems today a tractable target. The up-coming possibility of genome-wide association analysis will enhance pedigree based and pedigree-independent analyses and will shed light into the evolution and adaptive value of genetic variation. First approaches have already been performed with more traditional genetic and genomic approaches and keep great promise for an appropriate management of captive and natural populations of Mediterranean fish species.

Using metal, protein and DNA profiles in *Labeo rohita* as indicators of freshwater pollution

Chaudhry, A.S.[1] and Jabeen, F.[1,2], [1]Newcastle University, Agriculture, Food and Rural Development, Agriculture Building, Newcastle upon Tyne, NE1 7RU, United Kingdom, [2]GC University Faisalabad, Zoology Department, Faisalabad City, Faisalabad, Pakistan; a.s.chaudhry@ncl.ac.uk

This completely randomised study assessed the amounts of metals in water and different tissues of Labeo rohita and their effects on DNA and protein profiles, as biomarkers, of gills and muscles of these fish from three selected sites (one reference site and two polluted sites) of the Indus River, Pakistan. The Mn, Pb, Cu, Zn, Hg and Cr levels in water as well as gills, liver, muscles and skin of Labeo rohita from these sites were compared with the internationally acceptable levels of these metals. With the exception of Pb and Hg, all other metals in water were within the acceptable limits of drinking water. However, the levels of Mn, Hg and Cr in the fish tissues were greater than their acceptable limits in fish as a human food. Here the gills had greater metal contents than the other tissues. The biomarker patterns of these fish varied between the reference and the polluted sites. Although, the gills appeared to have lost four protein bands (16-55 kDa), the muscles showed four newly acquired bands (20-100 kDa) in fish from the polluted sites in comparison to the reference site of this study. The fish from the polluted sites of this River contained low molecular weight DNA in their gills but high molecular weight DNA in their muscles as compared to the fish from the reference site of this River. The variations between these sites for the DNA and protein profiles of fish could be attributed to the changes in metal profiles of water and its inhabiting fish. It appears from this study that the protein and DNA profiles in Labeo rohita could serve as potential biomarkers to assess the impact of multiple pollutions as the environmental stressors on the fresh water river systems and their fish populations.

Effect of partial substitution of fishmeal by silkworm meal on growth and physiological status of rainbow trout (*Oncorhynchus mykiss*) using recirculated water system

Papoutsoglou, E.S.[1], Orfanos, G.[1], Karakatsouli, N.[1], Papadomichelakis, G.[2] and Papoutsoglou, S.E.[1], [1]Agricultural University of Athens, Faculty of Animal Science and Aquaculture, Department of Applied Hydrobiology, Iera Odos 75, 118 55 Athens, Greece, [2]Agricultural University of Athens, Faculty of Animal Science and Aquaculture, Department of Nutritional Physiology and Feeding, Iera Odos 75, 118 55 Athens, Greece; stratospap@aua.gr

The increasing demand for fishmeal has (should) lead to the endeavour for high quality, low cost (preferably animal) protein sources for carnivorous fish feeds. In this context, there was an investigation on the effect of silkworm meal dietary inclusion on growth and physiological status of juvenile rainbow trout (Oncorhynchus mykiss) (64.5±0.54 g). Silkworm meal inclusion in diets replaced fishmeal at level 10% and 20% and the 8-week feeding trial included a control diet based on fishmeal. All experimental diets were isocaloric and isonitrogenous. Feeding level was adjusted to 2% body weight. No differences in growth, by means of final weight (146.9-150.2 g), SGR (1.37-1.40), percent weight gain (127.1-132.2%), FCR (1.24-1.27) or carcass chemical composition (water 72.1-72.4%, protein 18.6-19.1%, fat 6.4-6.8%, ash 2.43-2.56%) were demonstrated. Furthermore, no significant differences were detected among treatments for blood haematocrit, glucose, triacylglycerides, albumin or cholesterol. Present results support the dietary inclusion of silkworm meal in feeds for rainbow trout, even at 20% inclusion level. The importance of replacing fishmeal in diets for carnivorous fish with a by-product of the silk manufacturing industry, which is rich in protein and other nutrients, low-cost and abundant, can be profound for the aquaculture industry.

Session 24 Theatre 5

Is the extended use of fatty acid percentage in fish studies adequate and justified?

Karakatsouli, N., Agricultural University of Athens, Faculty of Animal Science and Aquaculture, Department of Applied Hydrobiology, Iera Odos 75, 118 55 Athens, Greece; nafsika@aua.gr

Fish are an excellent dietary source, among other nutrients, of n-3 long-chain highly-unsaturated fatty acids (FAs) of high nutritional value for human health. A detailed overview of papers related to fish FAs composition and published mainly in aquaculture journals during the last decade is revealing. In the overwhelming majority, FAs results are reported as percentage (weight or area percent of each FA to total FAs), in only few as content (weight of each FA per g sample e.g. lipid, muscle, whole body, carcass) and in even fewer FAs are presented with both expressions. This fact has two critical aspects. First, the FAs percentage is not informative enough as each FA value reported is depended on changes of the other FAs. However, percent FAs composition is related to fish health and permits comparisons among published data. When FAs are expressed as content then each FA value reported is independent of the other FAs and the actual amount of FAs in the sample is reported (which is directly related to product nutritional value), as well as studies on FAs metabolic pathways could be performed. Second, statistics based on FAs percentage do not always give the same result with statistics performed on FAs content, and thus divergent conclusions about treatment effects could be drawn. To support this statement, examples will be given (personal and from literature data) showing that according to FAs expression different interpretation of results could be made. Fish farming industry produces fish for human consumption and reporting FAs as absolute amounts in final product is the more relevant information from the consumer point of view. Fish farming industry is also concerned about fish health and welfare. The necessity of reporting both FAs percentage and content is strongly emphasized.

Epigenetic origins of litter phenotype and implications for post-natal performance

Foxcroft, G.R., University of Alberta, Agricultural, Food & Nutritional Science, 3-10A Agriculture-Forestry Centre, Edmonton, Alberta, T6G 2P5, Canada; george.foxcroft@ualberta.ca

Variation in litter growth performance after birth may be pre-programmed during embryonic and fetal development, yet may only express itself in the late grow-finish stages of production. Two hypotheses will be explored in the context of efficient pork production from contemporary sow populations: 1) Selection for increased litter size has resulted in indirect negative effects of intra-uterine crowding on placental development in early pregnancy, leading to reprogramming of fetal development, less efficient post-natal growth performance, and adverse effects on carcass quality at slaughter. 2) Sow metabolic state at breeding can also act through epigenetic mechanisms to affect the quality of litters born. Effects of prenatal programming on postnatal health and survivability are also important. Negative pre-natal programming effects on post-natal performance are consistently seen in hyper-prolific sows that produce total numbers of pigs born that exceed uterine capacity for optimal birth weight. Selection approaches that target pigs born live and surviving through the immediate post-farrowing period will partly correct the birth phenotype trend in hyper-prolific sows. In mature sow populations producing between 10 and 15 pigs per litter, differences between litters is the major source of variance in pig birth weight. There is also apparent repeatability of low and high litter birth weight phenotypes in these sow populations. Production strategies that might address variation in litter average birth weight and post-natal performance of litters include: 1) interventions in the farrowing house targeted at sows producing a low birth weight phenotype, 2) segregated management of litters through the nursery, grow-finish and marketing stages of production based on observed or predicted birth weight phenotype, and 3) nutritional strategies in gestation and lactation targeted to specific sub-populations of sows.

The influence of n-3 polyunsaturated fatty acids in the feed of the sow on parturition characteristics and piglet viability

Tanghe, S.[1], Missotten, J.A.M.[1], Vangeyte, J.[2], Claeys, E.[1] and De Smet, S.[1], [1]Laboratory for Animal Production and Animal Product Quality, Department of Animal Production, Ghent University, Proefhoevestraat 10, 9090 Melle, Belgium, [2]Institute for Agricultural and Fisheries Research, Technology and Food Science Unit, Burg. van Gansberghelaan 115 box 1, 9820 Merelbeke, Belgium; stanghe.tanghe@ugent.be

The aim of the present study was to examine the effect of n-3 polyunsaturated fatty acids (PUFA) in the maternal diet on parturition characteristics and on piglet viability. PUFA are important for the development and function of the nervous system, and dietary levels may affect animal's behaviour and performance. Two groups of twelve sows each were fed different diets from day 45 of pregnancy and during lactation on two commercial farms. On farm I, a control diet (palm oil; 20 and 23.5 g/kg during pregnancy and lactation respectively) and a PUFA diet containing both linseed oil (5 g/kg) and fish oil (10 g/kg) were fed. On farm II, the same diets were fed except for a lower level of fish oil (5 g/kg) in the PUFA diet. All diets contained equal amounts of linoleic acid (13 g/kg). Cameras were installed in the farrowing crates and parturition was recorded. For each piglet, the time of birth, the first attempt to stand up, and the time needed to reach the udder was recorded. On both farms, the PUFA diet resulted in a shorter time interval birth – udder reaching (significant on farm II). The average time interval between successive births and average duration of parturition were significantly longer for litters from sows receiving the PUFA diet compared to the control diet on farm II, but not on farm I. There was a great variability between litters. In conclusion, n-3 PUFA in the sow's diet may affect parturition traits, but in view of the great variability this needs further study.

Nutrition of the hyper prolific sow during lactation

Quiniou, N.[1], Noblet, J.[2] and Dourmad, J.Y.[2], [1]IFIP, BP35104, 35650 LeRheu, France, [2]INRA, UMR1079 SENAH, 35590 Saint-Gilles, France; nathalie.quiniou@ifip.asso.fr

Over the three last decades, a very important increase in the average litter size has been observed in many European countries. Such an evolution is associated with a higher milk output and consequently, increased nutritional needs. However, over the same period, spontaneous feed intake remained rather unchanged, leading to a deterioration of the nutritional status of the sow during lactation and at weaning. The definition of improved nutritional strategy for high producing sows relies on a good estimation of their nutrient requirements. Experimental results obtained during the last 20 years on energy, amino acid and phosphorus utilisation by the lactating sow have been integrated in the simulation model called InraPorc (http://www.rennes.inra.fr/inraporc). The amount of nutrients exported in milk during lactation can be predicted from litter size and average daily gain. This allows a precise determination of daily energy, ileal digestible amino acids and mineral requirements. This, combined with the knowledge of average feed intake, allows the nutritionist to define targets of dietary nutrient contents for a given herd. Moreover, the modelling approach also allows predicting the consequences on body condition of sows at weaning. Thereafter, solutions that help to increase spontaneous feed intake need to be investigated, such as a better control of ambient temperature or the formulation of low heat-increment diets, in order to reduce the energy deficiency and to preserve reproduction.

Negative energy balance during the transition period stimulates milk yield of lactating sows

Theil, P.K., Hansen, A.V., Lauridsen, C., Bach Knudsen, K.E. and Sørensen, M.T., Aarhus University, Dept. of Animal Health and Bioscience, Blichers allé 20, DK- 8830, Denmark; Peter.Theil@agrsci.dk

The experiment was carried out to evaluate whether sow nutrition during late gestation affects performance during the subsequent lactation. In total, 48 sows were fed one of four gestation diets until d 108 of gestation, and then sows were fed one of six transition diets until weaning. Three gestation diets were formulated to contain 35% DF (mainly from sugar beet pulp, pectin residue or potato pulp) and a low fiber control diet contained 17.5% DF from mainly wheat and barley. Transition diets contained either 3% supplemented animal fat +/- 2.5 g/d of hydroxyl methyl butyrate or 8% supplemented fat originating from coconut oil, sunflower oil, fish oil or fish oil/octanoic acid mixture (4+4%). Back fat depth was measured at d 108 of gestation and d 28 of lactation. Blood samples were collected from sows on d 108 and 113 of gestation and d 1, 10, 17 and 28 of lactation. Piglets were weighed at d 0, 7, 10, 14, 17 and 28 of lactation (weaning) and milk yield of sows were estimated on d 7-10, 14-17 and 17-28 (week 2, 3, and 4, respectively). The experiment was regarded as a complete balanced randomized factorial design and repeated measurements within a sow was taken into consideration in the statistical analysis. Sows lost an average of 2.7 mm back fat from d 108 of gestation to d 28 of lactation. The energy balance of sows at selected days of transition (d 115 of gestation and d 1 of lactation) was negatively correlated to milk yield of the sows at weeks 2, 3 and 4. Considerable changes in plasma metabolites (glucose, lactate, NEFA and SCFA) were observed during transition (d 108 and 113 of gestation and d 1 of lactation) but no significant correlations were found between plasma metabolites and subsequent milk yield of sows. The present study suggests that the feeding strategy at onset of lactation is important for the performance throughout lactation.

Nutrition and management of lactating sows

Kemp, B., Hoving, L.L., Van Leeuwen, J.J.J., Wientjes, J.G.M. and Soede, N.M., Wageningen University, Animal Sciences, Adaptation Physiology Group, Marijkeweg 40, 6709 PG Wageningen, Netherlands; bas.kemp@wur.nl

Normally during lactation the sow is in anoestrus. However, during the lactation period the hypothalamic/pituitary system has to restore its ability to fire high frequency/low amplitude pulses of LH to recruit follicles to grow out to ovulatory sizes and to be able to mount a sufficient preovulatory LH surge. Follicle growth also has to restore during lactation to have good quality antral follicles at the end of lactation. Low feeding levels (both in terms of protein and energy) during lactation reduces restoration of LH levels and pulse frequency during lactation, resulting in extended weaning to oestrus intervals. Low feeding levels during lactation also impair follicle growth during lactation, resulting in smaller, less quality follicles at the end of lactation. These follicles result in lower ovulation rates and lower embryo survival. An adequate feed intake during lactation, preventing high losses of body stores is therefore important. Feed intake can be stimulated through good management in which attention should be given to gilt development, feed intake during pregnancy, water intake, ambient temperatures in the farrowing stable, feeding systems and feeding pattern during lactation. Reducing the number of piglets during (part of) the lactation can be successful in improving reproductive results after lactation, but a risk is the occurrence of lactational oestrus. Post weaning feeding of insulin stimulating diets or management can partly restore the negative effects of lactational weight loss on reproductive output. Also extending the time of first service after weaning by using Regumate TM or skipping a heat improves subsequent pregnancy rate and litter size.

Cross suckling from a sow perspective: which sow nurses alien piglets?

Wallenbeck, A., Alanko, T. and Rydhmer, L., Swedish University of Agricultural Sciences, Department of Animal Breeding and Genetics, Box 7023, 75007 Uppsala, Sweden; Anna.Wallenbeck@hgen.slu.se

The study is based on the first 4 parities of 40 LxY sows. Sows were single housed the first 2 weeks (w) post partum (pp) and group housed (4 or 5 sows per group) from w2 pp until weaning (w7 pp). Cross suckling was registered during 6 h of continuous observation at w4 and w6 pp, and analysed as a binary trait of the sow; either the sow nursed alien piglets or not. The proportion of sows nursing alien piglets was 44% w4 and 33% w6. The statistical analyses were performed using SAS software (PROC MIXED and GLIMMIX) and included estimation of residual correlations and analysis of variances. The results show that whether a sow nurses alien piglets or not is repeatable within lactation (w4-w6: r=0.36, P=0.001). However, it was not repeatable over lactations, i.e. when analysing the bivariate trait 'nursed alien piglets', none of the residual correlations estimated between subsequent parities were significant and the variance estimated for the random effect of the sow was not significantly different from 0. Sows nursing alien piglets w4 pp had lost less fat w2 to 7 pp, compared with sows not nursing alien piglets (-6 vs. -23 mm back fat, P=0.001) and the trend was the same when nursing was recorded w6 pp. Additionally, sows allowing alien piglets to suckle both w4 and 6 pp tended to gain weight from w2 to 7 pp, while sows not nursing alien piglets lost weight (n.s.). Sows nursing alien piglets had smaller litters (litter size w 2 pp) than sows not nursing alien piglets (w4: 9.4 vs. 10.5 piglets, P=0.011; w6: 9.3 vs. 10.3 piglets, P=0.029). In conclusion, the influence of the individual sow on the occurrence of cross suckling is repeatable within but not over lactations. Litter size and sows' use of body recourses during lactation influence the occurrence of cross suckling in that sows nursing alien piglets tended to have smaller litters and use less of their body fat during lactation.

PDS (Postpartum Dysgalactia Syndrome) in sows: application of decision tree technique for data analysis

Gerjets, I.[1], Traulsen, I.[1], Reiners, K.[2] and Kemper, N.[1], [1]Institute of Animal Breeding and Husbandry, Christian-Albrechts-University, Olshausenstr. 40, 24098 Kiel, Germany, [2]PIC Germany GmbH, Ratsteich 31, 24837 Schleswig, Germany; igerjets@tierzucht.uni-kiel.de

Postpartum dysgalactia syndrome (PDS), with mastitis as main symptome, is an important disease in sows after farrowing associated with serious economic losses. As a multifactorial disease, PDS is influenced i.e. by management, feeding and hygiene, and, moreover, by bacterial pathogens. The aim of the study was to investigate the ability of decision tree technique for analysing datasets of PDS-infected animals. Decision trees may be effective tools for making large datasets accessible and different sow herd information comparable. Milk samples of 759 sows with PDS and 762 non-infected sows of different age were obtained on six piglet rearing and fattening units. Bacteria and virulence genes of *Escherichia* (E.) *coli* involved in pathogenesis were identified by advanced bacteriological analysis of this milk including molecular techniques like multiplex PCR. Top-down algorithms (C4.5 decision tree classifiers) were used for analysis of all gathered possible decision parameters and for bacteriological data. Besides fever (rectal temperature above 39.5 °C) and appearance of the mammary glands as most important decision parameters associated with PDS, other effecting parameters and their relations were detected and visualized by the decision tree algorithm. Graphical trees of the isolated bacteria were complex; however, specific patterns in the identified virulence genes of *E. coli* were detectable. This is the first time decision tree modeling was used to identify decision parameters and patterns in pathogen spectrum for PDS. Important parameters for the disease were illustrated with decision rules indicating that a decision tree approach can distinguish PDS-infected from healthy sows and predict outcome. This could prove beneficial in disease and herd management.

Feed consumption in lactating Norwegian loose housed hybrid sows

Thingnes, S.L.[1,2], Gaustad, A.H.[2] and Framstad, T.[1], [1]Norwegian School of Veterinary Science, Production Animals Clinical Sciences, P.O.box 8146 dep., 0033 Oslo, Norway, [2]Norsvin, P.O.box 504, 2304 Hamar, Norway; signe-lovise.thingnes@norsvin.no

The aim of this project was to investigate the feeding capacity of the Norwegian Landrace/Yorkshire (LY) sows and its correlation to weight loss and litter gain. Data were obtained from 148 dry fed lactating loose housed Norwegian LY-sows from two different commercial swine herds during three trials. Sows were fed using a pelleted lactation feed containing 9.86 MJ NE/kg feed and 8.26g Lysin/kg feed. Daily feed consumption of the sows was recorded. Sow and litter were weighed within 24-36 hours after parturition (day 1), at three weeks and at weaning. The data were analysed in SAS 9.1 2003 edition, using the GLM and CORR procedures. Trial and parity number were used as fixed effects and length of lactation as a covariate. Average length of lactation was 33.8 days. Average weight at day 1 was 219.3±14.2 kg for 1st parity, 251. 2±26.2 for 2nd parity and 291.2±28.7 kg for 3rd-7th parity sows. Average litter gain was 99.8±19.0 kg for 1st parity, 102.3±20.5 kg for 2nd parity and 105.7±18.0 kg for elder sows. 1st and 2nd parity sows have a lower feed intake capacity than elder sows ($P<0.0001$), and there is a tendency for 1st parity sows to eat less than 2nd parity sows ($P<0.1$). LS means and SE for feed consumption was 243.2±6.0 kg for 1st parity, 259.5±8.0 kg for 2nd parity and 294.0±3.5 for elder sows. A positive correlation was found between feed consumption and litter gain, regardless of parity number, with $r=0.45$, $P<0.0001$ (feed consumption 277.4±44.0 kg and litter gain 103.9±18.6 kg), and a negative correlation between feed consumption and weight loss; $r=-0.43$, $P<0.0001$ (feed consumption 277.4±44.0 kg and weight loss 32.7±19.1 kg). The data show that total feed consumption influence the weight loss of sows during lactation and the productive performance in terms of litter gain, and that feed consumption varies with parity number.

effect of dietary tryptophan supplementation during lactation on sow behaviour and performance

Muns, R.[1], Agostini, P.S.[1], Martín-Orue, S.M.[1], Perez, J.F.[1], Cirera, M.[2], Corrent, E.[2] and Gasa, J.[1], [1]Universtitat Autònoma de Barcelona, UAB, Barcelona, 08193 Bellaterra, Spain, [2]Indukern, S.A., Barcelona, 08820 El Prat de Llobregat, Spain; ramon.muns@uab.cat

Feed intake and management play an important role on sow productivity during lactation. Therefore, dietary tryptophan supplementation may become a useful tool to improve both, feed intake and behaviour during this period. The aim of the present study was to evaluate the effect of dietary tryptophan supplementation during lactation on sow's behaviour evaluated as individual position changes (lying, sitting or standing) and the time spent in each posture. A total of 28 hiperprolific LW x LD sows (primiparous and multiparous) from a commercial farm were fed the same maize-barley based lactation diet only differing in total tryptophan level: T-1; Trp = 0.2% (n=14) and T-2; Trp = 0.26% (n=14). The experimental diets were introduced on day 5 post-farrowing. Sows were video recorded on day 3, 6-7 and 12 of lactation in order to evaluate tryptophan effect on behaviour. Body condition scores (BCS), backfat thickness (BF) and piglet performance were measured. Sows' feed consumption was daily registered. On day 6-7 post-partum, T-2 sows showed lower number of position changes (P=0.034) than non supplemented sows. No differences between treatments were observed for time spent on each posture, BCS, BF loss or piglets' performance. However, the multiparous sows fed the tryptophan enriched diets showed numerically lower BCS and BF (8 and 15% respectively) than those fed T-1. Tryptophan supplementation at day 5 of lactation reduced the number of posture changes on day 6 and 7 post-partum but without affecting the time spent on each posture. Sow's performance was not affected by tryptophan supplementation.

Incidence of Endotoxins in MMA sows on Austrian pig farms

Schaumberger, S.[1], Ratzinger, C.[1], Krüger, L.[1], Masching, S.[2] and Schatzmayr, G.[1], [1]BIOMIN, Research Center, Technopark 1, 3430 Tulln, Austria, [2]BIOMIN Holding GmbH, Industriestrasse 21, 3130 Herzogenburg, Austria; simone.schaumberge@biomin.net

MMA in sows is a consistent problem in swine producing farms. In the last years an involvement of endotoxins in the complex of the symptoms was discussed repeatedly. Therefore the aim of this study was to evaluate the incidence of endotoxins on swine farms in Austria and the excretion of endotoxins after antibiotic treatment. For this reason 32 sows of 16 Austrian pig farms (two/farm) were sampled. Samples taken were milk, urine and feces of a treated and untreated sow in the first three weeks after farrowing. All samples were analysed with the kinetic chromogen Limulus-amoebycat-lysat (LAL) test (CRIVER) within 12 hours after sampling. Lab material which was used for analysis was proved pyrogen free. Before analysis all samples were serially diluted and heat inactivated. Feces was extracted in Tween20 for 1 hour before dilution and heat inactivation was conducted. Tests were valid with $r^2 > 0.97$ and a recovery of 50-200%. Regardless of lactation week mean values in milk (35.5 ng/ml; 32.4 ng/ml) and urine (17.04 ng/ml; 8.08 ng/ml) were lower in treated sows. Feces, however, showed a higher mean value in treated sows (66.3 µg/ml; 81.5 µg/ml). Regarding the lactation week a correlation between endotoxin values and treatment could be shown. Untreated sows showed higher values of endotoxin in milk and urine in the first week after farrowing compared to treated sows. Values of feces again behave the other way around and higher values are seen in the first days after farrowing. Our results indicate that endotoxins are present any time in excretions of sows and antibiotics do not eliminate them but move the excretion to a later point after farrowing. Moreover we could show that antibiotics lead to a more concentrated excretion of the intestine. Further studies will be performed to figure out consequences of endotoxin values for the piglets.

Turning science on robust cattle into improved genetic selection decision

Amer, P.R., AbacusBio Limited, P.O. Box 5585, Dunedin 9058, New Zealand; pamer@abacusbio.co.nz

More robust cattle have the potential to increase farm profitability, improve animal welfare, reduce the contribution of ruminant livestock to greenhouse gas emissions and decrease the risk of food shortages in the face of increased variability in the farm environment. This paper focuses on genetic improvement initiatives as a mechanism by which research into robust cattle can be put into practice. Genetic improvement initiatives can include breed choice, crossbreeding, more efficient breeding programs, and more appropriate selection of genestocks from the global gene pool. Cattle breeding programs have a long history of successful contributions to farm profitability, but there are also failures and inefficiencies. In developing countries, importation of potentially high performing dairy cattle has led to terrible failure and loss of food security for poor people, due to an inability to control the environment. In the western world, there has been too much focus on a narrow range of relatively easily recorded and economically important traits, such as milk yield in dairy cattle and growth rate in beef cattle. There are opportunities for genetic improvement initiatives to better achieve more robust cattle, but each faces potential conflicts. Crossbreeding is well known to contribute to robustness, but can lead to loss of genetic diversity, and pose a threat to traditional suppliers of improved breeding stocks. Breed substitution can contribute to loss of genetic diversity, and result in unexpected incompatibility with the new environment. For within-breed improvement programs, a major challenge remains to shift the focus from a narrow range of shorter term market driven breeding objectives. A case is made for industry and government initiatives to switch some of their investment effort back into incentivising new trait recording at breeding program and farm level. This, in conjunction with broader breeding goals, will lead to artificial evolution towards more robust cattle and dovetail synergistically with new technological developments in genomic prediction.

Robust dairy production systems in the USA

De Vries, A., University of Florida, Dept. of Animal Sciences, Bldg. 499, Shealy Drive, Gainesville, FL 32608, USA; devries@ufl.edu

Dairy production systems in the USA are very diverse and range from small, grass based family-run dairy farms to large, confined dairy businesses. The approximately 70,000 dairy farms take care of 9 million dairy cows. Although the average herd size is a little greater than 100 dairy cows per farm, 5% of the dairy farms house 50% of all dairy cows. There is no government implemented milk quota. Milk pricing is complex and varies by region, but generally follows international supply and demand of milk products. As a result, farms are encouraged to produce more milk when milk prices are relatively high. A major challenge for America's dairy farmers is the increasingly larger swings in milk prices. High milk prices are followed by low milk prices approximately 18 months later. Average milk prices in the good years of 2007 and 2008 were \$0.42/kg and \$0.40/kg, respectively, but the average milk price decreased dramatically to a low of \$0.28/kg in 2009. The cost of production remained high in 2009 which collectively led to major losses for almost all dairy farms with many going out of business. Robust dairy production system are therefore foremost those that can withstand these swings in milk prices. Anecdotal evidence and limited analyses suggest that dairy farms that have low debts, own plenty of land, grow their own forages, raise their own young stock, and manage input and output price risks are better able to withstand swings in prices and remain in business. These characteristics are mostly found in the smaller more traditional dairy farms in the Midwest and eastern part of the USA. Large dairy farms in the west, especially those with a limited land base who purchase most of their feeds were hit hard in 2009. In addition, high feed costs coupled with low milk prices have triggered a renewed interest in grazing dairy farms. Grazing not only provides some feed directly to the cows, but should be considered a low cost housing system. In conclusion, robust dairy production systems in the USA are able to control costs and limit price risks.

Session 26 Theatre 3

Selection for robustness and product quality in dairy cattle; an international effort

Berry, D.P.[1] and Veerkamp, R.F.[2], [1]Teagasc, Moorepark Dairy Research Center, Fermoy, Co. Cork, Ireland, [2]Animal Breeding and Genomics Centre, Wageningen UR Livestock Research, Lelystad, 6708WC, Netherlands; donagh.berry@teagasc.ie

The impact on profitability of reduced health, fertility and longevity in dairy cattle as a consequence of aggressive selection for milk production is unsustainable. Furthermore, there is a growing awareness among consumers of the increasing levels of animal wastage in modern day European dairy production systems as well as the quality of the milk produced. Exploration of the opportunities in animal breeding, to mediate against unfavourable trends in animal robustness and milk quality requires a concerted effort among international scientists, with expertise in different production systems. ROBUSTMILK is an EU Framework7 project (http://www.robustmilk.eu), bringing together geneticists from the Netherlands, the UK, Sweden, Belgium and Ireland with the objective of developing new practical tools to allow breeders to re-focus their selection to include milk quality and dairy cow robustness. Robustness is defined as (a) the ability of the animal to remain close to nutritional homeostasis, and (b) the ability of an animal to perform equally well in different environments. Milk quality relates to milk somatic cell count, fatty acid type and content and lactoferrin content. Phenotypic, statistical and genomic tools are proposed. Results to-date clearly demonstrate that mid-infrared spectroscopy analysis of milk can predict the major fatty acid component of milk across different breeds and production systems with a very high degree of accuracy. Prediction of energy balance from the mid-infrared spectrum is also possible. Large genetic variation has been identified in traits that better capture changes in somatic cell count and new statistical methods have been developed that can be used to select animals that have less residual variation and that avoid becoming sick but also recover more quickly if they become diseased. Genomic regions associated with milk production, somatic cell count and indicators of energy balance have been identified.

Session 26 Theatre 4

Ethical dilemmas in dairy farming

Stassen, E.N., wageningen university, animal sciences, marijkeweg 40, 6709 PG Wageningen, Netherlands; elsbeth.stassen@wur.nl

The introduction of loose housing systems and other developments in housing, feeding and treatment of dairy cattle have improved the efficacy of milk production resulting in less labour per cow and better working conditions for the farmer. At this moment, dairy farming is on the verge of major changes because liberal market principles will be introduced to the dairy sector. Despite an expected gradual increase in worldwide demand for milk, the costs of milk production will therefore become more important. As a consequence, the number of dairy farms will reduce in the coming ten years in combination with an increased number of dairy cattle per farm. In society there is concern that these developments will result in a growing objectification of the animals. Questions are raised about the longevity of dairy cattle and welfare problems such as claw disorders. It is of interest to understand the ethical dilemmas by giving consideration to the principles of wellbeing, autonomy and justice in relation to each stakeholder group: farmers, dairy cattle and citizens/ consumers. For that, different views on animal welfare including the relation to longevity and moral attitudes of people towards animals will be considered. These considerations will be further elaborated on the main cause of diminished welfare in dairy cattle, claw disorders. Despite considerable investment in research, technology and information transfer, no reduction in the prevalence of claw disorders and lameness has occurred. Insight into the key ethical dilemmas will help to bridge the gap between farmers and technical animal sciences on the one hand and society on the other hand.

Crossbreeding in dairy cows: effects on production, somatic cell counts and calving interval on Dutch organic dairy farms

Nauta, W.J.[1], Hoorneman, J.N.[2], Smolders, E.A.A.[3], Veerkamp, R.F.[3] and De Haas, Y.[3], [1]Louis Bolk Institute, Animal production, Hoofdstraat 24, NL-3972LA Driebergen, Netherlands, [2]Hendriks Genetics, Breeding, P.O. Box 114, NL-5830 AC Boxmeer, Netherlands, [3]Wageningen UR Livestock Research, Animal Breeding and Genomics Centre, P.O. Box 65, NL-8200 AB Lelystad, Netherlands; w.nauta@louisbolk.nl

Crossbreeding Holstein dairy cows with other native and foreign dual purpose breeds, like Brown Swiss, Montebèliarde, Dutch Friesian, Groninger White Headed and Meuse-Rhine-Yssel, is very popular in organic farms. Mainly because highly productive Holstein cows often cannot cope with the low input organic farming environment. However, until now there is no information available about how such breeds and crossbreeds perform in organic farms, with very diverse farm styles, based on the nature of organic farming. Data from 113 Dutch organic herds were analysed to determine the effect of breed composition, heterosis and recombination on milk production, udder health and fertility traits in different farm types, differentiated based on their soil type and housing system. Not surprisingly, crossing Holstein with dual purpose breeds always decreased 305d milk production, also when yields are corrected for milk fat and protein production. Crossing Holstein cows always improved fertility, and mostly improved udder health as well, except when crossing with White Headed bulls. Farm type affected the regression coefficients on breed components significantly for some breeds, indicating that differences in regression coefficients of some breeds on the analysed traits were larger when two farming systems were compared than for other breeds. For example, milk production was on average highest for all breeds on sandy soils, probably due to more home grown corn and cereals in the diets. This higher production has a negative correlation with udder health and fertility for most productive breeds.

Gasconne breed management in France and Spain

Guerrero, A.[1], Sañudo, C.[1], Mateos, J.A.[1], Campo, M.M.[1], Olleta, J.L.[1], Caillaud, S.[1], Toustou, J.[2], Gajan, J.P.[2] and Santolaria, P.[1], [1]Veterinary Faculty of Zaragoza, Animal Production and food science, Miguel Servet 177, 50013 Zaragoza, Spain, [2]Groupe Gascon, Villeneuve du Pareage, 09100, France; aguerre@unizar.es

The Gasconne is a cattle breed originating from the South of France, which from the 1980's has spread to the neighbouring territory of Spain. The aims of the study were to know the current situation of this breed in Spain and to compare it to that in France. A questionnaire was used, divided in 17 sections that included some reproductive, productive and breeding characteristics. Thirty two Spanish and twenty five French breeders were polled. The results showed an average of 23 head in Spain and 28 head in France at the beginning of the farming activity. The average number of head by farm (in the year 2009) was 85 and 118 respectively in these two countries. The typical dam's production system was extensive in both countries, with two people working in the farm, usually members of the family. The replacement index was 8.1% in Spain and 16.7% in France; however, regarding average productive life of the cows, the theoretical replacement index should be 6.9% and 7.9% respectively. The average age at first calving was 31.6 months for Spanish cows and 34.3 for French ones whereas the calving interval was one year in both countries. The number of farms that make its own fattening was 41% versus 64% respectively in each country. The main income came from calves (44% of the total) and yearlings (33%) in Spain and cows (40%), calves (21%) and steers (18%) in France. The most important criteria for animal replacement for Spanish breeders were the female origin (42%), followed by its conformation (33%). For the French's breeders the main criteria were conformation (38%) and the origin of the animal (22%). Docility was considered in third place in France, with 12% of the answers, and in fourth place in Spain, with 5%. Male replacement criteria were similar to those considered in females in both countries.

Crossbreeding in dairy cattle: results from national data in Germany

Bergk, N. and Swalve, H.H., Institute of Agricultural and Nutritional Sciences, Martin-Luther-University Halle-Wittenberg, Theodor-Lieser-Str. 11, 06120 Halle, Germany; nadine.bergk@landw.uni-halle.de

In the past years of dairy cattle breeding, breeding goals have focussed on improving production. Meanwhile, functional traits, e.g. fertility or resistance to disease have become more and more important. These traits are normally characterised by low heritabilities and therefore genetic gains achieved by selection are small. An alternative way of improving functionality is crossbreeding since this breeding method can exploit heterosis which should be beneficial for traits with low heritability. Production, fertility, calving and culling data of F1-crossbreds and their purebred Holstein herdmates were extracted from the national data recording base. The dataset included only dairy cows from herdbook herds that were born since 2001 and with records available at the genetic evaluation run as of April 2009. Crosses between Holstein and Fleckvieh (Fl x Hol), Jersey (J x Hol), Brown Swiss (BS x Hol), or Red breed (RB x Hol) were chosen for a comparison with purebred Holstein. Different datasets, one for each type of cross and group of traits, were created. The comparison of production traits showed that crossbreds are only partially able to produce as much milk as Holstein purebreds: Fl x Hol and J x Hol yielded less than Holstein, BS x Hol and RB x Hol had a milk yield comparable to Holstein. In herds with a high production level, purebred Holsteins were superior. Notable advantages for F1-crossbreds concerning fertility traits and stillbirth rate, especially in the first lactation, were observed. The differences between the various genotypes of crossbreds and their purebred contemporaries decreased with increasing parity number. Culling rate of crossbreds was comparable with the rate of purebred cows that were culled involuntarily. Compared to pure Holsteins, crossbred cows exhibited an increased rate of culling due to insufficient milk production.

Session 26 Theatre 8

Assessing the effect of heat stress in dairy cows by applying random regression test day models

Brügemann, K.[1], König, S.[1] and Gernand, E.[2], [1]University of Göttingen, Department of Animal Science, Albrecht-Thaer-Weg 3, 37075, Germany, [2]Thuringian State Institute of Agriculture, August-Bebel-Str. 2, 36433 Bad Salzungen, Germany; skoenig2@gwdg.de

Random regression models were applied to investigate the impact of heat stress, as measured by temperature-humidity indices (THI), on test day production records of dairy cows. Data comprised 1,094,601 first lactation protein yield records from 154,880 Holstein cows with calvings between 2003 and 2007 at large-scale dairy farms in East Germany. Test day data were merged with daily meteorological data from nearest weather stations located within 50 kilometres of the respective farms. Variance components for daily protein yield were estimated using animal models for REML. Fixed effects were herd-year, milking frequency and classes for days in milk. The modelling of regression curves for days in milk and THI was done by using Legendre polynomials for additive-genetic effects of order 4. Different residuals were allowed for the combined effects of days in milk and THI. Regression analysis revealed different reaction patterns of daily protein yield with increasing THI for different periods of days in milk. A THI threshold of 60 was proven to be a critical threshold for heat stress, i.e. it was generally associated with a decrease in daily protein yield. Cows were most sensitive to heat stress directly after calving, as indicated by highest rates of decline in protein yield. Daily additive genetic variances, daily permanent environmental variance, and heritabilities (0.04-0.19) were highest for low THI in combination with later stages of lactation. The paramount goal of each breeding program must be the accurate identification of genetically superior animals. This goal can be achieved through selecting optimal environments for progeny testing, so that animals can express their true genetic potential. Heat stress inhibited the full expression of the genetic potential for production traits, and has to be taken into account when creating optimal environments for progeny testing.

Effects of *in vitro* rumen protection of fish-meal coated with supplemental fats on digestibility and gas production rate using gas test technique

Palizdar, M.H.[1], Sadeghipanah, H.[2], Amanlou, H.[3], Nazer Adl, K.[1], Mirhadi, A.[2], Mohammadian-Tabrizi, H.[1] and Aghashahi, A.[2], [1]Department of Animal Science,faculty of agriculture, Islamic Azad University of Shabestar, Shabestar, Tabriz, 31757, Iran, [2]Animal Science research Institute, Animal Nutrition, Karaj, 31757, Iran, [3] Zanjan University,faculty of agriculture, Department of Animal Science, Zanjan, 31757, Iran; araghashahi@yahoo.com

The aim of this study was to investigate gas production rate and digestibility of fish-meal (FM) coated with different types [hydrogenated tallow (T) and hydrogenated palm oil (P)] and levels (0, 20, 40, 60 and 80%) of fat to decrease rumen digestibility of protein and gas production. Approximately 200 mg (DM basis) of sample is weighed and inserted in glass syringes, and then gas production was recorded after 0, 2, 4, 6, 8, 12, 24, 48, 72 and 96 h. The average of treatments were computed, the equation of $p= a+b\ (1-e^{-ct})$ and the coefficients of a, b and c predicted and compared using SAS. The result of the present study showed that supplemental fat which was mixed with fish-meal to protect it from the fermentation process, reduced *in vitro* digestibility and gas production during the time of incubation. Addition of supplemental fat (T and P fat) to fish-meal could significantly decrease gas production during incubation times ($P<0.01$). In comparison to P fat, coating fish-meal with T fat resulted in a significant reduced gas production ($P<0.01$). Furthermore, the values of b (the fraction which needs more time to degradation) and a+b reduced significantly as fish-meal was coated with these two types of fat and along with increasing the level of supplemental fat used in this study ($P<0.01$). It seems that if we could use a part of supplemental fats as a coating factor to protect protein and amino acids sources from rumen destruction, it could lead to an increase in the rumen un-degradable protein and amino acids fraction. Accordingly, a reduction in the fermentation rate of proteins could reduce the gas production (NH_3 and CH_4) in the rumen.

Economic importance of production, functional and carcass traits of Pinzgau cattle bred in dairy and cow-calf system

Krupa, E.[1], Krupová, Z.[1], Huba, J.[1], Polák, P.[1], Tomka, J.[1] and Wolfová, M.[2], [1]Animal Production Research Centre Nitra, Hlohovecká 2, 95141 Lužianky, Slovakia (Slovak Republic), [2]Institute of Animal Science, Přátelství 815, 104 00 Praha Uhříněves, Czech Republic; tomka@cvzv.sk

The economic weights of 14 production (dairy and growth), functional and carcass traits for Slovak Pinzgau cattle raised in dairy (A) and cow-calf (B) system were calculated. The production of breeding heifers for own herd replacement with ten reproduction cycles at maximum, and the sale of surplus male and female calves were assumed for both systems. Base price per milk value was corrected according to fat and protein content and somatic cells count. The marginal economic weights were calculated as the numeric derivation of the profit function using a bio-economic approach. Marginal values were standardised (multiplied by the genetic standard deviation of appropriate trait) and expressed as relative values (percentage proportion). In both systems, carcass traits reached the lowest marginal economic values. The highest marginal importance was found for production lifetime of cows in system A (88.44 € per year and cow), and also in system B (70.17 € per year and cow), respectively. Marginal values for functional traits achieved the highest values in both systems. Nevertheless, relative economic values for functional traits together represent 72.36% in system B, but only 39.20% in system A, respectively. Production traits cover 60.80% in system A, and 27.64% in system B, respectively. The highest relative importance achieved the 305 day milk production (34.06%) in system A and yearling weight (26.19%) in system B, respectively.

Investigations on pathogenic microorganisms sources in cows milk production chain

Konosonoka, I.H., Jemeljanovs, A., Valdovska, A. and Ikauniece, D., Latvia University of Agriculture, Research Institute of Biotechnology and Veterinary Medicine Sigra, Instituta street No.1, LV 2150, Sigulda, Latvia; biolab.sigra@lis.lv

For obtaining qualitative and safe cows' milk and further milk products, it is essential to gather information on critical microbial hazards in milk production environment. The objective of the current study was to investigate sources of *Listeria monocytogenes*, *Salmonella* spp., *Staphylococcus aureus*, *Bacillus cereus*, *Clostridium* spp. and moulds in cowshed environment. 130 feed, 21 water, 5 manure, 4 air samples from four dairy farms were bacteriologically and mycologically examined at the Scientific Laboratory of Biochemistry and Microbiology of the Research Institute 'Sigra'. Complex and selective culture media were used for the isolation and differentiation of bacteria. Acquired results showed that Listeria spp. were isolated from 14.3% grass, 11.1% fodder and corn, and 6.3% of silage samples. 12.5% of all feed samples were contaminated with Listeria monocytogenes. Most contaminated with *Clostridium* spp. spores and vegetative cells were silage samples. *Clostridium* spp. spores were detected in 62.5% of samples. 32,0% of feed samples were contaminated with *B. cereus* vegetative cells. *B. cereus* spores were detected in 26.0% of feed samples. There were detected 11 moulds' genera and species in feed samples. The greatest number of moulds has been detected in the hay samples: $8.5x10^8$ cfu g^{-1}. Water used in dairy farms is the source for raw milk contamination with *Escherichia coli*, *Enterococcus faecium*, *Enterococcus faecalis*, *B. cereus*, *Clostridium* spp. and coliform bacteria. Air in dairy farms is the source for contamination with bacteria of *Staphylococcus* genus. Manure is the source for raw milk contamination with *B. cereus*, *Listeria* spp., *Salmonella* spp., *E. coli*. Our investigations let us conclude that feed and water used in dairy farms and cows environment are the main sources for raw milk contamination with pathogenic microorganisms.

Health programs for small ruminants in Switzerland

Luechinger Wueest, R., Extension and health service for small ruminants, P.O. box 399, 3360 Herzogenbuchsee, Switzerland; rita.luechinger@caprovis.ch

The Swiss population of small ruminants is 446,000 sheep and 81,000 goats. Animal-movement and contact within flocks on alpine pastures are sources of high risk for infections. The 'health service for small ruminants' offers programs to prevent severe health problems. In 1980, about 80% of the Swiss goat population was infected with caprine arthritis and encephalitis (CAE). Economical losses due to lower milk production and an average age at culling of only three years were the main reasons to start with an eradication program in 1984. By eradicating seropositive animals less than 1% of the seropositive goats were achieved. Similar symptoms, due to Maedi-Visna, caused damages to the milk-sheep population. Using the same scheme as for CAE eradication, 42% of milk-sheep are now Maedi-Visna free. Because of high losses of lambs during summer due to parasite infection, the treatment-strategy has been expanded with a parasite survey program. Faecal egg counts from different groups of animals help to decide whether a treatment is necessary or not. The results allow to chose an efficient and appropriate product for the treatment and, with additional examinations, resistances can be detected. At present, around 500 flocks are included in the survey program. This program allows saving on average one treatment per year. Beside parasite-infection, the main problem in sheep is foot-rot. Hoofs are periodically inspected to detect clinical signs, infected and non infected animals are separated. Infected animals are treated by clipping and foot bathing. After two negative controls with no animals showing clinical signs, a flock is declared foot-rot free. Today around 20% of the Swiss sheep population are free of foot-rot. Abscesses due to infection by pseudotuberculosis are a problem of hygiene and a risk for product safety. Since 1999, a survey-program (regularly scanning the lymph-nods) is used to recognize flocks with pseudotuberculosis, followed by blood sampling and eradication of the positive animals.

A cost-benefit analysis of the 'Healthier goats' program in Norwegian dairy goats

Nagel-Alne, G.E.[1,2], Valle, P.S.[2], Asheim, L.J.[3] and Sølverød, L.S.[1], [1]TINE BA, TINE Extension Services, Christian Frederiks plass 6, 0051 Oslo, Norway, [2]Norwegian School of Veterinary Science, Institute of Production Animal Clinincal Science, Ullevålsv. 72, 0454 Oslo, Norway, [3]Norwegian Agricultural Economics Research Institute, P.O 8024 Dep., 0030 Oslo, Norway; gunvor.elise.nagel_alne@nvh.no

In 2001 the Goat Health Services of Norway initiated a program to sanitize for caprine arthritis encephalitis (CAE), paratuberculosis (Johne´s disease) and caseous lymphadenitis (CLA) in dairy goat herds ('Healthier goats'). Since its start about 250 out of 450 herds have completed the sanitation at a program cost of about NOK 2500 per goat. In the paper the net present value of the investments for farmers and the processing industry, will be presented based on farm data for a random sample of 24 sanitized and 21 control farms. Farmers costs, gathered in an inquiry, will be added to the industry costs to arrive at total sanitation investment costs. Data from a previous study show average milk yield increased by 21 percent (from 627 to 756 kg) per goat in sanitized herds, after adjusting for increase in a control group. This calculation assumes similar age distribution and will be reinforced if sanitized goats last longer. In that case there will also be lower costs of replacement. Benefit due to increased milk yield will however be restricted due to the quota system. Farm labour input, costs of buildings and use of feed decrease, however, so do headage payments and the goats may require more concentrated feeds. The price of milk increases due to improvements in its quality. The improved processing properties will lower milk losses and increase industry productivity. Finally the paper discusses the importance of preventive veterinary medicine and animal disease control in a strategy for a future oriented, robust and competitive Norwegian dairy goat industry, producing healthy and safe food.

Factors affecting lamb mortality in Cretan (Greece) sheep farms

Stefanakis, A.[1], Volanis, M.[2] and Sotiraki, S.[1], [1]NAGREF, VRI, Thermi, 57001, Greece, [2]NAGREF, Research Station Asomaton, Amari, 74061 Rethymno, Greece; stefanakisa@yahoo.gr

High mortality reduces the profitability of lamb production worldwide, and is an important welfare consideration. Data on neonatal lamb mortality and flock management in different lambing seasons were recorded on 15 sheep farms in Crete during 2006, 2007 and 2008. The major causes associated with lamb losses were watery mouth disease (33,2%), hypothermia & starvation (22.6%), digestive disorders (13.4%), pneumonia (10.3%), septicaemia and toxaemia (9.1%), endoparasitism (2.20%), accidental (2.1%) and undetermined causes (7.1%). The highest lamb mortalities occurred within the first 15 days of life. The lamb birth weight and ewe weight at lambing were highly significant ($P<0.01$) with lamb mortality rate during pre- and post-weaning stages. The sire of the lamb also had a significant ($P<0.01$) effect on lamb mortality rate at all ages. Intensive rearing systems appear to be associated with increased perinatal and postnatal mortality, although housing ewes at lambing was associated with a decreased risk of stillbirth. High perinatal mortality also was associated with poor hygiene, flocks that foster more lambs, and failure to provide appropriate nursing for sick lambs. Larger flocks, poor ewe condition at breeding and flocks with higher ewe-replacement rates were associated with higher postnatal mortality. Extensively managed animals, often living in harsh and unfavourable environments, need specific adaptations that promote survivability. This is particularly important at parturition and during the neonatal period, when lamb mortality is highest. Lambs <1 week old are at greatest risk, and tend to die of exposure hypothermia, starvation, septicaemia consequent upon inadequate colostrum intake. The findings of this study revealed causes of pre-weaning lamb mortalities to be mainly due to a low birth weight and non-parasitic diseases (predominantly watery mouth disease, hypothermia & starvation and pneumonia). These factors must be considered in any sheep production system.

Session 27 Theatre 4

Population dynamics of ovine coccidia in lambs reared under different management conditions in Greece

Saratsis, A.[1], Stefanakis, A.[1], Joachim, A.[2] and Sotiraki, S.[1], [1]NAGREF, VRI, Thermi, 57001, Greece, [2]University of Veterinary Medicine, Institute of Parasitology and Zoology, Vienna, 1210, Austria; smaro_sotiraki@yahoo.gr

Coccidiosis in lambs is caused by Eimeria spp and represents an important problem in sheep production. Infected lambs develop diarrhoea early in life and even though mortality is low economic losses can be remarkable due to weight gain losses and impaired performances in the subsequent production phases. The epidemiology of coccidiosis in dairy sheep systems has not been widely studied leading to excess anti-coccidial drugs use to prevent diarrhoea. This is costly for the producer, against consumer's demands and may lead to development of drug resistance. In this study, we assessed population dynamics and the effect of flock level factors on the risk and level of Eimeria infection. Data were collected from 7 (intensive and semi-intensive) flocks in different sites including lambs born in both early and late lambing period in each flock. Faecal consistency and oocyst excretion were recorded from faecal samples taken from in total 240 lambs starting at day 8 after lambing and subsequently every 6 days for 5 times. Eimeria oocysts were detected from samples coming from all flocks during the study. 11 Eimeria species were isolated including the most pathogenic E. ovinoidalis and E. crandallis. The lambs age at onset excretion ranged from 14 to 38 days and the cumulative incidence of infection per flock until the end of the study ranged from 60% to 100% (5 out of 7 flocks;both lambing periods). Oocyst excretion peaked at 25 to 33 days of life irrespective of the lambing period. Survival analyses showed that the onset of oocyst excretion was associated with the lambing period (i.e. higher risk to excrete oocysts earlier in life during the late lambing period, $P<0.05$). The results confirm that the spread of Eimeria infection among lambs is wide regardless of flock location and management and strongly related to environment contamination. Thus, appropriate hygiene measures could prevent or limit infection.

Session 27 Theatre 5

Use of European grown protein sources to control gastrointestinal parasitism in small ruminants

Sakkas, P.[1], Houdijk, J.G.M.[1], Athanasiadou, S.[1] and Kyriazakis, I.[2], [1]SAC, Animal Health, Edinburgh, EH9 3JG, United Kingdom, [2]University of Thessaly, Veterinary Faculty, Karditsa, 43100, Greece; www.panossak@gmail.com

Periparturient relaxation of immunity (PPRI) plays a key role in small ruminant parasite epidemiology, as nematode eggs excreted by ewes and goats are a main source of infection for their young. Increased metabolizable protein (MP) supply reduces PPRI, although its extent differs between studies. We suggest that this is due to variation in MP quality, defined as its amino acid (AA) profile. We address this hypothesis through both literature review and experimental studies. PPRI has been reduced using cottonseed meal, fishmeal, urea, soybean meal and protected soybean meal (SoyPass), leading to increased MP supply with variable digestible undegradable protein (DUP) proportion. Our study shows that the higher the DUP to MP ratio, the stronger the effect on PPRI, which suggests that effects depend on the AA profile of the extra MP. Cottonseed meal, soybean meal and urea result in low DUP levels, so that extra MP consists mainly of microbial protein. The relatively unbalanced AA profile of the latter, compared to that of animal protein in general and immune-proteins in particular, may account for their limited effect on reducing PPRI, as parasitism increases specific AA requirements to synthesize proteins to maintain homeostasis (e.g. albumin) and to mount immune responses (e.g. inflammatory agents, mucins and antibodies). In contrast, fishmeal and SoyPass have high levels of DUP, with especially for fishmeal a much more balanced AA profile, which combined may result in larger effects on PPRI. The current finding that methionine deficiency increases PPRI is consistent with this view. Whether this means that European grown pulses like field beans are less effective than SoyPass to reduce ovine PPRI is being assessed. In conclusion, the degree of PPRI is sensitive to both MP supply and MP quality, suggesting that protein source is an important issue to consider in non-chemical parasite control strategies.

A screening tool for monitoring scrapie in sheep flocks

Psifidi, A.[1], Dovas, C.I.[2], Basdagianni, Z.[1], Valergakis, G.E.[1], Arsenos, G.[1], Papanastassopoulou, M.[2] and Banos, G.[1], [1]Faculty of Veterinary Medicine, Animal Husbandry Laboratory, Aristotle University, 54124 Thessaloniki, Greece, [2]Faculty of Veterinary Medicine, Microbiology Laboratory, Aristotle University, 54124 Thessaloniki, Greece; geval@vet.auth.gr

Scrapie in sheep is controlled by genetic factors, with polymorphisms at codons 136, 154 and 171 of the PrP gene being the determining parameters. Breeding programs have been implemented in Europe, to control scrapie by eliminating VRQ-carriers and increasing the ARR allele frequency. The aim of this study was to develop a practical method for assessing the presence of VRQ-allele within each flock using bulk milk. Sixty milk samples were taken in 50 ml tubes from as many ewes of the Chios breed. These samples were used for the creation of two artificial bulk milk sample sets each containing VRQ at 8 different frequencies (0.5%-64%). Three samples were collected from each bulk and genomic DNA was isolated using a modified commercial kit (Nucleospin® Blood, Mackerey-Nagel). Polymorphisms of the PrP gene at codon 136 were identified by Restriction Fragment Length Polymorphism (RFLP) analysis of Polymerase Chain Reaction products using the enzyme BspHI. Moreover, standard samples were constructed by plasmid DNA, in order to determine the detection limit (LOD) of VRQ with RFLP analysis. DNA pools with predefined VRQ allele concentrations of 50%, 32%, 16%, 8%, 4%, 2%, 1%, 0.5%, 0.25% were generated by serial dilutions of VRQ plasmid DNA with ARQ plasmid DNA. Every dilution was analysed in triplicates. Experiments with plasmid DNA pools indicated that the LOD of VRQ was 0.5%. In bulk milk samples the allele was detected at 1%. It is concluded that the presence of one heterozygous or one homozygousVRQ ewe can be detected by analysing bulk milk of maximum 50 or 100 ewes, respectively. The proposed method can provide a useful tool for fast screening of flocks for the presence of VRQ allele. In case VRQ-carriers are detected, individual animal genotyping could be performed to exclude these undesirable animals from flocks

Health control programme in local Lipe sheep

Jovanovic, S.J.[1], Savic, M.S.[1], Katic Radivojevic, S.K.[2] and Vegara, M.V.[3], [1]Faculty of veterinary medicine, Animal breeding and genetics, Bul.oslobodjenja 18, 11000 Belgrade, Serbia, [2]Faculty of veterinary medicine, Department for Parasitology, Bul.oslobodjenja 18, 11000 Belgrade, Serbia, [3]Norwegian University of life Sciences (UMB), Department of International Environment and Development Studies, NORAGRIC, P.O. Box 5003, N-1432 Aas, Norway; maxi_et_commy@yahoo.com

The Lipe sheep is type of Zackel breed, traditionally reared in village Lipe. At Sorbonne University, in 1935, professor Slobodan Pavlovic, in his Doctoral Dissertation has described Lipe sheep as a robust sheep with excellent cheese making characteristics. At that time the size of Lipe sheep population was about 40,000. The size of original Lipe sheep population has been restricted as a result of crossing. Nowadays, Lipe sheep is an endangered population, registered at The Endangered – Maintained breed list. The active conservation programme has been applied according to the FAO strategies for managing animal genetic resources. Lipe sheep is a triple purpose, late maturing sheep, raised under semi-intensive management conditions. The average estimated body weight of ewes is 50 kg and rams 70 kg. Coat colour is white, with black head and dark legs. Horns are large and spiral, triangularly shaped in rams. Average milk yield is 150 kg. The lambing rate is 120%. Health control program was carried out on 3 flocks. Functional traits, such as adaptive traits, fertility, lambing ease and longevity have received more attention. Priority areas of the program included breeding for disease resistance. Parasite infections were the main health problem in Lipe sheep populations. Poor weight gains, anemia, weakness and diarrhea were noticed. The infection rate of Trichostrongilidae, (*Haemonchus* spp, *Trichostrongylus* spp. *Nematodirus* spp, *Cooperia* spp, *Ostertagia* spp.), *Eimeria* spp., *Fasciola hepatica*, *Paramphistomum* spp. and *Moniezia* spp. were evaluated. The results showed a great variability of host responses to parasite infections, variability which can enable improvement of the Lipe sheep breeding programme.

Session 27 Poster 8

Haematological and blood biochemical parameters of goats after experimental invasion with Ostertagia/Teladorsagia circumcincta third development stady (L3) larvae

Birģele, E., Keidāne, D. and Ilgaža, A., Preclinical institute, Latvia University of agriculture, Faculty of veterinary medicine, Helmaņa - 8, LV - 3002, Latvia; edite.birgele@llu.lv

The aim of the study is to investigate the hematological and blood biochemical parameters 2-3 months old kids and 10-14 months old goats in association with the 5,000 and 10,000 *O. circumcincta* infection. Parasites *O. circumcincta* larvae at the development stage L_3 were administered in animals during the morning feeding time through the animal fistula. The mean arithmetical value and standart deviation were calculated for the blood hematological and biochemical parameters. To compare and assess their changes between the experimental groups of goats, t-test was applied for comparison of two sample means with different dispersions. Hematological changes in infected goats depend on the animal age and level of infection: in kids infected with 5,000 larvae of *O. circumcincta* oesinophilic and neutrophilic segmented nuclear leucocyte count significantly increases and exceeds the physiological norm as well as the monocyte count in the blood, but in kids infected with 10,000 larvae of L_3 development stage only the lymphocyte and eosinophilic leucocyte count. In adult goats in both cases of infection, the eosinophilic leucocyte count is exceeded many times, and segmented nuclear leucocyte count increases significantly, but the lymphocyte count in the blood, on the contrary, decreases significantly. In kids and adult goats artificially infected with *O. circumcincta* larvae of L_3 development stage, significant changes are typical in some blood biochemical parameters: the total amount of protein and amount of albumin in the blood increase statistically confidently, and the albumin/globulin coefficient tends to increase.

Session 27 Poster 9

Serological survey of abortifacient pathogens in organic sheep and goat farms of western Greece

Fragkiadaki, E.[1], Ntafis, V.[1], Xylouri, E.[1], Bellacicco, A.L.[2], Buonavoglia, C.[2], Vretou, P.[3] and Pappa, A.[4], [1]Agricultural University of Athens, Faculty of Animal Science and Aquaculture, Iera Odos 75, 11855, Athens, Greece, [2]University of Bari, Dept. of Veterinary Public Health, Bari, Bari, Italy, [3]Institute Pasteur Athens, Lab. of Biotechnology, Institute Pasteur Athens, Athens, Greece, [4]Aristotle University of Thessaloniki, Dep. of Microbiology, School of Medicine, Thessaloniki, Thessaloniki, Greece; irfrag@yahoo.com

The aim of the study was to investigate serologically the abortions in organic small ruminants, for *Chlamydophila abortus*, *Brucella* spp., Caprine Herpes Virus (CpHV) and West Nile virus (WNV). The objective was to estimate the pathogens prevalence and their potential abortifacient impact. 427 non vaccinated sheep and goats from 36 organic flocks having massive abortions during two successive reproductive periods (2004-2006) were tested. The first year, preliminary study was performed only to aborted animals, while both affected and normal animals were sampled the year after. The serological methods used per pathogen were: C– ELISA (C. abortus), seroneutralization (CpHV), IFA (WNV) and both an agglutination assay and an ELISA test (*Brucella* spp.). Results from the first year show positive serological titres, for *C. abortus* in 50% ewes and 84.87% goats tested. The percentage recorded for CpHV was 63.06% for the goats. At the second year 66.67% of the affected and 64.63% of the normal ewes were positive for *C. abortus* together with 93.88% of the affected and 97.56% of the normal goats. Regarding CpHV, 52.94% of the affected goats and 82.35% of the normal were positive. No antibodies to WNV and *Brucella* sp were detected. Chlamydiosis and CpHV infection are enzootic in small ruminants', while no WNV or brucellosis was observed. For the first two pathogens, seroprevalence is higher than that reported in conventional farms. However, in order to safely conclude on the main etiological factors of abortions in organic small ruminants, more parameters should be studied.

Comparing environmental impacts of livestock production systems of varying intensity

Corson, M.S., Aubin, J. and Van Der Werf, H.M.G., INRA UMR SAS, 65 rue de Saint Brieuc - CS 84215, 35042 Rennes, France; michael.corson@rennes.inra.fr

Increasing human population and affluence is increasing the global demand for animal products. At the same time, societies are calling for a decrease in negative environmental impacts of agriculture, particularly livestock production. Unfortunately, increasing animal production via intensification has been implicated in increasing environmental impacts. In contrast, extensive or organic systems, often considered as having lower environmental impacts, usually have lower production per unit area. Designing innovative livestock production systems that can attempt to meet this double challenge requires the use of assessment methods than can quantify a variety of environmental impacts and identify specific management practices or production stages of systems with lower impacts. Life Cycle Assessment (LCA) is a multicriteria assessment method that estimates potential environmental impacts of products as a function of all the raw materials and energy used to produce them and pollutant emissions at all production stages. We present recent results comparing various production intensities in aquaculture, pork, and dairy systems estimated with LCA. In all systems, feed production contributed the most to nearly all impact categories. Less intensive systems tended to have lower environmental impacts than more intensive ones, but these differences often were not statistically significant or varied with the functional unit of analysis (e.g., impacts per kg of product vs. per ha of agricultural area). The environmental impacts of changes in intensification can vary depending upon the land use of a livestock system. For example, implementing less-intensive production modes in confinement systems may require larger changes to system design than doing so in pasture-based systems, which may influence relative environmental impacts.

Greenhouse gases emissions and energy consumption on a panel of French meat sheep production systems: what variability and what factors to explain?

Benoit, M. and Laignel, G., INRA, Unité Recherches Herbivores 1213, Centre de Theix, 63122 St Genes-Champanelle, France; marc.benoit@clermont.inra.fr

This study is under questioning about the environmental impact of livestock farming systems. This environmental impact is studied here through greenhouse gases (GHG) emissions and consumption of nonrenewable energy. We have adjusted the levels of methane emissions according to the latest scientific knowledge and we take into account the carbon sequestration in grasslands to be able to present also the net emissions of GHG per kilogram carcass produced. This work is made by modeling ten contrasted farming systems studied elsewhere (private farms or experimental stations). One objective is to identify key factors affecting the levels of GHG emissions. We study in particular the impacts on GHG emissions according to 1 / the ewe productivity level, 2 / the level of forage self sufficiency, 3 / the use of large forage areas (extensive systems). An analysis of economic performance is made in parallel. The study is not fully realized. We assume that there is no systematic correlation between low levels of GHG emissions and low consumption of nonrenewable energy. In addition, interesting correlations may emerge between the economic output (gross margin per ewe) and low levels of GHG emissions per kilogram of carcass produced. The correlation is a priori lower between the economic results and a low use of nonrenewable energy. Unlike in suckler cow systems, there is significant variability in the levels of GHG emissions (and nonrenewable energy) per kilogram of carcass produced depending on the system studied. Forage self-sufficiency is a key factor for a low non renewable energy consumption; the ewe productivity and the fodder area per ewe are the first factors affecting GHG emissions (after carbon sequestration). So, overall, the best performing systems on these two approaches (energy and GHG) are those that have maximum ewe productivity with the best use of forage, in extensive systems.

Sustainable pig production in the EU: an example of Erasmus intensive teaching programme

Montagne, L.[1], Bosi, P.[2], Boudry, C.[3], Cain, P.J.[4], Couvreur, S.[5], Franks, J.R.[4], Guy, J.[4] and Lundeheim, N.[6], [1]Agrocampus Ouest, 65 rue de St-Brieuc, 35042 Rennes, France, [2]Bologna Univ, 107 via Rosselli, 40100 Bologna, Italy, [3]Agro-Bio Tech, 2 passage des Déportés, 5030 Gembloux, Belgium, [4]Newcatsle Univ, 10 Kensington Terrace, NE17RU Newcastle Upon Tyne, United Kingdom, [5]ESA, 55 rue Rabelais, 49007 Angers, France, [6]Swedish Univ of Agricultural Sciences, box 7023, SE75007 Uppsala, Sweden; montagne@agrocampus-ouest.fr

Development of sustainable farming systems is crucial for the survival of the pig production in the EU. To reach this, relevant research projects, as well as education of students need to be initiated. The aim of this communication is to present the Erasmus Intensive programme 'Sustainable pig production in the EU'. This 2-wk training period was organised by a consortium of 11 universities. It took place at Agrocampus Ouest and was each year followed by 20+ students, most of them preparing for a Master degree in animal sciences. Pedagogic objectives were to make students sensitive to the scope and difficulties and to provide them knowledge and tools to understand the concept and to evaluate the sustainability of pig farming systems. This multi-disciplinary course consists of lessons given by the lecturers from the different universities and researchers from the French National Institute for Agricultural Research. A problem-based learning approach was taken where students worked in small groups. The task was to evaluate the sustainability of pigs farms which they visited. This encouraged the students to integrate knowledge from lessons, to be actively involved in the course and to promote fruitful exchange of ideas. Student's perceptions of sustainability relating to farming and pig production specifically were broadened as shown by the results of a short survey taken both pre- and post-programme. To conclude, this training period has been shown to be an enriching part of student's curriculum and one small step in the creation of the common European Higher Education Area encouraged by the Bologna declaration.

Modelling the influence of increasing milk yield upon greenhouse gas emission and profitability considering different co-product allocation

Zehetmeier, M. and Hoffmann, H., Technische Universität München, Institute of Agricultural Economics and Farm Management, Alte Akademie 14, 85350 Freising-Weihenstephan, Germany; monika.zehetmeier@tum.de

A continuous increase in milk yields can be observed in many countries in the last decades. Besides economic advantages for farmers this is often considered an important method to decrease greenhouse gas (GHG) emissions per kilogram (kg) of production. It should be considered that milk and beef production systems are often closely interlinked. Surplus calves and culled dairy cows play an important role in the beef production of many countries. The aim of this study is to identify the influence of increasing milk yield of dairy cows on GHG emissions, beef production and economic parameters. Models were developed incorporating several indoor dairy systems differing in milk yield and breed, bull fattening, heifer fattening and beef cow systems. The models stimulate on farm and off farm emission sources of methane, nitrous oxide and carbon dioxide. Different methods for allocating GHG emissions to milk and co-products (surplus calves and meat from culled cows) were used referring to the standard for life cycle inventory analysis - ISO 14041. The study demonstrates that generally profit per cow increases and the GHG efficiency of production improves with increasing milk yield. However when GHG emissions (kg CO_2eq) of the dairy system are allocated to milk and co-products according to their economic value the emission relative to farm profit displays only initial improvement in response to increased milk yield. In another system scenario the fattening of surplus calves and beef breeding are added to the dairy system. This model enables comparison of various levels of milk production corrected to maintain a constant level of meat output. This system displays increasing GHG emissions with increasing milk yields. This investigation demonstrates that the environmental (GHG) and economic evaluation of increasing milk yields in dairy farming is different depending upon the handling and value of co-products.

Mixed crop-livestock farming: an economical and environmental-friendly way to intensify production?

Ryschawy, J., Choisis, J.P., Choisis, N. and Gibon, A., INRA, UMR 1201 Dynafor, BP 52627, 31326 Castanet-Tolosan, France; julie.ryschawy@toulouse.inra.fr

Intensification and specialization of agriculture allowed increasing productivity but induced detrimental impacts on the environment and challenges for the economical viability of numerous farms. Association between livestock and crops, which was common in the past, is given consideration worldwide. It is regarded as a possible way for improving nutrient cycling while reducing chemical inputs, increasing sustainability of natural resources management, and also generating economies of scope. Mixed crop-livestock farming systems could therefore be a favourable alternative to specialized livestock systems. In this paper we address their environmental and economic advantages from a comparative assessment of real-life specialized and mixed farms in the region 'Coteaux de Gascogne' in Southwestern France. In this hilly region with frequent summer draughts, agriculture includes cattle and cash crop production. 47% of the farms have a mixed crop-livestock production, the remaining been specialized either in crop (14%) or cattle production (39%). We made an exhaustive survey of the farms working land in a small area of about 4,000 ha to assess local farming systems and farmers' practices and strategies as regards farm and land management. We analysed farm diversity in reference to technical, economic and environmental criteria. Our results pointed out an intensification gradient from low to high input farming systems amongst local farms. Beyond some general trends (e.g. dairy production more intensive than beef production), a wide range of management practices and intensification degrees was observed amongst farming systems of a similar orientation. Mixed crop-livestock farms appeared very heterogeneous as regards the use of external inputs. We compared intensification degrees in local livestock and mixed crop-livestock farms and discussed their respective advantages in reference to the feed cost per cow, the income per work unit and the local challenges attached to land-use change.

First insights into the relationship between production intensity and sustainability: application to the whole gradient of beef and dairy production systems in France

Teillard D'evry, F.[1], Doyen, L.[2], Jiguet, F.[2] and Tichit, M.[1], [1]INRA, SAD Science for Action and Development, AgroParistech, Paris, France, [2]CNRS, MNHN, rue Cuvier, Paris, France; teillard@agroparistech.fr

One of the key components of livestock farming system sustainability is their ability to ensure efficient production while preserving biodiversity. The pattern of biodiversity change with production intensity remains largely unknown. Most studies linking agricultural production with biodiversity focused on yield as a surrogate for production intensity. Yet, livestock production relies on complex technical choices which cannot be reduced to a simplistic opposition between management techniques generating more or less yield. This paper reports the results of a study aimed at (i) developing production intensity indicators based on input/output ratios (ii) modelling their relationship with biodiversity. Two France-scaled databases were used in order to account for the whole intensity gradient of livestock farming systems. With the Rural Development Watch we built geo-referenced indicators of production intensity. With the French Breeding Bird Survey database we constituted a study set of 25 species of farmland birds. Indicators derived from both databases were combined to model the response of birds to production intensity. The indicator of production intensity articulated input levels (feedstuffs, agrochemicals, energy) with outputs (income). It had contrasting effects on birds among regions and livestock farming systems. In contrast with theoretical bird responses already proposed by different authors, non-linear responses and threshold functions were expected i.e. birds' populations benefited from optimal intensity levels or threshold intensities induced severe detrimental or beneficial effects. On the basis of these relationships, we discussed the strengths and weaknesses of the production intensity indicator in relation with its use for the environmental certification of livestock farming systems.

Evaluation of beef production systems in Greece

Vessalas, S., Tsiplakou, E., Tsiboukas, K. and Zervas, G., Agricultural University of Athens, Nutritional Physiology and Feeding, Iera odos 75, Gr 11855, Greece; eltsiplakou@aua.gr

The objective of work was to describe and evaluate the beef meat production enterprises of Greece, their efficiency and their profitability. The self sufficiency of Greece in beef meat is only 30%. This meat is produced partly by fattening calves of dairy and suckler cows breeds and partly by imported live young beef stock which is kept up to 600-700 kg body weight and then slaughtered. Twenty one representative fattening calves enterprises were selected at random all over Greece, which agreed to fill in a questionnaire by giving information on breeds, type and size of farms, feeding and management of the farm, costs of feed, labour etc., revenues, marketing problems etc. The analysis of these data shows that: a. the practiced beef fattening, farming system is intensive where the animals are kept indoors and fed mainly with purchased concentrates and farm grown forages (alfalfa hay and corn silage). Intake from pasture is limited and concerns only the suckler cows calves from calving to September. The sector is heavily dependent on subsidies, since the Greek beef meat is sold in more or less similar price with the imported one. The more efficient and profitable enterprises are those which can manage to retail their production.

TM-QTL and MyoMAX® effects in Texel x Welsh Mountain lambs

Masri, A.Y.[1], Macfarlane, J.M.[1], Lambe, N.R.[1], Haresign, W.[2], Brotherstone, S.[3] and Bunger, L.[1], [1]SAC, Kings Buildings, EH93JG, United Kingdom, [2]IBERS, Aberystwyth, SY233AL, United Kingdom, [3]IEB, Kings Buildings, EH93JG, United Kingdom; amer.masri@sac.ac.uk

TM-QTL and MyoMAX® (MM) are two muscle-enhancing polymorphisms on Chr 18 and Chr 2, respectively, in Texel sheep. Earlier studies on crossbred lambs out of Mule ewes showed that heterozygous TM-QTL lambs had increased loin muscle depth and weight and heterozygous MM lambs had greater carcass muscling and were less fat. Since Welsh Mountain (WM) ewes make a significant contribution to UK finished lamb supplies, this study evaluated direct TM-QTL and MM effects in Texel x WM lambs. Lambs (n=177), sired by 4 Texel sires (all MM carriers, 2 TM-QTL carriers) and out of WM ewes, were ultrasound scanned at ~23w and slaughtered at ~24w. Carcasses were graded for conformation and fat class then video image analysis (VIA) scanned to predict weight of carcass primal (P) and trimmed primal cuts (TP), loin muscle area (MLL-A), depth (MLL-D) and width (MLL-W), and muscle volume and muscularity of the hind leg and loin regions. Of the 177 lambs, 132 could be classified for TM-QTL genotype [109 non-carriers (+/+), 23 heterozygous (TM/+)] and 174 for MM genotype [15 non-carriers (+/+), 129 heterozygous (MM/+), 30 homozygous (MM/MM)]. Carcass weight did not differ significantly between genotypes. TM-QTL significantly increased ultrasound muscle depth (5.7%), P-leg weight (2.2%), MLL-A (4.1%) and MLL-W (2.3%). Ultrasound fat depth and carcass fat class were significantly lower for MM/MM (25% and 20%, respectively) and MM/+ (11.6% and 8%, respectively) than +/+. MM significantly increased several VIA-predicted lean traits (P-Saleable Meat Yield, P-leg, TP-leg, TP-chump, MLL-D, MLL-A, hind leg muscle volume and hind leg and loin muscularity), mainly due to the superiority of MM/MM compared to +/+. These results agree with other reports that TM-QTL increases muscling in the loin only, whereas MM reduces fatness and increases overall carcass muscling.

Identification and evaluation of β-defensin polymorphisms in Valle del Belice dairy sheep

Monteleone, G.[1], Calascibetta, D.[1,2], Scaturro, M.[2], Galluzzo, P.[1], Palmeri, M.[2] and Portolano, B.[1,2], [1]Università di Palermo, Dip. S.En.Fi.Mi.Zo.-Sez. Produzioni Animali, Viale delle Scienze, 90128 Palermo, Italy, [2]Consorzio Regionale di Ricerca Bioevoluzione Sicilia, Via P.L. da Palestrina, 2, 92018 S. Margherita di Belìce (AG), Italy; giusi.monteleone@senfimizo.unipa.it

Defensins are a class of small peptides belonging to the antimicrobial peptides family, involved in the innate immunity mechanisms and acting directly against bacteria, viruses, and fungi. Due to their role in the immune response, β-defensin genes have been characterized in several domestic species like cattle, pig, and goat. The aim of this study was to identify, validate, and analyze SNPs on exon 1 and 2 of β-Defensin 1 (SBD1) and β-Defensin 2 (SBD2) genes in Valle del Belice dairy sheep. The study was conducted on 400 randomly selected animals belonging to 4 flocks. A total of 7 SNPs were identified, 2 of which in SBD1, i.e. a transition A>G and a transition T>C at position 1747 and 1757, respectively; whereas 5 SNPs were identified in SBD2: a transition C>T at position 89 and 4 transitions G>A at positions 1659, 1667, 1750, and 1761. SNP at position 1659 determines an amino acid change Arg^{42}>Lys^{42}, whereas the transition at position 1667 determines the switch Gly^{45}>Arg^{45}. Prediction of the functional effects of SNPs, obtained using Panther and SIFT programs, showed that amino acid substitutions do not affect SBD2 protein function. For their location in the 3'-UTR, SBD1 1747-1757 and SBD2 1750-1761 could have an effect on post-transcriptional events. However, our results, obtained by using RNA structure program, demonstrated that only SBD2 shows changes in mRNA secondary structures. Association between SNPs was estimated and haplotypes identified using Phase software. A total of 15 haplotypes were detected, 4 of which with a frequency higher than 5%. These preliminary results suggest that identified SNPs could play a role in the modulation of immune response. Further studies will be necessary to verify if these SNPs compromise the function of the protein.

Association between polymorphism g.276T>G in the FTO gene and fatness: related traits in Krškopolje pig and hybrid 12

Flisar, T., Žemva, M., Kunej, T., Malovrh, Š., Dovč, P. and Kovač, M., Biotehnical Faculty, University of Ljubljana, Department of Animal Science, Groblje 3, Rodica, 1230 Domžale, Slovenia; milena.kovac@bf.uni-lj.si

The aim of this study was to analyze the effect of polymorphism g.276T>G in FTO gene on fatness traits and fatty acid composition in Slovenian local breed Krškopolje pig (KP) and hybrid 12 (H12), which is a cross between Slovenian Landrace line 11 and Slovenian Large White. Experiment involved 23 animals of KP and 24 animals of H12. Pigs were reared under same condition at the same location. The tissue samples were genotyped with PCR-RFLP method. As phenotypic fatness traits, backfat thickness, intramuscular fat (IMF) and fat thickness measured on slaughter line were used. The fatty acid composition was determined for IMF in M. longissimus dorsi and backfat. The statistical model for fatty acid composition included fixed effects of gene FTO, breed/hybrid and interaction between gene FTO and breed/hybrid. In the model for fatness traits, sex as fixed effect and carcass weight as covariate were also included. The most frequent genotype was GG (KP: 43.48%, H12: 58.33%). The frequency of TT differed: 8.33% of animals had genotype TT in H12 and 30.43% in KP. Ratio between alleles G:T was 3:1 in H12 and 1.3:1 in KP. Significant effect of FTO gene on fat thickness was found. In KP, animals with GG were the fattest, in H12 the fattest were GT and the leanest animals with TT. Strong association between gene FTO and saturated, n-3/n-6 polyunsaturated fatty acids and aterogenic index in IMF was confirmed. The significant effect was also on mono- and polyunsaturated fatty acid. According to healthiness, the favorable genotype is GT in KP, while GT showed the prevalence to the lowest percentage of n-3 and n-6 fatty acid in H12. The effect of FTO gene was confirmed on the majority of fatness-related traits, involved in this study. According to different influence of genotypes in KP and H12, g.276T>G could be used as marker, linked to locus with effect on fatness and related traits.

Session 29 — Theatre 4

Gene polymorphism in relation to transcription level of selected candidate genes for reproduction traits in sows

Goluch, D., Korwin-Kossakowska, A., Pierzchała, M. and Urbański, P., Institute of Genetics and Animal Breeding Polish Academy of Sciences, Animal Immunogenetics, Jastrzębiec ul. Postępu 1, 05-552 Wólka Kosowska, Poland; d.goluch@ighz.pl

The physical role of proteins and biological significance in reproduction was the reason why the three genes coding: the osteopontin (OPN), amphiregulin (AREG) and epidermal growth factor (EGF) were taken under consideration. The aim of the study was to find a possible polymorphisms in the promoter and exon regions as potential mutations affecting the expression level of genes in ovaries, uterus and oviduct of sows. Sixty sows of F1 generation of Polish Large White and Polish Landrace crossbred were slaughtered in early luteal phase. Samples of ovary, oviduct, uterus body and horn were taken and frozen immediately in liquid nitrogen. Total RNA for expression analysis was extracted from the frozen tissues (ovarian, uterus and oviduct tissues) of all sows. The complementary DNA (cDNA) was synthesized. Real-time PCR analysis to determine the expression dynamics of the OPN, AREG and EGF genes in examined tissues was performed using Sybr Green I format in relation to "housekeeping" genes (b-actin, GAPDH). High Resolution Melting and SSCP method was used to find new polymorphisms within selected genes and for genotyping. Until now a few polymorphisms in promoter region, exons of the selected gene were found. Some of them are SNPs, one is a microsatellite sequence. A relationship between the OPN, AREG and EGF genotype (based on the novel polymorphism) and the mRNA level was estimated. Collected data were also set against reproduction traits revealing different expression patterns among the specimens. Among the others, polymorphism in exon 6 of OPN (OPN/C499A) seems to be the most interesting, because genotype AA may correlate with elevated level of expression in ovary and uterus.

Session 29 — Theatre 5

Detection of QTLs for growth in turbot

Cerna, A.[1,2], Fernandez, J.[3], Toro, M.[2], Bouza, C.[1], Hermida, M.[1], Vega, M.[1] and Martinez, P.[1], [1]Universidad de Santiago de Compostela, Genetica, Facultad de Veterinaria, 27002 Lugo, Spain, [2]Universidad Politecnica de Madrid, Produccion Animal, Ciudad Universitaria s/n, 28040, Spain, [3]INIA, Mejora Genetica Animal, Carretera La Coruna km 7, 28040 Madrid, Spain; miguel.toro@upm.es

The turbot is a highly appreciated European aquaculture species, whose production is expected to increase from the currently 9,000 tons up to more than 15,000 tons in 2012. Growth is an economically important trait in this species, as for most domesticated fish species. The search for QTLs is, therefore, of obvious interest. In the present study, five full-sib families obtained from the breeding programs of Stolt Sea Farm SA (SSF) and one from Pescanova SA in Northwestern Spain were used to search for growth-associated QTLs. Overall about 450 individuals were genotyped for nearly one hundred of homogeneously distributed microsatellites. Length and weight of fishes were measured when they were close to reach 100 g. Using the Gridqtl software we have detected seven QTLs for body weight and nine QTls for body length. Four and two of them, respectively, showed a chromosome-wide significance above 1%.

Association of DGAT1 gene with breeding value of Iranian Holstein bulls

Hosseinpour Mashhadi, M.[1], Nassiry, M.R.[2], Emam Jome Kashan, N.[3], Vaez Torshizi, R.[4], Aminafshar, M.[3] and Tabasi, N.[5], [1]Islamic Azad University Mashhad Branch, Animal Sci Dept, Emamie 59., Mashhad, Iran, [2]Ferdowsi University of Mashhad, Animal Sci Dept, Park Square, Mashhad, Iran, [3]Islamic Azad University Science and Research Branch, Animal Sci Dept, Ponak Square, Tehran, Iran, [4]Tarbiat Modares University, Animal Sci Dept, Peykan Shahre, Tehran, Iran, [5]BuAli - Immunology Research Center, Bu Ali Square, Mashhad, Iran; mojtaba_h_m@yahoo.com

The DGAT1 catalyzes the final step of triglyceride synthesis and is located on centromic end of bovine chromosome 14. The substitution AA to GC in exon VIII changes Lysine to Alanine. The allele K affects fat yield, percent of fat and protein, allele A affects the yields of milk and protein. Samples of semen from 103 bulls were genotyped. A 411bp fragment from exon VIII was amplified and digested by AcoI with PCR-RFLP method. The frequencies of KK, KA and AA were 0.592, 0.408 and zero respectively. The frequency of alleles K and A were 0.7961 and 0.2039. The PCR product of KK genotype was sequenced. The sequence of allele K of DGAT1 gene was registered in NCBI gene bank with accession number EU077528. The first lactation records of 43303 dairy cattle from 1999 to 2006 were used for prediction BV under Animal Model with DFREML software. The average BV for yield of milk, fat and protein, percent of fat and protein were 180.2 (±28.8) kg, 3.7 (±1.26) kg, 2.3 (±1.06) kg, -0.036% (±0.014) and -0.03% (±0.01), respectively. The association between BV of the bulls and the genotypes for DGAT1 gene were studied with GLM procedure of SAS package. The average BV of milk yield for KA and KK genotype were 288.8 and 109.6 kg ($P<0.01$) and for fat yield were 5.6 kg (KK) and 0.91 kg (KA) ($P>0.05$). The difference between BV of protein yield for KK (5.5 kg) and KA (0.025 kg) was significant ($P<0.05$). The average BV of fat percent were -0.009% (for KK genotype) and -0.067% (for KA genotype) and the difference of means was significant ($P<0.05$). The differences of these values for percent of protein for KK (-0.0158%) and KA (-0.0597%) were also significant ($P<0.05$).

Genes differently expressed in adipose tissue of dairy cows carrying 'fertil+' or 'fertil-' haplotype for one QTL of female fertility located on the chromosome 3

Coyral-Castel, S.[1], Ramé, C.[1], Elis, S.[1], Cognié, J.[1], Fritz, S.[2], Lecardonnel, J.[3], Hennequet-Antier, C.[4], Marthey, S.[3], Esquerré, D.[3] and Dupont, J.[1], [1]INRA PRC, centre de Tours, 37380 Nouzilly, France, [2]UNCEIA, 149 rue de Bercy, 75595 Paris, France, [3]INRA CRB GADIE, centre de Jouy, 78350 Jouy-en-Josas, France, [4]INRA URA, centre de Tours, 37380 Nouzilly, France; stephanie.coyral@tours.inra.fr

The fertility of high producing dairy cows has continuously decreased since several years, especially in Holstein breed. In order to better understand the genetic determinism, we studied Holstein cows carrying the favorable haplotype 'fertil+' or the unfavorable haplotype 'fertil-' for one QTL of female fertility located on the chromosome 3 (QTL-F-Fert-BTA3). After the first calving, 'fertil-' cows lost more weight than 'fertil+' cows during the first eight weeks. This result suggests a higher mobilization of body fat reserve in 'fertil-' cows than in 'fertil+' cows and a potential difference of gene expression in adipose tissue. To check this latter hypothesis, we made biopsies of adipose tissue in 9 'fertil+' and 9 'fertil-' cows, one or two weeks after calving, when non-esterified fatty acids levels were the highest. Total RNAs were extracted and after cDNA synthesis, samples were labeled with Cy3 and hybridized on a 385K array containing oligonucleotides of genes and non-coding sequence of the QTL-F-Fert-BTA3 (designed by Roche Nimblegen). Results were analysed with two different approaches to detect differentially expressed genes. First we performed a structural mixed model for variances on each probe. With this first method, 2 known genes are differently expressed between the two haplotypes. Then we performed a hierarchical model taking into account that many probes are available per gene, on two datasets containing only known genes. In this case, 40 genes (23 forward and 17 reverse) are differently expressed between 'fertil+' and 'fertil-' cows. These results will be confirmed by real-time RT-PCR and western blot for protein expression.

Bovine lactoferrin gene polymorphism and its association with prevalence of sub-clinical mastitis caused by *Staphylococcus* species

Sender, G.[1], Pawlik, A.[1], Galal Abdel Hameed, K.[2], Korwin-Kossakowska, A.[1] and Oprzadek, J.[1], [1]Institute of Genetics and Animal Breeding, Polish Academy of Sciences, Animal Sciences, Jastrzebiec, 05-552 Wólka Kosowska, Poland, [2]South Valley University, Faculty of Veterinary Medicine, Department of Food Hygiene and Control, Qena, 00-000, Egypt; g.sender@ighz.pl

Several studies have attempted to investigate associations between lactoferrin genotypes and somatic cell count. However, there are no studies examining sub-clinical mastitis caused by Staphylococcus species. The present study was designed in order to develop a better understanding of associations between lactoferrin polymorphism and sub-clinical mastitis caused by Staphylococcus aureus and coagulase-negative staphylococcus species (CNS). The objective this study was to verify hypothesis that polymorphism, occurring in intron 6 of bovine lactoferrin gene is associated with prevalence of sub-clinical mastitis caused by Staphylococcus species (Staphylococcus aureus and CNS) in Polish dairy cattle. Data on 680 composite milk samples were collected from 216 lactating Polish Holstein cows. The bacteriological status of the mammary gland was determined by collecting and culturing duplicate composite milk samples. A third sample was cultured in case the results from the first two samples differed. The polymorphism of lactoferrin gene has been analyzed using PCR–RFLP method. Analysis of variance was used to evaluate the associations between lactoferrin genotype and sub-clinical mastitis caused by Staphylococcus aureus and CNS. There was a decrease in the prevalence of sub-clinical mastitis caused by Staphylococcus species in animals having BB lactoferrin genotype but differences were not significant probably due to very low number of animals having BB genotype.

Session 29 Theatre 9

Leptin gene polymorphisms in wild and captive American mink

Arju, I. and Farid, A., Nova Scotia Agricultural College, Plant and Animal Sciences, 58 River Road, Truro, Nova Scotia, B2N 5E3, Canada; hfariod@nsac.ca

Leptin hormone plays important roles in the regulation of appetite and energy balance and signals nutritional status to the central reproductive axis. The roles that this hormone plays are particularly important in the mink where hormonal control of reproduction is tightly linked with body fat reserves and availability of food. We bi-directionally sequenced a 4738 bp segment of the leptin gene in two mink from each of four captive populations (black, brown, pastel and sapphire) and mink that were trapped in the wild. This segment included exon 2 (144 bp), intron 2 (1867 bp) and exon 3 (2727 bp). Multiple alignments of the sequences showed no polymorphism in the coding region of the gene but 17 single nucleotide polymorphisms (SNP) and an 11 bp deletion/insertion were found in intron 2 and the 3'-UTR of the gene. SNPs in the 3'-UTR may influence mRNA stability and the rate of translation of the gene by creating or destroying microRNA target sites. Genotypes of 20 mink from each of the five populations were determined for 10 of SNPs by High Resolution Melt using real-time PCR or RFLP-PCR. Captive mink were from different ranches, and wild mink were trapped in north-eastern Canada where there has been no history of mink ranching. Pairwise comparison of the mink color types for allele frequency distributions at every polymorphic site showed that 32 of the 101 comparisons were significantly different, and the greatest differences were between the wild and captive mink as 24 of the 32 significant differences involved the wild mink. In addition to the 11 bp deletion, which was detected only in the wild mink, the allele frequencies at seven of the SNPs were significantly different between wild and the combined data of the four captive mink populations. The results suggested that the leptin locus has possibly been under selection in captive mink populations.

Session 29 Poster 10

Genetic diversity in native Greek horses

Cothran, E.G.[1], Kostaras, N.[2], Juras, R.[1] and Conant, E.[1], [1]Texas A&M University, VIBS, TAMU4458, College Station, TX 77843, USA, [2]Amaltheia (Greek Society for the Protection of Indigenous Breeds of Domestic Animals), Argyrokastrou 51, 15669 Papagos, Athens, Greece; gcothran@cvm.tamu.edu

Genetic diversity was examined in eight native Greek horse breeds using 15 microsatellite loci. The breeds examined were the Andravidas, Crete Native Horse, Pindos, Pinias, Rhodes, Skyros Horse, Thessalias and Zakynthos. For the Rhodes horse, which is nearly extinct, only six animals were typed. Variability levels for the Crete and Pinias were near the average for domestic horse breeds. Values for the Andravidas, Pindus, Thessalias and Zakynthos were somewhat higher than the domestic mean. The Skyros had a relatively low level of variability (Ho of 0.602 compared to a domestic horse mean of 0.713) while the Rhodes horse had one of the lowest levels of heterozygosity seen for any horse population (0.264). Phylogenetic analysis placed the Crete, Pindos and Pinias breeds within the cluster of horses with an oriental origin. The Skyros showed no close relationship to any group of horses, possibly due to the low variation and its isolation on an island. The Andravidas, Thessalias and Zakynthos breeds fell into the cluster that contained the Thoroughbred and its allies. There is known crossing of these three breeds to Thoroughbreds. The horses of Greece illustrate a common problem for conservation of rare breeds. The Greek horses are named for the regions where they are found and for these breeds there are no organized breed associations and the preservation of the horses is left strictly in the hands of those who use them. In some cases, the circumstances are such that the breed characters are maintained but in others, the economics of horse breeding leads to crossing with breeds that will improve the price of the animals. It is of course necessary for the animal breeders to make money to continue to raise the animals; however, homogenization has become a major threat to livestock diversity. It is important that outcrossing be recognized to develop plans for the preservation of native breeds.

Session 29 Poster 11

The GT gene polymorphism as a potential marker for fatty acid profile changes in beef cattle

Urtnowski, P.S., Oprzadek, J.M., Pawlik, A. and Sender, G., Institute of Genetics and Animal Breeding, Jastrzebiec, 05-552 Wolka Kosowska, Poland; p.urtnowski@ighz.pl

Consumer demands are focused on GMO-free, healthy products, obtained with high animal welfare. Moreover, especially young people, are interested in healthy life style, including functional food. Red meat is a good source of high value protein, rich in available iron and vitamin (B12). Furthermore, grazed animals are characterized by favorable n-6/n-3 fatty acid ratio and CLA in muscles. The high-throughput DNA techniques such as PCR-RFLP, sequencing, SNP panels are good methods for better understanding of the molecular basis of metabolic reactions that are involved in meat production. The thyroglobulin gene (TG) was mapped in centromeric region of chromosome BTA14. TG encodes a glycoprotein that is thyroid hormone's precursor and the SNP as a potential predictor of fat content variability was reported. The PCR-RFPL method, modified by using HotStarTaq polymerase and optimized reaction conditions, according to Barendse (2001) was used. The aim of our study was to assess TG polymorphism (C/T) and its association with beef quality, as well as fat content in beef. PCR product was digested with BstYI endonuclease and restriction patterns were separated in 3.5% agarose gel. 180 animals of Polish Friesian young bulls were genotyped. Animals were kept in experimental barn of Institute of Genetics and Animal Breeding (IGAB), Jastrzebiec, Poland. Yearling bulls were killed in experimental slaughter house in IGAB and dressing percentages were recorded. Blood samples were taken from jugular vein to S-Monovette tubes with 1.6 mg K-EDTA/ml of blood. Alleles frequency were C – 0.919 and T – 0.081. The CC animals made up 83.8% and heterozygous ones CT- 16.2% of the herd. Among investigated animals we didn't find homozygous TT.

Expression pattern in the endometrium of genes involved in calcium signaling pathway during bovine estrous cycle

Heidt, H., Salilew-Wondim, D., Gad, A., Grosse-Brinkhaus, C., Tesfaye, D., Hoelker, M., Looft, C., Tholen, E. and Schellander, K., Institute of Animal Science, University of Bonn, Department of Animal Genetics, Endenicher Allee 15, 53115 Bonn, Germany; hhei@itw.uni-bonn.de

Calcium signaling is a fundamental intracellular pathway that mediates a variety of physiological functions including the resumption of meiotic maturation in oocytes. However, the role and expression level of key regulatory genes regarding the calcium signaling pathway during bovine estrous cycle is poorly understood. Therefore, the aim of the study was to investigate the expression profiling of genes involved in calcium signaling pathway in the endometrium during the bovine estrous cycle. For this, 12 Simmental heifers were estrus synchronized and slaughtered at day 0, 3, 7 and 14 of the estrous cycle. Endometrium samples were taken at the ipsilateral and contralateral regions of the uterus and used for mRNA expression profiling. The mRNA level of 11 genes (ADORA2B, ATP2B1, CALM3, HTR4, ITPR1, ITPR2, PLCD1, PLCE1, PLN, PRKCA and SLC8A1) were quantified at day 0, 3, 7 and 14 of the estrous cycle using quantitative Real-Time PCR. The statistical data analysis was done using SAS (version 9.2). Furthermore, the protein localization for ITPR1 was performed with immunohistochemistry. The results showed that the mRNA expression of all genes were lowest at day 0 of the estrous cycle. However, at day 7 the highest gene expression was observed for the genes ADORA2B, ATP2B1, CALM3, ITPR1, ITPR2, PLCD1, PLCE1, PLN, PRKCA and SLC8A1, followed by day 14 of the estrous cycle. Moreover the expression patterns of those genes were not significant different between the ipsilateral and contralateral side of the uterus. These observations were confirmed by the protein localization of ITPR1. The protein level of this gene showed the same trend as the mRNA profiles. Based on these results, the dioestrus was identified as the stage of the oestrus cycle with the highest expression level of genes involved in the calcium signaling pathway.

Interaction of LEP and SCD genes influencing basic carcass traits in crossbred bulls

Riha, J.[1], Kaplanova, K.[2] and Vrtkova, I.[2], [1]Research Institute for Cattle Breeding, Ltd., Vyzkumniku 267, Vikyrovice, 78812, Czech Republic, [2]Mendel University in Brno, UMFGZ, Zemedelska 1, Brno, 61300, Czech Republic; irenav@mendelu.cz

We analyzed 143 crossbred (Czech Spotted Cattle, Holstein, Ayshire, Blonde d´Aquitaine, Galloway, Charolais) bulls for 2 genes (LEP - leptin and SCD - stearoyl-coA desaturase) polymorphisms which are reported as candidate genes influencing fat deposition in cattle. Net daily gain, portion of beef fat in half of carcass and portions of first and second class meat were recorded for each animal. Animals were slaughtered in average age 560 days according to ČSN 57 6510 norm. PCR-RFLP method was performed to detect the genotypes in candidate genes. PCR primers were assumed from previously reported studies. GLM with two fixed effects of genotypes, effect of interaction, and two continous effects of slaughter age and weight was used for statistical analysis. After model fit, we used contrast analysis for a priori hypothesis testing. In netto daily gain (r^2=0.701, P≤0.001), there were observed no significant differencies in the interaction. In portion of separable beef fat (r^2=0.234, P≤0.001), effect of genotypes had a significant influence (P=0.0142). We observed significant differencies (P≤0.001) within LEP TT group: SCD VV(4.068%)>SCD AV(2.052%). Within LEP CT group, there was revealed significant difference between SCD AV (2.661%)>SCD AA(1.788%) genotypes. Also relationships, LEP TT, SCD VV>LEP CT, SCD AV (P≤0.05); LEP TT, SCD VV>LEP CT, SCD AA (P<0.001) and LEP TT, SCD VV> LEP CC, SCD (VV, AA, AV) are valid and significant (P≤0.001). For portion of first class meat (r^2=0.306), we observed only one significant difference (P≤0.05) within LEP CC group: SCD AA(31.787%)>SCD VV(30.741%). Antagonistic relationship were observed in portion of second class meat (r^2=0.384, P≤0.001) within LEP CC group (P≤0.05): SCD VV(34.458%)>SCD AV(33.353%), AA(33.199%). Especially, different expression in interaction of analysed candidate genes in separable beef fat seems to be useful for beef quality breeding.

Denaturing gradient gel electrophoresis as a prion protein gene screening method in goats for mutation detection at codons 142, 146, 151, 154, 211, 222 and 240

Fragkiadaki, E.[1,2], Ekateriniadou, L.[2], Kominakis, A.[1] and Rogdakis, E.[1], [1]Agricultural University of Athens, Faculty of Animal Science and Aquaculture, IERA ODOS 75, 11855, Athens, Greece, [2]National Agricultural Research Foundation, Veterinary Research Institute, Campus Of Nagref, Thermi, Greece; irfrag@yahoo.com

Denaturing Gradient Gel Electrophoresis (DGGE) has been applied for prion protein (PrP)genotyping in human, sheep and goats. The present DGGE protocol aimed to expand the mutation detection in more PrP gene codons and particularly at 142, 211 and 222 that are possibly associated with scrapie in goats. Genomic DNA extraction was performed from goat EDTA-treated blood and brain tissue. Melting domains of two overlapping PrP gene fragments, referring to the whole ORF were determined by MELT94 software and primer design was optimised. DGGE analyzed sequence referred to codons 106-154 (462bp fragment) and 178-256 (444bp fragment). DGGE analysis was performed in 6,5% polyacrylamide (37.5:1 acrylamide:bisacrilamide) gels with a linear gradient concentration of 20% to 80% denaturant in 0,5X TAE. RFLP-PCR with BspHI enzyme was used to distinguish polymorphisms at codons 151 and 154. DGGE analysis of the 462bp fragment revealed five distinct band patterns referring to polymorphisms in codons 138, 142, 146, 151 and 154. DGGE analysis of the 444bp fragment gave also five distinct band patterns based on the mutation presence at codons 211, 222 and 240. By combining DGGE results of these two overlapping PrP gene fragments, goat's genotype was determined based on codons 138, 142, 146, 151, 154, 211, 222 and 240. DGGE allows samples grouping based on their pattern and subsequently based on their allele-specific polymorphisms, by using a relative inexpensive and less sophisticated sequencing method. Representative samples from each group could be further direct-sequenced by a reference sequencing method extrapolating data to all group's samples.

Authentication of swine derived products by means of SNP and STR markers

Matassino, D., Blasi, M., Bongioni, G., Incoronato, C., Occidente, M., Pane, F., Pasquariello, R. and Negrini, R., consdabi@consdabi.org

In the last decade market awareness about 'food safety' issue increased significantly, fostered also by critical sanitary episodes. In this context, the major efforts are addressed to develop methods for product authentication and to guarantee consumers from frauds. The development of new molecular assay makes the DNA-based methods of large application. Currently, the most used molecular markers are SNPs and STRs. Within the frame of the national SELMOL project, the aim of this contribute is to evaluate the efficiency of a SNP panel and of a set of microsatellites to authenticate pig mono-breed products by means of 'population assignment test'. Two reference dataset were produced, genotyping 177 subjects belonging to 4 cosmopolite (Duroc, Landrace, Large White, Pietrain) and two Italian autochthonous breeds (Apulo-Calabrese and Casertana) with a panel of 47 SNPs selected in silico (www.ncbi.nlm.nih.gov.), and with a set of 20 microsatellites. Field experiments were carried out genotyping, with the same sets of markers, DNA extracted from 25 pig mono-individual (fresh and/or seasoned) products from Casertana ancient autochthonous genetic type (AAGT). The assignment of Casertana products to the reference set of populations, was performed using STRUCTURE 2.2 (pritch.bsd.uchicago.edu/software) and GeneClass ver. 2.0 (http://www.ensam.inra.fr/URLB/geneclass/geneclass.html). Within the limits of the observation field, the results obtained confirmed the suitability of both SNPs and STRs markers as a tool for implementing a DNA based traceability (authentication) of mono-breed and mono-individual products.

Comparison of Vascular Endothelial Growth Factor C (VEGF-C) expression in canine breast cancer by using quantitative real-time PCR technique

Doosti, M.[1], Nassiri, M.R.[1,2], Movasaghi, A.R.[1,2,3], Tahmoorespur, M.[1,2], Ghovvati, S.[1], Kazemi, H.[1,3] and Solatani, M.[1], [1]Ferdowsi University of Mashhad, Animal Science Department, Azadi Sq., 91775-1163, Iran, [2]Ferdowsi University of Mashhad, Institute of Biotechnology, Azadi Sq., 9177948974, Iran, [3]Ferdowsi University of Mashhad, Department of Veterinary Pathology, Azadi Sq., 9177948975, Iran; Ghovvati@stu-mail.um.ir

Recently, many studies have shown that vascular endothelial growth factor (VEGF) has a key role in tumor angiogenesis. The expression of VEGF-C is widely related to development of lymphatic vessels garnishment in cell. The aim of this study was to compare VEGF-C expression in canine mammary gland adenocarcinomas and its normal tissue. For this purpose, 314 sick dogs were screened for 10 adenocarcinomas samples and 10 normal samples. Total RNA was extracted and VEGF-C and B-Actine genes were reverse-transcribed using cDNA synthesis kit. Concentration and purity of RNA and cDNA were determined using Nano-Drop ND 2000 Spectrophotometer. Quantitative Real-time PCR approach based on SYBR Green was used to measure relative expression levels VEGF-C gene in non-malignant and malignant specimens with triplicates of each cDNA sample. Specificity of amplification products was confirmed by melting curve analysis. ABI 7300 SDS v1.4 software was used for analyzing the obtained CT values. Statistical analysis was performed using ANOVA, ANCOVA and t-test approaches of SAS software (Version 9.1). Results indicated that expression of VEGF-C in breast adenocarcinomas was higher than non-malignant breast tissue samples ($P<0.001$). There was a 34-fold overexpression of VEGF-C gene, in breast malignant tissue relative to non-malignant breast tissue specimens. Results demonstrated that VEGF-C gene overexpression in the mammary gland is a good marker to detect breast carcinoma in suspected dogs and it can be hypothesized that VEGF-C gene overexpression can increase the proliferation rate of tumor cells.

Effect of the Leptin gene polymorphism on the intramuscular fat deposition in Hungarian beef cattle

Anton, I.[1], Kovács, K.[1], Holló, G.[2], Farkas, V.[3] and Zsolnai, A.[1], [1]Research Institute for Animal Breeding and Nutrition, Genetics, Gesztenyés u. 1., 2053 Herceghalom, Hungary, [2]University of Kaposvár, Guba Sándor u. 40., 7400 Kaposvár, Hungary, [3]University of Pannonia, Georgikon Faculty of Agriculture, Deák F. u. 16., 8360 Keszthely, Hungary; istvan.anton@atk.hu

Intramuscular fat content, also known as marbling of meat, represents a valuable beef quality trait. Leptin is the hormone product of the obese gene synthesized and secreted predominantly by white adipocytes. This protein is supposed to be involved in the regulation of body weight by transmission of a lipostatic signal from adipocytes to the leptin receptor in hypothalamus resulting in appetite suppression and increased thermogenesis. The leptin gene has been mapped to bovine chromosome 4. Polymorphisms in the leptin gene have been associated with serum leptin concentration, feed intake, milk yield and body fatness. The objective of this study was to evaluate the effect of the leptin locus on beef quality traits in some cattle breeds in Hungary. 80 blood samples have been collected from different beef cattle breeds and genotypes were determined by PCR-RFLP assay. Animals had been slaughtered, beef quality traits data were determined and statistical analyses have been carried out to find association between genotypes and intramuscular fat deposition. Leptin TT animals showed the highest fat percentage values in the musculus longissimus dorsi and musculus semitendinosus, the difference between CC and TT genotypes was significant($P<0.05$). Concerning the herd genetic structure analyses, differences between the observed and expected genotype frequency values were not significant. This project was supported by the Hungarian Scientific Research Fund (Project 78174).

Ruminal infusions of soya and partially hydrogenated vegetable oils affect ruminal fermentation parameters and plasma fatty acid profile

Vargas-Bello-Pérez, E., Salter, A.M. and Garnsworthy, P.C., The University of Nottingham, Animal Sciences, Sutton Bonington Campus. School of Biosciences, LE12 5RD Loughborough, United Kingdom; Phil.Garnsworthy@nottingham.ac.uk

Dietary oils increase trans fatty acids (FA) production in the rumen, which can interfere with rumen fermentation and milk fat synthesis. The objective of this study was to compare effects of ruminal infusions of soya oil (SO, which increases 18:1 trans-11 by biohydrogenation) and partially-hydrogenated vegetable oil (PHVO, which supplies a range of preformed trans FA) on fermentation parameters and plasma trans fatty acid profile. Three non-lactating Holstein cows (Live weight 773±63 kg), fitted with rumen cannulae, were fed on grass hay (7 kg/d) and concentrate (2 kg/d) and treated in a 3 x 3 Latin square design with 3 daily bolus ruminal infusions of: 1) skim milk (SM; control; 500 ml/d); 2) SO (250 g/d in 500 ml SM); and 3) PHVO (250 g/d in 500 ml SM). Ruminal pH and proportions of individual volatile fatty acids (VFA) were not affected by treatment. Compared with SM, SO and PHVO lowered total VFA concentrations (89.7, 71.9 and 72.2 mM respectively) and NH_3-N (14.8, 8.3 and 12.9 mM respectively). Compared with SM, both SO and PHVO increased plasma trans fatty acids, but SO resulted in higher concentrations of 18:1 trans-10 and trans-11, and lower concentrations of 18:1 trans-4, -5, -9, -12 and cis-9 than PHVO. Plasma HDL and LDL lipoprotein cholesterol fractions were not affected by treatment. It is concluded that rumen microorganisms are equally susceptible to trans FA of rumen and dietary origin because both treatments lowered NH_3-N and total VFA concentrations, and that plasma trans FA concentrations differ according to source of trans FA.

Prediction of pasture dry matter intake for Holstein-Friesian dairy cows grazing ryegrass-based pasture

Baudracco, J.[1], Lopez-Villalobos, N.[1], Holmes, C.W.[1], Horan, B.[2] and Dillon, P.[2], [1]Institute of Veterinary, Animal and Biomedical Sciences, Massey University, Private Bag 11-222, Palmerston North, New Zealand, [2]Teagasc, Dairy Production Research Centre, Moorepark, Fermoy, Cork, Ireland; jbaudracco@yahoo.com

A model developed to predict pasture dry matter intake (DMI) for Holstein-Friesian (HF) cows grazing ryegrass-based pastures was adjusted and validated for grazing conditions in Ireland. The model had originally been developed for conditions in New Zealand, where pasture allowance is measured at ground level, while pasture allowance in Ireland is usually measured at a cutting height above ground level.The model combines theoretical and empirical equations. An upper limit to potential pasture dry matter (DM) intake at grazing is set, which is the lower of three limits set by either physical (rumen fill), metabolic (energy demand) or grazing restrictions. The cow's potential pasture DMI and the pasture allowance are then used to predict pasture DMI of un-supplemented cows using an empirical algorithm. This algorithm was obtained from analysis of data of 13 experiments in which cows grazed ryegrass-based pastures and pasture allowance was measured at a sampling height >3 cm above ground level. If supplements are fed, substitution rate is predicted to calculate actual pasture DMI. An independent dataset from two years of an Irish trial, with individual pasture DMI measurements (n=858) of three strains of lactating HF cows, was used to validate the model. Data within strains were averaged for every month of lactation, allowing 30 data points for validation. The fitness of the model was satisfactory, with a relative prediction error (RPE) of 6.4%, a concordance correlation coefficient (CCC) of 0.87 and a mean bias of +0.28 kg DM. The RPE were 6.1, 5.7 and 6.3%, and the CCC were 0.85, 0.87 and 0.87 for HF strains with 90, 80 and 13% of North American genes, respectively. The model successfully predicted pasture DMI of grazing cows under different combinations of nutritional, physiological and genetic variables.

Extent of biohydrogenation by rumen micro-organisms of n-3 PUFA in two protected fat sources incubated in-vitro

Estuty, N., Chikunya, S. and Scaife, J., Writtle College, Essex, CM1 3RR, United Kingdom; naserestuty@yahoo.com

The potential benefits of increasing the PUFA in ruminant products (meat and milk) are well recognised. Several strategies for protecting ingested PUFA from rumen biohydrogenation have been reported in the literature. However, few studies have investigated the kinetics of PUFA biohydrogenation in protected fat supplements. This study investigated the rate and extent of biohydrogenation in relation to time over a 48-hour period. A 70:30 grass hay:concentrate diet was incubated using the *in vitro* gas production procedure. The basal diet was then supplemented with four PUFA sources: 1) Megalac (MEG, control); 2) Protected fish oil (FO, rich in eicosapentaenoic acid, [EPA, C20:5 n−3]; and docosahexaenoic acid [DHA, C22:6n-3]); 3) Protected linseed (LIN, high in linolenic acid, C18:3n-3); and 4) 50:50 mixture of linseed and fish oil (LINFO). The extent of PUFA biohydrogenation was measured after 6, 12, 24 and 48 hours. Biohydrogenation (%) of C18:3n-3 was extensive on all treatments with means of 53, 35, 47, and 68 after 6 hours ($P=0.03$) for MEG, FO, LIN and LINFO respectively. The biohydrogenation of C18:3n-3 increased to 88, 72, 83, and 93 after 48 hours ($P<0.001$) for MEG, FO, LIN and LINFO respectively. The mean biohydrogenation rates (%) of C18:3n-3 across all treatments at different time intervals were; 50, 59, 77 and 84 after 6, 12, 24 and 48 hours respectively. No significant differences between the two fish oil-containing supplements were observed in relation to biohydrogenation of EPA and DHA. The mean extent of biohydrogenation of EPA in the FO and LINFO diets was 48, 52, 63 and 68%, after 6, 12, 24 and 48 hours respectively, whilst that for DHA was 44, 47, 51 and 58%. The extent of biohydrogenation in this study is comparable to that reported when unprotected linseed and fish oil were used, suggesting that level of protection in the supplements used was marginal. The results also show that *in vitro* biohydrogenation of PUFA is most active in the first 6-12 hours of incubation.

Synchronisation of energy and protein degradation in the rumen to improve nitrogen utilisation by dairy cattle

Van Duinkerken, G.[1], Klop, A.[1], Goselink, R.M.A.[1] and Dijkstra, J.[2], [1]Wageningen UR Livestock Research, Edelhertweg 15, 8219 PH Lelystad, Netherlands, [2]Wageningen University, Animal Nutrition Group, Marijkeweg 40, 6709 PG Wageningen, Netherlands; gert.vanduinkerken@wur.nl

Synchronisation of the availability of energy and nitrogen (N) in the rumen can be achieved either by altering the feeding pattern or frequency, or by altering dietary composition i.e. by synchronising rumen degradation rates of proteins and carbohydrates. The latter track was evaluated by performing an experiment with 32 dairy cows. The 2007 version of the Dutch DVE/OEB system for protein evaluation in ruminants was used to evaluate the diets in this experiment, which had a double Latin-square design with fermentable organic matter in the first 2 h after feed ingestion (FOM2) set at 4 levels. The experiment counted 4 experimental periods and was replicated twice with different ratios of grass silage and corn silage: 3:2 (GC3:2) and 1:4 (GC1:4) on a DM basis, respectively. Each experimental period consisted of 2 weeks for adaptation and 1 week for measurements. Cows were assigned to 1 of 4 blocks of 8 cows. These blocks were formed on the basis of similarity in parity, days in milk and milk performance. The four levels of dietary FOM2 were achieved by feeding two concentrates with contrasting FOM2 concentrations (175 vs. 358 g/kg) which were mixed in different ratios. Average FOM2 intake per treatment group varied between 5121 and 6603 g/d for GC3:2 and 5768 and 7091 g/d for GC1:4. Dry matter intake was significantly ($P<0.01$) affected by dietary FOM2 level, with a reduced intake at the highest FOM2 level. Milk yield was not influenced by treatments. Milk protein percentage increased with increasing FOM2 level, while milk fat percentage was lowest for the highest FOM2 level. Milk urea N was highest for the lowest level of FOM2 and calculated N utilisation was highest for the highest level of FOM2. It was concluded that optimisation of dietary FOM2 level can improve N utilisation by dairy cows.

Effect of dietary inclusion of clinoptilolite on the prevention of rumen acidosis in dairy cows

Karatzia, M.A., Roubies, N., Taitzoglou, I., Panousis, N., Pourliotis, K. and Karatzias, H., Aristotle University of Thessaloniki, School of Veterinary Medicine, 11 St. Voutyra str., 546 27, Thessaloniki, Greece; mkaratz@vet.auth.gr

Sub-acute rumen acidosis is often encountered in dairy farms and is characterized by repeating periods of depressed ruminal pH below 5.5-5.6. Its main cause is diets rich in readily fermentable carbohydrates and poor in cellulose, administered without prior appropriate adjustment of the microbial flora. The inclusion of zeolites in cow rations has positive effects on its prevention by increasing pH values and affecting volatile fatty acids' concentration. 10 clinically healthy Holstein cows 3-6 years old were used in the present study and were administered a diet that causes SARA (wheat pellets and barley, in 20% additional quantity over the standard diet). In the 1st group (n=5) concentrate mixture supplemented with 2,5% of clinoptilolite was administered and the 2nd group served as control. The experiment lasted 90 days and every 3 days (except day 0) ruminal fluid and blood were sampled. PH value, acetic acid and VFA's concentration were evaluated in ruminal fluid, while haematocrit (HCT), haemoglobin (Hb), white blood cell (WBC) lymphocyte and neutrophil granulocyte percentage were evaluated in whole blood. Beta-Hydroxybutyric acid (β-HBA), acetoacetic acid (AcAc) and glucose were evaluated in blood serum. Daily clinical examination was conducted and feed intake was ensured in all animals, simultaneously with possible diarrhoea grading. Ruminal fluid pH was 4.96-5.52 in control group, significantly lower than 6.76-6.94 in 1st group. VFA's concentration was significantly higher in control group, whereas in 1st group it remained stable. Percentage ratio of acetic acid (mol%) was significantly lower in 2nd group, while % ratio of butyric and propionic acid were increased. HTC, Hb, WBC and neutrophil granulocyte percentage significantly increased in control animals, while lymphocyte percentage decreased. Glucose, β-HBA and AcAc concentrations were significantly higher in 2nd group.

The effect of forage source and supplementary rumen protected methionine on the efficiency of nitrogen use in autumn calving dairy cows

Whelan, S.J., Mulligan, F.J., Flynn, B. and Pierce, K.M., University College Dublin, Belfield, Dublin 4, Ireland; stephen.c.whelan@ucd.ie

In dairy cow production systems the efficiency at which ingested feed nitrogen (N) is converted to utilisable milk-N is poor. This results in large quantities of ingested N being excreted and potentially lost to the environment as ammonia, nitrous oxide and nitrate leachate. This trial evaluates the effect of (1) forage source (grass silage vs. maize silage) and (2) rumen protected methionine supplementation on the efficiency of nitrogen use (ENU) in dairy cows offered low crude protein (CP) diets (133±1 g CP). Four primiparous and 4 multiparous dairy cows were offered 1 of 4 dietary treatments (n=2) in a 2*2 factorial, latin square design of 24 days duration. The diets were grass silage (GS), grass silage with methionine (GSM), maize silage (MS) and maize silage with methionine (MSM). Diets were iso-energetic (0.98±0.013 UFL kg DM-1) and iso-nitrogenous (133±1 g CP kg DM-1). Animals were housed and milked in metabolic stalls to facilitate collection, weighing and sampling of milk, feed and excreta. Data was analysed using PROC GLM of the SAS institute. Milk yield (20.26±0.74 kg day-1) was not effected by dietary treatment (P>0.05). There was also no effect of dietary treatment (P>0.05) on milk protein yield (0.72±0.03 kg day-1) and therefore no effect of dietary treatment (P>0.05) on the efficiency of N use (24±0.85%). Total N excreted (0.32±.01 kg-1) and urinary N excreted (0.125±0.006 kg day-1) were not effected by diet (P>0.05), however urinary urea nitrogen (UUN) output (1.3 mol urea-N day-1, grass silage; 1.5 mol urea-N day-1, maize silage) was effected by forage type (P<0.05). Methionine supplementation did not affect the milk yield, milk protein yield or the ENU of the cows used on this trial suggesting that other factors (e.g. lysine) may be limiting with diets of this CP. The lower UUN output with grass silage diets observed in this trial may be environmentally advantageous.

Effect of live yeast on dairy cows under practical conditions

Kampf, D.[1], Dusel, G.[2] and Schreiner, M.[2], [1]Orffa Deutschland GmbH, Lübecker Str. 29, 46485 Wesel, Germany, [2]University of Applied Sciences, Department of Agriculture, Berlinstraße 109, 55411 Bingen, Germany; kampf@orffa.com

The objective of the trial was to investigate the impact of 50 x 10^9 CFU per animal and day of Saccharomyces cerevisiae (MUCL 39885) on feed intake, milk production and milk composition in an on-farm experiment. The animals were accommodated in separate feed groups in cubicle housing with slatted floors and were permanently housed indoors. In the trial, one group was used as control and one as experimental group. Statistical evaluation was effected using 'proc mixed' by SAS, by utilising the mixed model and special modelling of the lactation curve. Apart from the trial group, lactation number and lactation date were analysed as additional factors of influence. The trial involved a total of 985 cows. The administration of live yeast improved average milk yield by 1.1 kg, to 38.0 kg/h/d ($P<0.05$). It was noticeable that older animals, in particular, with ≥ 3 lactations, reacted positively to the supplementation. Furthermore, the effect of live yeast was ensured from the 51st to the 150th day of lactation. Overall, the live yeast supply was proven to significantly influence the energy-corrected milk volume (+ 0.79 to 34.89 kg/h/d, $P<0.05$). The reason is most likely the improved digestibility of the ration, as the feed intake in the trial had not increased (Ø 20.7 kg DM/h/d). Consequently, the higher milk yield resulted in slightly lower concentrations of fat and protein. Nonetheless, the higher milk volume led to an increase in the daily production of these constituents (fat +0.024 kg/h/d; protein +0.024 kg/h/d, $P<0.05$).

The effect of dietary crude protein, supplementary methionine and starch source on milk yield and composition in early lactation grazing cows

Whelan, S.J., Pierce, K.M., Callan, J.J. and Mulligan, F.J., University College Dublin, Belfield Dublin 4, 0, Ireland; stephen.c.whelan@ucd.ie

In Ireland grazed grass is the principal component in the diet of the dairy cow. However there is a requirement for supplements in early lactation to bridge the gap between nutrients required and those supplied by the diet. This experiment evaluates the effect of 4 concentrate supplements on milk yield and milk composition in early lactation dairy cows offered grazed grass. 48 cows of mixed parity were selected based on parity, milk yield, milk constituent yield and calving date and assigned to 1 of 4 dietary treatments in a complete randomised block design (n=12). Diets consisted of perennial ryegrass based pasture and one of the following dietary treatments; Hi-Pro, Lo-Pro, Lo-Pro+ Meth and Lo-Pro+ Maize. Hi- Pro was 18% crude protein (CP), all other diets were reduced to 14% CP by replacing soybean with cereals, Lo-Pro+ Meth contained additional methionine and in Lo-Pro+ Maize barley was replaced with maize. Concentrates were offered at 6 kg day-1 during milking time. Milk yield was recorded daily and sampled weekly for constituents. Feed was sampled daily. Data was analysed using Proc GLM of the SAS institute. Milk yield (kg day^{-1}±s.e.d) was higher ($P<0.05$) for animals offered Hi-Pro (28.45±0.37) vs. other dietary treatments. Lo-Pro + Meth (26.49±0.37 kg) and Lo-Pro + Maize (27.28±0.37) were not different ($P>0.05$), however had greater ($P<0.05$) milk yields vs. Lo-Pro (25.19±0.37). Yields of fat (kg day^{-1}±s.e.d) were greater ($P<0.1$) for Hi-Pro (0.98±0.03) and Lo-Pro+ Meth (1.02 kg day±0.03) vs. Lo-Pro (0.95±0.03) and Lo-Pro+ Maize (0.94±0.3). Protein yield (kg day^{-1}±s.e.d) was greater ($P<0.05$) for Hi-Pro (0.97±0.02) vs. Lo-Pro (0.88±0.02 kg) and Lo-Pro+ Meth (0.91±0.02) but not Lo-Pro+ Maize (0.94±0.02 kg). These results indicate that reducing concentrate CP effects milk yield. However where low CP concentrate is fed, methionine supplementation and replacing barley with ground maize ameliorates reductions in yields of fat and protein respectively.

Effect of organic minerals in dry cow and lactating diets on health and fertility in Jersey cows

Wilde, D.[1] and Warren, H.[2], [1]Alltech UK Ltd, Rhyall Road, Stamford, United Kingdom, [2]Alltech Biotechnology Centre, Summerhill Road, Dunboyne, Co. Meath, Ireland; hwarren@alltech.com

Research carried out examining the inclusion of organic forms of trace minerals noted beneficial effects on dairy cow health and production. This trial was set up to evaluate effects of replacement of inorganic minerals with organic forms on health and fertility in a commercial, UK dairy herd. Pedigree Jersey cows (n=207) were fed a basal TMR plus minerals (inc. Cu 600; Mn 800; Zn 1800; Se 8 mg/d) from 1st April 08 - 31st March 09. Dry cows were fed the basal TMR plus wheat straw, dry cow minerals (inc. Cu 300; Mn 400; Zn 700; Se 7 mg/d). Animals were housed in cubicle sheds until 100-120d post-calving then moved to straw sheds. From 1st August 08, minerals were reformulated to totally replace inorganic Mn, Zn and Se and partially replace inorganic Cu in both lactation (Cu 600; Mn 150; Zn 600; Se 6 mg/d) and the dry period (Cu 300; Mn 100; Zn 400; Se 5 mg/d) with the respective organic form (Bioplex®, Alltech Inc., KY) for Cu, Zn and Mn. Inorganic Se was replaced by selenised yeast (Sel-Plex®, Alltech Inc., KY). Vitamin E remained constant at 1000IU/d. Incidence of mastitis, somatic cell count (SCC), days to 1st service and services per conception were measured during the trial. Data were analysed using ANOVA. Use of organic Zn, Mn, Se and Cu resulted in fewer ($P<0.05$) cases of mastitis based on month of calving and fewer ($P<0.05$) cases as a percentage of cows calved each month. There was no effect on herd average SCC or number of animals with high SCC (>400, 000). Days to 1st service were reduced ($P<0.05$) from 72 to 64 for cows receiving organic vs. inorganic minerals. Compared with the herd average, services per conception were numerically reduced from 2.01 to 1.37 for animals calving September 08 - February 09. These data demonstrate the benefits of organic mineral supplementation during both the dry period and lactation and support the growing trend towards nutrient management where bioavailabilty is of greater concern than total supply.

The effects of different silage additives on *in vitro* gas production, digestibility and energy values of sugar beet pulp silage

Kilic, U. and Saricicek, B.Z., Ondokuz Mayis University, Department of Animal Science, Faculty of Agriculture, 55139, Turkey; unalk@omu.edu.tr

The aim of this study was to investigate the effects of different silage additives during ensiling on silage quality, gas production, gas production parameters, energy values, organic matter digestibility and dry matter digestibility of sugar beet pulp silage (SBPS). A total of eight different silages were prepared from sugar beet pulp. Silage additive treatments were no additives (Control; CONT), the Artturi Imarın Virtanen (AIV:1 part H_2SO_4, 1 part HCl and 6 part water; 80 g/kg); urea (UREA; 1% of fresh weight material); formic acid (FAS; 2.2-2.5 lt/ton), microbial inoculants which were obtained from Alltech-Pioneer Maize All (MAL; 10 g/t) and Sil All (SAL; 10 g/t); F silofarm sodium formiat dry (SFD; 0.5 kg/ton) and F silofarm liquid (formic acid, sodium format and water; SLI 5-7 kg/ton). The effects of different silage additives were determined using chemical composition, cellulase method and *in vitro* gas production technique. Gas production of the silages was determined at 0, 3, 6, 9, 12, 24, 48, 72 and 96 h incubation times and their gas production kinetics were described using the equation $y = a + b\ (1-e^{-ct})$. The silage additives significantly influenced the nutrient composition of SBPS ($P<0.01$). The highest crude protein content was found in UREA added SBPS. AIV treatment resulted in the lowest *in vitro* gas production values at 24, 48, 72 and 96 h of fermentation and the values of AIV treatment were significantly different from those of other treatments ($P<0.01$). Highest energy values and gas productions were observed for MAL, SAL, SLI and SFD. The use of *in vitro* gas production technique can be recommended for the estimation of metabolisable energy and net energy lactation values of SBPS since this technique provides more reliable estimates as compared to cellulase method. In conclusion, suitable additive should be selected with the consideration of the rate and amount of roughage and concentrate feeds offered to animals.

Performance of male goats feed biological treated wheat straw

El-Bordeny, N.E.[1], El- Shafie, M.H.[2], Mahrous, A.A.[2] and Abdel-Khalek, T.M.M.[2], [1]Ain Shams University, Animal production, 68 Hadayeq Shoubra, Cairo, Egypt, 11241, Egypt, [2]Agriculture research center, Animal production research institute, Dokki, Giza, Egypt., 11111, Egypt; nasr_elbordeny@yahoo.com

Biological treatments for crop residues increase its protein and decrease CF contents consequently improve its nutritive value. Two experiments were conducted to evaluate effect of inclusion biological treated wheat straw in balanced diets (isonitrogenus isocaloric) on small ruminant performance. The first experiment was metabolic trials, which nine Ossimi rams were divided into 3 groups three animal each Three complete rations were formulated with commercial concentrate Feed mixture (CFM contain 16% CP and 63% TDN), Berseem hay, untreated wheat straw(UWS) and biological treated wheat straw(TWS). The first was control ration (T1) and contain UWS, hay and CFM. The treated wheat straw included in the second ration (T2) and the third ration (T3) to cover about 20 and 40% of total protein content of the ration. Biological treatmrnt of wheat straw with fungus increase protein and ash content and decrease OM, CF,NDF, ADF, hemicelluloses and cellulose content. Inclusion of treated wheat straw in the ration to cover 20 and 40% of the total protein has adverse effect on all nutrient digestibilities consequently the feeding value. pH value, NH3-N and total VFA's concentration as well as Acetic, propionic and butyric concentration and acetic :propionic ratio decreased ($P<0.05$) in the groups received treated wheat straw at different time. The same trend was observed for total rumen fungal and microbial protein concentration. And the second was feeding and growth experiment, which 30 male goats were divided into 3 groups 10 animal each. Inclusion TWS in the ration decrease ($P<0.05$) average daily gain, total gain and final weight in T3 compared to the control group and T2. DM, TDN and CP conversion was biter for T1 compared to the other groups. It could be concluded that using biologically treated wheat straw in animal ration can't improve animal performance and need to more studies.

Effect of stage of maturity at harvest and chop length on faecal particle size in dairy cows fed grass silage

Jalali, A.R.[1], Nørgaard, P.[1] and Randby, Å.T.[2], [1]Dept. of Animal and Veterinary Basic Sciences, Faculty of Life Sciences, University of Copenhagen, 1870, Frederiksberg C, Denmark, [2]Norwegian University of Life Sciences, Animal and Aquacultural Sciences, P.O. Box 5003, N-1432 Ås, Norway; pen@life.ku.dk

The particle size of faeces depends on the type of feed ingested, and may be indicative of good or poor rumen function. The objective of this study was to evaluate particle size distribution in faeces from dairy cows fed grass silage harvested at two stages of maturity and three chop lengths. The experimental design included ad libitum feeding with grass silage supplemented with 6 kg of concentrate to six lactating Norwegian Red dairy cows (27±9 DIM) during a three weeks period incorporating a 6×6 Latin square design with two stages of maturity and three physical forms of grass silage. The grass was harvested at an early (D-value 76%, 20% CP and 42% NDF per DM basis) or normal (D-value 70%, 15% CP and 52% NDF per DM basis) stage of maturity and fed unchopped (170 mm median particle length; MPL), medium chopped to 55 mm MPL or finely chopped to 24 mm MPL. Faeces were collected for 7 days, washed in nylon bags with a pore size of 10 μm and freeze dried before being sorted into six sieving fractions with square holes of 2.36 (O), 1.0 (M), 0.5 (S), 0.212 (D), 0.106 (C) mm and a bottom bowl (B). The faecal arithmetic mean particle size (APS), the geometric mean particle size (GPS), the most frequent particle size and median particle size values were significantly higher for normal stage of maturity at harvest ($P<0.001$). The APS and GPS values were significantly higher for finely chopped compared with unchopped silage ($P<0.05$). High frequencies of long faecal particles (>10 mm), which are indicative of poor rumen function due to lack of structural fibre, were not found for any of the rations. In conclusion, both the stage of crop maturity at harvest and its physical form when fed, affects particle size distribution in washed faeces from dairy cows to a high degree.

Efficacy of tree leaves extracts on controlling some gastrointestinal parasites species in growing lambs

Salem, A.Z.M.[1], Mejía, H.P.[1], Ammar, H.[2], Tinoco, J.L.[1], Camacho, L.M.[1] and Rebollar, S.R.[1], [1]Universidad Autónoma del Estado de México, Centro Universitario UAEM-Temascaltepec, 51300, Mexico, [2]Ecole Supérieure d'Agriculture de Mograne, Zaghouan, 1121 Mograne, Tunisia; asalem70@yahoo.com

The anthelmintic effects of *Salix babilónica* (SB) and *Leucaena leucocephala* (LL) or their mixture (SBLL), extracts-rich secondary compounds (ESC, 10 g dry leaves/80 ml of solvent) against naturally occurring infections of mixed gastrointestinal parasites were assessed in lambs under the pastoral field conditions of central Mexico. Thirty two crossbreed male (Katahdin X Pelibuey) lambs (LW 24±0.3 kg) that had not been treated previously with anthelmintics, were randomly allocated into four experimental groups (8 animals/group) in a factorial design. Lambs of control group (CTR) were not received ESC, while the other groups were submitted to a daily oral dose of 30 ml extracts of either SB, LL or the mixture (SBLL, 1:1 v/v) for 50 d. Animals of the different groups were fed ad libitum on a TMR (18% CP). Fecal samples from all animal groups were collected in three consecutive periods: P1 (d 15); P2 (d 35) and P3 (d 50). Nine nematode genera (i.e. *Haemonchus* sp.; *Oesophagostomum* spp.; *Ostertagia* sp.; *Cooperia* sp.; *Bonostomum* sp.; *Nematodirus battus*; *Chaberita* sp.; *Strongiloides papillosus*; *Nematodirus spathiger*) were identified in fecal samples, by the microscopic examination. Anthelmintic efficacy was estimated using percent faecal parasite (EPCR). There was a significant reduction ($P<0.05$) of the ESC on parasite counts per gram for all nematodes. Based on EPCR data, the most effective ESC ($P<0.05$) against nematodes was in LL (47 and 46%) than SB (40 and 26%) or SBLL (29 and 35%) lambs, respectively. Amongst the experimental period, nematode parasites followed a continuous decline being more pronounced in SB than LL or SBLL groups. Our results suggest that secondary compounds extracted from *L. leucocephala* and drenched to lambs were more efficient than *S. babilonica* or the mixture of both, in controlling gastrointestinal parasites.

The effect of calcium level in the diet of dry dairy cows on soluble faecal phosphorus

Nordqvist, M., Spörndly, R., Holtenius, K. and Kronqvist, C., Swedish Univ. Agric. Sci., Dept. of Animal Nutr. and Management, 75323 Uppsala, Sweden; maria.nordqvist@huv.slu.se

Phosphorus (P) in dairy cow faeces is an environmental problem causing eutrophication. The soluble fraction (sP) of total phosphorus in faeces can be used as a tool to estimate overfeeding of P. However, earlier studies have indicated an effect of dietary calcium (Ca) level on sP in faeces. The aim of this study was to evaluate the effect of different levels of Ca on P excretion and the proportion of sP in faeces. Twenty nine dry, pregnant dairy cows of the Swedish Red breed were fed a forage and concentrate diet with 3.7 g P/kg DM. Three weeks before expected calving, the cows were assigned to three treatments consisting of different levels of limestone, resulting in total concentrations of 4.9, 9.3 and 13.6 g Ca/kg DM. After one week of adaptation, faecal grab samples were collected twice daily during five days. Spot samples of urine were collected twice. All faecal samples from each cow were pooled and total faecal P was determined by plasma emission spectrometry. Soluble P was determined by adding 95 ml of water to 5 g of faeces. The mixture was transferred to a shaker for 1 h, after which the solution was centrifuged and filtered. The P content in the filtrate was measured using a commercially available kit. Apparent digestibility was determined using acid-insoluble ash as a marker. Data was analysed with a mixed model (SAS 9.2). Results showed a marked difference in sP among treatments. Addition of Ca decreased the sP fraction from 49% in cows fed 4.9 g Ca/kg DM to 14% and 10% for the cows fed 9.3 and 13.6 g Ca/kg DM, respectively. This may be explained by binding of sP to the surplus Ca in the gastro-intestinal tract. No differences in apparent digestibility, total P in faeces or urinary P was found. Addition of Ca to dairy cows resulted in a smaller fraction of the faecal P in soluble form. No negative effect of additional Ca was discovered, since the apparent digestibility as well as total excretion of P in faeces and urine was unchanged.

Polyunsaturated fatty acid supplementation reduces methane emissions

Rowntree, J.D.[1], Pierce, K.P.[1], Buckley, F.[2], Kenny, D.A.[1] and Boland, T.M.[1], [1]University College Dublin, S.A.F.V.M, Belfield, Dublin 4, Ireland, [2]Teagasc, Moorepark, Cork, Ireland; jason.rowntree@ucd.ie

Methane (CH_4) accounts for approximately 50% of the total greenhouse gas emissions from the average Irish dairy farm and a loss of up to 8.5% of the cow's gross energy intake (GEI). The aim of this study was to assess the impact of oils rich in PUFA on CH_4 output from grazing dairy cows compared to an unsaturated fatty acid supplement. Forty five Holstein Friesian cows were blocked on parity and allocated to one of three dietary treatments, balanced for days in milk and pre-experimental milk yield in a randomised block design. All treatments were allocated 17 kg grazed grass DM per day per cow and 4 kg (DM) of concentrates containing 160 g/kg (FW) of stearic acid (Control), soya oil (SO) or linseed oil (LO), daily. Individual CH_4 emissions were measured using the SF_6 technique at 17 (PI) and 44 (PII) days post diet introduction. After PII all cows were taken off the treatment diets and fed 3 kg per day of an average dairy nut (24% CP) and 28 days after PII, individual CH_4 emissions were measured from the LO and Control treatment groups (CO). Statistical analysis was performed using the mixed procedure of SAS with terms included for treatment, period and their interaction. Both treatment and period affected all CH_4 variables measured, and there was a significant interaction between treatment x period for SO ($P<0.001$) in that during PI SO and LO reduced daily CH_4 output (per cow) compared to the control but during PII only LO reduced CH_4 variables compared to the control. During PI both LO and SO showed reduced grams of CH_4 output per kg of milk solids ($P<0.001$), however during PII only LO showed a reduction ($P<0.001$). During the carryover period there was no effect of treatment on daily CH_4 output ($P>0.05$). Both SO and LO show the potential to reduce enteric CH_4 emissions from grazing dairy cows however the effects of LO appear to have a greater persistency over time. Once removed from the diet the mitigation effects of LO are lost within 4 weeks.

Reducing crude protein of calf starter considering difference amounts of methionin and lysine

Maddahi, H.[1,2], Mohammadi, H.[1] and Saremi, B.[3], [1]Education center of Jihad Agriculture, Animal science department, Shahid Kalantari BLV, Mashhad, Iran, [2]Taliseh Dairy farms, Mashhad, 91879, Iran, [3]Animal Science institute, Physiology and Hygiene, Katzenbergweg 7-9, 53115, Bonn, Germany; bsaremi@uni-bonn.de

Using high levels of protein (19-22% CP) in calf starter by adding considerable amounts of high quality protein sources (Soybean meal) is a common practice in modern dairy farms Although it had shown that it's not digestible more than 50%. On the other hand, first and second imitating amino acids (Lysin and Methionin) in cattle are well established and caused reducing total diet protein content which had commercial and environmental advantages. To our knowledge less works had been done on calf starter. The aim of this study was to reducing calf starter CP content from 20 to 19 and 18 using 6+2 g and 9+3 g rumen protected Lysin+Methionin. 29 female Holstein Calves were allocated to 3 treatments immediately after birth in individual stalls and received calf starter and water ad-libitum. Milk was fed 10% of body weight after colostrum. Daily dry matter intake and weekly body weight (W), body length (BL), heart girth (HG) and Stomach size (ST) was measured. Data were analyzed using GLM method of sas 9.2 ($P<0.05$). Data showed that none of the production traits was affected by 1 or 2 percent reduction in calf starter CP content. Indeed all the production traits were in upper level of normal ranges. Calves average growth rate was 600 g/day up to 60 days of age with a 1000 g/day calf starter intake results in feed to gain value of 1.5. It shows that by considering the level of amino acids in calf starter, CP content of diet could be reduced up to 2% without negative effects of production traits of calves that could be of interest from economic point of view. Later works is needed to review the metabolic advantages and less wasts in the calf and farm.

The use of polyethyleneglycol as a marker of fecal output in cows feed hay at two feeding levels

Casasús, I.[1] and Albanell, E.[2], [1]CITA-Aragón, Avda Montañana, Zaragoza, Spain, [2]Universitat Autònoma de Barcelona, Bellaterra, Barcelona, Spain; icasasus@aragon.es

Polyethyleneglycol 6000 (PEG) was selected as a potential indigestible estimator of fecal output in cattle, due to its simple detection by NIRS. Calibration equations of PEG fecal content were obtained adding PEG at 0.5% increments to feces from 2 cows fed a meadow hay (9.3% CP, 68.0% NDF), to final PEG concentrations from 0 to 10%, in duplicates. Spectra were NIRS-scanned (Foss NIRSystems 5000), and predictive equations were derived. Then a digestibility trial was conducted with 8 Parda de Montaña dry cows (622 kg), placed in digestibility cages for two 3-wk periods, during each of which half the cows received either a High (12 kg hay, as fed) or Low (9 kg hay) feeding level (FL). During the last 10 d cows were dosed either 175 (Low) or 235 (High FL) g PEG/d, and in the last 4 d total feces were collected. Fecal samples were scanned to determine fecal PEG content with the prediction equations. Individual data for actual and estimated fecal PEG content, fecal output and feed intake, fecal PEG recovery and diet DM digestibility were analyzed using PROC MIXED, and actual and estimated values were compared by paired T-tests. PEG recovery in feces was 95.7%, with differences ($P<0.001$) between actual values (5.71 vs. 6.11% in the Low and High FL, NS) and NIRS estimates (5.68 vs. 6.04%, $P<0.05$). Therefore, fecal output was slightly overestimated (+5.2%) and the difference between actual (3.05 vs. 3.81 kg DM/d in the Low and High FL, $P<0.001$) and estimated values (3.09 vs. 3.91 kg DM/d, $P<0.001$) was significant ($P<0.001$). Feed intake calculated from fecal output and average DM digestibility (61.3%) was overestimated, but the difference between actual (8.11 vs. 10.18 kg DM/d in the Low and High FL, $P<0.001$) and estimated values (7.98 vs. 10.11 kg DM/d, $P<0.001$) was only significant in the High FL, while estimates were accurate in the Low FL. In conclusion, PEG can be used as a marker of fecal output in dry cows fed hay diets, the accuracy of intake prediction depending on feeding level.

Detailed description of the in situ kinetics and synchronism of feed fractions degradation

Chapoutot, P. and Sauvant, D., AgroParisTech-INRA-AFZ, 16 rue Claude Bernard, 75231 Paris Cedex 05, France; patrick.chapoutot@agroparistech.fr

The in situ kinetics of dry matter (DM) component degradation were measured with 13 incubation points during 72 h for ten feedstuffs: barley (BAR), maize (MAI), pea (PEA), dehydrated lucerne (DLU), and several by-products: dried brewer's grains (BRG), corn gluten meal (CGM), dehydrated sugar beet pulp (SBP), palm kernel meal (PKM) and soyabean meal, untreated (SBM) or formaldehyde-treated (TSBM). The degradation patterns of DM, crude protein (CP), structural carbohydrates (SC=NDF) and cytoplasmic constituents (CC=DM-CP-NDF) were described and compared in order to display analogies or particular degradation behaviours between feeds and components. DM degradability was closely related to soluble nitrogen fraction in the short term ($R=0.94$, at 1 h), to the CC fraction during the 3 h-48 h period ($R=0.83$, at 12 h), and to the cell-wall content in the long-term ($R=0.86$, at 72 h). CP degradation kinetics were fairly parallel to the DM ones, except for SBP and MAI which exhibited very low CP degradation levels during 10 or 20 h, and BRG which showed a rather high proportion of undegradable CP. SC degradation presented a quite long lag phase (almost 16 h) for TSBM and PKM. The quantity of undegraded SC (expressed in g/kg initial DM) at 72 h varied mainly according to the ADL content with a quadratic effect ($R=0.97$, $RSD=22.3$). A cumulative synchronism index (CSI) was calculated using the actual kinetics to quantify the synchronicity of nitrogen and carbohydrate availability for micro-organisms, taking into account the particle outflow from the rumen. DLU and PKM appeared fairly harmonious, while BAR, SBP and MAI showed a deficit of nitrogen compared to their carbohydrate content of respectively -6, -10 and -12 g N/kg feed. At the opposite, an excess of N was observed for BRG, TSBM and PEA (respectively 7, 12 and 15 g N/kg) and above all CGM (35 g N/kg) and SBM (45 g N/kg). In conclusion, feeds presented very different patterns of their relative N/carbohydrate supply for micro-organisms over time.

Effect of feeding two different forms of supplemental methionine on the lactation performance and health status of dairy cows

Kudrna, V. and Cermakova, J., Institute of Animal Science, Pratelstvi 815, 104 00 Praha Uhrineves, Czech Republic; cermakova.jana@vuzv.cz

The objective of this experiment was to assess the effect of a newly developed form of methionine - isopropylester2-hydroxy-4 (methylthio) butanoic acid, called MetaSmart^{TM} on the lactation performance, in particular the production of milk protein, and on the physiological status of dairy cows. At the same time, the efficiency of this supplement was compared with the ruminally protected form of methionine Smartamine^{MT} M. The experiment of 3x3 Latin square design was conducted with a total of 30 high-yielding dairy cows, divided into three well balanced groups. Each period lasted four weeks; three weeks of the preliminary period and one week of experimental period during which the samples of milk, blood and rumen fluid were collected. A total mixed ration (TMR) based on maize silage, lucerne silage, lucerne hay, brewer's grains and concentrate mixture was offered ad libitum four times a day. The experimental diets contained MetaSmart^{TM} (170 g/d), and Smartamine^{MT} M (190 g/d), respectively. The control diet was supplemented with soybean meal to achieve the same concentration of crude protein. The highest average daily milk yields (31.34 kg) and simultaneously the highest production of milk protein and milk fat were found for the diet with MetaSmart^{TM}. Inclusion of Smartamine^{MT} M and MetaSmart^{TM} in the diet increased milk protein content by 0.11% (3.45%) and 0.07% (3.41%), respectively, in comparison with the control group. But no significant differences (P>0.05) were found for milk production, and between the values of of basic parameters of rumen fluid and blood plasma. However, the concentrations of methionine in blood plasma were 7.2 and 7.4 μmol/l higher for diets with Smartamine^{MT} M and with MetaSmart^{TM}, respectively, than for the control group. The inclusion of both Smartamine^{MT} M and MetaSmart^{TM} in the diet could increase milk production and improve the methionine status of dairy cows.

Effects of copper sources and levels on performance, Cu status, ruminal fermentation, metabolism and lipids oxidation in cattle

Correa, L.B., Zanetti, M.A., Melo, M.P. and Silva, J.S., University of Sao Paulo, Animal Science, Rua Duque de Caxias Norte, 225, CEP 13 635-900, Brazil; mzanetti@usp.br

Copper is associated with lipid metabolism, cholesterol reduction and oxidative stability of meat. The aim of this study was to determine the supplementation effect of two levels and two copper sources, during 84 days, on the performance, liver, muscle and blood copper concentration, ruminal fermentation, oxidative parameters and lipids and cholesterol metabolism. Thirty-five Nelore cattle were allocated in 7 feedlot treatment, as described: 1) C: control diet, without additional Cu supplementation; 2) I10: 10 mg Cu/kg DM, as Cu sulphate; 3) I40: 40 mg Cu/kg DM, as Cu sulphate; 4) O10: 10 mg Cu/kg DM, as Cu proteinate; 5) O40: 40 mg Cu/kg DM, as Cu proteinate. Copper supplementation provided higher Cu liver concentration (P<0.05), with the highest mean observed for the O40 treatment, nevertheless, the Cu status in muscle and serum were not affected by treatments (P>0.05). The highest ceruloplasmin activity (P<0.05) was observed for the I40 treatment. There was no significant effect (P>0.05) for daily weight gain, dry matter intake, feed efficiency, hot and cold carcass yield, cold loss, backfat and loin eye area of cattle, among control treatment and supplementation with inorganic or organic copper. Ruminal pH, volatile fatty acids and ammonia were not influenced by treatments (P>0.05). In general, copper supplementation altered the meat fatty acid profile (P<0.05), with an increase in the proportion of unsaturated fatty acids over saturated fatty acids. Cu supplementation (treatment I40 and O40) increased the activity of SOD (superoxide dismutase) related to control treatment (P<0.05), but did not influence the GSH-Px (glutathione peroxidase) activity (P>0.05). There was no effect of copper supplementation on triglycerides and cholesterol in blood; however, there was a cholesterol reduction in L. dorsi muscle related to the control treatment (P<0.05), by reducing the GSH concentrations.

Rumen protected methionine in reduced protein diet for lactating mediterranean buffaloes

Pace, V.[1], Carfì, F.[1], Contò, G.[1], Di Giovanni, S.[1], Mazzi, M.[1], Boselli, C.[2], Terzano, G.M.[1] and Terramoccia, S.[1], [1]CRA-PCM, Via Salaria 31, 00015 Monterotondo (RM), Italy, [2]IZS Lazio e Toscana, Via Appia Nuova 1411, 00178 Roma, Italy; giuseppinamaria.terzano@entecra.it

The aim of the research was to evaluate the effect of supplementing rumen protected methionine (RPM) on milk yield and quality in lactating buffaloes fed on a diet containing a reduced amount of crude protein (CP). Sixteen multiparous Mediterranean buffaloes (Bubalus bubalis L.), homogeneous for number and lactation stage, milk production traits and body condition score, were divided in two groups (A and B) and fed for 120 days on two isoenergetic diets (0.90 MilkFU/kg DM) containing, on DM basis, 44% corn silage,13% soybean meal, 15% corn meal, 26% alfalfa hay (group A, CP=15.53%) and 44% corn silage, 9.5% soybean meal, 18,5% corn meal 26%, alfalfa hay, 12 g/head/d of RPM (group B, CP=14.16%). Milk samples of each animal were collected every two weeks in the morning and afternoon; crude protein, casein, fat and urea percentage were determined and milk yield was also recorded. The differences between groups were tested using a monofactorial model. Milk yield and CP content resulted similar in the two groups (7.74 kg/head/d in A vs 7.97 kg/head/d in B and 4.71%.in A vs 4.60% in B respectively). No differences were found either in casein (3.99% in A vs 4.00% in B) or in fat percentage (9.39% in A vs 9.29% in B); the urea level in milk was lower in RPM supplemented diet (36.87 mg/100 ml in B vs 42.88 mg/100 ml in A) but the difference was not statistically significant. These results seem to indicate that a reduction of crude protein level in RPM supplemented diet does not affect the milk yield and quality. Particularly the content of casein, the most relevant parameter, with the fat, for the production of mozzarella cheese, could suggest that milk protein synthesis is improved when the limiting AA methionine is supplied in a protected form. In order to decrease the nitrogen excretion a further reduction of CP level of diets could be studied.

In vivo digestibility of dry matter and organic matter of a total mixed ration based on concentrate supplementation

Homolka, P., Koukolová, V. and Výborná, A., Institute of Animal Science, Department of Nutrition and Feeding of Farm Animals, Pratelstvi 815, 104 00 Prague, Czech Republic; homolka.petr@vuzv.cz

The current study (project No. NAZV QH81309 and MZE0002701404) aimed to determinate the *in vivo* digestibility of dry matter (DM) and organic matter (OM) of a total mixed ration (TMR) with and without a concentrate supplement. *In vivo* trials were performed with seven wethers (Romanovské breed, live weight 74 + 9 kg) in metabolism crates. The control TMR, containing maize and alfalfa silages and without the concentrate supplement (fed at 6.2 kg of TMR/animal/day), was initially evaluated with all animals. Thereafter the experimental TMR, supplemented with 170 g/animal/day concentrate, was fed at 5.3 kg of TMR/animal/day to the same group of animals. This was followed by a complete duplication of the sequence. TMR were offered at 6 a.m. and 6 p.m. each day. Animals had free access to water. Feed intake and the amount of residual feed and feces were measured on a daily basis. Samples were analyzed for contents of DM, crude protein (CP), ether extract (EE), ash, crude fibre (CF), neutral detergent fibre (NDF) and acid detergent fibre (ADF). The chemical composition (g/kg of DM) were 158 and 156.5 CP, 19 and 17.1 EE, 89.5 and 88.1 ash, 292.2 and 259.1 CF, 450.8 and 429.2 NDF and 349.8 and 311.8 ADF, for the control and experimental TMR, respectively. *In vivo* digestibility of DM and OM averaged 66.8 and 70.9%, respectively, for the control TMR, and 68.1 and 71.2%, respectively, for the experimental TMR. No significant differences ($P<0.05$) in *in vivo* DM and OM digestibility were found between the control and experimental TMR. Feed value as describe by chemical composition and *in vivo* sheep digestibility analysis are essential for feed quality evaluation.

Session 30 Poster 23

Bioavailability of methionine from two different levels of an hydroxyl-analog supplement

Migliorati, L.[1], Masoero, F.[2], Abeni, F.[1], Capelletti, M.[1], Giordano, D.[1], Cerciello, M.[1], Gallo, A.[2] and Pirlo, G.[1], [1]CRA, CRA-FLC, Via Porcellasco N°7, 26100 Cremona, Italy, [2]Universita Cattolica Del Sacro Cuore, Istituto Di Scienze Degli Alimenti E Della Nutrizione, Via Emilia Parmense N°84, 29122 Piacenza, Italy; luciano.migliorati@entecra.it

Bioavailability of the isopropyl ester of the 2-hydroxy-4-(methylthio)-butanoic acid (HMBi; MetasmartTM, Adisseo Inc., Antony, France) was assessed in nine lactating dairy cows by a standardized blood plasma test. The HMBi supplement was orally administered by a single dose to 9 cows divided in three groups: 1) control group with placebo treatment; 2) 21,3 g/cow HMBi fed group; 3) 85 g/cow HMBi fed group. Blood samples were obtained by jugular venipunctures at 8:00, just before HMBi administration, and at 1, 2, 3, 6, 9, 12, and 24 h after administration of placebo or HMBi treatments. Plasma concentrations of Met and Lys were determined. Statistical analysis was performed to consider repeated measurement over time for each cow, to evidence difference between treatments at each sampling. The model seemed to indicate that the kinetic of Met concentration in plasma comprised 2 phases. Plasma Met concentration increased during the first 3 h following supplementation in group 2, whereas the increase lasted 6 h after supplementation in group 3, reaching approximately 4′ the basal level, then decreasing in both groups to restore the basal values 24 h after supplementation. Only 85 g/cow Met fed group treatment determined a different plasma Met concentration in the first 12 h after administration.

Session 30 Poster 24

Effect of different levels of crude protein and methionine or methionine + lysine supplementation on performance of dairy cows

Migliorati, L.[1], Masoero, F.[2], Speroni, M.[1], Abeni, F.[1], Giordano, D.[1], Cerciello, M.[1], Fiorentini, L.[2] and Pirlo, G.[1], [1]CRA, CRA-FLC, Via Porcellasco 7, 26100 Cremona, Italy, [2]UCSC, Istituto Di Scienze Degli Alimenti E Della Nutrizione, Via Emilia Parmense 84, 29122 Piacenza, Italy; luciano.migliorati@entecra.it

Methionine (Met) and lysine (Lys) are the first two limiting aminoacids for lactating dairy cows fed corn-based diets. Aim of two trials was to evaluate the hypothesis that a low crude protein (CP) diet supplemented with the isopropyl ester of the 2-hydroxy-4-(methylthio)-butanoic acid (HMBi; MetasmartTM) (Trial 1) or HMBi + rumen protected L-Lys HCl (Relys®) (Trial 2) would support milk production as much as a high CP diet while reducing N excretion. In Trial 1, 20 early lactation Italian Friesian cows were used to compare three diets with different CP content and Met supplementation: 14% CP on dry matter (DM) without aminoacids supplementation (LCP); 14% CP on DM plus 10 g/d of bio-available Met (LCPM); 16.5% CP on dry matter without aminoacids supplementation (HCP). All diets were chosen in order to fulfil Lys and Met requirements, with a Lys:Met ratios in the diets that were respectively: 3.2:1; 2.7:1; 3.2:1. In Trial 2, 24 early lactation Italian Friesian cows were used to compare three diets with the same CP level (14% on DM); treatments were: no supplementation (C); 10 g/d of bio-available Met (M); 10 g/d of bio-available Met and 30 g/d of RP Lys (ML). The Lys:Met ratio in the diets were respectively: 3.2:1; 2.6:1; 3:1. N excretion was estimated utilizing milk urea content, milk yield and milk protein content. In Trial 1, no differences were found between treatments on milk yield, fat, protein, and lactose contents, and milk fat and protein yield, but milk urea N concentration was significantly lower in LCP and LCPM. Estimate N excretion was reduced by 15% in LCP and LCPM in comparison to HCP. In Trial 2, protein concentration was higher and milk urea N was lower in ML than the other two groups. In Trial 2, estimate N excretion was reduced by 8% in ML in comparison to C and M.

Changes of sodium and potassium balance in ruminant calves fed cation-anion diets with different proportions of roughage and concentrate

Salles, M.S.V.[1], Zanetti, M.A.[2], Saran Netto, A.[2] and Salles, F.A.[1], [1]supported by FAPESP, APTA, Ribeirão Preto, SP, Brazil, [2]USP, FZEA, SP, Brazil; mzanetti@usp.br

The function of cation-anionic system is to keep the body homeostasis and also is important due its influence on animal metabolism and performance. This research aims to investigate the influence of DCAD (Dietary Cation-Anion Difference) in diets with different proportions of roughage on Na and K balance. Twenty four male holstein calves were employed in two completely randomized block design experiments. In the first experiment, calves (90±12 kg LW) were fed a diet containing 60% of roughage (corn silage – 60R) and concentrates with DCAD -100, +200 or +400 mEq/kg of DM; in the second experiment, calves (117±20 kg LW) received diets with 40% roughage (40R) and concentrates with the same DCAD levels. Ammonium sulphate and sodium bicarbonate were added to reach the required DCAD levels. Ration, faeces and urine were collected for five days for mineral balance. Data were submitted to analysis of variance (SAS PROC GLM) followed by orthogonal polynomial contrasts. On both experiments, increased DCAD was followed by a linear increase of Na ingested, excreted on urine and faeces, absorbed and retained. The Na retention rate increased linearly on 60R. Na retention at -100 DCAD on experiment 60R (0.41) was higher than experiment 40R (0.21). Animals with 60R had a linear increase on K ingestion, K absorbed and retained. K retention increased according to DCAD increase (0.56, 0.63 and 0.62 at -100, +200 and +400, respectively). Animals with 40R showed a quadratic response on K ingestion and K on faeces (highest at +200). K absorbed and retained increased linearly following increased DCAD on these same animals. In conclusion, animals fed +200 or +400 DCAD had similar values of Na retention, irrespectively of roughage concentration. Regarding cationic diets, Na retention was higher on 40R (0.55) than on 60R (0.37) and the highest K retention was achieved on cationic diets with higher roughage proportion.

Session 30 Poster 26

effect of protein source on nutrients digestibility of sheep rations based on alfalfa hay

Zagorakis, K.[1], Gourdouvelis, D.[1], Abas, Z.[2], Liamadis, D.[1] and Dotas, D.[1], [1]Aristotle University Of Thessaloniki, Faculty Of Agriculture, University Campus, 54124 Thessaloniki, Greece, [2]Dimocritius University Of Thrace, Agricultural Development, Orestiada, 68200, Greece; konstantinoszagorakis@yahoo.gr

Introduction. The aim of this study was to evaluate the effects of replacing Soybean meal (SBM) with ground Lupin seeds (LS), Faba bean seeds (FBS) and Chickpea seeds (CPS), on nutrients digestibility. Material and Methods. An *in vivo* digestibility trial using a latin square 4x4 experimental design with castrated Chios rams, was conducted to evaluate the above effects. Rams were fed four rations containing alfalfa hay, ground corn grains plus one of the previous mentioned protein sources (LS, FBS and CPS), which were submitted their energy requirements for maintenance. Each treatment lasted fifteen days (eight days for adjustment and seven days for fecal collection). Statistical analysis was performed by the use of Tukey test and statistically significant differences was determined at $P<0.05$. Results. Digestibility coefficients of Dry Matter (DM), Organic Matter (OM), Crude Protein (CP), Ether Extract (EE), Crude Fibre (CF) and Nitrogen Free Extract (NFE) of the rations were not significantly affected by the protein source. Conclusions. Lupin seeds, Chickpea seeds and Faba bean seeds are valuable alternative protein sources for sheeps without having any adverse effects on nutrients digestibility of the rations. The cultivation of these legumes could be used for the exploitation of marginal and unfavorable Greek areas, according to the new revised Common Agricultural Policy. Further investigation is needed to verify the results of current study.

Session 30 Poster 27

Organically bound or inorganic Selenium for sheep nutrition

Van De Vyver, W.F.J.[1], Esterhuyse, J.[1], Cruywagen, C.W.[1] and Van Ryssen, J.B.J.[2], [1]Stellenbosch University, PBag X1, Matieland, 7602, South Africa, [2]University of Pretoria, UP, 0002, South Africa; wvdv@sun.ac.za

Selenium has been recognized as essential for domestic animals since the 1950's and is involved with cell membrane integrity, has anti-oxidative properties, supports the immune response and is important for reproductive performance of animals. Inorganic selenium supplementation is still the norm, but evidence exist that the organic form has additional benefits in alleviating deficiency symptoms. A trial consisting of 40 sheep randomly allocated to four treatments were performed at Stellenbosch University (ethical clearance: 2007B03006). Treatments were a low Se diet (Control, 0.11 mg/kg Se) or three Se enriched diets (0.3 mg/kg Se); inorganic Sodium selenite or organically bound Se YB or Se YA. Sheep were fed the respective diets for 90 days. Blood samples were collected at the onset of the trial and every 30 days to determine blood Se concentration and Glutathione peroxidase activity. At termination of the trial sheep were slaughtered by stunning and exsanguination and samples of L. dorsi, liver and kidney collected and analyzed for Se content and Glutathione peroxidase. Wool samples were also collected and analyzed for Se content. Data were analyzed using ANOVA and significance declared at $P<0.05$. Se YA had a higher blood Se concentration than the control group at day 30 (113.7±8.51 ng/ml, $P<0.05$) but blood Se concentration did not differ between the inorganic or organically bound Se treatments. Glutathione peroxidase (serum) followed a similar pattern but no significant differences were detected. Both organic Se treatments resulted in significantly higher Se concentration in L.dorsi, liver and wool samples ($P<0.05$), compared to the Control group. The Se concentration was higher for the organically bound Se treatments than the inorganic Se treatment for L.dorsi and wool samples. Selenium YA had a higher Se concentration in the kidney than the other treatments ($P<0.05$). It is concluded that scientific merit exists for the use of organically bound Se in sheep nutrition.

Session 30 Poster 28

Maize supplementation during last 12 days of pregnancy improves the metabolic state of the goats at partum

Ramírez, S.[1], Delgadillo, J.A.[1], Terrazas, A.[2], Flores, J.A.[1], Serafín, N.[3] and Hernández, H.[1], [1]Universidad Autónoma Agraria Antonio Narro, CIRCA, Periférico Raúl López Sánchez S/N, Torreón Coahuila, 27054, Mexico, [2]UNAM, Secretaría de Posgrado, FESC, Cuautitlán Izcalli, Edo de México, 54714, Mexico, [3]UNAM, Instituto de Neurobiología, Juriquilla, Querétaro Qro, 76001, Mexico; sarave2@hotmail.com

Prepartum fed supplementation in ewes maintained under good pastures improves the metabolic state of the animals. Goats grazed under semi-arid conditions posses a low body condition score (BC) due to that in some occasions the quality of the vegetation is poor. The objective of this study were to determine if flaked maize supplementation improves the metabolic state of goats assessed by measuring BC and blood glucose concentrations (BGC) around parturition. A group of 11 control goats (C) were maintained under traditional extensive management, in which the animals were taken out daily to graze the available vegetation until parturition. Another group of 14 supplemented goats (S) were managed in the same conditions as the C, but received in addition 0.6 kg of flaked maize/mother/day during the last 12 days of pregnancy. BGC were determined at day 14 before partum and at partum using a blood glucose meter. BC was recorded at day 14 prepartum and at partum. Data from two variables were analyzed between groups in each period using independent t test. At day 14 prepartum BGC did not differ ($P>0.05$), between goats from C (35±2 mg/dl) and S (34±1 mg/dl) groups. However, at partum BGC was higher ($P<0.05$) in goats from S (158±13 mg/dl) than in C (116±12 mg/dl) group. BC recorded at day 14 prepartum did not differ ($P>0.05$) between goats from C (1.8±0.1 points) and those from S (1.7±0.1 points) group. However, BC recorded at partum was higher ($P<0.05$) in goats from S (1.8±0.1 points), that in C (1.4±0.1 points) groups. We concluded that in goats maintained under semi-arid conditions, a supplementation with flaked maize during last 12 days of pregnancy maintain a good metabolic state of the females at partum.

The effect of different levels of monensin on Ghezel lambs performance by step-up feeding program

Safaei, K.[1] and Tahmasebi, A.M.[2], [1]Ministry Of Agricultural Jahad, Management of Pastoral Nomads of Kermanshah, Near Moalem park- Mostafa emami Blvd, 6714748845 Kermanshah, Iran, [2]Mashhad university, Animal science, 9177948974 Mashhad, Iran; kh.safaei@gmail.com

For evaluation of different levels of monensin on Ghezel lambs performance by step-up feeding program, 20 wether Ghezel lambs (mean live weight 24.5±4.9 kg) were allocated to individual cages and experiment carried out by completely randomized block design with five treatment (control diet and rations contain 0, 10, 20 and 30 mg monensin per kg DMI) and 4 replications. After 3 weeks of adaptation period, animals fed by experimental diets. At the start of experiment Forage: Concentrate ratio was 55:45 that was fixed for control treatment up to end of experiment but, it increased 5% in concentrate weekly on other treatments. So, at the end of experiment (9th week) F: C ratio was 10:90 with different levels of monensin. Diets composed of alfa alfa hay, wheat straw, barley, soy bean meal, vitamin-mineral supplement, salt and offered morning and afternoon every day. During the experiment, dry matter intake (DMI) and body weight gain (BWG) were recorded weekly. Results of this experiment indicated that increasing level of concentrate and monensin significantly decreased DMI, improved BWG and feed conversion ratio ($P<0.01$). However, this effect was more significant by adding 30 mg/kg DMI monensin in the ration. Data from this study show that monensin had no any effect on blood metabolites ($P>0.05$). Total rumen acidity ($P<0.05$) and pH ($P<0.01$) was affected by monensin and level of concentrate in the ration. In conclusion, addition of monensin to rations contain high level of concentrate has beneficial effect to improving of animal performance due to the changes in the rumen ecosystem and pattern of fermentation.

Influence of exogenous enzyme on dairy buffaloes performance

El-Bordeny, N.E.[1], Gado, H.M.[1], Kholif, S.M.[2], Abedo, A.A.[2] and Morsy, T.A.[2], [1]Ain Shams University, Animal production, 68 Hadayeq Shoubra, Cairo, Egypt, 11241, Egypt, [2]National Research Center, Animal production, Dokki, Giza, Egypt., 11111, Egypt; nasr_elbordeny@yahoo.com

Supplementing dairy animal ration can enhance nutrients digestibility and milk production. So this study was conducted to investigate effect of mixture of exogenous enzymes (ZADO) from anaerobic bacteria on performance of dairy buffaloes. Fourteen lactating multiparous Egyptian buffaloes (575±15.5 kg live weight) were randomly assigned into two experimental groups of 7 immediately after calving and fed ration containing 75.5% berseem, 6% rice straw and 18.5 concentrate feed mixture with or without of 40 g exogenous enzymes (ZADO) /head/d of for 12 weeks. Intake of Dry matter (DM) and organic matter (OM) were increased ($P>0.05$) by adding enzymes. Digestibility of all nutrient in the total tract was higher ($P<0.05$) in supplemented dairy buffaloes. Supplementation of enzymes insignificant affects blood parameters. Milk and 4% FCM yield were increased ($P>0.05$) for buffaloes fed ration supplemented by ZADO. Milk fat, SNF and protein % and its yield were increased ($P<0.05$) for supplemented animal. Supplemented lactating buffaloes ration by exogenous enzyme (ZADO) increase milk yield and milk fat and protein content and yield due to positive effect on nutrient digestibility.

Session 31 Theatre 1

Rearing entire pigs in barren or enriched housing: consequences on the human-animal relationship

Tallet, C., Brillouët, A., Paulmier, V., Meunier-Salaün, M.-C. and Prunier, A., INRA-Agrocampus Ouest, UMR1079 Systèmes d'Elevage Nutrition Animale et Humaine, domaine de la prise, 35590 Saint-Gilles, France; celine.tallet@rennes.inra.fr

Castration is believed to decrease the aggressiveness of entire pigs toward humans, but no scientific work has been done on the consequences of rearing entire males on human-animal relationship. This is why we compared entire and castrate males both in barren and enriched housing. Eighty males (groups of 10) were studied: 40 castrated (surgical) at 5-6 days of age and 40 left entire, half of each reared in a barren (1 m^2/animal, slatted floor) and half in an enriched (2.5 m^2/animal, straw bedding and outdoor run) housing. We evaluated their relationship to humans at 80 and 150 days of age by measuring the time taken to be transferred individually to a corridor (manageability) and the reaction to the presence and departure of an unfamiliar human after isolation in a test pen (1 m x 6 m). Data were analysed with ANOVA. Manageability was not affected by sex nor housing. No entire pig was aggressive toward the human. At 80 days entire pigs approached (<1.5 m) more rapidly and spent more time near the human than castrated ones, and reacted to his departure by increasing their locomotor activity whereas castrated pigs did not move more. Whatever the age, pigs from the barren housing approached/contacted the human sooner than pigs from the enriched housing and spent more time near/in contact with him. They also vocalised less in his presence than pigs from the enriched housing. At 150 days, pigs from the barren housing moved less in the pen during human presence than pigs from the enriched housing. These results do not confirm that entire males are aggressive towards humans and difficult to handle. At young age, they are even more (positively) attracted by humans than castrated pigs. Enriching the environment does not modulate the effect of the absence of castration on the human-animal relationship, but leads to fewer interactions with the human.

Session 31 Theatre 2

Impact of surgical castration and housing environment on the immune function of fattening male pigs

Merlot, E., Thomas, F., Mounier, A.-M. and Prunier, A., INRA, UMR1079 SENAH, domaine de la Prise, F-35590 Saint-Gilles, France; elodie.merlot@rennes.inra.fr

Androgens are known to be negative regulators of the immune function in adult mammals but the consequences of their deprivation since childhood, as it is the case for surgically castrated pigs, remain unexplored. The present study aimed to study the effects of castration on porcine immune function and health. Male pigs were surgically castrated or not at one week of age. Because the welfare problems related to entire male breeding might be improved by enrichment of the environment, and because stress response can also impact immune function, the effects of castration were evaluated on growing pigs housed in two different systems (slatted floor versus deep litter with an outdoor area, 4 experimental groups of 20 pigs each). Blood samples were drawn for measurements of blood leukocyte counts (at 3, 4 and 5 months (M) of age) and lymphocyte proliferation in response to concanavaline A (at M3 and M5). Organ weights and respiratory tract health status were recorded at slaughter (M5). Pigs housed on litter had lower lymphocyte proliferation at M3 ($P=0.08$) and M5 ($P<0.01$) and higher relative spleen weight at slaughter ($P<0.05$) than pigs housed on slatted floor. Entire pigs had higher total white blood cell numbers, higher proportions of lymphocytes ($P<0.05$), and lower proportions of granulocytes ($P<0.05$) than castrated pigs since at M3, and heavier thymus ($P<0.001$) at slaughter (M5). Housing interacted with sex for lymphocyte proportions at M4 ($P<0.10$). At that age, entire males displayed higher proportions of lymphocytes than castrated animals on deep litter only ($P<0.05$). Health status was similar among experimental groups. In conclusion, castration increased haematopoiesis, blood lymphocyte numbers and thymus weight, suggesting that androgens might have a stimulating effect on the maturing immune system of juvenile pigs, different from their well-known inhibitory effect in adults. Housing characteristics did not modulate significantly this influence.

Transmission of highly infectious animal diseases in a pig contact network

Traulsen, I. and Krieter, J., Institute of Animal Breeding and Husbandry, CAU, Olshausenstr. 40, 24098 Kiel, Germany; itraulsen@tierzucht.uni-kiel.de

To assess the impact of contact patterns in the pork production chain on the spread of animal-mediated infectious diseases, contact data from a producer community in Northern Germany are analyzed. The data contains information on 15,652 animal movements between 469 different premises in 2006 to 2009, in detail supplier, purchaser, as well as number and type of delivered livestock. Network theory is applied to set up a directed network of contacts for every calendar week. In the network, an edge represents a livestock movement (gilts, finishing pigs, boars) and a node indicates a premise (multiplier, farrowing or finishing farm, and abattoir). Network characteristics, such as in-, out-degree and closeness centrality, are used to describe the network. Starting from an initially infected premise in the first calendar week, the spread of an infectious disease through the network is determined using a dynamical percolation. In every calendar week, on average 112 premises are involved in 98 animal movements. The size of the network slightly changes from one week to another (110-124 premises, 95-109 movements). The average degree (number of contacts) of a premise is 0.88, and the range of in-going movements is much higher compared to out-going (55 vs. 15). Premises with few contacts (<50) in the whole observation period have on average the same in- and out-degree (0.54, 0.61). Premises with 50-100 contacts have much more out-going (0.27, 0.89), premises with more than 100 contacts much more in-going contacts (2.49, 1.31). On average 3.1 premises participate in disease spread and 3.0 infected premises with contacts in one week also have contacts in the subsequent week. In total, a mean number of 7.8 premises is infected, but in the worst case this quantity increases up to 87.3. The application of network theory provides the opportunity to identify key premises with an important role concerning disease transmission. Hence, further simulations evaluate control measures that preferentially target these premises.

Session 31 Theatre 4

Vaccination strategies to control classical swine fever epidemics

Brosig, J., Traulsen, I. and Krieter, J., Institute of Animal Breeding and Husbrandy, Christian-Albrechts-University, Olshausenstr. 40, 24098 Kiel, Germany; jbrosig@tierzucht.uni-kiel.de

The effect of emergency vaccination to eradicate classical swine fever (CSF) was compared to pre-emptive culling. A spatial and temporal Monte-Carlo simulation model described the spread and control of classical swine fever virus between individual farms on a daily basis. The virus could spread by animal, person and vehicle contact, as well as by local spread. Farm data of a region in Northern Germany were used, consisting of 1108 farms with a farm density of 1.22 farms/km^2. A farrowing farm in the centre of the region was set as index case. Four different control strategy combinations were simulated with 100 replications each: (1) no control measures, (2) establishment of protection (3 km) and surveillance zones (10 km), (3) establishment of protection and surveillance zones plus pre-emptive culling (500 m, 1000 m) and (4) establishment of protection and surveillance zones plus emergency vaccination (1-10 km). The establishment of protection and surveillance zones with a strict movement ban reduced the mean number of infected farms by approximately 95% from 1077 (scenario 1) to 60 farms. Adding pre-emptive culling (1000 m) resulted in a further reduction of the number of infected farms (35 farms), simultaneously the number of culled farms increased. Emergency vaccinations within an area of 10 km around an infected farm led to the same dimension of infected farms as pre-emptive culling, on average 37 farms were infected. The number of culled farms could approximately be reduced from 69 to 36. 422 farms need to be vaccinated. Due to the actual definition of freedom from CSF (EU legislation) vaccinated animals need to get culled. The real time RT-PCR is a diagnostic approach that is based on the detection of virus instead of antibodies. Freedom of CSF could be checked by rapid PCR testing so that culling could be avoided. Additional to emergency vaccination rapid testing will be investigated in further simulations.

Genetic resistance to natural helminth infections in two chicken layer lines

Kaufmann, F.[1], Das, G.[1], Preisinger, R.[2], König, S.[1] and Gauly, M.[1], [1]Livestock Production Systems, Animal Science, Albrecht Thaer Weg 3, 37075 Göttingen, Germany, [2]Lohmann Tierzucht GmbH, Postbox 460, 21789 Cuxhaven, Germany; mgauly@gwdg.de

Groups of Lohmann Brown (LB) and Lohmann Selected Leghorn (LSL) hens were reared under helminth-free conditions and kept afterwards in a free range system. Mortality rate, body weight development, laying performance and faecal egg counts (FEC) were recorded during a laying period of 12 months. At the end of the laying period, 246 LSL and 197 LB hens were harvested and worms counted following the World Association for the Advancement of Veterinary Parasitology (W.A.A.V.P.) guidelines. In addition adult Heterakis gallinarum and Ascaridia galli were sexed and measured for length. Significant ($P<0.01$) differences were observed in mortality rates between LSL and LB animals (12.9 vs. 5.7%). LSL hens showed significantly higher FEC when compared with LB hens at almost all dates of monitoring. Almost all animals became infected with at least one of the various helminth species. The most prevalent species were H. gallinarum, Capillaria spp. and A. galli. LB hens showed a significantly ($P<0.05$) higher average number of adult H. gallinarum, Capillaria spp. and tapeworms when compared with LSL animals. However, number of adult A. galli was in tendency lower in this genotype. In total, LB had a significantly ($P<0.05$) higher worm burden than LSL (192.3 vs. 94.3). The numbers of all different helminth species were positively correlated., whereas sex ratio of H. gallinarum and A. galli and average worm lengths were not significantly different between genotypes. There was no significant phenotypic correlation between body weight and worm burden in LSL, whereas it was in LB ($r=0.17$, $P<0.05$). The estimated heritabilities for total worm burden were 0.23 (SE ± 0.12) in LSL and 0.75 (SE ± 0.21) in LB, respectively. Based on the estimated heritabilities it is possible to breed for helminth resistance in both genotypes.

Effect of dietary non-starch polysaccharides (NSP) on *Heterakis gallinarum* infection in chicks

Das, G., Abel, H.J., Humburg, J. and Gauly, M., University of Goettingen, Department of Animal Science, Albrecht-Thaer-Weg 3, 37075, Goettingen, Germany; gdas@gwdg.de

Dietary NSP do influence intestinal parasite infections in pigs. The present study aimed to evaluate effects of NSP supplementation in *H. gallinarum*-infected chicks. Three consecutive experiments were conducted with one-day-old female layer chicks until an age of 11 weeks (wk). The birds were divided into 3 feeding groups receiving ad libitum (1) a basal diet (CON), (2) a basal diet plus insoluble NSP (I-NSP), or (3) a basal diet plus soluble NSP (S-NSP). At the end of wk 3 each feeding group was subdivided into an uninfected and a *H. gallinarum* infected group (200 embryonated eggs per bird). The CON diet led to greater slaughter weights than the NSP diets ($P<0.001$). Infection impaired body weight (BW) development, particularly in those animals being fed S-NSP ($P<0.05$). Higher incidence of infection and elevated worm burdens were observed in birds on the NSP-supplemented diets than in those on CON ($P<0.001$). Both NSP diets caused higher faecal egg counts (FEC) when compared with CON ($P<0.001$). The S-NSP diet resulted in higher FEC than I-NSP ($P<0.05$). The S-NSP diet increased size of small intestine and caeca in comparison to CON and I-NSP ($P<0.001$). Both, NSP and infection increased the relative pancreas weight referred to BW ($P<0.001$). Infection also increased caecal weight ($P<0.001$). The concentration of short chain fatty acids (SCFA) in the caeca contents was not influenced by diet ($P>0.05$). However, greater amounts of SCFA were produced in the caeca by feeding S-NSP ($P<0.001$). Infection with *H. gallinarum* decreased the concentration and total amount of SCFA and caused a higher value of pH in the caeca ($P<0.001$). The results indicate that both, I-NSP and S-NSP favour *H. gallinarum* infection in chicks. The S-NSP increased FEC and fecundity more than I-NSP. It is concluded that *H. gallinarum* benefits from NSP supplementation of host animal diets.

Session 31 Theatre 7

The effect of physical exercise during the dry period on the performance of dairy cows in early lactation

Goselink, R.M.A., Lenssinck, F.A.J., Van Duinkerken, G. and Gosselink, J.M.J., Wageningen UR Livestock Research, Edelhertweg 15, 8219PH Lelystad, Netherlands; roselinde.goselink@wur.nl

In dairy cattle, a gradual transition from the anabolic state during dry period to the catabolic state of early lactation will prevent metabolic disorders like ketosis. The effect of physical exercise during the dry period on health and performance of dairy cows during the transition period was studied in an experiment with 34 multiparous dairy cows. The experiment had a randomized block design. Cows were paired in blocks of 2 cows each, based on similarity in parity, body condition score, expected calving date and milk yield. Cows within each pair were randomly allocated to one of two treatment groups. Rations were based on grass silage with additional concentrates according to Dutch feeding standards. The experiment lasted from 6 weeks before expected calving date until 6 weeks after calving. One group (WALK) received daily physical training in a horse walker during the dry period, by two sessions of 45 min at 3.4 km/h, resulting in 5 km/d. After calving, training was stopped. The other group (CON) did not receive physical training. All cows were housed indoors in four groups (WALK vs. CON × dry vs. lactating) in a free stall with restricted space. Milk yield was recorded daily during the first six weeks of lactation and milk composition was determined weekly during the first three weeks. Blood samples were taken in week -6, -2, 1, 2, 3 and 4 for glucose, nonesterified fatty acids (NEFA) and β-hydroxybutyrate (BHBA) analysis. First results showed that milk yield was not significantly different between groups (average 34.9 kg/d). Milk fat concentration tended to be higher for CON ($P<0.10$). Serum glucose concentration was not significantly different. Antepartum, NEFA concentration tended to be higher for group WALK. Postpartum, BHBA and NEFA concentrations tended to be lower for group WALK ($P<0.10$). These results suggest better metabolic adaptation of dairy cows in early lactation with physical exercise before calving.

Session 31 Theatre 8

Effect of TMR chemical composition on milk yield lactation curves using a random regression animal model

Caccamo, M.[1], Veerkamp, R.F.[2], Petriglieri, R.[1], La Terra, F.[1] and Licitra, G.[1,3], [1]CoRFiLaC, Regione Siciliana, S. P. 25 Km 5 Ragusa mare, 97100 Ragusa, Italy, [2]WageningenUR Livestock Research, Animal Breeding and Genomics Centre, P.O. Box 65, 8200 AB Lelystad, Netherlands, [3]Catania University, DACPA, via Valdisavoia, 5, 95123 Catania, Italy; caccamo@corfilac.it

Several studies reported influence of diet on milk production. As part of a larger project aiming to develop management evaluation tools based on results from test-day models, the objective of this study was to estimate the effect of TMR chemical composition on milk yield curves. A random regression animal model was fitted to a full dataset (134,579 testday records) to obtain variance components. Subsequently the same model with fixed variance components was used on a subset where chemical composition of TMRs were sampled immediately before the testday. The subset contained 46,531 test-day milk yield records from 3,554 cows in 27 herds recorded from 2006 through 2008. The model included parity, days in milk, age at calving, year and season at calving, days dry, calving interval and stage of pregnancy as fixed effects. Days in milk was modeled using 9-order Legendre polynomial. Animal, sire and maternal grand sire effects were modeled using 3-order Legendre polynomials. Model fitting was carried out using ASREML. Total mixed ration samples were collected every 3 months and analyzed for dry matter (DM), ash, crude protein (CP), soluble nitrogen (SN), acid detergent lignin (ADL), NDF, acid detergent fiber (ADF), and starch. All chemical parameters and their interaction with days in milk were included in the model as fixed effects. Conditional Wald F statistic on fixed effects revealed significant effects ($P<0.001$) for DM, CP, SN, NDF, ADF and starch and their interactions with days in milk on milk yield. Lactation shape can therefore be used as an indirect indicator of feed management adequacy. Deviations from an average lactation curve shape could, for example, warn on low CP content in the diet, being positively correlated with lactation peak and negatively correlated with persistency.

Session 31 Poster 9

Rearing entire pigs: consequences of enriching the housing condition on the social activity

Tallet, C., Brillouët, A., Paulmier, V., Meunier-Salaün, M.-C. and Prunier, A., INRA, Agrocampus Ouest, UMR1079 Systèmes d'Elevage Nutrition Animale et Humaine, domaine de la prise, 35590 Saint-Gilles, France; celine.tallet@rennes.inra.fr

Rearing entire pigs may lead to welfare problems due to sexual and aggressive behaviour. One way to improve this could be to enrich the housing as it favours exploration of the environment that could be done at the expense of undesired behaviours. Our objective was thus to investigate whether enriching the housing conditions during fattening can modulate the effects of non-castration on the behavioural activity of pigs. Eighty males (groups of 10) were studied: 40 castrated (surgical) at 5-6 days of age and 40 left entire, half of each reared in a barren (1 m^2/animal, slatted floor) and half in an enriched (2.5 m^2/animal, straw bedding and outdoor run) housing. We observed their social activity by continuous sampling three times a month for 1 hour from 3 to 5 months of age. This included agonistic behaviour (eg biting), massages and the calculated total social activity (agonistic+non-agonistic+sexual). We analysed the mean of each activity per month with ANOVA. In both types of males, the amount of agonistic and total social activity decreased between the 3 and 4 month of age. Entire males showed more agonistic and total social interactions than castrates but the difference decreased with age. Entire pigs also gave more massages but independently of their age. Enriching the housing induced a decrease in agonistic and social interactions of castrate but not of entire males. It had no effect on the massages and this independently of the type of males. The welfare troubles linked to agonistic behaviour of entire pigs were mainly present during the first month of fattening, a period when the general social activity (including positive interactions) is also high in those animals. Entire pigs presented more massages what can become negative for the receivers. In our conditions enriching the environment does not seem to be an efficient way to decrease the agonistic activity of young entire males as it is in castrate pigs.

Session 31 Poster 10

Influence of ventilation and genetics on pig's biting behavior

Van De Perre, V.[1], Driessen, B.[1,2], Van Thielen, J.[2] and Geers, R.[1], [1]K.U.Leuven, Laboratory for Quality Care in Animal Production, Zootechnical Centre, Bijzondere Weg 12, B-3360 Lovenjoel, Belgium, [2]K.H.Kempen, Kleinhoefstraat 4, B-2440 Geel, Belgium; vincent.vandeperre@biw.kuleuven.be

Frustration behavior, specifically tail and ear biting, is a severe problem in modern pig production which reduces animal welfare and productive performance. This abnormal behavior is mostly observed in intensive housing systems and has a multi-factorial origin. In this study the effects of room climate, stocking density and genetics on frustration behavior of pigs were observed. Two identical rooms with piglets of 26 days old were observed, every week till the pigs had an age of 16 weeks, on a farm with a severe biting problem. However, the surface of the air inlet was adjusted in one room so the air came in with a velocity of 1 m/s and the minimum ventilation rate of the ventilator was adjusted according to 0.25 m^3/kg.h, in order to reduce frustration behavior. Piglets originated from 2 different boars. Tails were docked completely. Every week biting behavior, stocking density, temperature, CO_2-concentration, ventilation pattern and relative humidity was observed for both rooms. After optimization of the room climate, pigs expressed almost 50% less biting behavior. Also stocking density played an important role in biting behavior. A boar effect was found. These findings show the complexity of biting behavior motives, making solutions for this problem not easy.

The impact of the sow for the spread of Salmonella within the pork production chain

Hotes, S., Traulsen, I. and Krieter, J., Institute of Animal Breeding and Husbandry/CAU, Olshausenstraße 40, 24098 Kiel, Germany; itraulsen@tierzucht.uni-kiel.de

An individual-based Monte-Carlo simulation model was developed to analyse the spread of Salmonella infection in the pork production chain. The present study focused on the role of the sow in transmission of Salmonella to their piglets and the subsequent spread among growers and fattening pigs. Starting from 20 farrowing farms with an average of 200 sows, production flow and the chain of infection were considered on a weekly basis, down to lairage at slaughterhouse. Sows were divided in shedder and non-shedder and assumed as exclusive initial source of infection for the piglets. Health of the pigs was described by one of four mutually exclusive health states: susceptible animals, seronegative shedder, seropositive shedder, and seropositive not shedding carrier. Carrier could restart shedding, especially due to stress, e.g. during transport by lorry. Within 20 scenarios the Salmonella shedding prevalence of the sows was varied between 0% and 100%, increasing by 1% (0-10%) and 10% (10%-100%), respectively. Each scenario regarded 52 weeks and was repeated 20 times. The rate of direct transmission at farrowing unit enhanced from 0% to 3.7% with increasing shedding prevalence at sows. A concomitant effect was the forward shift of infection from fattening stage to farrowing and growing period. Seroprevalence at slaughter reached 83.7%, if all sows were shedding. Taking 1%, 5%, and 10% shedding sows as basis, seroprevalence at lairage was 4.2%, 17.8% and 29.8%, respectively. Following the infection route it became clear that with increasing proportion of shedding sows the relevance of infections between pigs of the same pen or between pigs of adjacent pens increased. In conclusion, the shedding prevalence of the sows influenced the prevalence in slaughter pigs. The direct infection between sows and piglets was less important compared to transmission among growers or fattening pigs. However, the initial source of these infections was a Salmonella shedding sow.

The effect of flooring on skin lesions and lying behaviour of sows in farrowing crates

Boyle, L. and Lewis, E., Teagasc, Moorepark Research Centre, Fermoy, Co. Cork, Ireland; laura.boyle@teagasc.ie

In spite of concerns for the welfare of sows farrowing in confinement there is still no viable alternative to the farrowing crate. Within the environment of the crate improvements to sow welfare are possible through modifications to housing and management practices. The aim of this study was to evaluate slatted flooring options in terms of lesions and lying behaviour of sows in farrowing crates. Seventy two multiparous sows were introduced to crates with four floors (Slatted steel [SS], n=19; Slatted steel with checker plate panel [CP], n=18; Expanded cast iron [CI], n=18; Plastic coated woven wire [PL], n=17) between d103 and 110 of pregnancy in 9 replicates. Latency to lie on entry to the crate and the time taken to go from lowering the knees to sternal recumbency was measured from 3 h video recordings of the behaviour of 10 sows in each treatment from entry to the crate. Abnormal lying was also recorded. Lesions to the hind limbs were scored from 0 to 6 according to severity on -6, -5, 8, 15, 21 and 27 (day prior to weaning) days relative to farrowing (day 0). Data were analysed by non-parametric Kruskall-Wallis and Mann-Whitney tests using methods of SAS. On average sows took 22 m 51 s to lie down after entering the crate and 23 s to go from kneeling to sternal recumbency. PL sows had both the shortest latency to lie (15 m13 s) and the fastest lying times (17 s) but there were no significant differences between treatments ($P>0.05$). Furthermore, there were no treatment differences in the proportion of sows (10% in total) adopting abnormal lying sequences ($P>0.05$); 2 sows were still standing 3 hours after entering the crate. SS sows had higher limb lesion scores than sows in the other three treatments on days 8, 15 and 21 ($P<0.01$) post-farrowing. Sows can have difficulty lying down in farrowing crates and flooring influences this behaviour. In accordance with findings for piglets, slatted steel is one of the most injurious flooring options for sows in farrowing crates.

Impacts of growth development and age at first calving on first lactation yield in Holstein dairy cows

Picron, P., Froidmont, E., Turlot, A. and Bartiaux-Thill, N., Walloon Agricultural Research Centre, Production and Sectors Department, Animal Nutrition and Sustainability Unit, 8 rue de Liroux, 5030 Gembloux, Belgium; froidmont@cra.wallonie.be

In a context of economic constraints, reducing the herd replacement costs represents a way to improve the economic efficiency of the dairy sector. Studies reported that heifers calving between 23 and 24.5 mo of age achieved the highest economic return. However, the heifer has to be well developed to guarantee milk performance, calving ease and low stillbirth rate. Monitoring growth, by Heart Girth (HG) measurements, constitute a convenient way to evaluate the feasibility to inseminate heifers early. The aim of this study was to analyse the relationship between HG, age at first calving and milk production at first lactation. HG measurements of 704 Walloon (Belgium) Prim Holstein heifers have been recorded, every 3 months, during 2 years (2006-2008). The impact of development, at key-ages (from 3 to 24 mo, every 3 mo), on age at first calving and first lactation yield, was determined using analysis of variance. Mean total milk production in February 2010 (n=319) rose to 7,686±1,388 l/365 days of lactation, for a mean age at first calving (n=421) of 27.4±3.5 mo. These preliminary results suggest a positive correlation between growth performances and first calving age. Heifers with a better conformation (HG > 2.5% of the French reference) calved significantly earlier than smaller animals. For instance, a lack of conformation at 15 mo implied a delay at first calving of 4.1 mo of age. HG measurement should therefore constitute a preferential tool of management of heifers herd. First lactation yield was significantly higher (950 l/365 d) for heifers calving older (> 30 mo) comparatively to early calving animals (24 mo). These results do not correspond to those related to larger samples in Wallonia. No significant relation could state, at this time, a direct relation between growth development and first lactation yield. Lactation data are still collected to complete this study.

Serological investigation of IBR and MAP in dairy cows in Greece

Mpatziou, R., Ntafis, V., Liandris, E., Ikonomopoulos, J. and Xylouri, E., Agricultural University of Athens, Department of Anatomy and Physiology of Farm Animals, Iera Odos 75, 11855, Athens, Greece; efxil@aua.gr

Infectious Bovine Rhinotracheitis (IBR) is a viral disease of cattle caused by Bovine Herpes Virus 1 (BHV-1). It is characterised by fever, rhinotracheitis, abortions and stillbirths. Paratuberculosis is a chronic enteritis of ruminants caused by Mycobacterium avium subsp. paratuberculosis (MAP). The disease is characterized by intermittent chronic diarrhoea thats leads to weight-loss, emaciation and often death. Both IBR and paratuberculosis have a significant financial impact due to loss of productivity that depending on the disease may refer to decrease milk and meat production and fertility rates. The aim of the study was to define the level of seropositivity for IBR and paratuberculosis in dairy cows in Greece using ELISA. For this purpose, we tested 550 serum samples collected randomly from a part of Greece(Macedonia, Thrace, Thessaly, Epirus, Central Greece and Ionian Islands) that hosts 94% of the country's bovine population. Sample size was defined so it would be representative of the targeted area at a 5% level of confidence. Of the animals tested 336 (66.5%) and 20 samples (3.64%) reacted positive for IBR and paratuberculosis respectively. Of all the positive reactors 295 were not vaccinated, 47 were vaccinated against IBR with the conventional vaccines and 24 with a deleted. None of the tested animals were vaccinated against paratuberculosis. The recorded IBR seroprevalence was higher in Greece compared to many other European countries. The level of MAP seropositivity on the other hand indicates low prevalence of paratuberculosis, which however cannot be considered conclusive to the pathogenesis of the disease. The resuts of this study indicate the level of awareness for the spread of IBR and perhaps bovine paratuberculosis in Greece, in order to safeguard animal health, preserve quality and safety standards of animal products and increase profit in dairy farms.

Factors affecting on open days and calving intervals in a training Brown Swiss cattle farm

Zakizadeh, S. and Sabzali, A., Hasheminejad High Education Center, Animal Science, Kalantari Highway, 9176994767, Iran; sonia_zaki@yahoo.com

In order to study the factors affecting open days and calving interval in Brown Swiss, 700 records related to 260 cows were used. Cows had all given birth during 1979-2009. Two linear regression models were applied to evaluate the effect of factors on open days and calving interval. The first one used for open days and included previous milk production, days in milk, age at artificial insemination, year and month of insemination, and service numbers. The effects of parity, open days, service numbers, year and season of calving were considered on the second model for calving interval. Results showed that year of insemination, service number and days in milk had highly significant effect on open days. Previous milk production had significant effect on the next open days, but the effect of age and month of insemination was not significant. There was also a negative correlation between milk production and open days. However, increasing of days in milk and number of insemination leaded the more number of open days. Conception ration per insemination increased with age and was the lowest at the third calving. Calving interval was significantly affected by open days and parity and increasing of service numbers. Year of calving had significant effect on calving interval, but not the season of calving. Although lactation number had significant effect on calving interval, it did not show any specific trend, because of unbalanced distribution samples per lactation and combining of the 7th parity and higher ones. Milk production had no significant effect; hence it was not included in this model. As calving interval is a management criterion, it seems that it might be influenced by milk production indirectly. This old training farm has a long history during the past 40 years with three different managements. Size of farm was different in these years, which all cows were pure and registered. Results showed that open days and calving interval tended to be decrease during these years, averagely

---

Effect of the first lactation age on longevity of Holstein cattle

Zakizadeh, S. and Asadi, I., Hasheminejad High Education Center, Animal Science, Kalantari Highway, 9176994767, Iran; sonia_zaki@yahoo.com

Reproductive performance has also been found to deteriorate as milk yield increased. Reduction in reproductive performance could affect culling rates and herd life and reduce the genetic gain from primary traits. Fertility, which is often measured by age at first calving, open days, calving interval, or number of services, is an important measure of reproductive performance. Conformation traits have also associated with herd life. Udder traits had largest absolute genetic correlations with herd-life traits, followed by body traits and feet and leg traits. The objective of this study was to investigate the effect of first lactation age on longevity in Yazd province of Iran. 1493 true herd life records used for the study were from 5 herds in Yazd province in Iran between 1992- 2004. The model included herd-year-season of the first lactation, parity, age at first lactation, as fixed effects. Results showed all independent factors had highly significant effect on longevity except the season of first lactation. Average longevity calculated and it was about 64 months in all herds. Herd factor had significant effect on longevity because of separate managements. Longevity decreased during 13 years. Year effect could be mainly based on climate and availability of feeding stuffs or different management decisions during these years. Longevity showed a positive linear relation with parity. Culling rates were 22.2 for the first lactation to 11.2 for the sixth lactation. The highest culling rates related to the first and second lactation. Age at the first of lactation had also significant effect on longevity. Although the positive linear relation between increasing of age at first lactation and higher longevity observed, but calving interval has a relatively high economic weight and a reduction in CI could be described as one of the outcomes of improved fertility. Furthermore, CI is open to management bias such as decisions to extend the lactation length of individual high-yielding cows within herds.

Session 31 Poster 17

The effect of seasonal and herd size on milk somatic cell counts in dairy cattle farms in Tehran province

Jamali, N., Moeini, M.M. and Souri, M., razi university, animal science, imam khomeini St. college of agriculture, animal science dep., 67155, Iran; mmoeini2008@gmail.com

Total bacterial count or Somatic Cell Count (SCC) is one of the biggest problems of dairy farmers whom had a contract with dairy factories. The effects of season and herd size on cow milk SCC were determined in dairy farms from January 2009 to December 2009 in Tehran province. The limit of SCC in this survey was 300,000 cells per Ml row milk. Data were recorded in six industrials and six traditional dairy farms. The mean of bulk milk tank and individual cow milk SCC recorded. The GLM methods of SAS statistical software was used for data analyzing. The results indicated that the mean milk SCC in the industrial dairy farm was 281,800±11,200 and in traditional dairy farm was obtained 10,000±7,020 cell/Ml. A general liner models was significant for fixed effect of herd size on mean of SCC ($P<0.05$). It was also same order for quarter milk samples (QMS) too that the highest infected cows were belonging to fall. The SCC was higher in spring than in fall but it was not different significantly ($P>0.05$). The SCC was lowers than our previous studies. It is concluded that the control of environmental mastitis is best achieved by using proper housing system, clean sand and bedding, dry cow treatment, post milking treat disinfection and antibiotic treatment of clinical mastitis, clipping and adequate ventilation also overcrowding should be avoided.

Session 31 Poster 18

Comparative assessment of Fumonisin B1 and hydrogen peroxide *in vitro* exposure on bovine lymphocytes proliferative response

Danieli, P.P., Catalani, E., Lacetera, N. and Ronchi, B., University of Tuscia, Dept. of Animal Science, Faculty of Agriculture, Via S. C. de Lellis snc, 01100, Italy; danieli@unitus.it

Fumonisin B_1 (FB_1) in feed and food has been associated with various diseases in animals and humans. FB_1 exposure is thought to cause impairment of the immune response. A possible mechanism trough which FB_1 exerts its immune-modulating effects, has been identified in the oxidative stress induction. The aim of the present *in vitro* study was to assess the effect of FB_1 on proliferation of peripheral blood mononuclear cells (PBMC) compared to that exerted by the reference oxidant hydogen peroxide (H_2O_2). Three healthy female Holstein calves (301±5 days of age) were used as blood donors. After isolation, PBMC (1 x 106 cells/ml) were incubate at 39 °C in an atmosphere of 5% CO_2 in the presence of 70 µg/ml of FB_1, 100 µM of H_2O_2 or 0.05% DMSO (negative control) for 2 and 7 days. At the end of the exposures, proliferative response of PBMC to mytogens (ConA, LPS, and PWM) was evaluated after further 66 hours incubation. Data were analysed by a full factorial ANOVA and differences were declared significant at $P<0.05$. After 2 days of exposure, FB_1 and H_2O_2 exerted a strong depressive effect ($P<0.01$) on proliferative response of PBMC using PWM and ConA as mytogens, while a moderate ($P<0.05$) to high ($P<0.01$) PBMC stimulation was observed with LPS. After 7 days, only FB_1 treated PBMC did not exhibit a proliferative response significantly different from the control, using ConA and PWM as mytogens. PBMC exposed *in vitro* to FB_1 and H_2O_2 exhibited a time and mytogen dependent proliferative response. Further studies are in progress to ascertain the involvement of oxidative-stress mechanism following the FB_1 exposure and to explain the observed cellular mytogen-dependent response.

Session 31 Poster 19

Effect of organic acids on Arcobacter butzleri attached to broiler skin

Skrivanova, E.[1], Molatova, Z.[1], Skrivanova, V.[1], Houf, K.[2] and Marounek, M.[1], [1]Institute of Animal Science, Pratelstvi 815, 10401 Prague, Czech Republic, [2]Ghent University, Faculty of Veterinary Medicine, Salisburylaan 133, B-9820 Merelbeke, Belgium; skrivanova.eva@vuzv.cz

Arcobacters have been considered as potential zoonotic food-borne and water-borne pathogens. Chicken meat is the main source of human infection with Arcobacter. It has been assumed that contamination of meat occurs during processing. The aim of this study was first to evaluate the antimicrobial activity of 17 organic acids on Arcobacter butzleri CCM 7051. The antibacterial activity of organic acids against A. butzleri was determined by real-time PCR and expressed as IC_{50} (a concentration which caused the 50% reduction of the target sequence synthesis). Acids with $IC_{50} < 1$ mg/ml were further tested for their effect against A. butzleri attached to broiler skin to assess their potential use as the food preservatives. Chilled broiler carcasses were inoculated at the surface with A. butzleri (6 Log_{10} CFU/ml) and treated with 5 mg/ml of organic acids for 1 min. Surviving bacteria were enumerated by the plating technique on the day 0, 1, 2 and 3 after the treatment. Treatments causing the highest microbial reductions were chosen for sensory evaluation. Results were analyzed using the SAS statistical program, procedure GLM followed by the Tukey´s test. Arcobacter butzleri was sensitive to all organic acids tested. The IC_{50} of 8 out of the 17 acids was bellow 1 mg/ml. The least effective was palmitic acid, with IC_{50} of 2.6 mg/ml. Evaluation of organic acids on their inhibitory effect against A. butzleri attached to broiler skin showed that all 8 acids were effective. The highest antibacterial activity was observed for the benzoic, citric and malic acid. Subsequent sensory analysis revealed benzoic acid as the most suitable organic acid for the chicken skin treatment. In conclusion, organic acids were shown to have a strong antibacterial activity against A. butzleri suggesting their potencial use to decrease the surface contamination of broiler skin. Supported by MZe 0002701404.

Session 32 Theatre 1

Postprandial evolution of the oxidative status in plasma of sheep

Wullepit, N.[1,2], Ginneberge, C.[2], Fremaut, D.[2] and De Smet, S.[1], [1]Ghent University, Lanupro, Proefhoevestraat 10, 9090 Melle, Belgium, [2]University College Ghent, Department of Animal Production, Schoonmeersstraat 52, 9000 Ghent, Belgium; nicolas.wullepit@hogent.be

The aim of the present study was to investigate what, if any, postprandial changes occur in the oxidative status in plasma of sheep. Three young male Zwartbles sheep were used and penned individually. They were fed at maintenance 800 g grass hay and 400 g concentrate per day in two equal portions at 09:00 and at 18:00 h. This resulted in a daily supply per animal of 0.06 g α-tocopherol. After two weeks of adaptation, blood samples were collected at 0.5 h before, and 0.5, 1, 2, 3, 5 and 8 h after the morning meal and 0.5 h after the evening meal. This was repeated per sheep on three separate days. The following oxidative status parameters were measured in plasma: ferric reducing ability of plasma (FRAP), α-tocopherol level, glutathione peroxidase activity (GSH-Px) and thiobarbituric acid reactive substances (TBARS, measure of lipid oxidation products). Data were analysed with a general linear model including sheep as random effect and time as fixed effect, and using Bonferonni as post hoc comparison of means test. FRAP values, plasma α-tocopherol concentrations and plasma GSH-Px activity were unaffected by time relative to feeding. Only TBARS were significantly affected by time postprandial. The mean TBARS level increased after the morning meal with a peak at 2 hours after feeding. Thereafter it decreased and returned to pre-feeding levels, and was again significantly higher after the evening meal. In conclusion, there were no short term postprandial effects on FRAP, α-tocopherol and GSH-Px in sheep plasma. On the contrary, TBARS levels did experience short term postprandial effects. The results indicate increased lipid peroxidation in plasma after feeding, probably due to higher blood fatty acid levels. Hence, sampling time relative to feeding should be taken into account and standardised in assessing plasma TBARS.

Metabolic hormones during the peripubertal period in sheep

Chadio, S., Pazalos, A., Kotsampasi, B., Kalogiannis, D., Deligeorgis, S. and Menegatos, I., Agricultural University of Athens, Anatomy and Physiology of Domestic Animlas, 75 Iera odos, 11855 Athens, Greece; shad@aua.gr

Aim of the study was to characterize the changes in certain metabolic hormones levels during the peribubertal period and to determine their role as possible mediators between energy status and onset of puberty in sheep. Seven male and six female lambs of the Chios breed were used. Blood samples were withdrawn at two week intervals between 105 and 300 days of age and plasma concentrations of leptin, insulin, IGF-1 and T3 were measured by radioimmunoassay. Puberty was defined as the first increase ≥ 1 ng/ml, followed by at least two more consecutive high values in plasma testosterone or progesterone concentration, measured twice weekly. Data were normalized to the date of puberty (day 0) and subjected to ANOVA. Pearson correlation coefficients were used to evaluate hormone associations. Male lambs showed a significant increase in all hormones at 30 days before day 0. Insulin and leptin levels were further increased and remained elevated up to 45 days after day 0. IGF-1 and T3 levels remained stable throughout the whole period. In female lambs leptin concentrations increased gradually reaching a peak at 30 days before day 0 and remained elevated thereafter. Insulin concentrations increased 45 days before and decreased 90 days after day 0, while IGF-1 and T3 showed subtle fluctuations over the whole period. Before day 0 a positive correlation was detected between leptin-insulin, leptin-IGF-1, insulin-IGF-1 and insulin-T3 in males and between leptin - insulin in females. After day 0 males continued to exhibit a positive correlation between leptin-insulin and leptin-IGF-1, while in females these correlations were less apparent. In conclusion these results indicate that metabolic hormones levels change during the peripubertal period. In particular the similar plasma profiles of leptin and insulin along with their correlation detected before day 0 implies that these hormones may act as permissive factors for the initiation of puberty in sheep.

Characterization of gastric intralumenal pH and development of the proventricular deep glands during perinatal period in ostrich

Duritis, I. and Mugurevics, A., Faculty of Veterinary Medicine, Preclinical Department, Kr.Helmana - 8, Jelgava, LV-3004, Latvia; ilmars.duritis@llu.lv

In ostrich, deep glands (gll. proventriculares profundi) that produce gastric juice are localized to a specific region within proventricular mucosa. The goal of this study was to determine gastric intralumenal pH and development of deep glands in ostrich chicks during perinatal ontogeny. Six embryos (38th day of development) and 37 chickens (neonatal; 3; 7; 14; 30; and 60 days old) of both sexes from African ostrich farm in Latvia were used. Intralumenal pH was measured in the proventriculus above deep glandular region and within ventriculus. Samples were collected from proventricular deep glandular region for histological characterization using sections stained with H&E and PAS. Results were analyzed with ANOVA Post Hoc test and Pearson correlation analysis (SPSS 11.5). Proventricular pH was significantly higher ($P<0.05$) on the 38th day of embryonal development (3.93±0.51) than on the day of hatching (2.47±0.11). Similarly, ventricular pH on the 38th day of embryonal was significantly higher (4.56±0.41; $P<0.05$) than ventricular pH in 14 day of chicks (3.00±0.21). There was positive correlation between pH in the proventriculus and ventriculus ($r=0.79$; $P<0.01$) and negative correlation between gastric pH and age of chickens ($r= -0.64$; $P<0.01$). On the 38th day of embryonal development deep glandular cells were few and they formed indistinct tubules in the periphery of glandular lobules. Beginning from 7 days of age tubules were well developed with wide lumen and decreased amount of connective tissue between groups of glands and lobules. In 30 day old chickens glandular histology resembled that of adult ostrich. Gastric lumenal acidity indicates that hydrochloric acid is secreted in the deep glands of proventriculus as early as 38th day of embryonal development and intensity of secretion increases beginning from hatching day as evidenced by decreased pH. The development of deep glandular epithelium occurs at accelerated rate between 3rd and 7th day of age.

Impact of dietary omega-3 fatty acids on sperm quality parameters in the Holstein bulls

Gholami, H.[1], Chamani, M.[1], Towhidi, A.[2] and Fazeli, M.H.[3], [1]Science and Research Branch, Islamic Azad University, Department of Animal Science, Tehran- Iran, 3136935531-Karaj, Iran, [2]University of Tehran, Department of Animal Science, Karaj- Iran, 3136935531-Karaj, Iran, [3]Islamic Azad University of Shahrekord, Department of Veterinary Science, Shahrekord- Iran, 3136935531-Karaj, Iran; gholami_hd@yahoo.com

The aim of current study was to investigate the effect of feeding an omega-3-enriched nutriceutical on the quality of fresh bull semen. Samples were obtained from nineteen Holstein bulls (n=10 per control and 9 per treatment group) used for semen collection at Semen Production Center, Karaj, Iran. Control group were fed a standard concentrate feed while treatment group bulls had this standard feed top dressed with 100 g of a commercially available nutriceutical (Optomega 50, Optivide International Limited, Nottinghamshire, UK). The nutriceutical was composed of salmon oil and a carrier with a total of 25% omega-3 fatty acids. Semen quality was assessed on ejaculates collected after 10 weeks of supplementation. Computer-assisted assessment of sperm motility (Hamilton Thorne Biosciences, Beverly, MA, USA), viability (eosin-nigrosin) and hypo-osmotic swelling test (HOST) were conducted. Dietary supplementation of omega-3 enriched nutriceutical for 10 weeks was indeed found to affect sperm motility parameters. The treatment significantly increased total motility, progressive motility, HOST, average path velocity and straightness in the fresh semen of bulls (P<0.05). However dietary supplementation did not significantly affect viability, straight-line velocity, curvilinear velocity, amplitude of lateral head displacement and linearity, but there was an increasing tendency in the viability and sperm velocities evaluated (VSL and VCL). Consequently, it is concluded that dietary omega-3 treatments result in improvements of fresh semen quality parameters in Holstein bulls.

Buffalo heifers: body measurements and hormonal values at the onset of puberty

Terzano, G.M., Maschio, M., Razzano, M., Mazzi, M., Napolitano, F. and Catillo, G., Animal Production Research Centre, Animal Production, Via Salaria, 31, 00015 Monterotondo, Rome, Italy; giuseppinamaria.terzano@entecra.it

Twenty six prepubertal 8-mo-old buffalo heifers weighing a mean of 134 kg were used to determine some body measurements and hormonal values around the pubertal period. Thirteen heifers were bred in intensive feeding (IF), 13 heifers were bred on pasture (PS). Starting from the 13th till to 22th month of age, blood samples were collected: every 10 days to assay the concentration of plasma progesterone (P4); monthly to assay the plasma concentrations of leptin and insulin –like growth factor-I (IGF-1); every 20 days to assay the plasma concentrations of inhibin-A. At puberty, physical parameters (body weight, withers height and body lenght) were measured. Heifers achieved puberty when plasma P4 level exceeded 1 ng/ml. Data were analyzed by linear model with fixed effect of rearing systems (SAS/STAT). The puberty occurred at the same mean age in both groups (599 vs 610 days, for IF and PS, respectively) but rearing systems significantly affected puberty weight (P<0.05) and ADG (P<0.001), with about 250 g/d more in IF. At puberty, the average values of body weight and body mass index ((body weight(kg)/withers height(cm)/body length(cm)x103) were 462 kg - 29.15 in IF and 375 kg -23.64 in PS (P<0.05), respectively. Plasma leptin concentrations were mantained at lower levels until the onset of puberty in PS than in IF (2.80 vs 5.56 ng/ml, P<0.0001, respectively at puberty). Plasma IGF-1 concentrations exhibited a curvilinear profile relative to the onset of puberty and the values at puberty were 47.9 and 27.6 ng/ml in IF and PS, (P<0.03), respectively. Plasma Inhibin-A concentrations began to gradually increase four weeks before the onset of puberty in IF and PS (18.83 and 13.07 ng/ml, respectively at puberty) and this increase continued throughout the peripuberal period. These results imply that, in spite of the differences in all the studied parameters in IF and PS, Inhibin-A can be used as a novel marker of the onset of puberty in buffalo heifers.

Session 32 Poster 6

Heat stress affects adipokines secretion in mice

Morera, P., Basiricò, L., Papeschi, C., Lacetera, N. and Bernabucci, U., University of Tuscia, DIPA, via S.C. de Lellis, s.n.c, 01100 Viterbo, Italy; bernab@unitus.it

In a recent study, we suggested that heat shock inducing changes in the biology of adipokines may be a cofactor involved in acclimation of energy metabolism in animals exposed to hot environment.The aim of this *in vivo* study was to verify the effect of heat stress (HS) on leptin and adiponectin secretion and the possible role of HS in modulating glucose metabolism in HSed mice. Forty-eight male mice (Mus musculus) 7 weeks old were used. After an adaptation period of 14 days under thermoneutral (TN) conditions (20-24 °C, 50-60% relative humidity), animals were subdivided into two groups each of 24. One group was maintained under TN, the other was exposed to 35 °C ambient temperature (HS). The experimental period lasted 7 days. At the beginning and after 2, 5 and 7 days of the experimental period blood samples were taken and 6 animals of each group were used to perform a glucose tolerance test (GTT). Plasma concentration of leptin, adiponectin and insulin were determined by an indirect ELISA using commercial kits. The GTT was carried out on 15 hour starved animals. Mice received a glucose solution by intraperitonal injection (2 g/kg body weight) and blood glucose was determined at 0, 30, 60, 90 and 120 minutes after injection. Data were analyzed using the MIXED procedure by the Statistica-7 software package. Compared with TN group, plasma leptin, adiponectin and insulin levels were higher ($P<0.001$) in the HSed group after 2, 5 and 7 days of exposure. The GTT after different time of exposure (2 or 5 days) to 35 °C revealed lower ($P<0.05$) plasma glucose at 30, 60 and 90 minutes in HSed mice compared to their TN counterparts. This *in vivo* study provides evidence about a direct effect of HS on adiponectin and leptin secretion, suggesting a modulating role of these adipokynes in the metabolic adaptive responses observed in subjects exposed to hot environment. Data also indicate that heat stress is responsible for an increase in insulin sensitivity. The research was financially supported by the PRIN-Project.

Session 32 Poster 7

Influence of orally administered bovine lactoferrin on metabolic and hormonal disturbances in lipopolysaccharide-injected calves

Kushibiki, S., Shingu, H. and Moriya, N., National Institute of Livestock and Grassland Science, ikenodai-2, tsukuba, ibaraki, 305-0901, Japan; mendoza@affrc.go.jp

Lactoferrin (LF) is known to have a role in iron absorption and is believed to be an important component of host defense. Therefore, we report the modulatory effect of LF on the plasma concentrations of tumor necrosis factor-alpha (TNF), metabolites, and hormones in the lipopolysaccharide (LPS)-induced inflammatory response in preruminant calves. Thirty clinically healthy Holstein male calves were used in this study. At 4 days of age (day-10), each calf was randomly assigned to one of three treatment groups (n=10), matched to body weight. The treatments were the oral administration of LF 1 g/day, 3 g/day, or 10 ml of saline/day (control) for 10 days (day-10 to day-1). The LF or saline was added to whole milk daily for the LF or control group, respectively. The day after the end of LF feeding for 10 days (0 day), the calves received an i.v. injection of LPS (50 ng/kg BW, *E. coli* 055:B5) at 09.00 h. Plasma TNF concentrations at 2 h after LPS treatment were lower ($P<0.05$) in LF 1 g/day-fed claves compared with LF 0 g/day (control) calves. In LF groups, plasma haptoglobin concentrations slightly increased after LPS injection, but those levels at 6-24 h were lower ($P<0.05$) than in the control group. The LF treatment inhibited ($P<0.05$) the reduction of plasma ferrin concentration in calves following LPS challenge. The concentration of plasma insulin-like growth factor-1 (IGF-1) in all groups was decreased by LPS treatment while, in LF groups, the IGF-1 level was higher ($P<0.05$) than in the control group. Plasma insulin concentration in LF groups was lower ($P<0.05$) than in control calves at 2 h after LPS injection. These data suggest that LF has a substantial anti-inflammatory effect on the modulation of the host defense system in preruminant calves.

Session 32 Poster 8

Effect of n-3 fatty acids and vitamin E on freezing ability of goat semen

Towhidi, A.[1], Ansari, M.[1], Moradi Shahre Babak, M.[1] and Bahreini, M.[2], [1]University of Tehran, Department of Animal Science, Karaj, 3158777871, Iran, [2]Animal Breeding Center of Iran, Department of Biotechnology, Karaj, 31585963, Iran; atowhidi@ut.ac.ir

The aim of this study was to investigate the effect of adding a source of n-3 fatty acids accompanied by vitamin E to an egg-yolk free extender (Bioxcell) on freezing ability of goat semen. Six mature Mahabadi bucks were selected for semen collection using an artificial vagina. Semen samples were collected and pooled after quality evaluation. Semen sample (a total of 6 ml) was divided into 12 groups consisting of 4 n-3 fatty acid (0, 0.1, 1, 10 ng/ml) and 3 Vitamin E (0. 0.1, 0.2 mM,VE) levels. The semen characteristics including the percentage of motility, progressive motility, viability and abnormality were evaluated at 37 °C and after thawing. Recovery rate was also calculated. Data analysis was done using factorial design and GLM procedure in SAS program. Percentage of motility in 10 ng/ml n-3 FA, 0.2 mM VE and 0.1 ng/ml n-3 FA, 0.2 mM VE treatments after thawing was significantly higher than that the other groups. Percentage of progressive motility was significantly higher in 10 ng/ml n-3 FA, 0.2 mM VE treatment than that other treatment groups. The 10 ng/ml FA, 0.1 mM VE treatment had higher viability than other groups. Before freezing, the 10 ng/ml n-3 FA, 0.1 mM VE and after thawing the 1 ng/ml n-3 FA, 0.1 mM VE groups had the lowest percentage of abnormality compared to the other groups. The 10 ng/ml n-3 FA, 0.2 mM VE and 0.1 ng/ml n-3 FA, 0.2 mM VE treatments had significantly higher recovery rate compared to other treatments. Collectivelly, the treated group that received 0.2 mM vitamin E and 10 ng/ml n-3 fatty acid had the best semen parameters after thawing in comparison with the other treated groups. Results suggested that adding n-3 fatty acids accompanied by an antioxidant to extender Bioxcell could improve freezing ability of goat sperm cell.

Session 32 Poster 9

Effects of caponization on meat quality, lipid composition and selected physiological characteristics of broilers and male layers

Symeon, G.[1], Bizelis, I.[1], Mantis, F.[2], Sinanoglou, V.[2], Laliotis, G.[1], Charismiadou, M.[1] and Rogdakis, E.[1], [1]Agricultural University of Athens, Animal Breeding and Husbandry, 75, Iera Odos, GR 188 55, Athens, Greece, [2]Technological Educational Institution of Athens, Food Technology, EGALEO, GR 122 10, Greece; jmpiz@aua.gr

The aim of the study was the evaluation of the effects of caponization on meat quality, lipid composition and other physiological characteristics of broilers and male offspring of layer breeds. Three experiments were conducted. In the first two, a medium growth broiler hybrid (Redbro) was used while in the third experiment, males from laying breeds (Lohmann Silver) were used. For the evaluation of breast meat quality pH24, color, cook loss, shear values and intramuscular fat were measured along with lipid composition of intramuscular and abdominal fat. Also, abdominal adipose tissue cellularity was measured along with NADP-enzymes activity in the liver. Serum concentrations of cholesterol, triglycericides, HDL-cholesterol and testosterone were also evaluated. In addition, the expression of the NADP-enzymes was measured. The data were analyzed using GLM procedures. Meat quality was affected by the caponization mainly in terms of colour and shear values. In capons, adipocytes volume was greater in comparison to intact males but no difference was observed for adipocytes number. Caponization did not significantly affect the individual NL and PhL profile and content in intramuscular tissue, except for TL and TG content, but resulted in a significant ω-6/ω-3, PUFA and PUFA/SFA ratio reduction and a significant atherogenic and thrombogenic indices increase in intramuscular fat. A tendency for increased serum lipoproteins concentrations was observed due to caponization while malate dehydrogenase activity was increased. The respective gene expression followed a similar trend. An opposite trend followed the isocitrate gene expression in capons layers Conclusively, caponization could be applied in order to produce chicken meat of 'special quality'

Efficiency of guar gum, pectin and wheat bran for reduction of oxidative stress in laboratory rats

Trebušak, T., Rezar, V., Levart, A., Frankič, T., Salobir, J., Orešnik, A. and Pirman, T., University of Ljubljana, Biotechnical faculty, Department of Animal Science, Groblje 3, 1230 Domžale, Slovenia; tina.trebusak@bfro.uni-lj.si

The objective of the present study was to determine the effect of different sources of dietary fibre on oxidative stress induced by a high fat diet in laboratory rats. Thirty two laboratory rats were penned individually and divided into four groups: CONT (high fat diet), G (70 g/kg guar gum), P (70 g/kg apple pectin) and WB (155 g/kg wheat bran). After 11 or 13 days of treatment DNA damage of blood leukocytes was measured by Comet assay and lipid peroxidation was studied by malondialdehyde (MDA) concentration in liver and in urine. In comparison with the CONT group, the degree of DNA damage was significantly lower in the WB group. G and P group also reduce DNA damage but not significantly. Similar results were also obtained in the liver MDA concentration. All three studied groups reduced the liver MDA concentration but only WB significantly. In comparison with the CONT group, the WB and P group significantly reduced the 24-hour MDA excretion, but not the G group. The results of the experiments confirmed that a high wheat bran intake effectively reduced oxidative stress induced by high fat diet. We can also conclude that pectin and guar gum may reduce oxidative stress but less effectively.

Seasonal effects on the quality of stallion semen

Siukscius, A., Pileckas, V., Urbsys, A., Kutra, J. and Nainiene, R., Lithuanian Veterinary Acadamy, Instititute of Animal Science, R.Zebenkos 12, Baisogal, Radviliskis district, LT-82317, Lithuania; arturas@lgi.lt

A study was carried at the Animal Reproduction Department of the LVA Institute of Animal Science and Vilnius Stud to determine the effects of the season on the physiological responses of fresh and preserved stallion semen. Qualitative and morphological parameters of fresh diluted and cryopreserved stallion semen were estimated. Semen freezing qualities were determined using stallion semen that met the requirements for fresh semen, i.e. the semen with the sperm motility not lower than 6 points (60%) and concentration not lower than 0.15 mlrd./cm^3. The study indicated that the season had no significant influence on the pH-value of semen, yet influenced the ejaculate volume and sperm concentration. Semen pH values ranged from 7.28 to 7.21 throughout the year, however, these values corresponded to the physiological standard (6.8-7.7). The ejaculate volume was lowest in autumn (approx. 20.9 cm^3), but in other seasons it remained comparatively stable and ranged from 27.4 to 27.8 cm3. The lowest sperm concentration in fresh semen was determined from June to August. Sperm motility was lower in spring (38%) if compared with other seasons (54-59%). Visual estimation indicated that the number of normal spermatozoa accounted for approximately 60-65%. There was a significant difference in the number of spermatozoa with normal heads in different seasons: 21% in summer and 30-33% in other seasons. The season had no influence either on the number of live spermatozoa in a thawed semen dose (53-57 million) or on the post-thawed sperm motility (20-23% spermatozoa with progressive movement).

Evaluation of semen physiological parameters and fertility of the boars used in Lithuania

Siukscius, A., Nainiene, R., Kutra, J., Pileckas, V. and Urbsys, A., Lithuanian Veterinary Academy, Institute of Animal Science, R.Zebenkos 12, Baisogala, Radviliskis district, LT-82317, Lithuania

The study was designed to analyze the reproductive traits of various pig breeds kept in Lithuania with respect to fertilization and litter size. Besides, semen usage and fertilization data from separate boar breeds have been collected and the physiological parameters of semen and its suitability for cryopreservation determined. Semen usage analysis indicated that the usage of Large White boars semen was the highest and accounted for 31.1% (P<0.05). Pjetrain boars were second in this respect. The analysis of the reproductive traits indicated that the fertilization rate of pigs breeds in Lithuania accounts for on average 64.4%, the lowest pig fertilization rate being 44.4% and the highest up to 93.65%. The average litter size of the sows in Lithuania is 10.6±0.25 piglets. The study of the physiological parameters of semen indicated that the semen of Large White boars was most qualitative (ejaculate volume was 315.0±45.32 ml, concentration -0.24±0.09 milliard/ml. initial sperm motility -79.3±4.87%, pathologic spermatozoa count only 20.1±5.62% (P<0.05).

Calving body condition affects somatotropic axis response and ovarian cyclicity in beef cows

Álvarez-Rodríguez, J.[1], Tamanini, C.[2] and Sanz, A.[1], [1]CITA, Av. Montañana 930, 50059 Zaragoza, Spain, [2]Università di Bologna, Via Tolara di Sopra 50, 40064 Ozzano Emilia, Italy; asanz@aragon.es

The effect of maternal body reserves on somatotropic axis response and ovarian cyclicity were studied in two beef cattle breeds, Parda de Montaña (PA) (n=68) and Pirenaica (PI) (n=28). After calving, cows were stratified by body condition score (BC) (>2.5, [2.6-2.9] or ≤2.5, [2.2-2.5]) and fed a diet meeting maintenance requirements. Weekly live-weight (LW) and monthly milk yield were recorded (oxytocin technique). Blood samples were assayed for progesterone (twice weekly) and growth hormone (GH) (weekly) by RIA and for insulin-like growth factor-I (IGF-I) (fortnightly) by chemiluminiscence. The interactions among effects were not significant (P>0.10). Calving LW was greater in BC>2.5 than in BC≤2.5 cows (605 vs. 521 kg, P<0.001), but cow daily gain did not differ between groups (-0.10 kg/d, P>0.10). Birth calf LW was not affected by dam BC (40.9 kg, P>0.10) but calf daily gains were greater in BC>2.5 than in BC≤2.5 cows (0.83 vs. 0.74 kg/d, P=0.01). Energy-corrected milk yield (ECM) was greater in BC>2.5 than in BC≤2.5 cows (8.3 vs. 7.1 kg/d, P=0.02). Breed did not affect cow-calf LW or gains (P>0.10), but ECM was greater in PA than in PI (8.5 vs. 6.9 kg/d, P=0.01). ECM was greater throughout the first and second month post-partum than subsequently (8.5 and 7.8 vs. 6.7 kg/d, P<0.05). Post-partum anoestrus length was shorter in BCS>2.5 than in BCS≤2.5 cows (54 vs. 76 d, P<0.001). GH plasma levels were lower (2.05 vs. 2.53 ng/ml, P=0.01) whereas IGF-I concentration was higher (64 vs. 55 ng/ml, P=0.04) in BC>2.5 than in BC≤2.5 cows. Peripheral GH levels were highest at weeks 1 and 8 post-partum (2.77 vs. 2.20 ng/ml, P<0.05) whereas IGF-I concentration did not vary during lactation (60 ng/ml, P>0.10). Neither GH-IGF-I axis activity nor luteal function differed between breeds. It is likely that previous nutritional status influenced post-partum ovarian cyclicity through somatotropic axis response during lactation.

Effects of polyunsaturated fatty acids and prostaglandins on activation of the matrix metalloproteinase of placental fibroblast cells

Kamada, H.[1], Hayashi, M.[1] and Kadokawa, H.[2], [1]National Institute of Livestock and Grassland Science, Molecular Nutrition, Ikenodai-2, Tsukuba-shi, Ibaraki, 305-0901, Japan, [2]Yamaguchi University, Animal Reproduction, 1677-1 Yoshida, Yamaguchi-shi, Yamaguchi, 753-8511, Japan; kama8@affrc.go.jp

In order to obtain information about the mechanism of placenta discharge after delivery, we used an *in vitro* cell culture system. Our previous experiment showed that arachidonic acid (Ara) and its 12-LOX metabolites (12-oxoETE) activate the matrix metalloproteinase (MMP, a possible mediator of placenta separation at delivery) of placental fibroblast cells (*in vivo* source of MMP-2). In this experiment, the effects of other polyunsaturated acids (PUFA) and prostaglandins on the MMP of fibroblasts were investigated. Fibroblast cells were prepared from cotyledon tissue of pregnant cows. 200000 viable cells were seeded per well in medium 199 containing 10% fetal calf serum (FCS). After monolayer formation, FCS was removed, and test fatty acids were added. Saturated, mono- and di- unsaturated fatty acids had no activity, while fatty acids containing more than three double bonds showed cell exfoliation activity (MMP activation). These activities were observed in not only n-6 but also n-3 series of PUFA. The insertion of OH or OOH residues in this region resulted in the loss of activation activity, thus, the polyunsaturated structure in the carbon chain is essential for MMP activation. In addition, the activation of MMP by 12-oxoETE was inhibited by prostaglandin E2, but not F2α. This work was supported by the Programme for Promotion of Basic and Applied Researches for Innovations in Bio-oriented Industry (2008-2010)

A simulation model of dairy cows' herd with focus on the information system (SITEL)

Brun-Lafleur, L.[1,2,3], Rellier, J.P.[4], Martin-Clouaire, R.[4] and Faverdin, P.[1,2], [1]INRA, UMR1080 Production du Lait, 35590 St-Gilles, France, [2]Agrocampus-Ouest, UMR1080 Production du Lait, 35000 Rennes, France, [3]Institut de l Elevage, Monvoisin, 35652 Le Rheu, France, [4]INRA, UR875, Biométrie et Intelligence Artificielle, France; laure.brun@rennes.inra.fr

In an instable economic context, farmers need tools to anticipate the consequences of their management decisions. In practice, those decisions are taken from the incomplete, imperfect and delayed knowledge the farmer has of his herd. Currently, numerous technical solutions appear to increase information availability on dairy herd, but their benefit on herd performances is difficult to estimate. Our objective was to assess the effect of existence, dynamics and quality of information acquired by the farmer to take decisions on herd performances using a modelling approach. A dynamic, stochastic, individual-based herd simulation model named SITEL was built. SITEL is composed of a biotechnical subsystem (the herd), a decision and operating subsystem (the farmer), and an informational subsystem (the knowledge the farmer has of the herd). The herd is composed by individuals, mainly cows and heifers described by reproduction with stochastic events and lactation processes. The farmer takes management decisions according to his own strategy and the knowledge available in the informational subsystem. The farmer realizes two types of activities. First, information activities are used to complete the informational subsystem from the biotechnical one. Those are imperfect diagnosis or measurement activities, like oestrus monitoring or pregnancy diagnosis, but also discrete quantitative measurements (milk yield or composition). Second, technical activities are used to modify the biotechnical subsystem, like drying off, breeding or selling animals. SITEL makes it possible not only to analyse complex interactions between biophysical and decisional processes, but also to study the importance or limits of information quality on herd management and performances according to the farmers' objectives.

Session 33 Theatre 2

Analysis of stockbreeder hierarchies within the livestock farming domains (cattle and sheep)

Ingrand, S.[1]*, Pailleux, J.Y.*[1]*, Poupart, S.*[1]*, Magne, M.A.*[2] *and Mugnier, S.*[1]*,* [1]*INRA, SAD, UMR 1273 Metafort, 63122 Saint-Genes Champanelle, France,* [2]*ENFA, 2 route de Narbonne BP 22687, 31326 Castanet Tolosan Cedex, France; stephane.ingrand@clermont.inra.fr*

The objective in this study was to analyze the relative importance given by farmers to the livestock farming domains (LFD). Surveys were carried out in 30 meat sheep and 30 beef cattle farms. The farmers had to classify 9 LFD according to 3 criteria: the attractiveness, the importance for the survival of the farm and the satisfaction about the results obtained. The 9 LFD proposed were: composition of the herd, breeding, feeding, grazing, production of stored forage, production of concentrates, accounts, marketing, health management. Two LFD come systematically at the top: 'herd' and 'breeding'. The range for 'herd' is very close to 1 with sheep farmers for the 'attractiveness' criterion (specificity of the emotional side with this species). The best and the least well classified domains are quite similar in sheep and cattle. Worth noting is the desire in both species to progress on 'concentrates' and 'health'. The greatest difference between the two species is in the domains that are the best classified for the 'attractiveness' criterion: directed towards the animal in sheep farming (herd, breeding, feeding), and more on resource with cattle (forage, grazing). The least dispersed ranks are observed for the satisfaction criterion for both species. The factorial analysis makes a clear distinction between 4 situations: the farmers who place the animals in the forefront (herd, breeding) versus the farmers who put forward the resource (grazing, concentrates, forage); the farmers who place economics in the forefront (accounts, marketing) versus those who stress feeding and health management. The technical domains still have a high priority for farmers. The least well classified domains on the 'satisfaction' criterion could constitute priorities for advice. The work follow-up consists of connecting these results with the ways in which information is collected for each of the domains.

Session 33 Theatre 3

Effect of dehydrated alfalfa supplementation on milk fatty acid composition: a farm survey

Ballard, V.[1]*, Couvreur, S.*[2] *and Hurtaud, C.*[3]*,* [1]*Coopedom, 11 rue Louis Raison, 35113 Domagné, France,* [2]*Groupe ESA, Unité de Recherches sur les Systèmes d Elevage, 55 rue Rabelais, 49007 Angers, France,* [3]*INRA, UMR 1080 Production du Lait, Domaine de la prise, 35590 Saint-Gilles, France; s.couvreur@groupe-esa.com*

Nowadays, consumers ask for a higher nutritional value of milk fat. The effect of dehydrated alfalfa (DA) supplementation on milk fatty acid composition (FA) is poorly known. The aim of this study was to characterize, at the farm level, the effect of DA supplementation in different diets on milk FA level. A survey has been led on a network of 30 dairy farms. The farms were divided in 6 groups according to the characteristics of the diet: (i) maize+grass silage based diet, (ii) maize+grass silage supplemented with DA, (iii) maize+grass silage supplemented with DA and linseed, (iv) maize silage supplemented with DA, (v) maize silage supplemented with dehydrated grass and DA and (vi) DA and dehydrated maize. Farms were surveyed twice in winter 2009. At each visit, the diet composition (thanks to a closed questionnaire) and a milk sample (representative of one or two days of milking) were collected. Statistical analyses were made according to diet groups. The study showed that a DA supplementation in a maize (+grass) silage diet increased the C18:3 level (0.25% vs 0.38% of total FA in milk produced with a maize+grass diet or the same diet supplemented with DA, respectively). This effect is greatly improved in a silage diet supplied with DA+linseed and total dehydrated maize+DA diet (0.70% and 0.88% of total FA, respectively). This increase in C18:3 level improved the C18:2/C18:3 ratio and thus the balance between omega-6 and omega-3 in milk. Finally, the estimated transfer rate of C18:3 from diet to milk fat is greater when there is DA in the diet (4.7% vs 7.2% of total FA in milk produced with a maize+grass diet or the same diet supplemented with DA, respectively, $P<0.001$). Finally, the supplementation of DA improves the nutritional value of milk fat.

Session 33 Theatre 4

Economic conditions for organic pig production in Sweden

Mattsson, B.[1] and Johansson, A.[2], [1]Svenska Pig AB, Svenska Pig, S-532 89 Skara, Sweden, [2]LRF Konsult, Skaraborgsg 11, S-532 30 Skara, Sweden; barbro.mattsson@svenskapig.se

Today, the certified organic pig production constitutes less than 1% of the total Swedish pig production (approx 35 herds, 26 000 pigs per year). The governmental intention is to increase this proportion. However, the share has been almost constant for the last 15 years. The question is why the production concept has not increased in volume. Are the economic and biologic conditions satisfactory for a sustainable organic pig production, according to the rules set? In this study, the economic records, as well as production records for 12 certified organic pig productions units were analyzed. These results were compared with corresponding information from conventional pig production in Sweden. The results show that organic pig production is difficult to manage, resulting in high variation in productivity between farms. According to figures from a national database, the average production in conventional pig production was 22.4 piglets produced per sow and year. In organic pig production the average production was 18.2 piglets per sow and year, with a variation between 15.8 and 21.0. Costs for organic feed, energy, machinery and labour were higher in organic pig production than in conventional production, but payment per kg carcass was almost double. Results from fattening period showed about 10% lower feed efficiency in organic pig production. In conventional piglet production, the average labour use was 15 hours per sow and year and 0.2 hours per fattening pig produced. In organic herds, the corresponding figures were 30 hours per sow and year and 0.85 hours per fattening pig. Despite of higher costs, the average profitability was comparable between organic and conventional production. However, farmers experienced and uncertainty regarding market demands for organic pork production. Thus the willingness to invest and expand production is restricted. There is also a need for improvements regarding new housing systems and technological solutions for organic pig production.

Session 33 Theatre 5

Effect of winter nutritional managements during the postpartum of beef heifers calving in autumn, on milk production and calves performance

Quintans, G.[1], Velazco, J.I.[1], Scarsi, A.[1] and López-Mazz, C.[2], [1]National Institute for Agricultural Research, Ruta 8 km 281, Treinta y Tres, 33000, Uruguay, [2]School of Agronomy, Garzón 780, Montevideo, 12900, Uruguay; gquintans@tyt.inia.org.uy

In Uruguay the mating period of beef cows coincides with summer, high temperatures and frequent periods of draughts. Some farmers are changing their production systems to an autumn service period. However, heifers mated in autumn have the challenge to lactate during winter when native pastures (NP) have lowest availability. The aim of this study was to evaluate the effects of 3 different winter nutritional systems during lactation on milk production and calves performance. Forty one heifers with their calves were managed together on NP until 87±4 d postpartum (pp), when they were assigned to one treatment during 90 d of winter: i) grazing NP (NP, n=14); ii) grazing improved pastures with Lotus subbiflorus (IP; n=13), iii) cows supplemented with a concentrate on NP (S, n=14). Supplemented cows were offered a concentrate (16%CP) at 1% of body weight (BW). When the treatments finished, all cows were managed together on NP until weaning (Day 210±3.8 pp). Calves BW were recorded at calving and every 28 days. Milk production (MP) was assessed at day 40, 60 and every 30 d until weaning. At 90 d pp all cows produced 3.1±0.3 k/d. At the end of winter S cows produced more ($P<0.0005$) milk than IP (4.4±0.3 vs 2.8±0.3 k/d) and IP cows produced more ($P<0.05$) milk than NP (1.9±0.3 k/d). At the onset of treatments calves weighed 86±2.8k. During winter calves presented different ($P<0.05$) daily weight gain (0.279±0.02, 0.347±0.03 and 0.493±0.02 k/d for NP, IP and S, respectively) and BW at weaning was lower ($P<0.01$) in calves from NP than calves from IP and S (144±3.4, 161±6.7 and 164±4.7 k for NP, IP and S, respectively). Heifers grazing IP or being supplemented produced more milk and weaned heavier calves respect to cows grazing NP. These managements would be considered in commercial conditions, when lactation coincides with winter.

Effects of feedlot duration and sex on carcass characteristics of Kalkoohi dromedary camels

Asadzadeh, N., Sadeghipanah, H., Banabaz, M.H. and Aghashahi, A., Animal Science Research Institute of IRAN (ASRI), Animal Breeding and Genetics, First Dehghan Villa, Shahid Beheshti St., 3146618361 Karaj, Iran; n_asadzadeh@asri.ir

To determine the effects of feedlot duration and sex on carcass characteristics, 12 Kalkoohi dromedary camels (12-month-old; 6 male and 6 female) were used in a 2×2 factorial experiment. Main effects were levels of feedlot duration (5 or 8 months) and sex (male or female). Camels in individual pens were fed diet containing 25% alfalfa, 25% wheat straw, 50% concentrate as a total mixed ration (TMR) and ad libitum. At the end of feedlot period, camels slaughtered and carcass characteristics were recorded. Effects of feedlot duration and sex and also interaction effect of feedlot-duration×sex on dressing percentage were not significant ($P>0.05$). As increasing feedlot duration, percent thigh and shoulder significantly ($P<0.05$) decreased and percent hump increased especially in females. Percent neck in females was higher than males ($P=0.041$). Interaction effect of feedlot-duration×sex on percent bone of carcass was significant ($P=0.001$). In camels that fattened for 5 months, percent bone of carcass in females was higher than in males; conversely in camels that fattened for 8 months, percent bone of carcass in males was higher in females. These results suggested that fattening of one-year-old Kalkoohi dromedary camels for a 5-month-duration in comparison to a 8-month-duration produces carcasses with better quality.

Grassland classification by Remote Sensing (RS) data and Geographical Information Systems (GIS)

Bozkurt, Y.[1], Basayigit, L.[2] and Kaya, I.[3], [1]Suleyman Demirel University, Faculty of Agriculture, Department of Animal Science, Isparta, 32260, Turkey, [2]Suleyman Demirel University, Faculty of Agriculture, Department of Soil Science, Isparta, 32260, Turkey, [3]Kafkas University, Faculty of Veterinary Medicine, Department of Animal Science, Kars, 36600, Turkey; ybozkurt@ziraat.sdu.edu.tr

This study aimed to classify grassland types by using Remote Sensing (RS) data and Geographical Information System (GIS) technology in Kars province located in north-east region of Turkey. For this purpose, the LANDSAT 5 TM satellite images, taken in 2005, were used and grassland classification maps were produced using GIS. In order to determine the current status of grasslands, red (0.45-0.52 μm), near infra-red (0.52-0.60 μm) and infra-red (0.63-0.69 μm) bands of Landsat images were used and unsupervised classification was applied to produce the distribution map of grasslands showing the grassland types based on biomass quality. Grasslands in the study area were classified as Type I, II and III together with 3 sub-classes within each type according to biomass quality and vegetation cover and compared according to NDVI (Normalised Difference Vegetation Index). The results showed that the whole grassland area occupied 638968.3 ha and accounted for 66.7% of the total area of the province. Grassland types I, II and III occupied 216917.7, 326334.7 and 95716 ha and accounted for 22.6, 34.1 and 10% of the total grassland area respectively in the study region. The defined types will certainly contribute to grazing management decisions, such as determination of starting date of grazing season, stocking rate and the most suitable grazing systems to be applied in the region.

Session 33 Poster 8

Effect of breeding technology on somatic cell count in cow milk on mountain farms

Frelich, J. and Šlachta, M., University of South Bohemia, Faculty of Agriculture, Studentská 13, 37005 České Budějovice, Czech Republic; slachta@zf.jcu.cz

The aim of this study was to evaluate the effects of different breeding techniques applied on mountain farms on somatic cell count in cow milk. The somatic cell count (SCC) in individual cow milk samples were analysed in thirty-two herds in the Czech Republic in 2004-2008. The herds consisted of the two most common dairy breeds in the region, the Holstein and the Czech Fleckvieh. Three breeding techniques were distinguished, (1) seasonal pasturing and loose housing for the rest of a year, represented by five herds, (2) seasonal pasturing and tie stalls for the rest of a year, represented by ten herds, and (3) all-year-round loose housing, represented by seventeen herds. The General Linear Model analysis was conducted in order to evaluate the effects of the breeding technique on SCC in milk. The effects of breed, parity, year and days in milk (covariate) were also included in the analysis. The effect of all factors in the analysis was highly significant ($P<0.001$). Seasonal pasturing with loose housing (variant 1) revealed lower SCC in cow mik samples ($P<0.001$) than the other two variants. The geometric mean of SCC was 95, 126 and 128 thousands ml^{-1} in variant 1, 2 and 3. The interaction between breeding technique and breed ($P<0.001$) indicated a more pronounced difference in SCC between breeds in variant 1. The geometric mean of SCC was 112 thousands ml^{-1} in the Holstein and 78 thousands ml^{-1} in the Czech Fleckvieh in variant 1 herds. The SCC values were associated with a good udder health in all the examined herds. Breeding technique and the breed had significant effects on SCC. The lowest SCC was found on farms applying the seasonal pasturing and loose housing system, mainly with the Czech Fleckvieh breed.

Session 33 Poster 9

Piglet mortality in organic herds

Prunier, A., INRA, SENAH, 35590 St-Gilles, France; armelle.prunier@rennes.inra.fr

Productive performance of organic pig farms is lower compared to conventional farms, but only very few data exist. Better knowledge of the productivity of organic herds regarding litter size at birth, piglet losses around birth and during lactation, as well as housing and management conditions should help to identify critical points and hence to improve the situation. Therefore, a research project was initiated in 6 EU countries (Corepig). As part of this, farmers recorded production data during 3-11 months starting between January and July 2008. Farmers were asked to record the numbers of piglets born dead, born alive as well as the number of piglets at weaning. Taking into account the quality of the records and setting a threshold of $\geq$ 10 litters/farm, data from 38 farms in 4 countries (France: 14, Germany: 12, Austria: 7, Sweden: 5) were analyzed (mean: 69, 10 to 713 litters/farm). Most farmers were not present at farrowing, meaning the number of piglets that were classified as 'born dead' was probably greatly overestimated. Therefore, mean total litter size at birth (born dead + born alive, MTLS), its standard deviation (SDLS), litter size at weaning and percentage of total losses (born dead + lactation losses, pLOSS) were calculated at the farm level. Overall, MTLS was 12.9±1.6 piglets at birth, 9.2±1.1 piglets at weaning and pLOSS was 26.7±7.1% with a lactation duration of 45.3±5.9 days. Mortality of piglets increased with MTLS (2.1±0.7% additional loss per piglet, $P=0.004$) and with SDLS (3.9±1.6% additional loss per unit of SDLS, mean ± SEM, $P=0.021$). MTLS was correlated with SDLS ($r=0.44$, $P=0.006$). These data confirm the detrimental influence of large litter size at birth on piglet mortality. This is commonly observed in conventional pig production and related to a higher proportion of piglets with low birth weight and to increased competition for teats. High variability in litter size may exacerbate these problems, and in addition may be an indicator for other problems on the farm.

Dairy farms in La Pampa (Argentine): typologies according to livestock management and economic indicators; preliminary results

Perea, J.[1], Giorgis, A.[2], Larrea, A.[2], García, A.[1], Angón, E.[1] and Mata, H.[2], [1]University of Cordoba, Animal Production, Edificio Produccion Animal - Campus Rabanales, 14071, Spain, [2]National University of La Pampa, Animal Production, Veterinarian School - General Pico, 0021, Argentina; pa2pemuj@uco.es

The aim of this study was to characterise dairy farms through structural, technical, productive and economic aspects located in the region of La Pampa (Argentine). The area of study extends over 32,467 km2 and concentrates a population of 172 dairy farms and a census of 26,408 heads. Information on 96 representative variables was collected in 2007 through stratified random sampling with proportional allocation by department. The sample of farms comprised 57 dairy farms (33% of the official census). The principal components analysis revealed 5 factors explaining 77% of the original variability: the first factor defines the farm size and the intensification of the system; the second factor indicates the farm specialization in dairy farming and its technological level; the tird factor explains the relationship between reproductive management and gross margin; the fourth factor indicates the relationship between health and direct cost and family involvement in management and work; and the fifth factor explains the inverse relationship between milk productivity per cow and fat percentage. 5 systems were identified from cluster analysis. Group I concentrates the 27.6% of farms which are characterized by their small size and highly specialized in dairy farming. Group II (17.0% of farms) is formed by family mixed cattle-dairy farms with low productivity. Group III (27.6% of farms) concentrates family farms of small size and highly specialized in dairy farming. Group IV (10.6% of farms) is formed by large farms with high productivity and greater diversification (agriculture-dairy farming). Group V (12.7% of farms) develops the system of higher technological level and is characterized by medium-scale farms with high specialization in dairy farming.

Session 33 Poster 11

Management of dairy farms in La Pampa (Argentine): factors affecting economic performance; preliminary results

Perea, J.[1], Giorgis, A.[2], García, A.[1], Mata, H.[2], Acero, R.[1] and Larrea, A.[2], [1]University of Cordoba, Animal Production, Edificio Produccion Animal - Campus Rabanales, 14071, Spain, [2]National University of La Pampa, Animal Production, Veterinarian School - General Pico, 0024, Argentina; pa2pemuj@uco.es

The aim of this study was to determine a model to estimate the probability of improving the economic performance of dairy farms of La Pampa (Argentine) by acting on farm management. The area of study extends over 32,467 km^2 and concentrates a population of 172 dairy farms and a census of 26,408 heads. The sample of farms was obtained using a stratified sampling by departments and comprised 33% of the census (57 farms). The farms were classified according to their economic profits, positive or negative, and a multinomial logistic regression model was used to detect the variables that explain, with a greater likelihood, the economic profit of the farm. The results show that only 5 of the 30 variables initially selected are significant predictors of the economic performance of the farm ($P<0.05$). The model correctly predicts 83% of the cases studied. The global significance of the model was checked by maximum likelihood test, which takes a value of 29.98 with 5 degrees of freedom ($P<0.000$). The goodness of fit was checked by Pearson's test, which takes a value of 43.04 for 41 degrees of freedom ($P=0.384$), indicating that the quality of the prediction is correct. The farm size increases the probability of obtaining a positive economic result in 1.006 times per hectare. The systematic collection of internal information of the farm increases 5.928 times the probability of obtaining a positive economic result. If this information is also used in the decision making process, the probability of obtaining a positive economic result increases by 35.40 times. The use of any external information source (journals, technical seminars, etc.) increases the likelihood of economic success in 5.910 times. Finally, if the farm has more of one advisor increased the likelihood of success is 5.738 times.

Session 33 Poster 12

Productive response of certain laying hens hybrids, reared within various farming systems

Usturoi, M.G., Boisteanu, P.C., Radu-Rusu, R.M., Dolis, M.G., Pop, I.M. and Usturoi, A., University of Agricultural Sciences and Veterinary Medicine, Animal Science Faculty, 8 Mihail Sadoveanu Alley, 700490, Iasi, Romania; umg@uaiasi.ro

Laying hens accommodation in conventional cage batteries will be forbidden from 2012, the alternative husbandry systems being created for replacement. However, they were not tested under production conditions or, more important for certain actual genotypes. The researches focused on the production response given by two laying hybrids (Lohmann Brown-A and Hisex Brown-B), exploited under different systems (Lc=conventional cages rearing, in climate controlled hall; L1exp=rearing in opened cages batteries, in climate controlled hall; L2exp=rearing at ground, in climate controlled hall; L3exp= rearing at ground, in a hall providing access to external paddock). Although fowl weight was found within standard, at the end of experiments, in Lohmann hybrids, the conventional cages hens (Lc-A) proved to have 4.46-8.06% higher values than those reared under alternative systems conditions, while in Hisex hybrids, hens from classic batteries (Lc-B) were 0.80-2.66% heavier than those in experimental groups. Lohman hens accommodated in conventional cages (Lc-A) achieved an average yield of 325.05 eggs/hen, which meant 4.22-15.89% higher than other husbandry versions. In Hisex hens, fowl from Lc-B group laid and average amount of 324.17 eggs, thus 3.28-15.87% more than the hens from experimental group. During 60 exploitation weeks, casualties rate in Lohmann hens reached 11.66% in control group and 7.36-11.08% in experimental ones, while in Hisex hens, mortality was found between 12.08% in control and 8.08-9.17% in the experimental ones. Conclusion of research was that alternative systems decreased laying hens production performances, although it fulfilled welfare needs, inducing also reliability diminution. It was also find that analysed genotypes gave different responses to the new tested husbandry versions.

Session 33 Poster 13

A survey on animal production in the north of the Cacheu province (Guinea Bissau, West Africa)

Almeida, A.M.[1,2] and Cardoso, L.A.[1,2], [1]IICT - Instituto de Investigação Científica Tropical, Centro de Veterinária e Zootecnia, CVZ-FMV Av. Univ. Técnica, 1300-477 Lisboa, Portugal, [2]CIISA - Centro Interdisciplinar de Investigação em Sanidade Animal, CVZ-FMV Av. Univ. Técnica, 1300-477 Lisboa, Portugal; aalmeida@fmv.utl.pt

Guinea Bissau is one of the lowest income per capita countries in the world and the vast majority of the population is dedicated to very small scale subsistence farming where animal ownership has an important role in both food supply and ceremonial events. Despite such fact little is known about animal production and genetic resources in the country. The North of the Cacheu province comprehends two sectors, Bigene and São Domingos, considered one of the poorest regions of Guinea Bissau. We have conducted a survey and enquire in both sectors aiming to briefly characterize local animal genetic resources and animal production systems. Such characterization is intended to serve as a basis to possible agricultural and animal production development projects in this area. Cattle (N'Dama and West African horthorn breeds) are owned by the two larger ethnic groups in the area, the Felupes and the Balantas that have however very different management practices. Sheep are relatively scarce whereas goats (West African Dwarf) are the most important small ruminant species for both ethnic groups. Pigs (crioulo breed) and dwarf chickens play a very important role as they are the only species regularly used in trade at the village level.

Genetic analysis of several economic important disease traits in German Holsteins cows

Hinrichs, D., Stamer, E., W. Jung, W., Kalm, E. and Thaller, G., Christian-Albrechts-University, Institute of Animal Breeding and Husbandry, Hermann-Rodewald-Straße 6, 24118 Kiel, Germany; gthaller@tierzucht.uni-kiel.de

Diseases are one of the major problems in commercial milk production and cause major economic losses. Precondition for genetic improvement are clear trait definitions and appropriate recording schemes. In the present study several disease categories were investigated in three lager herds in North Germany. Diseases were grouped into fertility diseases, udder diseases, metabolic diseases, and claw and leg diseases. Periods of data recording were 50, 100, or 300 days in milk. The impact of increasing numbers of daughters per sire on reliabilities of breeding values was examined and genetic correlations between disease categories were estimated. In addition, all diseases were analysed simultaneously. Frequencies of the disease categories were moderate to high and varied between 7% and 78%. The most frequent disease categories were fertility diseases and udder diseases. Heritabilities for all diseases varied between 0.03 and 0.15, and were 0.02 to 0.05 for fertility diseases, 0.06 to 0.08 for udder diseases, 0.08 to 0.15 for metabolic diseases, and 0.01 to 0.03 for claw and leg diseases, respectively. The genetic correlations between disease categories ranged from -0.18 to 0.82. Data recording is especially important at the beginning of lactation because most diseases occurred during the first 50 days of lactation. However, as claw and leg diseases are distributed over the whole period of lactation data recording should be extended beyond the first 50 days in milk for these traits. Fertility diseases should be divided into ovarian problems and non ovarian problems. The frequencies of diseases are not affected by the average number of daughters per sire. The results of our study shows that enhanced effort on data recording makes sense and can be implemented on larger commercial dairy farms with an acceptable additional costs.

A dynamic-stochastic simulation model to determine the optimal voluntary waiting period for insemination of Dutch dairy cows

Inchaisri, C., Jorritsma, R., Vos, P.L.A.M., Van Weijden, G.C. and Hogeveen, H., Utrecht University, Farm Animal Health, P.O.Box 80151, 3584 CL, Utrecht, Netherlands; C.inchaisri@uu.nl

The voluntary waiting period (VWP) is defined as the time in weeks post-partum after which farmers are start to inseminate their cows. The optimal VWP is very difficult determine with field data. A Monte-Carlo dynamic-stochastic simulation model was created to calculate the economic effects of different VWP's. The model is dynamic and uses time steps of one week to simulate the reproductive cycle (ovulation, estrous detection and conception), the occurrence of postpartum disorders and lactation curve for individual cows. Input of the model was adjusted for Dutch dairy farm management. During the simulation process, cows with a given breed, parity, month of calving and 305-d milk yield were created. The variables of interest were randomly varied, based upon relevant distributions and were adjusted for cow statuses. The lactation curve was modeled using Wood's function. The peak milk yield, time of peak yield and milk persistency were determined with this function. The economic factors in the analysis were costs of lower milk production (€0.07-0.20 per kg), calf price (€50-200 per calf), insemination costs (€7-24 per insemination), calving management costs (€137-167 per calving) and culling costs (expressed as the retention pay-off; €-1,616-318). A partial budget approach was used to calculate the economic effect of VWP's varying from 7 to 15 wks pp, using a VWP of 6 wks as basis. Per iteration, the VWP with the maximum profit was determined as the optimal VWP. The optimal VWP of most cows (93%) was shorter than 10 wks. For only 7% of the iterations, a VWP extended beyond 10 wks pp was optimal. Every VWP gave, on average, economic losses larger than 6 wks. However, in the results of sensitivity analysis, for a cow with high milk persistency and milk production within market circumstances of a low milk price and high costs of insemination a VWP larger than 10 wks pp is optimal.

Development of a methodology to analyse reproduction problems linked to environmental factors in cattle herds in Wallonia with key indicators and decision making trees

Knapp, E.[1], Chapaux, P.[2], Istasse, L.[1], Dufrasne, I.[1] and Touati, K.[1], [1]Liège University, Veterinary Faculty, Bd de Colonster, 20, B-4000 LIEGE, Belgium, [2]Association Wallonne Elevage, Insemination, Champs Elysées, 4, B-5590 CINEY, Belgium; eknapp@ulg.ac.be

The impacts of decreased of the reproduction performance in cattle herds on the profitability are large. There are many environmental factors which can influence reproduction performance. The clinical signs (poor heat detection/quality, infertility or abortion) are almost unspecific. At farm level, the farmers and their partners are faced with two major difficulties: definition and quantification of the problem and diagnosis of the environmental factor(s) implicated. So two tools were developed to answer these questions. The first one is a computer program available on every veterinarian-inseminator PDA of the Association Wallonne de l'Elevage. This program calculates online relevant retrospective and prospective key indicators. So, at any time, the real reproduction performances of the herd can be compared with targets fixed by the farmer and the inseminator. Watch lists (anoestrus, calving, repeat breeders..) can be printed in the farm after every visit for the monitoring of the reproduction. When the targets are not obtained, a methodology of diagnosis of environmental problems linked to reproduction is available. This is the second tool. Three decision-making trees (one for each reproduction problem: heat detection, infertility or abortion) organised in four levels allow the user to progress with logic and method to adjust the diagnosis. The first level indicates the clinical signs to observe, the second one proposes the complementary exams, the third one shows the different possible diagnosis and the fourth one explains the physiological links to the reproduction problem. The user can highlight specific reproduction problems, prioritize the major causes and propose relevant solutions to the farmer. This methodology is actually tested in 27 dairy and/or beef farms in Wallonia.

Effect of temporary suckling restriction and a short-term supplementation on ovarian cyclicity and early pregnancy in beef cows in low body condition

Quintans, G.[1], Velazco, J.I.[1], López-Mazz, C.[2], Scarsi, A.[1] and Banchero, G.[1], [1]National Institute for Agricultural Research, Ruta 8 km 281 Treinta y tres, 33000, Uruguay, [2]School of Agronomy, Garzón 780, Montevideo, 12900, Uruguay; gquintans@tyt.inia.org.uy

Nutrition and suckling are the most important factors affecting the anoestrous period. The use of suckling restriction (SR) with nose plates (NPs) is a useful tool on range conditions but it does not increase consistently pregnancy rate when cows are in low body condition score (BCS). The aim of this study was to evaluate the effect of temporary SR and an increase in the nutritional level by a short-term supplementation, on ovarian cyclicity and pregnancy rate in low BCS cows. Thirty eight multiparous cows (3.6u BCS) were assigned to 3 treatments on 71±2 d postpartum (Day 0): i) suckling ad libitum (S, n=12), ii) calves fitted with NPs for 14 d remaining with their dams (NP, n=13), iii) same as ii) plus cows supplemented during this period (NP+S, n=13). Cows in NP+S were supplemented with rice bran at 0.7% of BW. All cows were in anoestrous at Day 0. Mating started 14 days before onset of treatments (Day -14) and lasted 60 days. BW and BCS were recorded at Day -14 and biweekly thereafter. Presence of corpus luteum (CL) was recorded at Day 14 and early pregnancy rate was considered when cows got pregnant during the first month of mating. Cows in S tended to present lower (P=0.08) BW than cows in NP group (441±4.0 vs 454±3.9k) but no differences were detected respect to NP+S cows (447±3.9k). There was no effect of treatment on BCS and in average cows presented 3.8±0.03u of BCS along the experimental period. By Day 14 more (P<0.0004) cows from NP and NP+S presented CL respect to S cows (62, 54 and 0% for NP, NP+S and S, respectively). During the first month of mating there was a tendency (P=0.08) that more NP+S cows got pregnant respect to S cows (85 vs 50%) but there was no difference respect to NP cows (77%). Fourteen days of supplementation would increase the reproductive response when NP are used in low BCS cows.

Strategies to improve the pregnancy rate of German Holstein Friesian cows based on a combined use of Ovsynch and progesterone supplementation protocols

Tsousis, G.[1], Forro, A.[2], Sharifi, A.R.[3] and Bollwein, H.[2], [1]Aristotle University of Thessaloniki, Clinic of Farm Animals, School of Veterinary Medicine, St. Voutyra 11 str., 54627, Thessaloniki, Greece, [2]University of Veterinary Medicine Hanover, Clinic for Cattle, Bischofsholer Damm 15, 30173, Hanover, Germany, [3]University of Göttingen, Institute of Animal Breeding and Genetics, Albrecht-Thaer-Weg 3, 37075, Göttingen, Germany; tsousis@vet.auth.gr

The aim of this study was to examine the effects of a progesterone (P4) supplementation post insemination (p.i.) on pregnancy rates (PR) after the application of two OvSynch protocols. German Holstein Friesian cows were randomly synchronized at Day 60 after parturition with either the classical OvSynch protocol (GnRH d0-PGF2α d7-GnRH 48 h after PGF2α) or with a modified OvSynch60 protocol (2nd GnRH 60 h after PGF2α). At Day 4 p.i., in each OvSynch group, one half of the animals was supplemented with P4 (P4+) by applying PRID® intravaginally for 14 days, while the other half remained untreated (P4-). At Day 33 p.i. a pregnancy diagnosis was performed with the use of an ultrasound device in 398 cows. A generalized linear model with the SAS Proc GLIMMIX was used to analyse the data. Variables showing $P<0.1$ remained in the model. There was no difference in PR between the two OvSynch protocols (38.4% vs. 44.1%, $P=0.21$). P4+ animals tended to have higher PR values compared to P4- cows (44.4% vs. 38.1%, $P=0.08$). The retention of fetal membranes and a low Body Condition Score at insemination (BCS60) were significant factors in the final model. Moreover, an interaction between BCS60 and P4 supplementation was apparent, i.e. the difference in PR between P4+ and P4- animals was significant from a BCS of 3.25 and above. In conclusion, the elongation of the time interval between the injections of prostaglandin and GnRH from 48 to 60 hours had no positive effects on pregnancy rate. A progesterone supplementation improved the PR of OvSynch protocols and this effect became more apparent in animals with BCS above 3.25.

Reproductive performances of Charolais cows: analysis of 18590 carriers from 124 farms on a 37-year period

Zsuppan, Z.[1], Lherm, M.[2] and Ingrand, S.[3], [1]PE-GMK, Animal science, Deàk Ferenc u 16, 8360 Keszthely, Hungary, [2]INRA, Phase, URH, 63122 Saint-Genes Champanelle, France, [3]INRA, SAD, UMR1273 Metafort, 63122 Saint-Genes Champanelle, France; zsuppanzs@yahoo.com

A french network of Charolais farms was established by Inra in the 70's. We built-up a database of 18 590 cows at least 3 calvings.Our objective was to analyze the variability of their reproductive performances. Eight variables were included in the analyses: nb of calvings (NbC), age at 1rst calving (AgC1), calving-to-calving interval (CCI),variability of CCI (CCstd), interval between 1rst and 2nd calving (C1C2), the shorter CCI (CCmin), the longer CCI (Ccmax),the average difference between 2 successive calvings (CCsucc). Two groups were distinguished:cows(n=3784)from one group (G1) have lower performances and shorter carrier than cows(n=14806) from the other group (G2).The two groups differ for all the variables ($P<0.001$), respectively for G1 and G2: AgC1 =1036 and 1058 d, CCI=411 and 374 d, CCstd=7 and 20, C1C2=440 and 386 d, CCmin=359 and 345, CCmax=489 and 410 d, CCsucc=-22 and -5. The same analysis was performed separately for the two groups. Respectively 4 and 7 sub-groups were obtained for G1 and G2. Within G1, the sub-groups differ in nbC: 3, 3.1, 4.8 and 6.1. For the two first ones, the difference is in CCstd (respectively 52 and 17 d). The 308 cows with nbCC=4.8 on average had the worst performances (CCI=476 d, CCstd=116 d, CCmax=677 d). They also had the lowest AgC1(1018 d). The cows with NbC=6.1 had both short and long CCI during their carrier (337 and 501 respectively for CCmin and CCmax). Within G2,one group is composed of 837 2-year-first-calving cows, with the lowest C1C2 interval (376 d)and the lowest CCI (369 d). One sub-group is made of cows with short carriers (NbC=3) and another with long carriers (NbC=8.7). The two groups with the highest CCI (380 and 388 d) have also the highest C1C2 (respectively 397 and 393 d).Cows in those two groups have more variable CCI, from respectively 311 to 423 and 338 to 438 d (std=30 d vs 13 to 19 in the other 5 groups.

Cystic ovaries, silent estrus and respiratory disease occurrence in Austrian Fleckvieh heifers

Fuerst-Waltl, B.[1], Koeck, A.[1] and Egger-Danner, C.[2], [1]University of Natural Resources and Applied Life Sciences Vienna, Department of Sustainable Agricultural Systems, Division Livestock Sciences, Gregor Mendel-Str. 33, A-1180 Vienn, Austria, [2]ZuchtData EDV Dienstleistungen GmbH, Dresdner Str. 89/19, A-1200 Vienna, Austria; birgit.fuerst-waltl@boku.ac.at

Continued non-consideration of diseases in replacement stock eventually leads to increased disease incidences. A routine registration system is an essential prerequisite for the reduction of disease frequencies by both, management and breeding. However, in Austria, where health registration started in 2006, but also in Scandinavian countries with a long history of health registration, the main focus is on lactating cows. While reports supporting management decisions are also provided for replacement stock, no routine breeding value estimation exists for heifers' diseases in either country. The aim of this study thus was to describe the current situation of recorded reproductive and respiratory diseases in Austrian Fleckvieh (dual purpose Simmental) heifers. Besides, first steps towards the estimation of genetic parameters were made. Data of 14,136 Fleckvieh heifers under health recording and aged 12 to 32 months were provided in order to investigate cystic ovaries, silent estrus/anovulation and respiratory diseases. Diseases were defined as binary traits (1 or 0) based on whether or not the heifer had at least one veterinary treatment within the given time span. Mean frequencies of cystic ovaries, silent estrus/anovulation and respiratory diseases were 3.1%, 3.6% and 0.8%, respectively. A first genetic analysis revealed low heritabilities not significantly different from zero for all traits. Further research and the improvement of the heifers' health data quantity and quality are recommended.

Morphological characterisation of Piemontese cows by using score card

Lazzaroni, C. and Biagini, D., University of Torino, Department of Animal Science, Via L. da Vinci 44, 10095 Grugliasco, Italy; carla.lazzaroni@unito.it

Score card based on linear description of different body areas of animals are widely used in animal judging, checking the subject conformity to the Breed Standards for selection purpose. For Piemontese cows the score card for the breeding stock allowed, until last year, to obtain: - data on subjects (age at judging, age at last calving); - data on farming (housing system and length, nutritional condition); - animal biometric data (withers height, trunk length); - animal evaluation (final score and size, thoracic capacity, meat potential, rump, limbs, lactiferous characters scores); - morphological description (- appearance: head, top line, fore limbs, hind limbs, pastern; - muscularity: withers, shoulders, loin width and thickness, thigh muscularity, buttock convexity; - fineness: skeleton, skin); - notes (defects). Data on 381 cows from 88 farms were collected and analysed to describe the animals reared in a low hills area near Torino. The housing system heavily used was the tied stall (60%), although the partially open pen was spreading (>25%), and the pasture was used for more than 6 months/years (>85%). The evaluated cows had 23-208 months of age, with the last calving at 21-198 months, showed a normal nutrition, a height at shoulder of 133±3.9 cm (range 125-139 cm) and a trunk length of 162±5.6 cm (range 155-169 cm). The cows were mainly evaluated between 80 and 89 point of final score (mean 83.9±2.8), values matching the breed standard, although some subjects deviating from them (range 71-91). Regarding the morphological description (scale 1-9), the subjects obtained values almost always slightly higher than the medium code (excluding limbs, only 20-30% corrected), especially for muscularity and finesse, very positive results for the productive attitude of the breed (meat production). Finally, only 18 subjects (4.7%) presented defects such as discoloration, atypical coat, brachignathism and prognathism.

Comparison of animal judging at show-ring and at farm-shed: preliminary results on Holstein cows

Lazzaroni, C. and Biagini, D., University of Torino, Department of Animal Science, Via L. da Vinci 44, 10095 Grugliasco, Italy; carla.lazzaroni@unito.it

The morphological evaluation of cattle could be performed analytically at farm level, by mean of detailed score card (20 traits examined by breed classifiers) using linear description to be used in genetic index, but also at livestock exhibitions during which the best animals present in each of the 15 provided classes are awarded. In this study the correspondence of the classification in the exhibition with the final scores obtained at the stable were analysed on about 200 Italian Holstein cows attending different shows in different age classes (7 classes: from less than 30 months to more than 6 years of age). As general rule, in the show-ring cows were classified according to the more detailed score obtained at farm-shed, but sometime this did not occurred, as in the provincial show for the 2 year senior, 3 year senior and 4 year cows, and in the regional show for the 2 years junior, 2 years senior and 3 years senior cows. This is because for the show-ring the cows are prepared in order to enhance and bring out their best morphological traits and conceal defects. So it is possible that animals with low scores in the stable are placed in higher positions than those with high scores, as at farm cows are evaluated as they are, while at show cows appear at the best, or because they are in top form, thanks to fitting or because their weak points and defect are hidden (by foot care, washing, clipping, brushing, etc.), or conversely (not optimal conditions, wrong milking time, limping or problems arising in transport, etc.). It should be pointed out also that going from younger to older age classes, the official scores of the cows are rising (from 84-85 to 90-91). In conclusion, there is a substantial difference between the two judging techniques, and even if the results achieved are often in agreement, the future of animal judging is in the experts, who should use the same criteria and assessed the animals taking in account all rules learned during their training.

Growth modelling of males Serrana Soriana and Charolais breed in Soria

Miguel, J.A., Calvo, J.L., Ciria, J. and Asenjo, B., E.U. Ingenierias Agrarias, Produccion Animal, Campus Universitario De Soria S/N, 42004, Spain; jangel@agro.uva.es

Alternative livestock productions, increasingly more importance, are based on the rational use of resources, in line with the environment and the use of native breeds that many cases are in danger of disappearance, since her calf is not economically viable. A way to predict and know the growth of animals is by using mathematical models. This work presents the results of estimates of growth according to the model of the local breed Serrana Soriana and the Charolais males Gompertz-Laird. 50 males Soriana Serrana breed animals and 50 Charolais breed were used, all remained with their mother in an extensive system of exploitation until weaning, so his power was based in breast milk and available in the areas of grazing grass. At weaning (28 weeks old), calves were housed in free-range system, with a density of 30 m2/animal. The power was based on a commercial feed (1 UFL and 88.63 g PDI/kg DM), provided ad libitum and with free access to barley straw. The calves are weighed individually each week from birth until weaning (28 weeks), every 4 weeks until the 20 postdestete, and each 5 until week 70 postdestete (98 weeks of life). The growth of the animals was modelling according to the formulation of Gompertz-Laird (1965), the data were adjusted to the model by the non-linear regression procedure of the computer program SPSS 11.5. The weight to the birth of the breed calfs Charolais (45.14 kg), it was superior to those of Serrana Soriana (40.13 kg), and presented a greater relative growth (9.9%) before of the inflection point of the growth curve; furthermore it was also superior the maximum decline relative after this point (6%), Serrana Soriana reached 7 days after the maximum growth point but with a smaller weight (315.05 kg vs 384.76 kg). Was calculated a weight to the maturity of 1076.90 kg for Charolais and 856.40 kg for Serrana Soriana.

Influence of the age and the season of the first calving on milk performances of dairy cows

Froidmont, E.[1], *Mayeres, P.*[2], *Bertozzi, C.*[3], *Picron, P.*[1], *Turlot, A.*[1] *and Bartiaux-Thill, N.*[1], [1]*Walloon Agricultural Research Centre (CRA-W), Production and Sectors Department, Animal Nutrition and Sustainability Unit, Bertrand Vissac building, rue de Liroux 8, 5030 Gembloux, Belgium,* [2]*Walloon Association of the Breeding (AWE), Technico-economic Service, Rue des Champs Elysées 4, 5590 Ciney, Belgium,* [3]*Walloon Association of the Breeding (AWE), Research & Development Unit, Rue des Champs Elysées 4, 5590 Ciney, Belgium; froidmont@cra.wallonie.be*

An initial dataset (n=65,592) from Holstein dairy heifers born between 2000 and 2007 was used to determine an optimal age and season for the first calving. The data have been sorted in order to keep only heifers having a first calving between 18 and 42 months, a milk production in first lactation > 2500 L, and a first lactation length between 45 and 704 days (n=62,969). Age classes at first calving were fixed as follows: 18-22, 22-26, 26-30, 30-34, 34-38 and 38-42 months (class 1 to 6 respectively). A GLM analysis was performed considering two effects (age class and season at first calving) and their interaction on dependent variables (305-d milk production and composition for the first two lactations). The results confirm that the 305-d milk, fat and protein production in first lactation were maximised from the classes 2 to 5 and lower for classes 1 and 6. A first calving around 24 months of age appears interesting: it does not induce a loss of milk production and minimizes the feeding costs during the breeding phase. Cows from class 2 had also a better 305-d milk, fat and protein production in second lactation compared to all other classes. This could result from a farmer effect: those having the highest milk performances have also a better calving management. The 305-d milk production was also influenced significantly by the first calving season (spring: 6,531 l, summer: 6,870 l, autumn: 7,011 l, winter: 6,680 l; $P<0.001$) and probably reflects different feeding strategies. The overall results suggest that an early calving in autumn is advised to dairy producers.

Rearing of calves in the period of milk nutrition in relation to the system of housing

Tousova, R., Stadnik, L. and Ruzickova, M., Czech University of Life Sciences Prague, Department of Animal Husbandry, Kamýcká 129, 16521, Prague 6 - Suchdol, Czech Republic; stadnik@af.czu.cz

Comparison of two housing systems of Holstein calves (n=150) in the period of milk nutrition and evaluation of their effect on growth and health traits were performed. A system of individual outside boxes was compared to the housing of calves in boxes in a cow barn. Calves raised in the outdoor system were not exposed to a dusty environment and undesirable values of the microclimate were not recorded in this system either, whereas the level of combustion gases observed exceeded the permissible. The CO_2 content of was 0.2733% (limit 0.25%), and the NH_3 content represented 0.00325% (limit 0.0025%). The intensity of light outside was 3 times higher (758-789 luxes) than inside (247-387 luxes). No changes in the calves´ health in relation to changing conditions of the environment were recorded in the outdoor system of housing. Sneezing attacks with signs of respiratory disease were determined in 20% of calves in the cow barn housing, and 46.67% calves breathed with difficulty depending on rainy weather and increased humidity. The average daily live weight gain was 0.726 kg in the stable and 0.841 kg in outdoor individual boxes ($P<0.05$). This work was funded by MSM6046070901.

Gen frequency distribution of the BoLA-DRB3 Locus in Polish Holstein Cattle

Oprzadek, J.M.[1], Urtnowski, P.[1], Sender, G.[1], Pawlik, A.[1] and Oprzadek, A.[2], [1]Institute Of Genetics And Animal Breeding, Animal Science, Jastrzebiec Ul. Postepu 1, 05-552, Poland, [2]Agricultural Property Agency, Dolanskiego 2, 00-215 Warszawa, Poland; j.oprzadek@ighz.pl

The objective of this study was to describe the allele frequency distribution of the DRB3.2 locus in chosen dairy cattle herds and the associations between alleles and functional and dairy performance traits. Van Eijk's methods (1992) were used. Allele patterns were described according to ISAG BoLA Nomenclature Committee. Observations were taken on two dairy herds. DRB3.2 allele frequency was similar to those found by other authors. There were similar differences DRB3.2*24 allele frequency in both herds (0,144 and 0,209). Second frequent allel was DRB3.2*08 (frequency in herd 1 - 0,130 in herd 2- 0,13). Some of alleles occurring in one of experimental herds were not present in the other. Analysis on the associations between BoLA DRB3.2 polymorphism and functional or performance traits has proved significant differences between genotypes in reference to milk, fat and protein yield. There has been found association between DRB3.2 genotypes and length of the calving interval. In conclusion, the results shown that there is an association between DRB3.2 polymorphism and both functional and performance traits. The results of this study demonstrated that BoLA-DRB3.2 is a highly polymorphism locus in Polish Holstein cattle. Supported by the State Committee for Scientific Research grant: N311 037 31/0685

Session 34 Poster 14

The effect of milk price on economic weight of some traits of dairy cattle

Szabó, F.[1], Fekete, Z.[1] and Wolfová, M.[2], [1]University of Pannonia, Animal Science and Husbandry, Deák 16., 8360 Keszthely, Hungary, [2]Institute of Animal Science, Uhrinéves, Prague, Czech Republic; szf@georgikon.hu

Marginal and relative economic weight of 7 traits was calculated for the dairy cattle population in Hungary in 2009, using a bioeconomic model based on the program package ECOWEIGHT (Wolf et al. 2005) considering a high (€ 0.3/kg) and a low (€ 0.2/kg) milk prices. The study was based on the typical dairy farm size of 330 Holstein-Friesian cows, with a production level of 7,000 kg annual milk yield. Cows were managed in a loose-housing system with parlour milking, representing current commercial dairy enterprises. A total mixed ration based on maize silage and concentrates, with some alfalfa hay, was offered to 4 groups (first-, second-, third-phase of the lactation and a dry group). Income came from milk, calves, culled cows and manure sale. About 50% of the total costs related to feed, with the remainder due to factors such as management, reproduction and health services, labor, interest and amortization. Annual revenues and costs were used for the economic calculations. Gross margin was taken as a difference between income and variable costs. Marginal economic value of a given trait was defined as the partial derivative of the profit function, which was standardized by multiplying by the genetic standard deviation of the trait. The relative economic weight for traits were expressed as a percentage of the standardized economic value of 305-d milk yield. The obtained relative economic weights in decreasing order of the evaluated traits in case of high price were as follows: 1st 305-d milk yield (100%), 2nd length of productive life (52%), 3rd conception rate of cows (35%), 4th 305-d protein yield (35%), 5th 305-d fat yield (21%), 6th stillbirth (13%), 7th pregnancy rate of replacements (3%). In case of low price the order of traits has changed: 1st length of productive life (163%), 2nd 305-d protein yield (139%), 3rd conception rate of cows (133%), 4th 305-d milk yield (100%), 5th 305-d fat yield (51%), 7th pregnancy rate of replacements (13)%.

Salt addition to reduce concentrate intake in young bulls

Blanco, M.[1], Villalba, D.[2], Casasús, I.[3], Sanz, A.[3] and Álvarez-Rodríguez, J.[3], [1]PCTAD, Avda. Montañana 930, 50059 Zaragoza, Spain, [2]UdL, Av. Rovira Roure 191, 25198 Lleida, Spain, [3]CITA, Avda. Montañana 930, 50059 Zaragoza, Spain; dvillalba@prodan.udl.es

Organic beef producers seek for fattening diets that comply with the 40:60 concentrate:forage (c:f) ratio imposed by EC Regulation 889/2008 without increasing labour consumption. Adding salt to concentrate may be interesting to reduce its intake provided it does impair animal health. Twenty-two young bulls (290 kg) were assigned to one of two feeding treatments during 42 days (d). Both groups received on ad libitum basis a concentrate (11.8 MJ ME/kg DM, 16.4% CP) that differed in sodium chloride content, 0.5% (Control) and 10% (Supplemented). Animals received lucerne hay (10.2 MJ ME/kg DM, 15.8% CP) and water on ad libitum basis. Concentrate intake and weight were controlled individually and water and hay intake on a group basis. Air temperature was recorded daily. Animals were bled biweekly to study the electrolyte balance (Na, K and Cl concentrations) and renal function (urea and creatinine concentrations). Salt addition reduced weight gains (1.58 vs. 1.76 kg/d for Supplemented and Control animals respectively, P<0.05) and concentrate intake (5.1 vs. 6.9 kg/d, P<0.001) while hay (3.7 vs. 1.5 kg/d) and water intake (73 vs. 50 l/d) increased. Concentrate:forage ratio was 58:42 and 82:18 for Supplemented and Control animals, respectively. Blood parameters remained within the normal reference ranges, except for Na and K in both groups on d 14 and 28. Salt addition increased plasma Cl concentration on d 0, 14 and 28, and K concentration on d 28 and 42, and it decreased creatinine at d 28. However, as water was freely available salt poisoning did not occur. Air temperature increased on d 28 above 25°C, causing an increase in plasma Na concentration and, concomitantly, water intake. However, plasma Na concentration was not affected by salt addition. Inclusion of 10% salt in the concentrate reduced its intake without impairing the metabolic homeostasis, and c:f ratio was reduced although the compulsory ratio was not achieved.

No influence of coarseness of grain and level of rumen by-pass starch of pelleted concentrates on performance, carcass quality, and rumen wall characteristics of rosé veal calves

Vestergaard, M., Eriksen, M. and Jarltoft, T.C., Aarhus University, Department of Animal Health and Bioscience, Foulum, DK-8830 Tjele, Denmark; mogens.vestergaard@agrsci.dk

In order to improve rumen environment and reduce the high incidence (16%) of liver abscesses in rosé veal calf production, two strategies were tested to improve the traditional (N) pelleted concentrates based on finely ground ingredients. In the R-concentrate, the cereal ingredients (barley and wheat) were the same as in N-concentrate but were coarsely ground resulting in a mean particle size of 1.5 compared with 0.6 mm in N before pelleting. In the S-concentrate, half the barley and wheat was replaced by finely ground sorghum and corn, which increased the theoretical by-pass starch to 68 g/kg compared with 25 g/kg in N and R. All three concentrates had the same total starch (345 g/kg), NDF (170 g/kg) and crude protein (15%) content and a pellet size of 4 mm. A total of 57 Holstein bull calves (n=19/treatment) were offered one of the three concentrates ad libitum from weaning (2 months) to slaughter (< 10 months). Intake was individually registered using Insentec feeders. Barley straw was available ad libitum. Daily gain (1430 g/d), feed utilization (4.2 kg concentrate/kg gain), LW at slaughter (386 kg), carcass weight (194 kg), and EUROP conformation (3.9) were not affected by type of concentrate (P>0.05). Rumen papillae length and shape evaluated in atrium and ventral rumen sac at slaughter was not affected by concentrate (P>0.05). Rumen wall condition showed degrees of clumping, hyperemia and necrotic areas in all treatment groups, but with no general differences between type of concentrate (P>0.05). Only for hyperemia in the ventral sac, S was slightly better than N (P<0.05). The results show that it was possible to obtain the same high level of production performance with all three types of concentrates but that neither more coarse ingredients nor more by-pass starch in a pelleted concentrate could improve rumen wall condition.

The Albanian buffalo: a case study of a successful conservation programe

Papa, L.[1], Kume, K.[2] and Tahiri, F.[2], [1]Agricultural University, Animal production, Kamez, Tirana, Albania, [2]CATT-Fushe Kruja, Rinas, Kruje, Albania; kkume@icc-al.org

Albanian buffalo is an authochonous breed, classified in the group of Mediterranean buffalo. Buffalos has been used mainly as draught power in Albania. However, Albanian farmers and consumers have been interested for their milk and meat products. Milk yield is around 450-600 kg in first lactation going up to 850-980 kg in third one. Its fat and protein content are respectively, 3-10.2% and 5.3-6.8%. Fertility rate is round 80-85% and days open 120-150 days. The mantel is black or dark grew and rarely with white spots. Horns arched and back and side inward bent. After political and economic system changes in Albania that happened during 1990 years, the population size of buffalo decreased rapidly. According to 2000 year census, the population size counted 110 animals, out of which 67 cows and 7 bulls. At that critical situation, flocks belonged to a few private farms; the biggest one had 36 buffalo animals, 22 buffalo cows and 3 buffalo bulls. The rest of 27 buffalos, 19 cows and 4 bulls were distributed to five small scale farms located at Divjaka region. The survival of Albanian buffalo breed is done just in Divjaka region. In the frame work of Albanian Action Plan for in-situ conservation and sustainable use of local breeds, measures are undertaken for buffalo breed also. As result the number of buffalo population is increasing. During 2009 year the buffalo population counted of 255 animals out of which 137 buffaloes, 27 bulls and 91 young stock. Two 'breeding nucleus' in two farms with 56 and 58 animals respectively have been established. Until now 12 male lines are selected. Nevertheless, the buffalo population went through the prolonged bottle-neck period, there is a need to control inbreeding level and provide breeders with sound and timely advice on exchange of the breeding stock. The current task is also to ensure that buffalo breed will have its place in farm production system. So, the breeders will continue to get profits from buffalo farming even when the subsidies system will not be available.

Comparison of organic and inorganic zinc supplements on some blood test parameters in early lactating Holstein cattle

Sobhanirad, S.[1], Bahari-Kashani, R.[1], Hosseinpour, M.[1] and Rabiei, E.[2], [1]Islamic Azad Unoversity Mashhad Branch, Animal science, Animal science dep.-Islamic Azad Unoversity-Golbahar-Mashhad, 91735-413, Iran, [2]Ferdowsi University of Mashhad, Agriculture Faculty-Ferdowsi University of Mashhad- Azadi sq. -Mashhad, 91775-1163, Iran; Sobhanirad@gmail.com

This experiment investigated two supplemental zinc (inorganic and organic) and basal diet of Holstein dairy cattle (N=54). Zinc content of the control diet was 42 mg/kg DM. The amount of 500 mg of Zn/kg of DM as zinc sulfate monohydrate (ZnS) and zinc methionine (ZnM) was added to the basal diet. The main ingredients of the basal diet (%) were alfalfa hay (15), corn silage (23), ground corn (25), soybean meal (18.5), beet pulp (14), protected fat (3), urea (0.2), dicalcium phosphate (0.3), salt (0.3), trace minerals and vitamins premix (0.7). This experiment was started in the first phase of lactation (35±3 days after parturition). Animals were housed in individual stalls and fed the experimental diet as a TMR at 08.00 and 20.00 h. Overall effect of treatment was tested using cow within treatment as the error term. Blood were taken on 0, 2, 4, 6, 8 and 10 weeks of experiment and in tubes containing no additive was allowed to clot at ambient temperature (15 to 21 °C). At the laboratory, they g for 10 min), and the serum stored at -20 °C until they were centrifuged (3500× further analysis. The data were analysed using the mixed procedure of SAS (9.1) for a block randomized design with repeated measures. There were no differences for cholesterol, triglyceride, BUN, HDL, and LDL. There were also no differences for albumin, glucose, and total protein among different sources of Zn, despite the lactating dairy cattles fed ZnM had numerical higher concentration of these parameters in blood serum. These data indicated that organically bound zinc had numerically higher effect on some blood parameters.

Analysis of the evaluation methodology for behavioral parameters established on Fighting-Bulls

Valera, M.[1], Daza, J.[2], Bartolomé, E.[1] and Molina, A.[2], [1]Department of Agro-forestry Sciences. University of Seville, Ctra. Utrera km1, 41013 Seville, Spain, [2]Department of Genetics. University of Cordoba., Rabanales Campus. Ctra. Madrid-Cordoba km 396a, 14071 Cordoba, Spain; ebartolome@us.es

During the last years, the MERAGEM research group has being working on the development of a reliable methodology for the evaluation of behavioral traits on fighting-bulls. This method would allow us to carry out a genetic evaluation for fighting ability in this breed. It was denominated 'Linear Standardized Evaluation of the Behavior' (CLEC) and tries to measure the most desirable behavior patterns for a fighting-bull (bravery, nobility, endurance, etc.). A panel of judges collected linear behavior parameters ('Maximum starting distance', 'Horse promptness', 'Speed during the attack', etc.). In order to increase the reliability of this study, environmental factors that could determine the behavior of the fighting-bull, were also included in the analysis (bulls' weight, temperature, wind, audience, etc.). CLEC methodology included 17 parameters (6 behavioral characters of the fighting-bull attitude against the horse; 9 behavioral characters against the muleta and 2 global evaluations (one for the bullfighter and one for the bull). We have analyzed the data attained from CLEC evaluation methodology, during the period 2008-2009. It has been computed the mean reproducibility and repeatability for each behavioral parameter. Deviations of these evaluations were also computed for each parameter. The values ranged from 87.3% ('Speed during the attack') and 96.1% ('Bull-ring location of the bullfight') for reproducibility. Regarding to mean repeatability, values ranged from 82.9% ('Promptness') to 98.6% ('Bull-ring location of the bullfight'). These results showed that CLEC methodology was suitable for behavior analysis on fighting-bulls

---

Public perceptions of dairy sector in northern Greece

Mitsopoulos, I.[1], Ragkos, A.[2] and Abas, Z.[3], [1]Technological Educational Institute of Thessaloniki, Department of Animal Production, Alexander TEI of Thessaloniki, P.O. Box 141, 574 00, Thessaloniki, Greece, [2]Technological Educational Institute of Thessaloniki, Department of Rural Development and Farm Management, Alexander TEI of Thessaloniki, 57400, Thessaloniki, Greece, [3]Democritus University of Thrace, Department or Agricultural Development, 193 Pantzidou Str., Orestiada, Greece; gmitsop@ap.teithe.gr

The dairy cattle sector in Greece is one of the most important sectors in farm animal breeding. Apart from its vital contribution to incomes and employment in rural areas, the dairy cattle sector is a multifunctional activity, as it sustains rural landscapes and poses threats on environmental quality, especially in protected areas. Despite its complex interactions with the environment, dairy farming in Greece is a traditional economic activity whose role in safeguarding cultural features and preventing depopulation in rural regions is incontestable. Such non-economic characteristics of the sector often provide additional arguments in favor of retaining protectionism in the greek dairy cattle farming sector. The purpose of this study is to examine the attitudes of the greek public towards the dairy cattle sector. Central to this investigation is the recognition of the main stakeholders, which are the groups that are directly or indirectly affected by the sector. Stakeholders' attitudes are thus examined by means of a questionnaire-based survey. The questionnaire includes questions regarding the economic, environmental and social functions of dairy cattle farming in Greece as well as respondents' social and economic characteristics. The results of the empirical analysis reveal stakeholders' awareness of environmental issues such as pollution of water resources (59.0%), farm animal welfare (95.0%) and rural landscapes (63.7%). Furthermore, respondents strongly acknowledge the dairy sector's socioeconomic role in rural areas (94.0%). Given the sector's multiple functions, the public endorses price policy measures (83.2%) as well measures that will improve market conditions.

Mothering-spring relationship in Guzerat (*Bos indicus*) under selection for milk production

Pires, M.F.A.P.[1], Praxedes, V.A.[1], Peixoto, M.G.C.D.[1], Braga, L.D.C.[2], Brito, L.F.[3], Freitas, L.S.[4], Pereira, M.C.[1], Bergmann, J.A.G.[4] and Verneque, R.S.[1], [1]Embrapa, Embrapa Gado de Leite, Rua Eugênio do Nascimento,610, 36038330, Brazil, [2]Faculdade de Ciências Agrárias da UFVJM, Diamantina, 39100-000, Brazil, [3]Universidade Federal de Viçosa, Departamento de Zootecnia da UFV, Viçosa - MG, 36570000, Brazil, [4]UFMG, Escola de Veterinária da UFMG, Belo Horizonte - MG, 31270-901, Brazil; fatinha@cnpgl.embrapa.br

Milk production systems using some Zebu (Bos indicus) breeds face restrictions in Brazil related to animal behavior. They are regarded bad temperament animals, mainly after calving, presenting risks to human, increasing production costs and reducing milk quality. The objective of this study was to understand the mothering-spring relationship of Guzerat cows in order to identify factors influencing this trait and to set up adequate management practices. Observations were carried out in two herds located in the southern (S) and eastern (E) Minas Gerais, Brazil. Indices of temperature and humidity were 67.40±3.72 (S) and 80.25±2.39 (E). Cows and calves were directly observed in grass paddocks since the beginning of the delivery and so during four hours after the end of calving. Only the daytime calvings were monitored, in a total of 12. Calving length averaged 56.33±29.68 minutes. The average of birth weight was 34.95±4.5 kg. Results of this initial study found latency to calves stand up of 52.33±18.09 minutes when it begins to look for teats (12.33±17.09 minutes). Latency from the calving to the first suckling averaged 93.60±53.71 minutes. First suckling has taken 3.20±2.16 minutes and during the observation period it has taken the average of 9.25±13.37 minutes. Latency from stand up to the first suckling was 41.40±35.75 min. Cow movements for stimulation and cleaning the calf lasted on average 64.83±38.26 minutes and time cow expended with other activities was 160.08±39.80 minutes. Along observation period, calves stayed lying down for 65±25% e stand up for 27±25%. Remained time was expended with the attempts to stand up.

The influence of feeding time and parturition induction on the time of calving in Japanese black beef cows

Sakurai, Y.S., Hokkaido Animal Research Center, Shintoku,Hokkaido, 081-0038, Japan; sakurai-yosie@hro.or.jp

The objective of this study was to evaluate the influence of feeding time and parturition induction by PGF2α on the time of calving. Twenty three multiparous Japanese black were devided into four groups with feeding time and calving method, 1) Free feeding (FF) and NP: free access to feed and natural parturition (NP), 2) Night feeding (NF) and NP: time access to feed from 17:00 to 9:00 on the next day and NP, 3) FF and parturition induction (PI): free access to feed and PI by PGF2α 50 mg administration at 19:00 a week before the expected date of calving, 4) NF and PI. The time of parturition was defined as the time of fetal expulsion and categorized into the two periods of day time (7:00-19:00) and night time (19:00-7:00 on the next day). One hundred % (4/4) of FF-NP cows and 50% (2/4) of NF-NP cows calved during the daytime. There was no treatment effect to increase the rate of daytime parturition; however, it was few cows used in these treatments. In the PG treatment cows, calving was induced 55.0±9.3 hours in FF and 58.2±19.9 hours in NF after PGF2α treatment. The rate of daytime parturition was 63% (5/8) in FF-IP and 86% (6/7) in NF-IP. This result suggested that the combination of night feeding and parturition induction increases the rate of daytime parturition. This work was supported by the Programme for Promotion of Basic and Applied Researches for Innovations in Bio-oriented Industry (2008-2010).

Animal welfare related to temperament and different pre slaughter procedures

Del Campo, M.[1], Manteca, X.[2], Soares De Lima, J.[1], Brito, G.[1], Hernández, P.[3] and Montossi, F.[1], [1]INIA Uruguay, Ruta 5 Km386, 45000 Tbó, Uruguay, [2]Univ Aut Barcelona, Bellaterra, 08193, Spain, [3]Univ Polit Valencia, Camí de Vera s/n, 46022, Spain; mdelcampo@tb.inia.org.uy

Sixty steers were finished on 1) native pasture + corn grain and 2) high quality pasture, to study the effect of diet, temperament (TMP) and lairage (LA) time on animal welfare. Animals were slaughtered the same day in two groups (n=30, 15 from 1 and 15 from diet 2) staying 15 and 3 hours in LA pens, respectively. TMP was assessed monthly by flight time and crush score tests, and health was daily monitored. Stress was evaluated after transport and LA, and immediately pre slaughter, by cortisol, creatin kinase and non esterified fatty acids concentration in blood. Behaviour was directly observed during LA using a scan sampling technique. Carcass pH was registered at 24 hours post mortem. Data was analised by parametric and non parametrics tests, using SAS package. Productivity and TMP did not differ between diets and health status was satisfactory. Calmer animals had higher daily gains and a lower stress response at all pre slaughter stages. LA and preslaughter handling increased the hypothalamus-hypophisis-adrenal axis in both slaughter groups, suggesting some grade of psychological stress. A high frequency of negative behaviour was detected in both groups during the first hour in LA, but animals became calmer afterwards. Noise generated by regular abattoir activity, was higher in the morning. This, added to the insufficient resting period in the short LA group, probably remained animals excitable before slaughter, contributing to glycogen depletion and to a higher carcass pH. The 15 hours group had a higher metabolic stress response but animals remained in LA overnight with greater opportunities to recovery, having a better rate of pH decline. This results suggest that more than 3 hours pre-slaughter time should be necessary, mainly depending on LA conditions. TMP is an important feature, considering its effect on productivity and also on stress response at different pre-slaughter stages.

Macroelements content in the milk of six dairy cattle breeds

Hódi, K.[1], Szendrei, Z.[1], Holcvart, M.[1], Kovács, B.[2], Mihók, S.[1] and Béri, B.[1], [1]University of Debrecen Centre for Agricultural Sciences and Engineering, Institute of Animal Husbandry, Böszörményi str. 138, 4032 Debrecen, Hungary, [2]University of Debrecen Centre for Agricultural Sciences and Engineering, Institute of Food Science, Quality Assurance and Microbiology, Böszörményi str. 138, 4032 Debrecen, Hungary; hodik@agr.unideb.hu

The level of mineral elements is an important factor regarding the quality of milk. The aim of our research study was to determine the content of mineral elements in the milk of Holstein, Jersey, Brown Swiss, Ayrshire, Norwegian-red, Swedish-red cows in the first stage of lactation. All cows were fed the same organic feed and were kept under the same conditions at the organic dairy farm. The concentration of macroelements (K, Na, Ca, P, Mg, S) in digested milk samples was determined by inductively coupled plasma optical emission spectrometry (ICP-OES). The data were analysed using one-way analysis of variance (ANOVA) to examine statistical significance of differences in the mean concentration of macroelements. Our examination has started in April of 2009. According to the results received so far we found that the Jersey's milk contained significantly more Na, Ca, P, Mg, S, while it contained less K than milk of other breeds.

Session 35 Theatre 1

Serum cortisol and plasma NEFA response to adrenocorticotropin hormone and insulin, respectively, is influenced by sire in Merino sheep

Preston, J.[1], Tilbrook, A.J.[2], Dunshea, F.R.[1] and Leury, B.J.[1], [1]The University of Melbourne, Department of Agriculture and Food Systems, Parkville, VIC 3010, Australia, [2]Monash University, Department of Physiology, Clayton, VIC 3800, Australia; brianjl@unimelb.edu.au

There is a significant relationship between an animal's serum cortisol response to exogenous administration of adrenocorticotropin hormone (ACTH) and the efficiency of feed utlization measured in unselected meat breed rams. However, the link between sheep metabolism and wool growth efficiency is poorly defined and the importance of sire is unknown. Therefore, this study investigated the influence of sire on serum cortisol and plasma metabolite responses to ACTH and insulin administration, respectively. Progeny from four sires (n=8 single and n=8 twin lambs per sire) with different breeding and production backgrounds were selected and managed identically during the study. There was no difference in wool yield between sires but singletons produced more wool than twin lambs (P<0.05). Following intramuscular ACTH administration, serum cortisol increased significantly by +30 mins (P<0.001); this response was significantly modified by sire (P<0.05) but not birth type. There was a weak positive correlation between wool yield and cortisol response 60 minutes post injection (P<0.05). Following insulin administration, plasma glucose and NEFA concentration decreased reaching nadir's at +60 and +30 mins (P<0.001), respectively. Plasma NEFA concentration rebounded significantly above baseline concentration by +90 mins and this was modified by sire (P<0.05) and birth type (P<0.001). In conclusion, ACTH and insulin induced stress responsiveness, indicated by changes in serum cortisol and plasma metabolites, is different between progeny from different Merino sires and this may impact production efficiency and metabolic adaptation to environmental challenges.

Session 35 Theatre 2

The beta3-adrenergic agonist (BRL 35135a) acutely increases oxygen consumption and plasma intermediate metabolites in sheep

Samadi, T.[1], Jois, M.[2], Leury, B.J.[1] and Dunshea, F.R.[1], [1]The University of Melbourne, Agriculture and Food Systems, Parkville, VIC 3010, Australia, [2]La Trobe University, Department of Agricultural Sciences, Bundoora, VIC 3083, Australia; fdunshea@unimelb.edu.au

An atypical adrenoreceptor subtype may be involved in mediating some of the physiological effects of catecholamines, particularly in some adipose tissue sites. Therefore, three experiments were conducted to determine the metabolic and energetic responses to oral dose of the beta$_3$-agonist BRL35135A in ruminant lambs. The post-prandial increase in O_2 consumption (0.109 v. 0.139 L/min) and CO_2 production (0.102 v. 0.127 L/min) at 30 min after feeding were greater (P<0.05) in the lambs receiving 5 mg of the BRL35135A. Treatment x time interactions over the period between -50 and 220 minutes indicate significant increases in plasma non-esterified fatty acids (NEFA) (P<0.001), glucose (P<0.001) and lactate (P=0.024) in lambs consuming a single oral dose of 5 mg BRL35135A. In a subsequent experiment there were similar interactions over the period between -120 and 1440 minutes for NEFA (P<0.001), glucose (P<0.001) and lactate (P<0.001) in lambs consuming 1 mg of BRL35135A. The effects of BRL35135A on plasma NEFA (P=0.95), glucose (P=0.84) and lactate (P=0.68) were not modified by the beta$_{1\&2}$-adrenergic antagonist alprenolol indicating that the effects were mediated via beta$_3$-adrenergic receptor subtypes. In conclusion, these experiments confirm that BRL35135A is acutely orally active in sheep.

Upregulation of plasminogen activator-related genes in ovine macrophages and neutrophils during mastitis

Theodorou, G.[1], Daskalopoulou, M.[1], Chronopoulou, R.[1], Baldi, A.[2], Dell'orto, V.[2], Rogdakis, E.[1] and Politis, I.[1], [1]Agricultural University of Athens, Animal Sciences and Aquaculture, 75 Iera odos Str, 11855 Athens, Greece, [2]University of Milan, Veterinary Sciences and Technology for Food Safety, via Celoria, 10, 20133 Milan, Italy; gtheod@aua.gr

Migration of neutrophils and monocytes towards the mammary gland is of undoubted importance during mastitis. The objective of the present study was to examine changes in expression of plasminogen activator (PA) related genes in ovine monocytes and neutrophils isolated from healthy and diseased (mastitic) animals. A total of 48 blood samples were collected from 16 animals (8 healthy and 8 mastitic) belonging to the breeds Chios, Boutsiko and a synthetic breed (50% Boutsiko, 25% Arta, 25% Chios). All animals belonging to the mastitic group exhibited clinical symptoms. Additional criteria used to classify an animal as belonging to the mastitic group were SCC and bacteria numbers in milk. Monocytes – macrophages and neutrophils were isolated from all blood samples followed by RNA extraction. Expression of urokinase-type PA (u-PA), u-PA receptor (u-PAR) and PA inhibitors type 1 (PAI-1) and type 2 (PAI-2) was evaluated by Real Time PCR. Results indicated that expression of all four PA-related genes (u-PA, u-PAR, PAI-1 and PAI-2) was very low in monocytes and essentially non-existent in neutrophils isolated from healthy animals. In contrast, there was a 2-4 fold increase in expression of all four PA-related genes in monocytes isolated from mastitic animals. A dramatic increase in the expression of all four genes was observed in neutrophils isolated from mastitic animals. Upregulation of PA-related genes in monocytes and neutrophils, is thought to facilitate the rapid migration of these cells towards the mammary gland during a mastitic infection.

Session 35 Theatre 4

Effect of long day photoperiod on milk yield of dairy goats in early and late lactation

Russo, V.M.[1], Cameron, A.W.N.[2], Tilbrook, A.J.[2], Dunshea, F.R.[1] and Leury, B.J.[1], [1]The University of Melbourne, Department of Agriculture and Food Systems, Parkville, VIC 3010, Australia, [2]Monash University, Department of Physiology, Clayton, VIC 3800, Australia; brianjl@unimelb.edu.au

Short day length is associated with reduced milk yield from dairy animals. Dairy cows and sheep have been successfully housed under lights throughout winter to extend the day length and increase milk yield. Very little data exists on the effect of extended day length on the milk yield of dairy goats. The objective of this study was to examine the effect of long day photoperiod (LDPP) on milk production in dairy goats in early or late lactation. The study was conducted over an 8 week period from June to August (winter in the Southern hemisphere). The goats were kept in open sided sheds in which the Control treatment received natural lighting while the LDPP treatment received 16 hours of light, part of which was artificial. There was a difference ($P<0.001$) in milk yield between treatments, with a far greater response to LDPP occurring in late lactation animals. After 8 weeks of treatment the late lactation goats under LDPP produced 18.3% more milk than late lactation animals under control conditions, while those exposed to LDPP in early lactation produced 8.7% more milk than their controls. For both stages of lactation the effect of treatment was significant after only one week. Throughout the 8 week period no refractory effect was seen in late lactation LDPP and only a minor effect in early lactation. Milk composition was significantly affected by treatment, such that the total milk solids concentration and bulk milk cell count decreased with LDPP treatment. Plasma prolactin concentrations were significantly greater ($P<0.001$) in does housed under the LDPP compared with control, which is consistent with the reported effects of LDPP increasing plasma prolactin in lactating animals. Increasing photoperiod during winter can substantially increase milk yield in Australian dairy goats, especially lactation persistence and yield in late lactation.

Ration optimization of dairy sheep

Verkaik, J.C., Hindle, V.A. and Van Riel, J.W., Wagingen UR Livestock Research, Animal Welfare and Health, P.O. Box 65, 8200 AB Lelystad, Netherlands; jan.verkaik@wur.nl

Efficient use of protein and reductions in feeding costs are of particular relevance to organic diary sheep farmers in the Netherlands. In particular the protein standards are being questioned. On Dutch dairy farms Dynamic Linear Modelling (DLM) has proven to be a reliable tool for economic optimization. DLM is a self-adjusting model that calculates a daily economic saldo based on actual milk yield in response to concentrate intake. The current pilot study aimed to test the consequences of using DLM on sheep farms on flock performance and feed composition. A 16-week on-farm assessment was performed with 3 groups of ewes in early lactation at 3 organic diary farms. Every week DLM produced a flock level feeding advice based on daily feed intake and milk production, weekly averages of milk composition (fat and protein), milk price and feed costs and the estimated feeding value of concentrates and roughages. Body weight and condition was determined on four occasions during the trial period. A comparison was made between DLM and the standards based on milk yield, growth, intake of protein (DVE) and energy (VEM). No differences were observed between groups in terms of lambing performance. The starter rations fed during the last weeks of pregnancy were judged to be sufficient for the transition to early lactation. The DLM advice resulted in lower levels of protein being fed to the flocks compared to Dutch standards (15-20%). Overall milk production of the flocks was equal to the Dutch standard (2.41 kg/d). Growth and condition increased during the trial in all groups (0.08 kg/d and 0.4 respectively). The assessment demonstrates that DLM is a useful management tool for economic optimisation of dairy sheep nutrition on organic farms. In addition, it appears that a lower level of protein compared to conventional Dutch standards is acceptable from a performance point of view. This support recent recommendations lowering on protein standards for Dutch diary sheep.

Session 35 Theatre 6

Effect of dietary cation-anion difference in dairy ewes at mid-lactation

Schlageter, A., Caja, G., Ben Khedim, M., Salama, A.A.K., Carné, S. and Albanell, E., Universitat Autònoma de Barcelona, Grup de Recerca en Remugants (G2R), Campus universitari, Edifici V, 08193 Bellaterra, Barcelona, Spain; gerardo.caja@uab.es

A total of 40 ewes (Manchega, n=20, Lacaune, n=20) at 84±27 DIM were used to study the performance effects of diets with distinct dietary cation-anion difference (DCAD). The DCAD values were calculated in mEq/100 g DM as $(Na^+ + K^+) - (Cl^- + S^{2-})$. Ewes were allocated to 8 groups of 5 animals and blocked by breed, BW and milk yield (Manchega, 71.0±7.7 kg BW and 0.59±0.15 L/d; Lacaune, 69.2±8.0 kg BW and 0.97±0.17 L/d). Treatments were applied for 10 wk and consisted of total mixed rations in which the DCAD values were: 7, 26, 45 or 64 mEq/100 g DM. Individual dry matter intake (DMI) was measured at wk 5 and 10 using polyethylene glycol 6000 (PEG, 50 g/d for 14 d) as indigestible external marker. Milk yield and milk composition were recorded weekly and biweekly, respectively. Jugular blood samples for acid-base balance were taken and analyzed at wk 4, 8 and 10. Individual DMI and DMI/kg $BW^{0.75}$ showed a linear and quadratic response by effect of DCAD. Maximum intake was calculated between 40 and 45 mEq/100 g DM, for DMI and DMI/kg $BW^{0.75}$, respectively. Milk yield showed a linear response to DCAD in Lacaune ewes but it did not vary in the Manchega. There were no significant effects of treatment in milk composition. We also observed differences in acid-base blood indicators according to breed; Manchega ewes showed a neutral stage, while Lacaune ewes passed from metabolic acidosis (7 and 26 DCAD diets) to neutral (45 and 64 DCAD diets). Urine pH showed a linear response by effect of DCAD and was a representative indicator of the DCAD values in the ration. Blood Cl^- and K^+ showed linear and quadratic response by effect of DCAD in Lacaune ewes, increasing both ions in the anionic treatments. According to the obtained results, diets containing DCAD values in the range of 35 to 45 mEq/100 g DM are recommended for lactating dairy ewes. High yielding dairy ewes were more sensitive to DCAD than low yielding ewes.

Session 35 Theatre 7

Oral thiazolidinediones increase feed intake and growth in sheep

Fahri, F.T.[1,2], Clarke, I.J.[3], Pethick, D.W.[2], Warner, R.D.[4] and Dunshea, F.R.[1], [1]The University of Melbourne, Department of Agriculture and Food Systems, Parkville, VIC 3010, Australia, [2]Murdoch University, School of Veterinary and Biomedical Sciences, Murdoch, WA 6150, Australia, [3]Monash University, Department of Physiology, Clayton, VIC 3800, Australia, [4]Department of Primary Industries, Sneydes Road, Werribee, VIC 3030, Australia; ffahri@unimelb.edu.au

Feed efficiency, growth and body composition are all key drivers of productivity and profitability in livestock production systems. Thiazolidinediones (TZD) are synthetic compounds currently used clinically as oral antidiabetic drugs.Although the mode of action is still not well defined, research suggests that TZD interacts with specific genes that regulate fat cell development, energy balance, and lipid metabolism in fat tissue and skeletal muscle. Thirty Poll Dorset x Merino x Border Leicester 5 month old lambs of mixed sex (15 ewes and 15 wethers) were blocked on live weight and body fat (as evaluated by Dual X-Ray absorptiometry (DXA)) and allocated to a 2x3 factorial design with the respective factors being sex (ewe or wether) and dose of TZD (0, 8 or 24 mg/d of rosiglitazone maleate). Placebo or TZD tablets were administered via gavage daily for the duration of the study. Lambs were fed commercial pellets (Rumevite™) ad libitum, and 100 g of lucerne chaff daily for 8 weeks. Lambs were DXA scanned live at Day 0, 26, and as carcase at day 55 where meat quality measurements were carried out. Lambs treated with TZD gained more than control (215 vs. 266 g/d, P=0.024) and had higher FCE (0.136 vs. 0.158, P=0.017). Treated animals had greater back fat at the 8th rib (9.4 vs. 12.3 mm, P=0.015) and carcase fat (27.5 vs. 29.7%, P=0.027). These data suggest that TZD promotes weight gain but that some of the weight gain is as fat.

Session 35 Theatre 8

Conditioned aversion to olive tree leaves by lithium chloride in sheep and goats

Manuelian, C.L., Albanell, E., Salama, A.A.K. and Caja, G., Universitat Autònoma de Barcelona, Grup de Recerca en Remugants (G2R), Campus universitari, Edifici V, 08193 Bellaterra, Barcelona, Spain; ahmed.salama@uab.es

Lithium chloride (LiCl) was used to induce aversion to olive (*Olea europaea* L.) tree leaves in 10 Murciano-Granadina does and 10 Manchega ewes. Two parallel experiments were done to compare olive leaf intake in averted vs. control animals. Animals had no previous contact with olive leaves. All does and ewes were dry and non-pregnant and were randomly allocated into 2 experimental groups of 5 animals each per specie. For aversion induction, all animals were penned in individual boxes during 6 d, fed tall fescue hay ad libitum, and offered the olive leaves for 5 min (does) or 1 h (ewes). Aversion was induced using a drenching gun for giving a solution of LiCl in water (0.2 g/kg BW) after olive leaf consumption in the averted groups. In the control groups only water was drenched. Does learnt faster to eat the olive leaves and developed a stronger aversion to the LiCl treatment than ewes. After the aversion induction period, all the animals joined the respective herd or flock, grazed a cultivated pasture during the day and were complemented with tall fescue hay in the shelter. Aversion memory was evaluated by fortnightly tests of 5 min lasting 144 d in does, and 10 min lasting 130 d in the ewes. No LiCl was given during the memory test period. No olive tree leaf consumption was detected until d 53 and 23, for does and ewes, respectively (P<0.05). Consumption of the averted groups was lower than in the control groups throughout the experiment (P<0.05). Does and ewes behavior differed between averted and control groups during learning and memory test periods. Animals in the control group avidly ate the olive leaves, whereas animals in the averted group strongly rejected the olive leaves. In conclusion, the LiCl conditioned aversion to olive tree leaves proved to be an efficient short-term method in sheep and goat. The method may be of special interest for selective grazing among olive trees and for avoiding the use of herbicides in organic agriculture.

Effect of tick-borne fever on weaning weight in sheep

Grøva, L.[1,2], Olesen, I.[1,3], Steinshamn, H.[2] and Stuen, S.[4], [1]University of Life Sciences, Animal and Aquacultural Sciences, Pb 5003, 1432 Ås, Norway, [2]Bioforsk Organic Food and Farming Division, Gunnarsv 6, 6630 Tingvoll, Norway, [3]Nofima Marin, Pb 5010, 1432 Ås, Norway, [4]Norwegian School of Veterinary Science, Section for small ruminant research, Kyrkjev 332/334, 4325 Sandnes, Norway; lise.grova@bioforsk.no

A main scourge in Norwegian sheep farming is tick-borne fever (TBF) caused by the bacteria *Anaplasma phagocytophilum* and transmitted by the tick *Ixodes ricinus*. The main consequence of *A. phagocytophilum* infection in sheep is the ensuing immunosuppression that may lead to secondary infections. Sheep flocks on tick pasture may suffer heavy losses due to mortality, crippling conditions in affected lambs, but also impaired growth rate. It is expected that more than 300,000 lambs (ca 20%) are affected annually by TBF, but the extent of production losses due *A. phagocytophilum* infection is less known. Thus, the objective of the present work is to quantify indirect production losses due to *A. phagocytophilum* infection on tick pasture. Blood from lambs was sampled during autumn 2007 and 2008 from a total of 1,220 lambs from 12 different farms of one ram circle in the Sunndal, Ålvundeid and Todal municipalities on the west coast of Norway. Autumn body weight was recorded at an average of 137 days. Serum was analyzed for antibodies to *A. phagocytophilum* by an indirect fluorescent antibody assay (IFAT). Body weight of antibody positive and negative lambs was compared using ANOVA and MIXED procedure in SAS. Antibody results showed that 55% of lambs were seropositive to *A. phagocytophilum*. Furthermore, antibody positive lambs obtained a 1.34 kg (±0,412) (or 3,0%) lower autumn bodyweight than antibody negative lambs. The present study shows that *A. phagocytophilum* infection is widespread in the sampled area. The moderate weight reduction of 1.34 kg in infected lambs compared to 3.8 kg observed in a previous less extensive study might be explained by the presence of different genetic variants of *A. phagocytophilum*.

Research regarding the situation of goat size exploitations and goat breed structure in Romania

Raducuta, I.[1] and Ghita, E.[2], [1]University of Agronomical Sciences and Veterinary Medecine, Faculty of Animal Sciences, Str. Marasti no.59,district 1, 011464, Bucharest, Romania, [2]National Research-Development Institute for Animal Biology and Nutrition, Animal Biology, Calea Bucuresti no.1, 077015, Balotesti, district Ilfov, Romania; ghita.elena@ibna.ro

The purpose of this work is to investigate the situation of goat size farms and goat breed structure in Romania after the integration in EU. At present, in Romania the total number of goat exploitations is still high (131,795 units) compared to other EU member states with tradition of rearing this species (France, Greece, Spain, etc.), which is due to the fact that there are still a few large farms (1,789 units with over 50 head/unit) and many units where the number per herd is very small (121,779 units with 1-10 head/unit). In these last units, goats are kept only for family self-consumption. However the total number of goat exploitations was significantly reduced in 2008 compared to 2003, respectively 44%. As regards the goat breed structure there are five breed classes in Romania which detain in order the following percentage from the globally goat livestock: Carpathian (91.30%), White Banat (7.37%), Saanen (0.30%), French Alpine (0.07%) and Crossbreeds (0.96%). In recent years were imported dairy breeds such as Saanen and French Alpine to improve the milk production of local Carpathian breed. Crossbreeds resulting from these crosses are characterized by a better body conformation and a double milk production versus local Carpathian breed.

Prevalence of natural gastrointestinal nematode infections of sheep in Germany

Idirs, A., Moors, E. and Gauly, M., Livestock Production Systems, Animal Science, Albrecht Thaer Weg 3, 37075 Goettingen, Germany; mgauly@gwdg.de

The objective of the present study was to determine the prevalence and intensity of nematode infections of lambs based on individual faecal eggs counts (FEC) from various regions in Germany. Therefore, a total of 3924 lambs (3 to 15 month-old) with different genetic backgrounds (Merinoland, Blackhead Mutton, Rhoen, Texel and Merinolangwoll) were sampled during the grazing season. Pooled faecal samples from each farm/ breed were cultured for third-stage-larval differentiation of the nematodes spp.. Intensity of coccidia infection was semi quantitatively scored via a four score scaling system (0 = coccidia free to class 3 > 6,000 oocysts per gram of faeces (OPG)). 62.8% of lambs were infected with at least one species of gastrointestinal nematodes. Mean FEC was 315 (±777). The infections were, in most cases, mixed and involved several different species. The eggs of Nematodirus spp. were detected in 13.0% of all samples. The larval cultures revealed Trichostrongylus spp. 54.1% of all larvae counts, Haemonchus contortus 32.2%, Teladorsagia circumcincta 13.3% and Cooperia curticei 0.4%. The samples of only 11.4% of lambs were *Eimeria* oocyst free. Tapeworm eggs were encountered in 13.2% of all samples. Lambs with more than 6,000 OPG had significantly higher log (FEC+10) in comparison to lambs that were either not infected with coccidia or had less than 1,800 OPG (Log-EPG= 4.61, 4.27 and 4.03, respectively; $P<0.05$). The present study shows that natural infections with gastrointestinal nematodes of lambs in Germany are common. The intensity of infections was moderate and involved multi-species infections. The *Trichostrongylus* spp. were the predominant species.

Session 35 Poster 12

Space allowance and housing conditions influence the welfare and production performance of dairy ewes

Caroprese, M.[1], Santillo, A.[1], Marino, R.[1], Bruno, A.[2], Muscio, A.[1] and Sevi, A.[1], [1]University of Foggia, PrIME, Via Napoli, 25, 71100 Foggia, Italy, [2]CRA, Istituto Sperimentale per la Zootecnia, Via NApoli, Segezia, Foggia, Italy; m.caroprese@unifg.it

Intensive farming systems are characterized by a large number of animals per unit of space to maximize the production levels. The reduction in space allowance and in outdoor areas can markedly affect the welfare of farmed animals. The experiment was conducted to determine the effects of 2 different stocking densities and 2 different housing conditions on welfare, and on production performance of dairy ewes; 45 Comisana ewes were divided into 3 groups of 15. The stocking densities tested were: high stocking density (1.5 m^2/ewe, HD group) and low stocking density (3 m^2/ewe, LD group); the 2 housing conditions tested were: ewes housed indoor (LD group, 3 m^2/ewe) and ewes allowed to use an outdoor area (LDP group, 3 m^2/ewe divided in 1.5 m^2/ewe indoor and 1.5 m^2/ewe outdoor). The cell mediated and humoral immune responses to antigens were evaluated. Behavioral activities of ewes were monitored. After lamb weaning, individual milk yields were measured and milk composition analyzed weekly. Housing conditions affected cell mediated response, which was higher in LDP than in LD ewes ($P<0.001$). Concentrations of anti-OVA IgG were mainly influenced by space allowance, with higher IgG in LD than in HD ewes ($P<0.001$). Both housing conditions and space allowance affected sheep behavioral activities; a greater proportion of LDP ewes displayed standing and drinking behaviors than LD ewes ($P<0.001$ and $P<0.01$, respectively), and a greater proportion of LD ewes was observed walking than HD ewes ($P<0.01$). LDP ewes had a higher protein content ($P<0.05$) and lower somatic cell count in their milk ($P<0.05$), whereas reduced space allowance led to a reduction in milk yield and an increase in somatic cell count of milk ($P<0.05$). Results indicate that a space allowance of 3 m^2/ewe and the availability of outdoor areas can improve the welfare and production performance of the lactating ewe.

Activities of milk fat globule membrane enzymes in ewe's milk during lactation

Salari, F.[1], Altomonte, I.[1], Pesi, R.[2], Tozzi, M.G.[2] and Martini, M.[1], [1]Università di Pisa, Dipartimento di Produzioni Animali, Viale delle Piagge 2, 56124, Pisa, Italy, [2]Università di Pisa, Dipartimento di Biologia, Via S. Zeno 51, 56127, Pisa, Italy; mmartini@vet.unipi.it

Several components of milk fat globule membranes have been recently identified as being beneficial for human health. The milk fat globule membrane contains several enzymes, some acting as protective factors for the newborn. Although the enzymology of some milk fat globule membrane enzymes is documented, the physiological role of most of them in milk is unclear. Moreover only a few studies evaluate enzyme activities in ewe's milk fat globule membrane. With the aim of improving the knowledge about milk fat globules membranes we focused on the activities of 5 membrane enzymes: xanthine oxidase, xanthine dehydrogenase, γ-glutamyltranspeptidase, alkaline phosphatase and 5'-nucleotidase. A trial was carried out on 7 Massese ewes, homogeneus in terms of parity and feed, kept indoors one week before partum. The experiment lasted from 12 hours to 120 days *post partum*. During the experiment 77 individual milk samples were collected. The enzyme activities were evaluated on a milk fat globule membrane extract from colostrum and milk. The activity of xanthine oxidase increased significantly during lactation and reached the climax of activity on 45th day, then decreased on 120th day. On the contrary xanthine dehydrogenase activity did not change. The activity of alkaline phosphatase significantly increased as the lactation phase progressed, while 5'-nucleotidase increased after the 45th day and decreased on the 120th day. As reported on previously studies about bovine colostrum and milk, γ-glutamyltranspeptidase showed a higher activity during the first 10 days of lactation, a decrease between 20th and 45th day and a subsequent increase. Membrane proteins (mg/ml) showed higher values at 12 hour post partum and then decreased.

Effects of lateral presentation of food on dairy ewes

Acciaro, M.[1], Decandia, M.[1], Molle, G.[1], Cabiddu, A.[1], Rassu, P.G.[2], Pulina, G.[2] and Vallortigara, G.[3], [1]AGRIS Sardegna, Department of Animal Production Research, Olmedo, 07040, Italy, [2]University of Sassari, Department of Animal Sciences, Via De Nicola 2, Sassari, 07100, Italy, [3]University of Trento, Centre for Mind-Brain Sciences, Rovereto, 38068, Italy; macciaro@agrisricerca.it

Right and left brain hemispheres are different in the ability to regulate autonomic processes in the organism. Direct unilateral stimulation of the brain causes side-dependent endocrine, immune and other visceral reactions. Dairy cows showed differences in milk yield and reproductive performance when submitted to unilateral presentation of food. An experiment was undertaken to study the effect of lateral presentation of food on the milk production of dairy ewes. Thirty-six primiparous Sarda ewes were blocked for milk yield (MY), milk composition, body weight (BW) and body condition score (BCS) in 2 homogeneous groups. The two groups were stall-fed alfalfa hay and commercial concentrate from left (group L) and right (group R) side, respectively. Each group included 2 sub-groups starting the experimental treatment successively according to their lambing period. Nutrient intake of feedstuff (on a sub-group basis) was measured daily. Individual MY and milk composition (fat, protein, lactose, urea and somatic cell count), were measured weekly while BW and BCS every 3 weeks. Data of nutrient intake were tested by GLM using side of feed presentation, lambing period and their interaction as fixed effects. Data of BW, BCS, MY and milk composition were tested by MIXED statement using side of feed presentation, lambing period and their interaction as fixed effects and animal as random effect. Total DM intake was higher in L than R group (2.36±0.01 and 2.33±0.01 kg/head, $P<0.02$) as well as NDF intake (1.21±0.01 and 1.16±0.01 kg/head, $P<0.01$). No differences were found on milk yield 1301±36 ml (L) and 1318±36 (R), milk composition, BW and BCS. Thus, lateral presentation did affect food intake although at a mild level but contrary to results obtained in dairy cows it did not affect milk performance in dairy sheep.

Parameterisation of a dynamic rumen model in order to simulate the voluntary intake of Latxa sheep during lactation

Díez-Unquera, B.[1], Ruiz, R.[1], Silveira, V.[2], Mandaluniz, N.[1], Bernués, A.[2] and Villalba, D.[3], [1]NEIKER-Tecnalia, Animal Production, P.O. Box 46, E-01080 Vitoria, Spain, [2]CITA, Av. Montañana 930, 50059 Zaragoza, Spain, [3]Univ. Lleida, Av. Rovira Roure 191, 25198 Lleida, Spain; bdiez@neiker.net

A key step when assessing the accuracy of an animal simulation model is to predict voluntary intake in a reliable manner. Since the relation between forage intake and degradability is well known, the most common approach assumes that the rumen is the main site of the digestive tract controlling intake and models that predict intake as a function of animal requirements and forage quality are numerous. In order to build a model that simulates animal production and development, a simple model of rumen dynamics was developed derived from the works of Illius & Gordon (1991), Sniffen et al., (1992) and Herrero (1997). Visual Basic programming language was chosen for the implementation and parameterisation of the model. Main inputs related to the animal traits (live weight, physiological status, and activity) and feed characteristics (cell content, degradable cell wall, indigestible cell wall, degradation rate, protein fractions and crude protein content) were used. Validation of the model was conducted with published data containing feed characteristics and sheep voluntary intake of certain forages offered ad libitum and results were compared with model predictions. The results showed a sub-estimation of around 10%. Then, data collected within an experimental flock of Latxa ewes were used in a second validation. The data were collected from 12 ewes fed on a basis of 0.7 kg DM of a commercial concentrate, plus 0.5 kg DM of lucerne hay and herbage silage offered ad libitum. The observed silage intake was recorded during 5 weeks and compared with the predictions of the model. Initially the model showed less reliability when more than one food was included. The suitability of the model for assessing voluntary intake in more complex diets is discussed and options for improving fitness of accuracy are proposed.

Session 35 Poster 16

The effect of roughage quality and dried distillers grains as protein component on feed intake and milk performance in the diet of dairy goats

Ringdorfer, F. and Huber, R., LFZ Raumberg-Gumpenstein, Sheep and goats, Raumberg 38, 8952 Irdning, Austria; ferdinand.ringdorfer@raumberg-gumpenstein.at

Optimal diets for dairy goats are very important for high performance and healthy animals. For high milk yield the animals must be fed according their requirement with best quality feed stuff. Because of high cost of concentrate the amount in the ration must be considered well. High quality roughage can reduce the amount of concentrate. The feeding value of roughage mostly depends on the time of harvesting. Early cutted grass has a higher nutritive value than late harvested grass. With the building of a factory for ethanol production corn dried distillers grains (DDG) are available for feeding ruminants. The evaluate the use of DDG in the diet for dairy goats a feeding experiment was carried out. The aim was to replace soybeans as protein component in the concentrate with DDG. Three different concentrates was offered: group K-0 with 11.8% soybeans and no DDG, group K-50 with 5.9% soybeans and 9.4% DDG and group K-100 without soybeans and 18.7% DDG. The protein content was the same in all 3 groups, 15.6%, the energy content was 12.26, 12.15 and 12.04 MJ ME/kgDM. Additionally to the concentrate the animals had ad libitum access to hay with good (3 cut meadow) and best (4 cut meadow) quality. The experiment was carried out with 36 female Saanen goats for the first lactation. Animals where housed in pens on straw and was fed individually. Water was available for free intake. Body weight, daily hay and concentrate intake and milk performance where recorded. During the first 22 weeks of lactation there where no significant effect of DDG in concentrate on daily dry matter intake (1.95, 1.86 and 1.88 kg), on daily milk yield (2.34, 2.17 and 2.24 kg) and on milk fat (2.74, 2.86 and 2.79%) and milk protein (2.82, 2.91 and 2.84%) content. Hay quality had a significant effect on daily hay intake, the intake of 4-cut hay was higher than that of 3-cut hay (1.5 vs. 1.3 kg DM).

Effect of creep mixture diet on growth performance and carcass characteristics of pre-weaned lambs

Souri, M., Panah, M., Varahzardi, S. and Moeini, M., razi university, animal science, imam khomeiniSt. college of agriculture, animal science dep., 67155, Iran; M.souri@razi.ac.ir

The effect of *ad libitum* creep mixture (estimated metabolisable energy and crude protein concentration of 3.11 Mcal and 160 g per kg DM) on dry matter intake, liveweight gain and carcass characteristics of suckling lambs were studied. Ten male and ten female Sanjabi lambs (20-30 days of age) with initial liveweight of 13.4±1.3 and 11.6±0.9 respectively were used in a 100 day study. The lambs were blocked within sex according to their liveweight and randomly allocated to one of two treatments: control (free suckling and free choice of alfalfa leaves) or supplementary (*ad libitum* creep mixtures with free suckling and free choice of alfalfa leaves) in a 2x2 factorial arrangement. The growth study was continued up to approximately 120 days of lamb's age during which daily body weight changes of all lambs were recorded. Daily milk intake was recorded by lamb suckling method in daily throughout the study. At the end of the growth study, all lambs were slaughtered. Average daily milk intake of males was higher than the female lambs ($P<0.05$), whereas it was similar in the dietary treatments. Daily liveweight gain (gr/d) was significantly greater in males (177±0.015 for control group and 304±0.010 for treatment group) than female (103±0.013 for control group and 214±0.010 for treatment group) lambs ($P<0.05$). Both male and female lambs in supplementary deposited significantly ($P<0.01$) more fat tissue than control treatment. It could be concluded that the creep mixture diet affected growth rate but lead to the fatter lambs.

Evaluation the body composition of crossbred kids using computed tomography

Németh, T.[1], Nyisztor, J.[2], Kukovics, S.[1], Lengyel, A.[2] and Toldi, G.Y.[2], [1]Research Institute for Animal Breeding and Nutrition, Gesztenyés út 1., 2053 Herceghalom, Hungary, [2]Kaposvár University, Pf. 16., 7401 Kaposvár, Hungary; nemeth.timea@atk.hu

Sixteen female kids were selected from the same farm. The (Hungarian Milking Brown x Alpine) F_1 were 6 months old (n=8), while the (Hungarian Milking Brown x Boer) F_1 were 4 months old (n=8). The kids were scanned by Computed Tomography to evaluate and compare the body composition of Alpine and Boer crossbred progenies in order to estimate the quantity of meat. The area of fat, aquaeous, muscle and bone tissues were measured and recorded by the 'Australian method' with a 10 mm slice thickness and 20 mm slice distance. Forty-sixty pictures by individual, depending on the longness of vertebral column were evaluated by Medical Image Processing V1.0 software. The area of tissues were recorded in mm^2 and analysed by GLM (Generalized Linear Model) using live weight as a covariant (LSD-test; $P<0.05$) and partial correlation corrected for body weight. In Alpine crossbreds the average area of fat tissue was 64.6 thousands mm^2, while in Boer progenies it was 71.1 thousands mm^2. The average aquaeous tissue was much higher in Boer than Alpine crossbreds. The area muscle tissue in Alpine crossbred kids was 290.4 thousands and 372.0 thousands mm^2 in Boer crossbreds. The measured 81.6 thousands mm^2 difference was significant. The average area of bone tissue was similar in two crossbreds groups. The strongest partial correlation in Alpine F_1 was found between fat and aquaeous tissue, with a medium strongness as well as between muscle and aquaeous tissue, however, none of the relationships calculated became significant. In the case of Boer crossbred kids the correlations were medium, but two correlations (between fat and aquaeous tissue, and between muscle and bone tissue) were strong and significant. According to our conclusion the Boer goat had stronger and significant effect on body tissue composition comparing to Alpine breed concerning to meat production.

Correlation between ultrasound and direct measurements of carcass traits in Ripollesa, Lacaune and Ripollesa×Lacaune lambs

Esquivelzeta, C., Casellas, J., Fina, M. and Piedrafita, J., Universitat Autònoma de Barcelona, Departament de Ciència Animal i dels Aliments, Facultat de Veterinària, 08193 Bellaterra, Spain; joaquim.casellas@uab.cat

Real time ultrasound techniques were advocated as useful tools for predicting body composition in livestock. Their advantages were mainly linked to the assessment of carcass traits in live animals measured at relatively low cost in comparison with other techniques for carcass composition. The aim of this study was to evaluate the correlation between ultrasound and direct measurements of carcass traits in 130 lambs of the Ripollesa and Lacaune breeds, and their cross. Real time ultrasound images were taken perpendicular to the vertebral column one day prior to slaughter at around 24 kg, between 12th and 13th ribs and between 1st and 2nd lumbar vertebrae by means of a Sonosite 200 machine. Subcutaneous back-fat thickness (BFT), loin muscle depth (LMD), width (LMW) and area (LMA) were determined with the ImageJ 1.42q software. Averages of ultrasound measurements at the thoracic and lumbar level were very similar. Means for ultrasound and carcass measurements at the thoracic level were 0.16±0.004 and 0.23±0.01 cm for BFT, 2.40±0.02 cm and 2.47±0.03 cm for LMD, 4.53±0.05 cm and 4.50±0.04 cm for LMW, and 9.79±0.11 cm^2 and 9.97±0.12 cm^2 for LMA, respectively. Carcass cuts averages were 1.42±0.01 kg for right leg, 3.01±0.03 kg for rack, 1.08±0.01 kg for shoulder, and 0.82±0.01 kg for neck. Correlation coefficients between ultrasound and direct measurements were similar for both thoracic and lumbar locations, showing moderate estimates for BFT (0.55 and 0.61, respectively) and high for LMD (0.88 and 0.86), LMW (0.70 and 0.70) and LMA (0.72 and 0.78), all estimates reaching p-values smaller than 0.0001. Carcass pieces were also correlated with thoracic (0.21-0.35) and lumbar (0.18-0.39) ultrasonic measurements. In conclusion, ultrasound measurements are useful for predicting LMD, LMW, LMA and BFT.

Researches related to fattening performances achieved by hybrids issued from romanian sheep breeds crossed with meat type rams

Pascal, C., Gilca, I., Ivancia, M. and Nacu, G., University of Agricultural Sciences and Veterinary Medicine Iasi, Faculty of Animal Sciences, 3, Mihail Sadoveanu Alee, 700490, Romania; pascalc@uaiasi.ro

The researches aimed to assess the fattening aptitudes of sheep youth belonging to some different sheep groups. Control groups included local sheep breeds - Merinos of Palas (M1) and Ţigaie (M2), while the experimental treatments comprised weaned lambs, issued from Texel breed crossed with F1 hybrid females Bluefaced Leicester x Merinos de Palas (L1) and from Suffolk breed crossed with F1 hybrid females - Bluefaced L. x Tigaie (L2), respectively. Weaning has been done when lambs turned 85 days old. Intensive fattening technology was applied across a period of 90 days. The whole gain achieved by L1 lambs was 20.31% higher, compared to M1, while high degree of statistic significance ($P<0.01$) occurred for the average daily gain values. The same analysis, between M2 and L2 groups revealed 37.87% higher live weight in L2 group. All the differences found between L1 and M1, between L2 and M2 respectively were statistically significant for $P<0.01$ or $P<0.05$. L2 group had an average final weight of 39.242±0.258 kg, which was 43.65% cumulated during suckling period, respectively 56.35% during fattening. Feed conversion was found at 4.58 UFC for L1 and at 4.72 UFC in L2, these values being 21.44% better, respectively 20.13% better than M1 and M2. The data acquired when fattening ended suggested that hybrids synthesis and their rearing could be a more efficient way to improve sheep meat yield.

Basic carcass characteristics of light lambs of the Romanov breed and its Suffolk-sired crossbreeds

Kuchtik, J.[1], Dobes, I.[1] and Hegedusova, Z.[2], [1]Mendel University in Brno, Department of Animal Breeding (LA330, LA09031), Zemedelska 1, 613 00 Brno, Czech Republic, [2]Research Institute for Cattle Breeding, Vyzkumniku 267, 788 13 Vikyrovice, Czech Republic; kuchtik@mendelu.cz

The main aim of the study was to assess the basic parameters of carcass value of light lambs of Romanov breed (R, n=26) and the crossbreeds between Suffolk and Romanov breed (SF x R, n=29). All lambs under study were males and descended from three-year old ewes. The study was carried out on the organic sheep farm in Kuklík. All of the lambs were born in winter (January) and housed indoor till the slaughter. The daily feeding of lambs consisted of mother´s milk (ad libitum), meadow hay (ad libitum) and mineral lick (ad libitum). All lambs were slaughtered approximately at the same age (89 vs. 91 days) and live weight (14.06 vs. 13.86 kg). The evaluation of carcasses (colour of muscle and fattiness) were carried out according to SEUROP classification system for light lambs. The crossing had no significant effect on cold dressed carcass weight (5.67 vs. 5.84). However the genotype significantly affected the carcass yield (CY), while its higher value was found in SF x R (42.17 vs. 40.25%). The genotype had no significant effect on fattiness and colour of muscle, nevertheless the purebred Romanov lambs displayed a lower fattiness (1.32 vs. 1.53) and their carcasses were evaluated as lighter (1.15 vs. 1.24). On the other hand the SF x R crossbreeds showed non-significantly better conformation. Also the crossing significantly didn't influence the weight of skin (1.30 vs. 1.27 kg), proportion of kidney (0.39 vs. 0.43%) and kidney fat (0.31 vs. 0.33%).

The effect of a Suffolk-sired genotype on the fatty acid composition in extensively fattened lambs

Kuchtik, J.[1], Zapletal, D.[2], Dobes, I.[1] and Hegedusova, Z.[3], [1]Mendel University in Brno, Department of Animal Breeding (LA330, LA09031), Zemedelska 1, 613 00 Brno, Czech Republic, [2]University of Veterinary and Pharmaceutical Sciences Brno, Department of Nutrition, Animal Husbandry and Animal Hygiene, Palackeho 1-3, 612 42 Brno, Czech Republic, [3]Research Institute for Cattle Breeding, Vyzkumniku 267, 788 13 Vikyrovice, Czech Republic; kuchtik@mendelu.cz

The aim of the study was to evaluate the effect of a Suffolk-sired genotype on meat quality of quadriceps femoris muscle with respect to the fatty acid profile. Three different genotypes of Suffolk-sired crossbreeds were included in the study: F1 Suffolk-Charollais (SF 50 CH 50, n=10), F11 Suffolk-Charollais (SF 75 CH 25, n=10) and Suffolk-Improved Walachian (SF 50 IW 50, n=10). All lambs were singles and males. During the experiment, all of the lambs were reared in one flock under identical conditions without any differences in nutrition or management. All lambs were slaughtered in approximately the same age; however SF 75 CH 25 lambs had a significantly lower live weight at slaughter. Meat of SF 75 CH 25 lambs contained a significantly higher proportion of lauric acid, myristic acid and palmitic acid than other genotypes. By contrast, a higher proportion of stearic acid was observed in the SF 50 CH 50 lambs. The total content of saturated fatty acids was significantly higher in meat of the SF 75 CH 25 lambs. The genotype also affected the content of palmitoleic acid and oleic acid. In addition, the meat of IW type lambs had a lower saturated fatty acids proportion, lower values of atherogenic and thrombogenic indexes and a higher P/S ratio than meat of the Charollais genotypes. This resulted in better nutritional characteristics of meat from IW type lambs with regard to human health. Concerning crossing between the Suffolk and Charollais breeds, a favorable fatty acids profile of meat was found in the SF 50 CH 50 lambs as compared to the SF 75 CH 25 lambs. The SF 50 CH 50 genotype also showed a lower level of delta9-desaturase (16) index.

Physic-chemical and sensorial characteristics of meat from lambs finished with diets containing sugar cane or corn silage on two levels of concentrate

Silva Sobrinho, A.G., Leão, A.G., Moreno, G.M.B., Souza, H.S.B.A., Rossi, R.C. and Giampietro, A., FCAV, Unesp, Zootecnia, Via de Acesso Prof. PAulo Donato Castellane, s/n, 14884-900, Brazil; americo@fcav.unesp.br

With the aim to evaluate physic-chemical and sensorial characteristics of meat from lambs finished on feedlot with diets containing sugar cane or corn silage on two roughage:concentrate ratios, 60:40 or 40:60, it was used 32 Ile de France lambs, non castrated, with 15 kg of corporal weight. The animals were confined on individual stalls and were slaughtered at 32 kg of corporal weight. Studied diets and muscles didn't influenced on pH at 45 minutes (6.56) and 24 hours (5.62) after slaughter, water holding capacity (58.38%) and cooking losses (34.04%). Meat and subcutaneous fat colour didn't differ among diets, but the meat colour was affected by muscle type. Longissimus lumborum and Triceps brachii muscles had the followed values of L*, a* and b*: 34.64 and 36.86; 13.54 and 11.97; and 2.40 and 1.82 at 45 minutes; and 43.62 and 47.74; 16.12 and 14.21; and 5.36 and 4.49 at 24 hours after slaughter. Shearing force (1.85 kgf/cm^2) wasn't affected by diets, but it was different between muscles, with values 1.41 and 2.28 kgf/cm^2 for Longissimus lumborum and Triceps brachii, respectively. On sensorial analysis of ovine loin and shoulder, the lambs fed up with sugar cane and higher quantity of concentrate, had higher scores for flavor (8.07 and 8.26), texture (8.53 and 8.53), preference (8.20 and 8.46) and acceptance (8.33 and 8.26), respectively. It was possible to conclude that sugar cane on feedlot lambs food maintained physic-chemical quality of meat. In this case, it can be used on this production phase and when it was associated at higher quantity of concentrate in diet, it improved sensorial quality of lamb meat.

Structural characteristics of meat from lambs finished with diets containing sugar cane or corn silage on two levels of concentrate

Silva Sobrinho, A.G., Leão, A.G., Boleli, I.C., Alves, M.R.F., Moreno, G.M.B. and Souza, H.S.B.A., FCAV, Unesp, Zootecnia, Via de Acesso Prof. Paulo Donato Castellane, s/n, 14884-900, Brazil; americo@fcav.unesp.br

With the aim to evaluate structural characteristics of meat from lambs finished on feedlot with diets containing sugar cane or corn silage on two roughage:concentrate ratios, 60:40 or 40:60, it was used 32 Ile de France lambs, non castrated, with 15 kg of corporal weight. The animals were confined on individual stalls and were slaughtered at 32 kg of corporal weight. Experimental design was the totally random on split-plot, having on plots a factorial scheme 2 x 2 (two roughage:concentrate ratios and two roughages), and on subplots the evaluated muscles (Longissimus lumborum e Triceps brachii). The averages were compared by Tukey test at 5% of probability. The diets didn't change frequency of slow oxidative (SO), fast oxidative glycolytic (FOG) and fast glycolytic (FG) fibers. However, those containing 60% of concentrate propitiated larger diameter, area and total area relative at the same. About muscle type, frequency of SO and FG fibers was higher on Triceps brachii, and frequency of FOG fibers was higher on Longissimus lumborum. For all muscle fiber types, Triceps brachii muscle had areas and diameters larger than Longissimus lumborum. Independently of diet and evaluated muscle, the contents of total and soluble collagen of meat not differed, with average values of 2.45 and 0.30 mg/g of muscle, respectively. It was possible to conclude that characteristics of lamb muscle fibers change more in function of muscle type than diet. Diets containing higher quantity of concentrate (60%) increase muscle hypertrophy on lambs. In feedlot lambs, contents of meat total and soluble collagen were not affected by diets and muscle type.

Nutritional characteristics of meat from lambs finished with diets containing sugar cane or corn silage on two levels of concentrate

Silva Sobrinho, A.G., Leão, A.G., Moreno, G.M.B., Souza, H.S.B.A., Perez, H.L. and Loureiro, C.M.B., FCAV, Unesp, Zootecnia, Via de Acesso Prof. Paulo Donato Castellane, s/n, 14884-900, Brazil; americo@fcav.unesp.br

With the aim to evaluate nutritional characteristics of meat from lambs finished on feedlot with diets containing sugar cane or corn silage on two roughage:concentrate ratios, 60:40 or 40:60, it was used 32 Ile de France lambs, non castrated, with 15 kg of body weight. The animals were confined on individual stalls and were slaughtered at 32 kg of body weight. Experimental design was the totally random on factorial scheme 2 x 2 (two roughage:concentrate ratios and two roughages), being the averages compared by Tukey test at 5% of probability. Lamb meat had 74.55% of moisture, 19.61% of crude protein, 1.04% of ash and 51.28 mg/100 g of cholesterol, being ethereal extract content, highest on meat of lambs fed up with diets containing corn silage (3.97%) and on those of animals that received more concentrate food (4,02%). The fatty acids on highest concentration in lamb meat were: miristic (4.18%), palmitic (26.41%), stearic (17.09%), oleic (37.93%) and linoleic (4.00%). The meat of lambs fed up with diets containing sugar cane had higher contents of capric (0.47%) and arachidonic (4.17%) acids, and lower contents of palmitoleic (2.02%) and linoleic (0.25%) acids. The meat of lambs fed up with higher quantity of sugar cane (60%) had higher contents of pentadecanoic (0.68%), heptadecanoic (2.13%) and eicosadienoic (1.34%) acids too. These acids were reduced when sugar cane was used on lower quantity. It was possible to conclude that diets containing sugar cane and diets with higher quantity of roughage (60%) originates lamb meat with lower fat content. Roughage type had higher influence on fatty acid profile of meat than roughage: concentrate ratio of diets.

Effect of weaning age on feedlot performance of Kalekoohi male lambs

Karkoodi, K. and Motahari, M., Islamic Azad University, Saveh Branch, Saveh, Markazi, Iran; kkarkoodi@yahoo.com

A number of 16 male lambs; as 45, 60, 75 and 90 d old were selected randomly to study the effect of different weaning ages on feedlot performance of Kalekoohi male lambs. They weaned and after a 15-d adaptation period and they underwent vaccination for enterotoxaemia and parasite diseases. Lambs were fattened for 112 days and fed the same TMR diets. This study was carried out on a completely randomized design with four weaning ages (treatments) in four replicates. Results showed that increasing the weaning age significantly increased daily feed intake and warm carcass weight; but had no significant effect on carcass shrunk weight. Carcass efficiency of group 4 was numerically higher than group 3 (44.96 vs. 43.57%) but significantly higher than those other groups. No significant effects on total live weight changes and average daily gain was observed, but 60 d old group was numerically the highest of all (144 g/d). Total dry matter intake and feed conversion ratio (FCR) was significantly lower in 60-d weaned lambs compared to groups 75 and 90 d weaned lambs. However results showed that although earlier weaning of lambs can be economically beneficial, but environmental and ecological factors for lambs them selves and also ewes; weaning in 60-d age may be suggested for weaning age of Kalekoohi male lambs.

Session 35 Poster 27

The effect of BCS on reproduction parameters of Fat tail Afshari sheep

Moeini, M.[1], Aliyari, D.[1], Shaheer, S.[2] and Abdolmohammadi, A.[1], [1]razi university, animal science, imam khomeini St. college of agriculture, animal science dep., 67155, Iran, [2]zanjan university, Animal science Dep., Zanjan University college of agriculture, 1588, Iran; mmoeini2008@gmail.com

The effect of body condition score (BCS) and body weight of fat tail Afshari sheep before mating time on reproductive performance were studied. Total 162 fat tail Afshari ewes (3-8 years old) randomly selected from animal unit of Zanjan University. Ewes divided to 4 treatments group according BCS and weight (2, 2.5, 3, 3.5 and more). The reproductive parameters; rate ewe lambing, lambs born to ewes after mating, number of lambs born per lambing, kg lambs born per ewe mating, pregnancy period, lambs birth weight, weaning weight and some blood metabolites such as glucose, total protein, albumin and globulin were determined. The result of indicated that Body condition score had a significant effect (P<0.05) on the lambs born per ewe at mating. Ewes with BCS=3 had a better performance in lambing rate and kg lambs born per ewe. While the lambing rate reduced in ewes with BCS of 3.5 or more. Ewes with BCS=3 had more (P<0.05) normal estrous while ewes with low BCS had shorter estrous period. Birth weight of lambs was significantly affected by body weight of their dams (P<0.01) but BCS had no significant effect on lambs' weight at birth and weaning. There was no significant effect of BCS on blood metabolite parameters in this study. It is concluded that BCS had significant effect on lambing rate and lambs born per ewe. It's recommended flushing of ewes with low BCS and the BCS of 3 at mating could optimize profitability of Afshari ewes.

Session 35 Poster 28

Effect of different level of waste date palm on fattening performance of male lambs

Aghashahi, A.[1], Atashpanje, M.[2], Mirghazanfari, S.M., Hosseini, A. and Mahdavi, A., [1]Iran animal scienc research inst., ANIMAL nut, iran anmal science research inst.shahid beheshti ave., Karaj, 31585-1483, Iran, [2]jihad-keshavarzi org., jihade-keshavarzi org-zahedan-iran, 31585, Iran; araghashahi@yahoo.com

This experiment were conducted to evaluate growth performance, feed intake and feed conversion of Baluchi lambs, fed diets supplemented with low quality date palm(WDP).in this trial animal were distributed in to four groups, eight lambs each. The first group (c) was fed a control diet containing 40% alfalfa hay and 60% concentrate. Three level of WDP, 10% (T1), 20% (T2), 30% (T3) were used to substitute the concentrate portion. Animals were subjected to experimental diets for 88 days. At the end of this period lambs were slauthered and different part of body and carcasses (meat, fat, bones) were recorded. on the basis of results of this experiment, supplementing of fattening rations with 20% and 30% WDP had no adverse effect on daily gain,(196 and 192 gr per day respectively), but feed conversion for T3 was better than T2 (P>0.05). Meat percent of carcass of T3 lambs (58.7) were more than T2 (57.52) and C groups lambs (P>0.05). From an economic point of view diet supplemented with up to 30% WDP could be used efficiently in fattening ration of Baluchi lambs.

Nitrogen and carbon dioxide mixtures for stunning pigs: effect on aversion and meat quality

Llonch, P.[1], Rodríguez, P.[1], Manteca, X.[2] and Velarde, A.[1], [1]IRTA, (17121), Girona, Spain, [2]UAB, Facultat veterinària, (08193) Barcelona, Spain; pol.llonch@irta.cat

Stunning pigs with high concentration of CO_2 produces aversion before the loss of consciousness, but inhalation below 30% is not aversive. The aim of this study was to assess the aversion during the inhalation of gas mixtures with N_2 and CO_2: 70%+30% (70N30C), 80%+20% (80N20C), 85%+15% (85N15C), and the effect on meat quality in comparison to 90% CO_2 (90C). Sixty-eight female pigs (92.6±1.19 kg) were divided in 4 groups and stunned by exposure to different gas mixtures. Retreat attempts, escape attempts, gasping, muscular excitation and vocalisations were scored to determine aversiveness. The time to loss posture and the duration of muscular excitation were also recorded. The pH (pH45) was determined in the semimembranosus (SM) and longissimus dorsi (LD) muscles at 45 min *post mortem*. Electrical conductivity (EC), water holding capacity (WHC) and presence of petechiae were assessed at 24 h. Statistical analyses was performed with SAS using MIXED (for continual data), GENMOD (for binary data) and CORR statements. Pigs showed less escape attempts with nitrogen mixtures in comparison to 90% CO_2. Time to loss posture was not different between treatments. The duration of muscular excitation was higher in pigs stunned with nitrogen mixtures than 90C (P=0.05). The EC of LD and SM was higher in 85N15C than in 70N30C and 90C treatments. The duration of muscular excitation was correlated with pH45 (r=-0.39 in SM and r=-0.42 in LD) and EC (r=-0.66 in SM and r=-0.50 in LD). WHC was lower in 90C than 85N15C (P<0.05). Petechiae were found on average in 26% of hams from pigs stunned with 70N30C, 80N20C and 85N15C while they were not found in 90C. Aversive behaviour was seen in all 4 treatments. However, nitrogen gas mixtures reduced the incidence of escape attempts. The pH45 decreased with longer muscular excitation. The low pH45 in 80N20C and 85N15C may reduce the WHC which is confirmed by the EC. The duration of muscular excitation could affect the number of petechiae and also affected meat quality.

Enthalpy as a possible index of thermal load for animals in transit?

Mitchell, M.A.[1], Kettlewell, P.J.[2], Villarroel-Robinson, M.R.[3], Barreiro, P.[3] and Farish, M.[1], [1]SAC, SLS, Bush Estate, Penicuik, Midlothian, EH26 0PH, United Kingdom, [2]ADAS, Drayton, Alcester Road, Stratford on Avon, Warwickshire, CV37 9RQ, United Kingdom, [3]University of Madrid, Escuela Técnica Superior de Ingenieros Agronomos, Madrid, Spain; malcolm.mitchell@sac.ac.uk

Elevated thermal loads imposed upon livestock during commercial transportation are a major cause of reduced welfare and production losses. Thermal loads experienced by animals in transit are commonly defined in terms of temperature alone and limits prescribed in legislation do not use thermal indices incorporating air temperature and water vapour content. In the present study the application of enthalpy to quantify thermal loads upon pigs in transit was investigated. Temperature-humidity sensors were installed on commercial vehicles on 7 long distance livestock journeys from the UK to Spain using a commercial livestock transport vehicle carrying 90 pigs. These were undertaken during the summer period to impose a range of thermal loads upon the animals. The psychrometric data from the temperature-humidity measurements were used to calculate the psychrometric properties of the air during transport. A temperature derivative was computed using the Savitzky-Golay algorithm to smooth one-dimensional, tabulated data and compute the numerical derivatives using the Savgol routine. The phase space for temperature has been applied and its corresponding gradient (dT/dt). Pig behaviours during a 3 hour period at the end of the journey were utilised to assess the transport stress and were correlated with the temperature and enthalpy analyses. These indicated that the higher values for the relative changes in temperature (°C/second) and humidity (%RH/second) and enthalpy parameters were correlated (P<0.01) with behavioural indicators of increased transport stress. It is suggested that relative changes in enthalpy may constitute an integrated index of thermal load in transit which is more useful and sensitive than absolute values of temperature or relative humidity.

Current enrichment materials fail to address the cognitive potential of pigs

Van Weeghel, H.J.E., Driessen, C.P.G., Gieling, E.T. and Bracke, M.B.M., Wageningen UR Livestock Research, P.O. Box 65, 8200 AB Lelystad, Netherlands; ellen.vanweeghel@wur.nl

Under free-range conditions pigs experience numerous challenges, such as finding food in all seasons and fleeing from threats. In order to successfully master these challenges, pigs possess a wide array of behavioural strategies and cognitive skills. However, modern intensive pig husbandry systems are characterised by a predominant focus on production efficiency, resulting in stimulus-poor housing environments that offer few opportunities for the pig to display species-specific behaviours and to utilise their cognitive abilities. In the EU legislation has been developed to mitigate the main negative welfare consequences. EC Directive 2001/93 prescribes that pigs must have permanent access to a sufficient quantity of material to enable proper investigation and manipulation activities. In most EU countries this has been implemented as providing a chain with a piece of plastic. In Denmark natural materials are obligatory that reach till floor level, often a piece of wood hanging on a metal chain. Organic pigs generally receive straw. The objective of this contribution is to describe the implementation of this prescription about enrichment materials in relation to the pigs' cognitive capacities. Most enrichment materials provide some kind of stimulation to the pigs, mostly emphasising the physical expression of exploration (animal-material interaction time). In addition, most or all materials trigger operant and classical learning processes to a very limited degree. This is true, even for the 'best' materials such as straw. We conclude that at present pigs in both intensive and organic farming do not get what they need to express their cognitive capacities. Further improvements towards sustainable pig farming requires more attention to the cognitive functioning of pigs and the design of enrichment materials, either through (re-)creating natural conditions or through more artificial ways, in order to provide in the pigs' need for experiencing cognitive challenges.

Session 36 Theatre 4

Infrared thermography of the skin to assess pain in piglets

Kluivers, M., Reimert, H. and Lambooij, E., Wageningen UR Livestock Research, Edelhertweg 15, 8219 PH Lelystad, Netherlands; marion.kluivers@wur.nl

Piglets are subjected to several painful procedures shortly after birth. To assess the pain inflicted by tail docking and ear tagging, objective and valid measures are needed. Preferably, noninvasive measures should be used to prevent the effect of extensive handling. The objective of this study was to assess Infrared Thermography (IRT) of the skin as a measure of acute pain in piglets. An experiment was conducted with 64 piglets from 16 litters, in 2 batches. Per litter, 4 female piglets were enrolled in the experiment. Treatments were: 1. SHAM (only handling), 2. DOCK (tail docking), 3. MELDOCK (tail docking after meloxicam), 4. TAG (ear tagging). In order to take IRT pictures, piglets were lightly restrained in a hammock for 20 min. A 5 min period was used to establish a pretreatment baseline level, after that tail docking or ear tagging was performed. MELDOCK piglets received meloxicam (a NSAID) 35 minutes before being docked. During a 20 min period IRT pictures were taken of the eye area; every 10 (batch 1) or 30 seconds (batch 2). For the analyses, maximum temperature of a skin area just mediodorsal of the eye was measured and a REML variance components analysis in Genstat was used. Results show that after restraining, skin temperature decreased within 2 minutes to a steady line of 37.7 °C in all treatments. Skin temperature decreased significantly in DOCK (-0.18 °C), MELDOCK (-0.14 °C) and TAG (-0.22 °C), compared to SHAM (-0.01 °C). Lowest skin temperature was reached between 4 to 6 minutes after the procedure. After that, skin temperature rose slowly, but had not yet reached pretreatment levels 15 minutes after. No significant differences were found between DOCK, MELDOCK and TAG in skin temperature. It can be concluded that IRT of the eye area is suitable for measuring acute pain in young piglets. On the basis of skin temperature it can be concluded that tail docking and ear tagging inflict a similar amount of pain and that meloxicam has no immediate pain relieving properties during tail docking.

Session 36 Theatre 5

Breed and sex effect on pork meat quality

Rodrigues, S., Lourenço, M.C., Pereira, E. and Teixeira, A., CIMO, Escola Superior Agrária - Instituto Politécnico de Bragança, Campus Sta Apolónia Aptd 1172, 5301-855, Portugal; srodrigues@ipb.pt

This work had as objective to evaluate the physical-chemical and sensory quality of two categories of pork meat from, a commercial meat pork and a selected meat from the Portuguese black pork (Preto Alentejano breed). The Preto Alentejano is a local non improved swine breed which survived during the last years owing to a demand increasing of Iberian products and the protection of origin designation products. Commercial pig breeds have great prolificacy and precocity, raised purely on an intensive way, using a more advanced technology that translates into a possible improvement in terms of carcass yield. Sixteen animals were used, 4 females and 4 males from each breed. Animals had 80-100 kg of live weight. The longissimus muscle between the 5th thoracic vertebra and the 10th lumbar vertebra was used in the analysis. Regarding meat physical-chemical quality, samples were analyzed for protein, fat, pigments, ashes, dried materials, water-holding capacity, and texture. Results of fat and pigments contents indicate significant differences for all treatments. For protein, ashes, dried materials, water-holding capacity and texture no significant differences were found. In the analysis of fatty acids composition, ten were detected, being the main ones C16:0, C18:0, C16:1, C18:1, C18:2. There was a predominance of monounsaturated fatty acids, followed by saturated and polyunsaturated. Differences were significant for sex and breed. Preto Alentejano breed and females presented the higher percentages of saturated and monoinsaturated fatty acids. The taste panel found differences, mainly between breeds. The panellists scored Preto Alentejano meat as being juicier, more tender, with richer taste and more acceptable than Commercial meat. The higher juiciness score of Preto Alentejano meat were probably attributable to the higher intramuscular fat content compared to Commercial meat. The Commercial pork was characterized mainly by high toughness.

Session 36 Theatre 6

Outputs and economic assessment of RFID technology and molecular biology (STRs) as innovative tools for traceability and origin protection in 'Suinetto di Sardegna'

Cappai, M.G.[1], Baglieri, V.[2], Nieddu, G.[1] and Pinna, W.[1], [1]University of Sassari, Animal Biology Dept., Via Vienna 2, 07100 Sassari, Italy, [2]SDA Bocconi School of Management, Via Bocconi 8, 20136 Milan, Italy; mgcappai@uniss.it

The suckling piglet is a typical meat product of Sardinia, resulting from the traditional regional breeding of Swine within a scattered value chain (9.17% of national production). About 9,300 small family-run farms rear a limited number of sows that give birth to piglets bred into a strictly traditional farming practice. Piglets are normally slaughtered around one month of age: whole or half carcasses are the usual ways the product is presented to consumers on the market. Beyond retailers, the suckling piglets meat supplies also agritourisms and restaurants for typical dishes and local specialties. This product is highly appreciated both by local and foreign consumers for meat grilled following the traditional recipe and the added value due to the typical farming practice. Potentials of such production may lead the suckling piglet meat to achieve 'a niche' within the market: in this light, it needs to be protected from frequent commercial frauds, by importers who pass foreign pork meat off as local products. In previous work, safety, efficacy and efficiency for piglets identification and traceability of products from farm to slaughterhouse were tested within the suckling piglet of Sardinia: a total of 355 suckling piglets from 6 farms were electronically identified (injectable transponders HDX 32.5×3.8 mm, 134.2 kHz) and double-sampled *in vivo* and post mortem for a 6 microsatellites (FAO/ISAG panel) PCR on DNA from auricle tissue. EID showed: accidents (2.54%); deaths (0.56%); *in vivo* transponder readability 99.15%; post mortem readability 100%; reader reliability 99.93%; transponder recovery at abattoir 99.15%. The DNA analysis on 42 random animals showed: amplifications anomalies (10.7%); none no-identity. The economic assessment of the integrated system RFID+STRs showed a 27.1% of total costs/carcass produced.

Session 36 Theatre 7

Effect of weaning age and rearing systems on Iberian piglets

González, F., Robledo, J., Andrada, J.A., Vargas, J.D. and Aparicio, M.A., School of Veterinary Science, University of Extremadura, Avda. de la Universidad S/N (Facultad de Veterinaria), 10003 Cáceres, Spain; fgonzalej@alumnos.unex.es

Weaning is a stressful time for piglets. The adaptability of piglets at this critical time is essential to restore their productive balance. The aim of these experiments was to assess the effect of weaning age and rearing systems, at post-weanig period, on body weight (W), daily average gain (DAG) and Conversion Ratio (CR). A total of 85 iberian piglet were weighed weekly from 21 day of age to 63. The piglets were reared on three different systems: Intensive, Traditional and Out-doors. Piglets were weaned at 28 and 42 days of age, and were placed into six different groups according their weaning age and rearing system (two groups for each rearing systems and a weaning ages for each). Data were analysed using Spss® (T-test, ANOVA, Tukey) and statistical significance were accepted at $P<0.05$. Comparing the two weaning age in each rearing system, at 42 day of age were significant differences in W, DAG and CR between intensive piglets groups, and between Out-doors piglets groups. However there are not differences between Traditional piglets groups. At 63 day of age, intensive groups and traditional groups had not differences, but out-doors groups differences remained. Therefore productivity of early weaned piglets were lower than piglets weaned late due to post-weaning stress Nevertheless, the intensive early weaned piglets were able to recover the previous differences at 63 day of age, but out-doors were unable. Comparing the three rearing systems in each weaning age, from weaning to 63 day of age, were significant differences between early weaning (out-doors (a) intensive (b) traditional (b)), but were not differences in late weaning. These results show that rearing system has a greater influence on early weaning. In conclusion, the rearing systems influence on early weaning Iberian piglets, and late weaning are better than early weaning in out-doors Iberian piglets.

Session 36 Theatre 8

Birds as source of important bacterial pathogens in sows in outdoor production

Kemper, N. and Ziegler, I., Christian-Albrechts-University Kiel, Institute of Animal Breeding and Husbandry, Olshausenstr. 40, 24105 Kiel, Germany; nkemper@tierzucht.uni-kiel.de

Sows in outdoor production are exposed to pathogens originating from infection sources different from those in indoor production. Due to the open environment, a transmission via birds is possible, but detailed studies are lacking. This examination aims at the qualitative analysis of important bacteria in the faeces of sows and their piglets, in their environment and in the faeces of seagulls. The examined farm was located in proximity to the Baltic Sea and seagulls defecated regularly on the paddock. On four consecutive sampling dates every three weeks between May and July 2009, faecal samples from sows (n=20) and piglets (n=40), and pooled faecal samples from seagulls (n=20) were taken. Additionally, soil samples from the paddocks (n=20) and the huts (n=20) were collected. All samples were analysed bacteriologically with routine and molecular methods. Faecal indicators were the predominant species: *Escherichia coli* was isolated in 96%, and Enterococcus spp. (species) in 57% of all samples. Not only in the faecal samples, but also in the soil samples, these indicators were found in high percentages. Neither *Salmonella* spp. nor *Yersinia* spp. were detected. *Campylobacter* (C.) *coli/jejuni* were present both in sow samples (30%), piglet samples (38%) and bird samples (5%). Moreover, *Vibrio parahaemolyticus* was isolated from 5% of the sow samples, 5% of the piglet samples, 10% of the bird samples and 20% of the soil samples. These results confirm the role of both pigs and birds as natural reservoirs for *Campylobacter* spp. This is the first detection of the halophil *V. parahaemolyticus* in porcine faeces. A transmission via seagulls can be assumed, because the natural habitat of this pathogen is salt water and the occurrence was only sporadic on one single sampling date. The transmission of bacteria via birds to outdoor sows is possible, and especially for *Campylobacter* spp., further studies are required because of the importance of this pathogen as possible zoonotic agent.

Seasonal infertility in swine caused by high ambient temperatures

Vatzias, G.[1], Asmini, E.[2], Maglaras, G.[1], Papavasiliou, D.[1] and Markantonatos, X.[3], [1]Technological Educational Institute of Epirus, Animal Production, Kostakii, Arta, 47100, Arta, Greece, [2]Technological Educational Institute of Larissa, Project Management, Trikalon Ring Rd, 41110, Larissa, Greece, [3]Region of Epirus, Agricultural Development, V. Epirou 20, 45445, Ioannina, Greece; vatzias@teiep.gr

A major problem in the swine industry is seasonal infertility. High atmospheric temperature may cause physiological stress leading to a series of adverse effects on several reproductive parameters, including poor sperm quality, lower conception and farrowing rates. The objective of the present study was to determine the effects of seasonal temperature variations on the weaning to estrus interval and on farrowing rates. Data were obtained from the breeding and farrowing records from a commercial farm with a breeding stock of 500 sows, for a period of four years. The results indicated that sows weaned and bred during the warmest months of the year (June - September, with an average monthly atmospheric temperature ranging between 21 °C - 26 °C) exhibited a prolonged weaning to estrus interval ($P<0.05$) and a lower farrowing rate (($P<0.003$), compared to sows weaned and bred during the coldest months of the year (December-March with an average monthly atmospheric temperature ranging between 4.7 °C - 9.6 °C). The above results indicate that environmental factors, such as high ambient temperature may play a crucial role in swine seasonal infertility. Facilities with controlled environment and proper reproductive management may improve fertility and reduce the negative impact of the heat stress.

Seminal plasma and spermatozoa quality in the boar: identification of novel biomarkers

Dewaele, L.[1], Tsikis, G.[2], Ferchaud, S.[3], Labas, V.[2], Spina, L.[2], Teixeira, A.-P.[1], Druart, X.[2] and Gérard, N.[2], [1]INRA IASP, Centre de Recherche de Tours, 37380 Nouzilly, France, [2]INRA UMR 85, CNRS UMR 6175, Université de Tours, Haras Nationaux, 37380 Nouzilly, France, [3]INRA UEICP, Centre de Poitou-Charentes, 86480 Rouillé, France; gerard@tours.inra.fr

Sperm fertility is progressively acquired during maturation in the genital tract, under the influence of biochemical interactions with paracrine factors secreted within the tract. Thus, it is characterized with modifications of membrane properties, such as fluidity and composition. It has been recently demonstrated that in the boar, hypotonic resistance of ejaculates is positively correlated with *in vivo* fertility. The aim of the present study was to analyze the effect of seminal plasma on the hypotonic resistance of boar spermatozoa, by using mass spectrometry. For this purpose, seminal plasma (SP) were collected from several boars with known fertility and stored. Epididymal spermatozoa (EpSpz: n=3 boars) were recovered from cauda epididymis by microperfusion. EpSpz were incubated in the presence of saline or SP for 10 min at 37 °C before being assessed for hypotonic resistance by flow cytometry, and analyzed by Intact Cells Matrix Mass Spectrometry (ICM-MS) that allow to achieve direct protein profiling. Our results showed that hypotonic resistance of the 3 populations of EpSpz displayed similar profiles, in relation to seminal plasma in which they had been incubated. Thus, two SP that induced the lowest, and two other that induced the best, hypotonic resistance to EpSpz were selected (SP low and SP high, respectively). EpSpz incubated with SP high or SP low displayed different ICM-MS profiles, some molecular species being increased and other decreased (or absent) between the two groups of SP. Identification of these peptides/proteins is under analysis. In conclusion, our results demonstrate that hypotonic resistance induced by SP to EpSpz is related to their ICM-MS profiles, and that this approach may be useful to identify novel biomarkers of fertility in the boar.

Effect of mutation at the MC4R locus on carcass quality of native Puławska breed

Szyndler-Nędza, M.[1], Tyra, M.[1], Blicharski, T.[2] and Piórkowska, K.[1], [1]National Research Institute of Animal Production, Department of Animal Genetics and Breeding, ul. Sarego 2, 31-047 Kraków, Poland, [2]Sciences Institute of Genetics and Animal Breeding, Department of Animal Immunogenetics, Jastrzębiec, 05-552 Wólka Kosowska, Poland; mszyndle@izoo.rakow.pl

C, linear measurements were made and carcasses dissected according to the DLG method. Analysis of genotypes in the pig group studied showed the highest frequency of MC4R A/G genotype animals and the lowest frequency of MC4R A/A genotype animals. Frequency of the MC4R A allele was only 15.2% lower than that of the MC4R G allele. In addition to highly productive breeds, Polish breeders keep local breeds involved in the conservation programme, which include Puławska pigs. Analysis of Puławska pig data spanning 20 years showed that carcass lean content increased considerably from 41.43% in 1983 to 45.68% in 2003, with a concurrent decrease in fat content. Considering the relatively high rate of changes in these traits, it would be appropriate to find parameters that serve as a criterion for evaluating the degree of heterozygosity in this breed. The aim of this study was to determine the melanocortin receptor gene polymorphism in Puławska pigs and its effect on carcass quality. The study involved 66 Puławska fatteners. After slaughter and 24-hour cooling of carcasses at 4 In Puławska pigs, the MC4R A allele had a significant effect on increasing backfat thickness, especially in the loin area. The MC4RA allele was significantly correlated to a greater amount of fat in the neck. Animals with the A allele at the MC4R locus were also characterized by a significantly lower amount of meat in this carcass cut. The results obtained for frequency of different genotypes in Puławska pigs could serve as reference values for selection-induced changes, thus reflecting the level of genetic variation in this breed.

Session 36 Poster 12

Characteristics of the Polish nucleus population of pigs in terms of intramuscular fat (IMF) content of *m. longissimus dorsi*

Tyra, M. and Żak, G., National Research Institute of Animal Production, Department of Animal Genetics and Breeding, ul. Sarego 2, 31-047 Kraków, Poland; mtyra@izoo.krakow.pl

Subjects were Polish Landrace (PL), Polish Large White (PLW), Puławska, Hampshire, Duroc, Pietrain and line 990 gilts tested in the Polish Pig Performance Testing Stations (SKURTCh). A total of 4430 animals were investigated. During the test, animals were kept individually and fed ad libitum. Intramuscular fat (IMF) content was determined as crude fat by Soxhlet extraction with fat solvents (Soxtherm SOX 406, Gerhardt). Samples for analysis were taken from the middle cross-sectional area of m. longissimus dorsi behind the last rib. The highest level of IMF was observed for Duroc animals (2.23%) and the lowest for the Pietrain breed (1.66%). In the two most popular breeds in Poland (PLW and PL), this parameter was below the level acceptable for good quality meat (1.84% and 1.76%, respectively). IMF content of m. longissimus dorsi showed high variation ranging from V=27.8% to V=41.8% according to the breed.

Meatness and fatness traits of Polish Large White and Polish Landrace pigs differing in fattening traits

Zak, G., Tyra, M. and Rozycki, M., National Research Institute of Animal Production, Sarego 2, 31047 Krakow, Poland; z1zak@cyf-kr.edu.pl

Rate of growth and feed conversion during the fattening period were studied in relation to carcass meatiness and fatness traits such as fat area over loin eye, loin eye area, fat to muscle ratio, carcass meat percentage, mean backfat thickness from 5 measurements, backfat thickness over shoulder, and backfat thickness over loin eye. The present study showed a low relationship between the analysed fattening traits and carcass slaughter parameters, with low coefficients of correlation. Breeding programmes aimed at improving growth rate and increasing feed conversion ratio (kg/kg) may lead to decreased meat content and increased carcass fatness. Polish Large White pigs are characterized by greater variation of carcass muscling and fatness traits compared to Polish Landrace pigs.

Session 36 Poster 14

Improving pork quality through genetics and nutrition

Zak, G. and Pieszka, M., National Research Institute of Animal Production, Sarego 2, 31047 Krakow, Poland; z1zak@cyf-kr.edu.pl

Meat quality is a complex issue that encompasses several physicochemical factors. Adequate levels of quality parameters determine meat palatability, appearance and processing suitability, i.e. attractiveness for the consumer. In addition, traits such as cholesterol content, composition and proportions of fatty acids, and vitamin and mineral content determine the health value of meat. In Poland, pork is the most popular meat. Improvement of the national pig population for increased lean content has negatively affected many parameters of meat quality. This problem is common to most pig producers in the world. It is therefore increasingly important to pay attention to improving meat quality, especially through breeding work. Considering the relationships between carcass meat quantity and quality, the importance of pork quality examination and monitoring should be stressed. Breeding programmes should be carefully formulated based on monitoring results to improve and consolidate the desired levels of quality parameters in the active population of pigs. The application of the results obtained using population genetics, coupled with efficient feeding and high health status of the farm should produce expected results in terms of good quality of pork.

Effect of dietary selenium on muscle fatty acid composition in pork

Okrouhlá, M., Stupka, R., Čítek, J. and Šprysl, M., Czech University of Agriculture, Faculty of Agrobiology, Food and Natural Resources, Department of Animal Husbandry, Kamýcká 129, 165 21, Czech Republic; citek@af.czu.cz

Intramuscular fat content and composition are important meat quality characteristics. The aim of this study was to determine the influence of dietary selenium on the fatty acid composition of pig muscle. A total of 40 pigs (HxPN) x (CLWxCL) were divided into four groups according to selenium diets. The first group was fed with organic selenium during of all test, second group was fed with organic selenium until 60 kg live weight, third group was fed with organic selenium when pigs got 60 kg live weight until the end of the test and last group had no organic selenium in diet. Amount of organic selenium was always 5 mg/kg diet. Representative muscle samples were taken from musculus longissimus lumborum et thoracis. They were then homogenized and submitted to FAME analysis. Palmitic, stearic and myristic acids were dominat acids from the saturated fatty acids (SAFA), oleic, palmitoleic and eicosenoic were dominant from monounsaturated fatty acids (MUFA) and linoleic, arachidonic and docosadienoic acids were dominant from polyunsaturated fatty acids (PUFA) in all groups. The highest values of SAFA, MUFA and PUFA and the lowest value of n-6/n-3 ratio were found in group supplemented with organic selenium during test. The oleic acid was dominat in all groups, it means 356.54, 221.35, 196.68 and 209.27 mg/100 g muscle. Statistically significant differences have been found in caprylic, capric, margarin, α-linolenic ($P<0.05$) and nervonic ($P<0.01$) acids. This search was supported by the Ministry of Education of Czech Republic (MSM 60460709).

Determination of the effect of gender and MYOG gene on histological characteristics muscle in pigs

Kratochvílová, H., Stupka, R., Čítek, J., Šprysl, M., Okrouhlá, M. and Dvořáková, V., Czech University of Life Sciences, Department of Animal Husbandry, Kamýcká 129, 165 21 Prague 6 Suchdol, Czech Republic; citek@af.czu.cz

The purpose of this study was determined differences in histological characteristics between gilts and barrows and differences between genotypes of MYOG gene. A total 30 pigs of Pietrain x (Czech Large White x Czech Landrace) were evaluated. The samples were obtained from loin (m. longissimus lumborum et thoracis), ham (musculus semimembranosus) and shoulder (m. serratus ventralis). The individuals with genotype BB had the highest weight of loin without skin (6.95 kg), the largest cross sectional area of muscle fibre in loin (7,250.83 µm2), the largest diameter of muscle fibre (91.59 µm) and the least number of fibre on 1 mm2 (114). The animals with AA genotype had the highest weight of shoulder (4.81 kg), the largest cross sectional area of fibre in ham (6,778.85 µm2), the largest diameter in ham (89.55 µm) and the least number of fibre on 1 mm2 (131). Pigs with AB genotype had the highest weight of ham without skin (11.75 kg), the least cross sectional of muscle fibre in loin (5,110.99 µm2) and ham (5,924.72 µm2), least diameter of fibre in loin (78 µm) and ham (83.95 µm). The greatest cross sectional area of muscle fibre in loin and shoulder was measured in barrows. The greatest diameter in loin and shoulder was in barrows too. The greatest cross sectional area and diameter of muscle fibre in ham was found in gilts.

Chemical composition, colour, oxidative and sensory attributes evaluation of national and imported hams

De Santis, D.[1], Danieli, P.P.[2], A. Bellincontro, A.[1], Bertini, D.[1] and Ronchi, B.[2], [1]University of Tuscia, Dept. of Food Science And Technology, Via. S. C. de Lellis snc, 01100, Italy, [2]University of Tuscia, Dept. of Animal Science, Via. S. C. de Lellis snc, 01100, Italy; danieli@unitus.it

The overall quality and the organoleptic attributes of the dry-cured ham are mainly correlated to meat origin. The aim of the present study was to assess the influence of meat origin on chemical composition and final sensory characteristics of dry-cured hams. Samples of national (NH) and imported (IH) hams (slices 2 cm thick, in double) were collected from a dry-cured ham making enterprise located in Viterbo (Italy) at 0 (fresh thighs), 8, 65, 120 and 210 days (marketable product). Moisture, pH, protein, lipids, ash, texture, colour, thiobarbituric acid reactive species (TBARs) and sensory profile were investigated. For some chemical parameters, Near Infrared Reflectance predictive PLS models were also developed. Data were analyzed using the GLM procedure of Statistica 7 (StatSoft Inc., USA); differences were declared significant at $P<0.05$. Chemical characterization and colour at different time of seasoning did not show difference between NH and IH samples apart for pH with IH samples always below the respective NH ones. I dry-cured hams were softer than NH ones. Aromatic profiling of dry-cured hams outlined a higher prevalence of aldehydic components in NH than in AH samples, the latter showing a slight but significant major content of TBARs. Sensorial evaluation by expert sensory assessors (ISO 5492) (experts) assigned a higher value of preference to the N seasoned hams than the I ones. NIR spectra was not sufficient different to discriminate N vs. I samples although good predictive PLS models were developed and cross-validated for water activity (RMSECV=0.007, R^2=0.97), ash% (RMSECV=0.68, R^2=0.96) and pH (RMSECV=0.05, R^2=0.91) The origin of meat influences several volatile compounds and numerous sensory attributes of dry-cured hams with implications on overall quality of marketable products.

Number of myofibers increases from birth to weaning in pigs

Rodríguez-López, J.M.[1], Pardo, C.[2] and Bee, G.[2], [1]Instituto de Nutrición Animal, Estación Experimental del Zaidín, CSIC, Granada, 18100, Spain, [2]Agroscope Liebefeld Posieux, Posieux, 1725, Switzerland; giuseppe.bee@alp.admin.ch

Currently, it is believed that total number of muscle fibers (TNF) is fixed at birth. However, there are indications that at birth between primary (P) and secondary (S) fibers very-small diameter fibers containing embryonic and fetal myosin heavy chain isoforms exist. They represent a different population of myotubes, designated tertiary myotubes and might contribute to hyperplastic growth after birth. The goal of the study was to establish if TNF changes from birth to weaning. For the trial 8 pairs of female littermates with a similar birth weight (1.41±0.113 kg; P=0.82) were used. One piglet of each pair was sacrificed either at birth or at weaning at d 28 of age (BW: 6.93±0.527 kg). Subsequently, internal organs and the semitendinosus (ST) muscles were collected and weighed. Histological analyses were performed on the ST using the mATPase staining procedure after pre-incubation at pH 10.2. This allowed identifying the muscle cross-sectional area, TNF, number of P and S fibers and the S/P ratio of the dark (STD) and the TNF of the light (STL) portion of the ST. Relative to slaughter weight, the spleen and ST were 90 and 26% heavier ($P<0.01$), respectively, whereas lungs, liver, heart and kidneys were 17, 16, 30 and 24% lighter ($P<0.06$) at weaning than at birth. From birth to d 28 of age TNF increased in the STD (151,020 vs. 235,191; $P<0.01$) but not in the STL (395,497 vs. 405,836; P=0.83). The increase resulted from both a greater number of P (4,597 vs. 6,605; $P<0.01$) and S fibers (146,423 vs. 228,586; $P<0.01$) with no changes in the S/P ratio (32 vs. 35; P=0.25). Overall the TNF of the ST was only numerically greater (546,517 vs. 641,028; P=013) in weaned than newborn piglets. This preliminary data suggest that the TNF of parts of muscles are not fixed at birth.

Effect of transport, unloading, lairage, pig handling and stunning on meat quality

Van De Perre, V.[1], Permentier, L.[1], De Bie, S.[2], Verbeke, G.[3] and Geers, R.[1], [1]K.U. Leuven, Laboratory for Quality Care in Animal Production, Zootechnical Centre, Bijzondere weg 12, B-3360 Lovenjoel, Belgium, [2]VLAM-BELPORK, Leuvensplein 4, B-1000 Brussel, Belgium, [3]K.U.Leuven, Interuniversity Institute for Biostatistics and statistical Bioinformatics, Kapucijnenvoer 35, B-3000 Leuven, Belgium; Vincent.VandePerre@biw.kuleuven.be

Appropriate pre slaughter handling of pigs is very important, not only from a welfare point of view, but it also affects pork quality and is consequently linked with economic implications. Stressful situations before slaughter such as loading, transportation, unloading, lairage and driving pigs to the stunning line could be responsible for the development of aberrant pork quality. Other important factors are season, stunning method, genetics, chilling and handling on farm. In this study the effect of different pre-slaughter parameters i.e. transport, unloading, lairage, pig handling, stunning method, genetics and season on meat quality based on pH measurements were examined. Between March and December 2009, a total of 95 transports were followed up in one to three different trials each time at the same 17 Belgian commercial slaughterhouses. The final model showed significant ($P<0.05$) effects for season, level of decibels produced during unloading, percentage of pigs that showed a high respiration frequency and the frequent use of electric goads on meat pH. CO_2 concentrations >80% gave a better stunning and meat quality than lower concentrations.These results showed that stress immediately before slaughter has more effect on meat quality than genetics.

Development of a new crossbred based evaluation for carcass quality traits of Piétrain boars in the Walloon Region of Belgium

Dufrasne, M.[1], Hammami, H.[1], Jaspart, V.[2], Wavreille, J.[3] and Gengler, N.[1,4], [1]University of Liege, Gembloux Agro-Bio Tech, Animal Science Unit, 2 Passage des Déportés, 5030 Gembloux, Belgium, [2]Walloon Pig Breeders Association, 4 rue des Champs-Elysées, 5590 Ciney, Belgium, [3]Walloon Agricultural Research Centre, 9 rue de Liroux, 5030 Gembloux, Belgium, [4]National Fund for Scientific Research, 5 rue d'Egmont, 1000 Brussels, Belgium; marie.dufrasne@ulg.ac.be

The aim of this study was to develop a genetic evaluation model to estimate the genetic merit of Piétrain boars for some carcass quality traits. These boars are now evaluated on performances recorded on their crossbred progeny fattened in a central test station. Data provided by the on-farm performance recording system (used on farm but also in the test station) were utilized in this study. Traits analysed were backfat thickness (BF) and lean meat percentage (%meat). The data file contains 60 546 records measured on pigs between 150 and 300 days of age. Model developed was a multitrait animal model. Fixed effects were sex, contemporary groups and heterosis, modeled as regression on heterozygosity. Random effects were additive genetic and permanent environment, modeled by random regressions using linear splines, and residual. Variance components were estimated by restricted maximum likelihood (REML) method on random samples of the total dataset and then confirmed by a Gibbs sampling algorithm on the total dataset. Fit of the model was tested by computing residuals from a BLUP (Best Linear Unbiased Prediction) evaluation. BF and %meat have a high heritability that increase with age. These two traits are also highly genetically correlated. Mean residuals are not significantly different from zero for both traits. Given that BF and %meat had high heritability, genetic improvement of carcass quality is possible by selection on these two traits. Like residuals are close to zero for both traits, it seems that model developed explain a great proportion the variance in each traits.

Seasonal influences on reproduction in sows

Citek, J., Stupka, R., Sprysl, M., Kratochvilova, H. and Likar, K., Czech University of Life Sciences, Animal Husbandry, Kamycka 129, 165 21 Prague 6-Suchdol, Czech Republic; citek@af.czu.cz

The purpose of this work was the examination of the season influence on the reproduction of sows. The reproduction indicators in the select stock were monitored in 10 634 litters during 3 years (2006-2008). These indicators were monitored: the number of total born, born alive and weaned piglets, the amount of alive born piglets, the number of weaned piglets, the weight of weaned piglets and the litter weight gain. The daily outside temperature in degree Celsius was monitored as well. The highest decline by total born piglets was demonstrated in hot seasons (July): 0.75-1.0 piglet in the litter in comparison with maximum in March-April. The decline of the number of weaned piglets of one sow was found out as a result of increasing outside temperature. At average were weaned 10.08 piglets in hot seasons June, 7.05 piglets (-30.1%), July 10.45 piglets (+3.7%), or, August 9.25 (-8.2%) piglets. The temperature influence on litter gain was evident. The average litter gain for one day was 2.46 kg in hot seasons: June 2.22 kg (-8.1%), July 2.19 kg (-11.0%) and August 2.33 kg (-5.3%). In cold months were registered gains, above-average gains were registered in the spring and autumn. The lower intensity of the growth in hot seasons we can explain by sows´ lower feed intake and their lower milk production. Experiments were performed within the framework of the research CMEPt No. MSM 6046070901.

Session 36 Poster 22

Effect of dietary creatine monohydrate supplementation on chemical composition, meat quality and oxidative status of pork

Bahelka, I., Polák, P., Tomka, J. and Peškovičová, D., Animal Production Research Centre, Hlohovecká 2, Lužianky, Slovakia (Slovak Republic); peskovic@cvzv.sk

Thirty hybrid pigs L x (HAxPN) with genotype NN (RYR1 gene) were involved in the experiment and divided into control and experimental groups. The pigs were fed finisher diet (control) or finisher feed supplemented with creatine monohydrate (CMH) 2.0 g/kg feed for 30 days prior to slaughter. The animals were slaughtered at 110±5 kg live weight and chemical composition of meat (musculus longissimus dorsi) and pork quality parameters 45 min., 24 hours and 5 days after slaughter were evaluated. To determine the stability of the skeletal muscle lipids against oxidation process, TBARS value was found. CMH supplementation had no effect on chemical composition of pork. Similarly, pH45 and pH24, meat colour after 24 h and 5 days were not influenced by creatine monohydrate. On the other hand, drip loss was statistically significantly influenced by CMH (4.34% - control vs. 3.36% - experimental group). Positive effect of CMH supplementation on oxidative stability of pork after 30 min incubation of muscle homogenate was determined (2.00 vs. 1.22 mg/kg). Results suggest that creatine monohydrate supplementation in pig diet can improve some meat quality parameters and oxidation status of pork muscles.

Session 36 Poster 23

Effect of liveweight and backfat depth at first mating on subsequent reproductive performance of sows

Sprysl, M., Citek, J., Stupka, R., Dvorakova, V. and Kratochvilova, H., Czech University of Life Sciences, Animal Husbandry, Kamycka 129, 165 21 Prague 6 Suchdol, Czech Republic; citek@af.czu.cz

This work is concerned with the quantification of the interaction of the body score of gilts expressed by their live weight and P2-backfat thickness during the first mating on subsequent reproduction productivity. The reproduction potential of 96 gilts of LWDxL genotype, originating from a production herd, were divided into groups in accordance with live weight and P2-backfat thickness. The division criterion in gilts was the standard division value of the monitored variables. The reproduction potential in accordance with parity (1st-4th litter) was assessed in particular groups. The following reproduction variables (number of total piglets, born alive and weaned piglets; their live resp. weaned weight; litter weight after parturition and the weaning and farrowing interval) were assessed using the GLM procedure of SAS (SAS Institute Inc., 2001). Fixed effects of live weight (Hi) and backfat thickness (Sj) were used, and differences between the means were tested by Tukey's method. The following model was used: $Yij = \mu + Hi + Sj + ei$. On the basis of the results obtained, it is possible to state that: 1) the live weight of gilts non-significantly affected the litter frequency; 2) the greatest litter frequency is in gilts farrowed at a live weight 141-145 kg; 3) the first litter frequency significantly affected the subsequent litter frequency; 4) in accordance with the first litter frequency, susequent litter frequency cannot be predicted; 5) gilts with 10-12 mm P2-backfat thickness showed the highest first litter frequency; 6) with a P2-backfat thickness decline, the average number of total and piglets born alive declined; 7) the number of weaned piglets and their live weight does not affect the body score in gilts. Experiments were performed within the framework of the research CMEPt No. MSM 6046070901.

Session 36 Poster 24

Effect of RBP4 (Retinol Binding Protein 4) genotype on reproductive traits and rearing performance of piglets

Mucha, A., Tyra, M. and Piórkowska, K., National Research Institute of Animal Production, Sarego 2, 31-047 Krakow, Poland; amucha@izoo.krakow.pl

During pregnancy, Retinol Binding Protein (RBP) is secreted from uterus. Along with retinal and retinoic acid, retinol is a form of vitamin A. This vitamin has a role in the development of the cardiovascular and nervous systems, the skeleton and limbs, and the genitourinary system, and is stored in fetal liver (as trans-retinol), especially in late pregnancy. For this reason, it can affect reproductive and rearing traits of piglets. A total of 122 Polish Large White and Polish Landrace piglets of similar reproductive performance were investigated. Number of piglets and their weight at birth and at 7, 14 and 21 days of age were recorded in all 1st, 2nd and 3rd parity animals. Genomic DNA was isolated from blood and RBP4 polymorphism was determined by PCR-RLFP. The largest number of piglets was found in 1st and 3rd parity sows of AA genotype (11.88 and 11.69 piglets, respectively) and in 2nd parity sows of BB genotype (11.57 piglets). However, rearing mortality was the lowest in all parities of BB genotype sows. These sows gave birth to lighter piglets compared to AA and AB genotype sows. In addition, statistical analysis showed significant ($P<0.05$) differences for the number of 2nd parity piglets aged 7 days between AA and BB genotype sows.

Session 36 Poster 25

Body score effect on the farrowing interval in sows

Stupka, R., Sprysl, M., Citek, J., Dvorakova, V. and Kratochvilova, H., Czech University of Life Sciences, Animal Husbandry, Kamycka 129, 165 21 Prague 6 Suchdol, Czech Republic; citek@af.czu.cz

This work is concerned with the influence of the body score in sows on the farrowing interval lenght. For this purpose, the P2-backfat thickness before parturition as well as the P2-backfat thickness decline level during lactation were monitored in 1017 sows of three genotypes. On the base of the results obtained, it can be stated that backfat thickness level before parturition significantly influenced the subsequent farrowing interval lenght, respectively the weaning-estrus interval. It was found that lean sows showed the longest farrowing interval. When the body score was better, a shorter weaning-estrus interval, respectively farrowing interval (from 159.61 days in the leanest sows up to 146.21 in animals with the highest P2-backfat thickness) was shown. Also, it was found that the farroving interval is influenced by the backfat thickness decline level during lactation. Those animals with marked fat decline showed a longer farrowing interval (151.89 days). Sows with no bacfat thickness decline subsequently demonstrated a shorter farrowing interval (148.99 days) as well as a shorter weaning-estrus interval. Experiments were performed within the framework of the research plan CMEPt No. MSM 6046070901.

Session 36 Poster 26

Reproductive characteristics of pig farms in the region of Epirus, Greece

Maglaras, G.[1], Papavasiliou, D.[1], Bizelis, J.[2], Vatzias, G.[1], Labidonis, D.[2] and Nikolaou, M.[3], [1]T.E.I. of Epirus, Animal Production, Kostakii, Arta, 47100, Greece, [2]Agricultural University of Athens, Animal Science and Aquaculture, Iera odos 75, 1185, Greece, [3]Local Centre for Rural Development, Ioannina, 45110, Greece; gmag@teiep.gr

The objective of this study was to collect, analyze and investigate the various parameters that determine reproductive efficiency of sows, in order to evaluate the most suitable breeding pigs and improve competitiveness of Greek pig meat and products.The dataset consists of 128,678 recordings concerning reproductive data of 2,144 sows throughout 12,678 farrowings from 22 herds. Environmental data were also obtained from 20 farms with a total of 14,600 observations. In advance, blood sampling of one hundred and sixty pigs from eight different farms took place in order to investigate the presence of the ryanodine receptor gene (detection of 'T' mutation), which is associated with the production of PSE pig meat and increased stress sensitivity. It was revealed that farmers used 6 different genotypes (Genotypes: G1,G2,G3,G4,G5,G6). They have a significant effect ($P<0.05$) on piglet live born numbers, farrowing intervals, age at first service and first farrowing and duration of reproductive life ($P<0.01$). Traits such as total sow production (total number of piglets weaned during whole reproductive life) and sow productivity (expressed as total production per duration of reproductive life) were significantly better ($P<0.01$) for genotypes G4,G2 and significantly lower for genotypes G3, respectively. A large proportion (29.4%) of the sows were heterozygous or carriers of the genotype for the Halothane gene (HAL^n/HAL^N) suggesting possible detrimental effects on meat quality.It was also apparent that farm imposes a significant effect ($P<0.05$-$P<0.001$) on all the above mentioned traits with relative economic importance. This in conjunction with the environmental data reveals the variation of management and its effect on reproductive and productive results of the studied genotypes.

Group housing of lactating sows with electronically controlled crates

Bohnenkamp, A.-L.[1], Traulsen, I.[1], Müller, K.[2], Tölle, K.-H.[2], Meyer, C.[2] and Krieter, J.[1], [1]Institute of Animal Breeding and Husbandry, Christian-Albrechts-University, Ohlshausenstraße 40, D-24098 Kiel, Germany, [2]Camber of Agriculture Schleswig-Holstein, LVZ Futterkamp, Gutshof, D-24327 Blekendorf, Germany; itraulsen@tierzucht.uni-kiel.de

The aim of the present study was to evaluate group housing (GH) of lactating sows in comparison to conventional single housing pens (SH). 96 cross-bred sows were compared in 8 batches. 6 sows were moved in each system 1 week before farrowing. The SH-sows were housed in pens with conventional farrowing crates while GH-sows had individual pens with a crate, electronically controlled gates and a running area (13 m^2) between the pens. The GH-sows could move freely and were fixed in the crates excluding 3 days ante partum (a.p.) until 1 day post partum (p.p.). The piglets stayed in the pens for five days p.p. Piglets were weaned at the age of 26 days. Behavioural parameters, litter information and sow condition were recorded. After farrowing, the sows spent 22.5 h/d with their piglets voluntary. The sows' duration of staying in the farrowing crates decreased daily from 20 h/d 5 days p.p. to 10.4 h/d 11 days p.p.. The piglets born alive, stillborn and mummified were not different among housing systems (SH vs. GH, 14.4 vs. 13.8, 1.0 vs. 1.2, 0.5 vs. 0.4). The piglet mortality rate was enhanced in the GH (15.7% vs. 14.4%) because more piglets were crushed within the first 3 days p.p.. The numbers of piglets weaned did not differ (SH vs. GH; 11.4 vs. 11.3). The average weaning weights were 7.9 kg in the SH and 7.5 kg in the GH. GH-sows lost more body weight and back fat during lactation (-18.8 kg vs. -14.0 kg, -3.2 mm vs. -2.5 m), although the feed intake was similar (SH vs. GH, 169 kg vs. 173 kg). In conclusion in both housing systems equal numbers of weaned piglets were observed with different weaning weights and mortality rates. The GH-sows lost more body condition during the lactation period.

Digestible lysine: metabolizable energy to barrows and gilts diets at 50 to 70 kg: performance and ultrasson parameters

Gandra, E.[1], Berto, D.[1], Gandra, J.[2], Donato, D.[3], Budiño, F.[4], Schammas, E.[4] and Trindade Neto, M.[2], [1]FMVZ / Unesp Botucatu, Animal Production and Nutrition, Distr. Rubião Jr, s/n, 18.618-970 / Botucatu - SP, Brazil, [2]FMVZ / USP Pirassununga, Animal Production and Nutrition, Rua Duque de Caxias Norte, 225, 13635-900 / Pirassununga - SP, Brazil, [3]FCAV / Unesp Jaboticabal, Animal Production and Nutrition, Via de Acesso Prof. Paulo Donato Castellane s/n, 14884-900 / Jaboticabal - SP, Brazil, [4]IZ / Nova Odessa, Rua Heitor Penteado, 56, 13460-000 / Nova Odessa - SP, Brazil; messiasn@usp.br

The effects of digestible lysine were studied in performance and ultrasson parameters (backfat and Longissimus muscle area) assay with specific line of barrows and gilts at 50 to 70 kg body weight alloted in randomized block design. In experiment were used 72 gilts and 72 barrows, with 6 replicates and 2 animals per pen. The treatments were 0.70, 0.80, 0.90, 1.00, 1.10 and 1.20% digestible lysine in diet with 16.18% CP and 3400 kcal/kg, in average. The control performance parameters occurred in 16 and 32 days of experiment. There was linear effect ($P \leq 0.05$) in crude protein intake (g), in crude protein eficience and backfat; and quadratic effect ($P \leq 0.05$) in feed conversion and Longissimus muscle area in gilts. In barrows there was linear effect ($P \leq 0.05$) in daily gain (kg), in relative gain (%), in feed conversion, in crude protein intake (g) and in crude protein eficience. The results suggests 1.06% digestible lysine the better level to gilts.

Digestible lysine: metabolizable energy to barrows and gilts diets at 50 to 70 kg: apparent digetibility and blood parameters

Gandra, E.[1], Berto, D.[1], Gandra, J.[2], Garcia, P.[2], Budiño, F.[3], Schammas, E.[3] and Trindade Neto, M.[2], [1]FMVZ / Unesp, Animal Production and Nutrition, Distr. Rubião Jr, s/n, 18.618-970 / Botucatu - SP, Brazil, [2]FMVZ / USP, Animal Production and Nutrition, Rua Duque de Caxias Norte, 225, 13.635.900 / Pirassununga/SP, Brazil, [3]IZ, Animal Science, Rua Heitor Penteado, 56, 13460-000 / Nova Odessa - SP, Brazil; messiasn@usp.br

The effects of digestible lysine were studied in apparent digestibility and blood parameters assays with specific line of barrows and gilts at 50 to 70 kg body weight allotted in randomized block design. In digetibility assay were used 20 pigs, 4 replications and one animal for experimental unit. In blood profile assay were used 24 barrows and 24 gilts, 4 replications for sex and one animal for experimental unit. The treatments were 0.70, 0.80, 0.90, 1.00, 1.10 and 1.20% digestible lysine in diet with 16.18% CP and 3400 kcal/kg, in average. In digestibility assay the feed intake based in dry matter for metabolic unit (kg 0.75). The control blood parameters occurred at first study (farm condition) using 4 gilts and 4 barrows and the final blood parameters occurred at finish study (segregate condition) using 24 gilts and 24 barrows. There was linear effect of lysine in neutrophil/limphocit. There was linear effect ($P \leq 0.05$) in digestible dry matter (%), crude protein in urine (g), nitrogen intake (g), nitrogen in urine (g); and quadratic effect ($P \leq 0.05$) in crude protein intake (g), nitrogen retention (g), nitrogen absorbed (g), nitrogen retention / absorbed (%), crude energy intake (kJ/g), digestible energy (kJ/g), metabolizable energy (kJ/g) and energy balance (kJ/g). There was sex effect in nitrogen metabolism. The gilts showed more efficiency in nitrogen retention than barrows. The results suggests 1.04% of digestible amino acid. The Dunnet test indicates significant differences between farm and segregate conditions.

Microbial and biochemical properties of wet pig feed on Swedish farms

Lyberg, K. and Lindberg, J.E., Swedish University of Agricultural Sciences, Department of Animal Nutrition and Management, Box 7024, SE-750 07 Uppsala, Sweden; Karin.Lyberg@huv.slu.se

Microbiological and biochemical properties of liquid diets for growing pigs were studied over time in different farms. Standard grower liquid diets were sampled as fed in triplicates during four subsequent weeks on four representative pig farms in the central region of Sweden. Yeast, lactic acid bacteria and Enterobacteriaceae were quantified and pH as well as organic acids was measured on fresh samples. Yeasts were quantified on Malt Extract Agar plates supplemented with chloramphenicol, Lactic acid bacteria were quantified on de Man Rogosa Sharp agar supplemented with Delvocid, and Enterobacteriaceae were quantified on Violet Red Bile Agar. Yeast and lactic acid bacteria (log cfu counts/g sample) differed in quantity and in diversity on each farm over time ($P<0.001$) and Enterobacteriaceae were only present when pH were above 4.5. Acetic acid levels were low (maximum 3.8 g/l), and lactic acid levels were connected to a lower pH and absence of Enterobcteriaceae. Succinic acid and ethanol differed also on each farm over time. Level of ethanol was connected to a low pH suggesting a more fermented wet feed. The varying microflora and biochemical properties of the liquid feeds on each farm, measured over such a short period as four weeks, indicate unstable conditions. Although the farms did not notice any measurable negative effects on production or pig health during time of the study, the current data suggests a need to develop more controlled liquid feeding systems.

Effect of roughage on the growth performance of intensively raised pigs under outdoor housing system

Juska, R., Juskiene, V., Leikus, R. and Norviliene, J., Institute of Animal Science of LVA, Department of Animal Hygiene and ecology, R. Zebenkos 12, LT-82317, Baisogala, Lithuania; violeta@lgi.lt

The experiment was performed to evaluate the effects of roughage (red clover grass) on the growth performance and behaviour of intensively raised highly muscular pigs under outdoor housing system. The study was conducted at the LVA Institute of Animal Science with two groups of 28 Norwegian Landrace and Norwegian Landrace x Pietrain crossbreed pigs. Two analogous groups – control and experimental – were formed. Pigs of both groups were raised outdoors in enclosures with fitted sheds on 9.0 areas. The pigs in both groups were given compound feed ad libitum. However, the diet of the experimental pigs was additionally supplemented with red clover. The results of experiment showed that both in the growing (under 60 kg of weight) and finishing (over 60 kg weight) periods and during the whole experiment, the pigs in the experimental group gained higher weights, respectively, by 14.3%, 5.1% and 7.9% than the pigs in the control group, but the red clover addition to the feed did not significantly affect the growth performance of pigs. Grass supplemented diets for pigs improved the intake of feeds, especially, in the growing period. In this case, the daily intake of compound feed was 5.3% higher and the intake of feeds per kg gain was 7.2% lower. The behaviour studies indicated that the pigs in the control group were a little more active and spent on the average 40 minutes more for rooting, yet the differences were insignificant. The pigs in the experimental group spent on the average 40.2% (P=0.052) more time for eating and drinking.

Relationships between performance test of gilts and their subsequent fatness, muscling and fertility

Mucha, A., Orzechowska, B. and Tyra, M., National Research Institute of Animal Production, Sarego 2, 31-047 Krakow, Poland; amucha@izoo.krakow.pl

The objective of this study was to determine if performance test results (obtained from animals aged between 150 and 210 days in Poland) can be used to predict subsequent body condition and reproductive performance of sows and rearing performance of piglets. A total of 82 Polish Large White and Polish Landrace gilts and sows, performance tested for fattening and slaughter traits, were studied. Over three successive litters, sows were weighed and measured at mating and at farrowing for backfat and longissmus muscle thickness using a PIGLOG 105 instrument. Number and body weight of piglets was recorded at birth and at 7, 14 and 21 days of age. Relationships were determined between body weight, backfat and longissimus muscle thickness measured during performance testing, the same three parameters measured at mating and at farrowing, and the number and body weight of piglets in subsequent weeks of life. The highest correlations were found between body weight measured at performance testing of gilts and that measured at first mating (r=0.343). Similar correlations were found for backfat thickness at the P2 position (behind the last rib, 3 cm off the midline) measured during the same time periods (r=0.330), backfat thickness at the P4 position (behind the last rib, 8 cm off the midline) (r=0.431), and height of longissimus muscle at the P4M position (r=0.209). It is concluded that body weight, fatness or muscling of sows in successive litters cannot be predicted from performance test measurements. It was also found that body weight, fatness and muscling measured on the day of performance testing are not correlated with the number and body weight of piglets.

Dietary preferences of piglets for mixtures of formic and lactic acids or their salts combined with or without phosphoric acid

Suárez, J.A.[1], Roura, E.[2] and Torrallardona, D.[1], [1]IRTA-Centre Mas de Bover, Ctra. Reus-El Morell km 3.8, 43210 Constantí, Spain, [2]Lucta S.A., Ctra. Granollers-Masnou km 12.4, 08170 Montornés del Vallès, Spain; david.torrallardona@irta.es

Weaning is a stressful process for piglets, in which the adaptation to the weaning diet is one of the most limiting factors. The palatability of the weaning diet ingredients may play a determinant role. A series of double choice trials with 180 pigs (20-25 kg BW) in 45 pens, was conducted to determine the preference of piglets, relative to a control (REF) diet, for diets containing a mixture of formic and lactic acids (FLa) or their corresponding Ca-salts (FLs) with or without partial replacement with phosphoric acid (PA). Each pen (4 pigs per pen) contained two feeding hoppers; one with the REF diet and one with the test diet with one of the products being studied. Each product was tested using 9 pens in three consecutive 4d-periods, in which low (0.5%), medium (1%) and high (1.5%) inclusion rates were tested. The diets tested were: REF (control); FLa; FLs; FLa+PA and FLs+PA. Preferences were calculated as the percent contribution of the test diet to total feed intake and are presented in brackets ordered by increasing doses. The preference values were compared to the neutral value of 50% with the Student t-test and values with an asterisk are significantly ($P<0.05$) different. Preferences for REF (48.2, 53.0, 51.3) and FLs (54.6, 46.5, 51.2) diets did not differ from 50%, whereas FLa significantly reduced feed preference (38.3*, 37.4*, 25.9*). The partial replacement of the mixtures with PA did not significantly affect preference for FLs (40.8, 43.0, 37.2), whereas it improved that for FLa particularly at the low and medium inclusion rates (49.0, 43.8, 33.9*). We conclude that the combination of lactic and formic acids but not that of their corresponding Ca-salts has a negative effect on piglet feed preference, and that this can partially be counteracted by the replacement of the acid mixture with phosphoric acid.

Interactive effect of dietary energy concentration and genotype on the growth rate, feed conversion and body composition of pigs in the growing-finishing period

Permentier, L.[1], Maenhout, D.[2], Broekman, K.[3], Deley, W.[4] and Geers, R.[1], [1]Laboratory for Quality Care in Animal Production, Zootechnical Centre, Catholic University Leuven, Bijzondere weg 12, B-3360 Lovenjoel, Belgium, [2]Hendrix Haeck, Zuidkaai 6, B-8770 Ingelmunster, Belgium, [3]Hypor B.V., Spoorstraat 69, NL-5831 CK Boxmeer, Netherlands, [4]Hypor Belgium NV, Leie Rechteroever 1, B-9870 Olsene, Belgium; Liesbet.Permentier@biw.kuleuven.be

In order to maximize the market value, the main focus of pig breeding is to optimize the production of lean meat on the one hand and to realize a high growth level on the other hand. Growth of pigs is influenced by several factors such as feeding or energy intake. Moreover, large differences in growth level appear between different genotypes. Piétrain crossbreds are commonly known as a slow growing genotype. The objective of this study is to investigate the interactive effect between genotype and dietary energy level. Two Piétrain crossbred types of pigs are selected according to their growth rate. Genotype 1 represents a slower growing genotype and genotype 2 a faster growing genotype. All the animals are fed either a standard energy diet or a high energy diet according to a three-phased ad libitum feeding scheme. The animals are weighted at the start of every phase, namely at 20, 40 and 70 kg. Daily growth per phase is analysed using the GLM-procedure (SAS). Preliminary results indicate a significant interactive effect, though depending on growth phase. During the first phase, a significant effect of genotype x diet reveals a higher daily growth for genotype 1 fed a high energy level diet in comparison to the standard diet. At a weight of 70 kg, an increased growth rate can be seen for genotype 2 compared to genotype 1, confirming the characteristics of Piétrain-crossbreds. Extra data on feed conversion and body composition will be added and further analysis involving the use of more sophisticated statistical models will be done in the near future.

Session 37 Theatre 1

Group housing horses under nordic conditions

Søndergaard, E.[1], Jørgensen, G.H.M.[2], Hartmann, E.[3], Hyyppä, S.[4], Mejdell, C.M.[5], Christensen, J.W.[6] and Keeling, L.[3], [1]AgroTech, Udkærsvej 15, Skejby, 8200 Århus N, Denmark, [2]Norwegian University of Life Sciences, P.O.Box 5003, 1432 Ås, Norway, [3]Swedish University of Agricultural Sciences, Box 7038, 75007 Uppsala, Sweden, [4]Agrifood Research, Opistontie 10 A 1, 32100 Ypäjä, Finland, [5]National Veterinary Institute, Ullevålsveien 68, 0454 Oslo, Norway, [6]University of Aarhus, P.O.Box 50, 8830 Tjele, Denmark; evs@agrotech.dk

Group housing satisfies the demands of horses with respect to regular movement and social contact. There are beneficial effects on e.g. trainability, and horses reared in socially complex environments develop better social skills useful for future contacts with different horses. Despite major improvements in horse husbandry over the last decades, many horses are still housed in single boxes due to concerns about group housing. For example, people are worried about an increased risk of injuries, the difficulty of separating a horse from the herd with potentially dangerous situations for humans, or the group composition itself is questioned. This project aimed to investigate issues related to group housing of horses and to provide practical solutions that could improve horse welfare and human safety. A second aim was to carry out basic experimental tests to study the behaviour of horses towards other horses and humans. During the 4-year experimental period we have investigated the effects of group composition in relation to sex, age and group stability, various elements of introducing horses to each other, effects of environmental enrichment, how separating a horse from a group is practiced, and we have validated a scoring system for injuries in horses that can be used in practical conditions. 381 horses in 67 groups have been included in the study and the data present a unique opportunity for investigating how various elements of group housing, like space, group size etc. affects the social behaviour of horses, their reactions in various situations, and the frequency of injuries. Results from the experiments included in this project will be presented.

Session 37 Theatre 2

Influences of husbandry and feeding management on welfare and stress of horses

Hoffmann, G.[1], Rose-Meierhöfer, S.[1], Niederhöfer, S.[2], Hohmann, T.[2] and Bohnet, W.[2], [1]Leibniz-Institute for Agricultural Engineering Potsdam-Bornim e.V. (ATB), Department of Engineering for livestock Management, Max-Eyth-Allee 100, 14469 Potsdam, Germany, [2]University of Veterinary Medicine Hannover, Institute for Animal Welfare and Behaviour, Bünteweg 2, 30559 Hannover, Germany; ghoffmann@atb-potsdam.de

The objective of these studies was to detect the occurrence of stress in warm blood horses under various keeping conditions. Different housing systems (single, double and group housing), different opportunities to move (horse walker, pasture, soil run), and different feeding situations (automatically and manual) were compared. Each of the four experimental groups consisted of six mares aged two to three years. To monitor treatment effects on stress and welfare, heart rate variability (HRV-parameters HF, SD1, SD2, pNN50) and faecal cortisol metabolite concentration were analyzed. After analyzing the stress parameters, the following order of the different housing systems according to stress level (from low to high) was established: group husbandry of six horses in a multi-room yards husbandry facility (GH) with non-divided lying area, GH with divided lying area, single box with freely accessible paddock, single box without paddock, and keeping of two horses in two shared boxes and one attached paddock. In group housed horses, the influence of offering different opportunities to move was investigated. The horses' stress level was lowest when offering either pasture or horse walker. Higher stress levels were found when offering non-grassy pasture land or no opportunity to move at all. Concerning concentrate feeding, lowest stress levels were observed when feeding horses by using automatic simultaneous feeding dispensers. The manual feeding induced the highest stress values and longer waiting periods resulted in a higher stress load. In addition to group-specific similarities, the observation of individual horses showed large inter-individual differences.

Session 37 Theatre 3

How could we measure horses' welfare: towards visible indicators

Fureix, C., Jego, P., Coste, C. and Hausberger, M., Université de Rennes 1, UMR CNRS 6552 Ethologie Animale et Humaine, Campus Beaulieu batiment 25 - 263 avenue Gl Leclerc, 35042 - Rennes cedex, France; carole.fureix@univ-rennes1.fr

A central issue of animal welfare studies has been, and continues to be, how can we measure animals' welfare? Among measures, health-related and physiological disorders appear as the most universally accepted indicators of altered welfare. However, some disorders could be difficult to assess directly, and consequently could be unappreciated. Moreover, even if good welfare starts with physical condition, it does not stop there and visible indicators reflecting the animals' internal state (mental well-being) must also be considered. Here we propose to present a critical review of welfare indicators in horses, then to present results from a first multidimensional study about horses' welfare performed in 59 horses coming from riding schools. The aim of this study was to identify reliable and visible indicators of the horses' welfare state, based on a multidimensional approach including health-related, physiological, behavioural and postural measures. Here we identified several visible postural and behavioural indicators that were clearly associated with altered welfare (i.e. presence of health-related disorders). Moreover, these indicators differed between riding schools, allowing us to identify riding schools' welfare profiles in addition to individual profiles. As a previous study revealed a relationship between horses' altered welfare and their aggressiveness towards humans, these visible postural and behaviours indicators are crucial to help professionals and non-professionals to identify their horses' welfare state and consequently to prevent accidents.

Session 37 Theatre 4

The hygro-thermal environments of horses during long-haul air transportation

Mitchell, M.A.[1], Kettlewell, P.J.[2] and Leadon, D.P.[3], [1]SAC, SLS, Bush Esatate, Penicuik, Midlothian, EH26 0PH, United Kingdom, [2]ADAS, Drayton, Alcester Road, Stratford on Avon, Warwickshire, CV37 9RQ, United Kingdom, [3]Irish Equine Centre, Equine Pathology, Johnstown, County Kildare, Ireland; malcolm.mitchell@sac.ac.uk

Thoroughbred horses are frequently transported by air over large distances. Such journeys may be associated with 'shipping fever' and the incidence and severity of the condition may be linked to the environmental conditions in transit. Detailed analysis of the hygrothermal transport micro-environment has not been reported. In particular, little information relating to absolute humidity is available. Therefore, 2 commercial flights (A-Sydney to New York and B-Shannon to Sydney) carrying groups of horses (A: 24 and B: 11) were monitored to characterize the variability in the thermal micro-environment around and within jet stalls. On each aircraft 20 data loggers, distributed throughout the transport space, were employed to record the air temperature and relative humidity at 1 minute intervals. The flights were of 19 (A) and 25 (B) hours duration. The mean temperature conditions on the two flights were: A: 20.9±3.8 °C, range 13.6-20.9 °C and B: 19.9±1.1 °C, range15.1-25.4 °C. The corresponding mean water vapor densities were A: 6.6±3.3 gm^{-3} and B: 5.4±3.0 gm^{-3} but with minima of A: 4.8 and B: 1.0 gm^{-3}. The lower VD on the longer flight B was associated with post transport pathology in 7 of the 11 horses. Comparison of the water content of the air during the flights with air at sea level and calculation of the water vapor density gradients available for evaporation between the upper respiratory tract and inspired air indicates a significant predisposition towards elevated drying rates of respiratory tract and associated dysfunction. It is proposed that a major predisposing factor to bacterial pleuropneumonia is the hygrothermal conditions prevailing in aircraft during long haul equine transportation.

Evaluation of physical activity and energy requirements in trotting horses using GPS technique during outdoor training

Giosmin, L., Bailoni, L. and Mantovani, R., University of Padua, Department of Animal Science, Viale Universita, 16 Legnaro (PD), 35020 Legnaro (PD), Italy; linda.giosmin@unipd.it

The aim of this study is to evaluate the actual physical activity and energy expenditure of trotting horses by GPS (Global Positioning System, Garmin®). The trial was carried out at Gina Biasuzzi stable (Mirano, Venice, Italy), using 7 GPS instruments simultaneously on 30 standardbred trotting horses in interval training program. Animals were divided in different classes of age (2, 3 and ≥4 years) and sex (males and females). The experiment lasted 5 weeks and the individual daily physical activity of horses was measured by GPS given to drivers or blocked on the sulky during the training. Overall 687 runs were measured. The energy expenditure was obtained on the basis of estimated oxygen consumption using equation from literature. BW and BCS were measured at the beginning and at the end of experiment. All data were analyzed using a mixed model for repeated measurements, accounting for the effect of age (A), sex (S), interaction AxS, random effect of horse, and period. BW and BCS changes during the whole trial were very low. As regard the GPS measures, none of the possible sources of variation affected the LSMEANS for speed, time and distance (on average 18 km/h, 28 min/d and 7.8 km/d respectively) on exercise, indicating a strong standardized method of training for the athletic horses. The energy allowances resulted greater than both energy requirements suggested by NRC-2007 (on average +6 Mcal/d) and energy expenditure estimated by GPS (on average +18 Mcal/d), pointing out some variations into age classes and sex. In conclusion, the use of GPS can provide quick useful data on physical activity of trotting horses, helping trainers to schedule athletic activity. About the differences between energy allowances and needs, more precise theoretical requirements could also be implemented by NRC for this kind of sporting horses and equations used to calculate the energy expenditure starting from GPS data could be developed.

Digestibility of extruded complete diet in horses

Etchichury, M., Tamas, W.T., Gonzaga, I.V.F., Wajnsztjen, H., Lopes, M.H., Ignacio, J.A. and Gobesso, A.A.O., Universidade de São Paulo, Nutrição e Produção Animal, Av. Duque de Caxias, Norte, 225, 13635-900 / Pirassununga/SP, Brazil; mariano@usp.br

Extruded complete diet (ECD) is a processed feed where hay and concentrated are mixed and extruded to form a long pellet. Feeding horses this single way presents an easier and more practical aspect in manage, avoiding supply additional grains, supplements, grass or hay, and facilitating transport, conservation, storage and labor. To compare ECD digestibility with a traditional diet, four mares aging 32±2 months and weighing 475±23 kg were fitted in a 4x4 Latin square model design during 4 periods of 11 days each, been the first 8 days to diet adaptation and the last 3 to faeces collection. Control diet (CD) was formulated with gramineous hay and concentrate containing ground corn, 50/50. Experimental diets (ED) consisted of replacing CD by ECD in 33% (ED1), 66% (ED2) and 100% (ED3). Mares were kept in stalls, and fed twice a day, at 7:00 and 19:00 h. To determine apparent digestibility of nutrients, total collection of faces was made 4 times a day, during 3 days. Resulting data was processed by ANOVA, and analyzed for simple polynomial regression using SAS program. During the trial, none nutritional disturbance or abdominal discomfort was noted in any horse. No significant differences (P>0.05) were observed among treatments (CD, ED1, ED2 and ED3) for digestibility of dry matter, organic matter, crude protein, ether extract, acid detergent fiber, neutral detergent fiber or starch. In terms of digestibility, this result shows ECD as a viable option for horse nutrition and fed industry.

Digestibility in horses fed maltodextrine in substitution of dietetic starch

Etchichury, M., Gil, P.C.N., Taran, F.M.P., Wajnsztejn, H., Gonzaga, I.V.F., Gandra, J.R., Tamas, W.T. and Gobesso, A.A.O., Universidade de São Paulo, Nutrição e Produção Animal, Avenida Duque de Caxias Norte, 225, 13635-900, Brazil; paulo.gil@ourofino.com

Maltodextrine (MD) is a polysaccharide obtained from cornstarch. Easily digested, its use in the nutrition of sport horses is presented as a potential alternative to replace more complex starches in diet. The aim of this study was to determine the effect of dietetic MD in substitution of starch (ST) on apparent digestibility. Four mares aging 30±2 months, and weighing 385±23 kg were randomly assigned to a 4x4 Latin square experimental design. Experimental diets were composed by 50% of concentrate and 50% forage. Concentrate contained no grains, but commercial starch, added to control concentrate (CC) at 50%, which was replaced by 33 (C1), 66 (C2) and 100% (C3) of MD in test diets. Mares were fed twice a day, at 7:00 and at 19:00 h. Collection of feces was made in a daily basis over 3 (three) days, with animals kept at stalls without bedding. Data obtained were submitted to ANOVA, and means compared by Tukey test. No significant difference ($P>0.05$) were observed in feces pH among diets. Apparent digestibility of C3 was higher ($P<0.05$) for dry matter, organic matter, crude protein, neutral detergent fiber and acid detergent fiber (ADF) than C1. No difference ($P>0.05$) was observed between CC and experimental concentrates, except in ADF, that was higher ($P<0.05$) in CC when compared to C1. The highest values of apparent digestibility were observed in C3. In this study, commercial starch was used aiming standardize experimental protocols, due to be easier digestible than cereal grains starch. Properly used, MD can be added to equine diets in replace of starch with good results in digestibility, avoiding some problems resulting from diets containing high quantities of complex starch, like colic or laminitis. More studies must be done to establish comparisons between dietetic MD and cereal grains starch.

***In vitro* fermentation parameters of equine cecal contents in a nitrogen deficient environment**

Santos, A.S.[1], Ferreira, L.M.M.[1], Guedes, C.M.[1], Cotovio, M.[1], Silva, F.[1], Bessa, R.J.B.[2] and Rodrigues, M.A.M.[1], [1]CECAV-UTAD, Po-Box 1013, 5001-801 Vila Real, Portugal, [2]CIISA-FMV, Technical University of Lisbon, 1300-477 Lisboa, Portugal; assantos@utad.pt

Microbial fermentation of protein in the equine hindgut is influenced by the type and amount of protein reaching these compartments. This microbial population is probably adapted to systematically low levels of nitrogen (N) as most of the protein is digested in pre-cecal compartments. In order to evaluate this hypothesis an *in vitro* study was conducted to evaluate fermentation parameters of cecal contents in an N deficient environment. Cecal contents were collected, two hours after the morning meal, from 3 horses fitted with permanent cecal canulas. In order to assure that N was limiting to microbial growth a pre-incubation was performed during 2 h with a buffer solution containing an excess of rapidly fermentable carbohydrates. It was assumed that all available N from cecal fluid was incorporated into bacterial N components. After the pre-incubation period, 20 ml of buffered cecal fluid were incubated in fermentation tubes and samples were collected at six incubation times (0, 2, 4, 8, 12 and 24 h after incubation). Samples were analyzed for pH, volatile fatty acids (VFA) and ammonia nitrogen ($N\text{-}NH_3$). Data were analyzed using one way Anova. $N\text{-}NH_3$ values at 0 h were of 1.13 mg/100 ml, and increased ($P=0.03$) to 1.82 mg/100 ml at 24 h, indicating that the pre-incubation period was sufficient to ensure that N was limiting. The results also indicated a minimum level of $N\text{-}NH_3$ in the environment since the lowest concentration was found at 0 h. As expected, an increase in total VFA ($P<0.001$) occurred from 0 h to 24 h, with a decrease ($P<0.001$) in [(C2 + C4)/C3] ratio, as an excess of rapidly fermentable carbohydrates were present in the incubations. Results for pH were in accordance with VFA evolution, and decrease ($P<0.001$) from 0 to 24 h. These data can be explained by the bacterial lyses to overcome the N-deficient environment.

Session 37 Theatre 9

Glycemic response in horses fed extruded complete diet

Gobesso, A.A.O., Etchichury, M., Tamas, W.T., Gonzaga, I.V.F., Wanjsztejn, H., Lopes, M.H. and Ignacio, J.A., Universidade de São Paulo, Nutrição e Produção Animal, Av. Duque de Caxias Norte, 225, 13635-900 / Pirassununga/SP, Brazil; gobesso.fmvz@usp.br

Extruded complete diet (ECD) is a processed feed where hay and concentrated are mixed and extruded to form a long pellet. Feeding horses this single way presents an easier and more practical aspect in manage, avoiding supply additional grains, supplements, grass or hay, and facilitating transport, conservation, storage and labor. To compare ECD glycemic response with that of a traditional diet, four mares aging 32±2 months and weighing 475±23 kg were fitted in a 4x4 Latin square model design during 4 periods of 11 days each. Control diet (CD) was formulated with gramineous hay and concentrate containing ground corn, 50/50. Experimental diets (ED) consisted of replacing CD by ECD in 33% (ED1), 66% (ED2) and 100% (ED3). Mares were kept in stalls, and fed twice a day, at 7:00 and 19:00 hs. At the last days of each period, blood samples were collected from jugular vein 30 min before (-30), and 30, 90, 150 and 210 min after 7:00 meal. Data obtained was processed by ANOVA test, and trapezoid area under curve (AUC) was calculated using SAS program. No change in behavior, such as irritability or hyperexitability, was noted in any horse during the trial. Results for AUC were 376.5 (CD), 384.94 (ED1), 396.25 (ED2) and 431.44 (ED3) mg/dl x min. A linear response ($P<0.05$) was observed through a linear regression equation, and an increase in 17.61 mg/dl x min of AUC every 1% replacement of CD by ECD was detect. Glycemic peak was reached at 90 min for all diets, been the higher one ($P<0.05$) that of ED3. Increase of glucose in the bloodstream of horses fed ECD diets is probably related to the processing method of ECD, in which heat generated by extrusion broke starch molecules allowing a higher absorption of glucose at small intestine. Substitution of traditional diet by ECD provides more glucose for longer, increasing energetic balance. In horses, it could mean an improvement in sport performance.

Session 37 Theatre 10

Glycemic and insulinemic responses in horses fed maltodextrine in substitution of dietetic starch

Gil, P.C.N., Taran, F.M.P., Wajnsztejn, H., Gonzaga, I.V.F., Gandra, J.R., Tamas, W.T., Etchichury, M. and Gobesso, A.A.O., Universidade de São Paulo, Produção e Nutrição Animal, Avenida Duque de Caxias Norte, 225, 13635-900 - Pirassununga/SP, Brazil; paulo.gil@ourofino.com

Maltodextrine (MD) is a polysaccharide obtained from cornstarch, widely used in human nutrition as energy source. Easily digested, its use in the nutrition of sport horses is presented as a potential alternative to replace more complex starches in diet. The aim of this study was to determine the effect of dietetic MD in substitution of starch (ST) on apparent digestibility. Four mares aging 30±2 months, and weighing 385±23 kg were randomly assigned to a 4x4 Latin square experimental design. Experimental diets were composed by 50% of concentrate and 50% forage. Concentrate contained no grains, but commercial starch, added to control concentrate (CC) at 50%, which was replaced by 33 (C1), 66 (C2) and 100% (C3) of MD in test diets. Mares were fed twice a day, at 7:00 and at 19:00 h. Blood samples were obtained from jugular vein 30 min before (-30), and 30, 90, 150 and 210 min after 7:00 meal at 11th day of each period. Data obtained were submitted to ANOVA, and means compared by Tukey test. No difference ($P>0.05$) among treatments were observed in glycemia at minutes 30 and 150. At minute 90, glycemia of C1, C2 and C3 were higher ($P<0.05$) than CC. At minute 210, glycemia of C2 and C3 were higher ($P<0.05$) than C1 and CC. Area Under Curve (AUC) of C3 was higher ($P<0.05$) than CC (439 vs 380 mg/dl). C3 Insulinemia at 30 min and 90 min was higher ($P<0.05$) than CC, C1 and C2. At 210 min, C2 an C3 were higher ($P<0.05$) than C1 and CC. No difference between treatments ($P>0.05$) were observed at 150 min. C3 insulinemic AUC was higher ($P<0.05$) than CC (43 vs 27 µUI/ml). Complete substitution of dietetic ST by MD provides more glucose for longer, increasing energetic balance. In horses, it could mean an improvement in sport performance.

Session 37 Theatre 11

Assessment of structural and mechanical properties of equine cortical bone with several techniques

Vale, A.C.[1], Fradinho, M.J.[2], Bernardes, N.[2], Pereira, M.F.C.[3], Mauricio, A.[3], Amaral, P.M.[1], Ferreira-Dias, G.[2] and Vaz, M.F.[1], [1]IDMEC, DEMat, IST, UTL, Av Rovisco Pais, 1049-001 Lisboa, Portugal, [2]C.I.I.S.A, FML, UTL, Pólo Universitário da Ajuda, 1300-477 Lisboa, Portugal, [3]Centro de Petrologia e Geoquímica, IST, 1049-001 Lisboa, Portugal; amjoaofradinho@fmv.utl.pt

This study describes some methods used to evaluate the properties of bone and it consists on a first approach to study structural and mechanical behaviour properties of equine third metacarpal bone. Samples were taken post mortem from the third metacarpal bone, on the dorsal and lateral regions, of Lusitano horses. Samples used in microscopy were mounted in a resin, polished and coated with gold. The mechanical compressive tests were performed in an Instron 8502 machine. The structure of the samples was observed under a field emission scanning electron microscope JEOL, JSM-7001F. The microscope is coupled with an energy dispersive x-ray spectroscopy (EDS) detector Oxford. Micro computed tomography (micro-CT) analysis was undertaken with a Skyscan 1172 – 1.3Mpix device. Stress-strain curves obtained from the compression tests allowed for the determination of the ultimate stress and elastic modulus, which evaluate respectively both mechanical strength and stiffness. Microscopic images allowed for the microstructural characterization with evaluation of the area of Haversian canals and osteons. Areas were measured using commercial image analysis software (SigmaScan Pro 5). EDS enables to detect the presence of C, O, P and Ca. Specimens were examined with 3-dimensional micro-CT, which generates sequential images of the samples interior enabling 3D reconstruction and morphometric parameters estimation. Both microscopy and mechanical results show the heterogeneity on bone properties. The techniques used were found to be adequate to fulfil the objectives of the work. Further tests will contribute for the establishment of the relationship between structure and mechanical properties.

Session 37 Theatre 12

Body measurements of one week old Lipizzaner foals

Potočnik, K., University of Ljubljana, Biotechnical Faculty, Department of Animal Science, Groblje 3, 1230 Domžale, Slovenia; klemen.potocnik@bfro.uni-lj.si

For analyzing body measurements and body weight of one week old Lipizzaner foals we used data from foals born in Lipica Stud farm between years 2003 and 2009. In the 7 year period 245 foals in Stud farm Lipica were born. All data from five foals have been missing. Data of 240 foals that were offsprings of 89 dams and 31 sires were used. The lowest number of foals (24) were born in the year 2007 and the highest number of foals (43) were born in the year 2005. Foals were measured between 5th and 7th day of age. The average body weight was 51.9 kg; height of withers – stick 96.4 cm; height of withers – band 101.6 cm; chest girth 84.5 cm and cannon bone girth 11.8 cm. We estimated high statistically significant correlations between both heights of withers with 0.92, between body weight and chest girth with 0.82. Estimated correlations between cannon bone girth and other traits were low and varied between 0.48 with chest girth and 0.58 with height of wither. For variance analysis mixed model was used. It included fixed effects season and sex of foal and sire and dam as random effects. The season had statistically significant effect on body weight, height of withers – stick, height of withers – band and chest girth, but not on cannon bone girth. The sex of foal had statistically significant effect on cannon bone girth, height of withers – stick, height of withers – band but not on chest girth and body weight. The dam had high statistically significant effect on all measurements and on body weight. The sire effect had statistically significant effect just on both heights of withers. For all traits maternal effect is much higher than paternal effect. In the future we shall study relationship between body measurements of foals and of adult horses.

Application of digital image equipment (DIE) for distance morphological evaluation of horses

Santos, A.S., Cotovio, M. and Silva, S.R., CECAV-UTAD, Po-Box 1013, 5001-801, Vila Real, Portugal; assantos@utad.pt

Morphological evaluation of horses is important for management and research purposes. However, obtaining morphological traits is time-consuming and requires trained labor. In addition, the procedure can be stressful, especially for feral animals. To overcome these difficulties, in recent years, studies with different species have involved the application of digital imaging for biometric examinations. As a result the aim of the present study was the development of an equipment to allow morphological measurements at a distance by analysis of digital images. The equipment consisted of a two parallel red laser with a 650 nm wavelength and a digital video camera (Sony DCR-SX31). The distance between the two lasers was 32 cm, and it was used as marker for image analysis. The equipment was calibrated by filming fixed and previously measured objects. The study involved 11 horses. In order to evaluate the equipment reliability three morphological measurements (H) (withers height - WH; rump height - RH and body length- BL) were obtained with hypometer (H) and with digital image equipment (DIE). The video imagery was divided into frames and the best frame from each horse was selected. From this frame a 1200 × 720 pixels JPEG image was generated. This image was analyzed using the NIH ImageJ software and WH, RH and BL were determined. Data were subjected to regression analysis to study relationships between H (dependent variable) and DIE (independent variable) measurements. All analyses were performed with JMP software and the simple regression equations were evaluated with the determination coefficient (R^2). All DIE measurements explained a large amount of the variation of the equivalent H measurements ($R^2 > 0.93$; $P<0.001$). Accuracies of predictions (in terms of R^2) were high for all measurements ($R^2=0.98$; 0.95 and 0.93 for WH, RH and BL, respectively). Findings of the current study show that DIE is able to accurately obtain morphological measurements in horses.

Changes in biochemical markers of bone metabolism in the young Lusitano horse

Fradinho, M.J.[1], Correia, M.J.[2], Beja, F.[2], Perestrello, F.[3], Mateus, L.[1], Bessa, R.J.B.[1], Caldeira, R.M.[1] and Ferreira-Dias, G.[1], [1]CIISA, Faculdade de Medicina Veterinária, TULisbon, 1300-477 Lisbon, Portugal, [2]FAR, Alter-do-Chão, 7441-909 Alter-do-Chão, Portugal, [3]Comp. Lezírias, Porto Alto, 2135-318 Samora Correia, Portugal; amjoaofradinho@fmv.utl.pt

Knowledge of equine bone metabolism is one of present concerns of horse production in order to improve sports performance. Previous research has shown that foal's bone formation can be assessed by biochemical markers like osteocalcin (OC) and bone alkaline phosphatase (BAP). The main objective of this study was to characterize changes in OC and BAP plasma concentrations from birth to two years of age in the Lusitano horse, under extensive management conditions. Twenty-eight foals from three stud farms (groups), born between February and May, were monitored from birth to two years. Blood samples were monthly collected from birth to one year and thereafter every two months up to 18 months of age and every three months until the end of the study. On the same days foals were weighed and withers height and cannon circumference were measured. Concentrations of OC and BAP were determined with specific competitive immunoassays (ELISA). Pearson's correlation coefficients were used to examine the variables relationship. A mixed model of analysis of variance for repeated measures was used to assess the effect of group and age on OC and BAP concentrations. For both markers, a significant decrease on plasma concentrations with age was observed ($P<0.0001$). The effect of group was not significant. Positive correlations were found between bone markers and foals average daily gain (0.59, $P<0.0001$ for OC and 0.32, $P<0.0001$ for BAP). When plasma concentrations of OC and BAP were plotted against month of the year, a transient increase was observed in both spring periods, suggesting a seasonal effect. These results and age-related changes were similar to others described on sport light breeds. This study provides valuable information on bone markers concentrations during the first two years of Lusitano horse.

On the way to a new defined freezing extender in the equine species

Pillet, E.[1], Batellier, F.[2], Fontaine, M.[1], Duchamp, G.[1], Desherces, S.[3], Schmitt, E.[3] and Magistrini, M.[1], [1]INRA, UMR85 Physiologie de la Reproduction et des Comportements, 37380 Nouzilly, France, [2]IFCE, Haras national de Blois, 62 avenue Maunoury - BP 14309, 41043 Blois, France, [3]IMV-Technologies, R&D, 10 rue Clemenceau BP 81, 61302 L Aigle, France; michele.magistrini@tours.inra.fr

In most domestic animal species, sperm freezing extenders are composed of animal products as milk and egg yolk (EY). However, these products are potential sources of bacterial contaminations. So, we developed 10 years ago an extender named INRA96®, containing the purified fraction of native milk caseins, for long-term sperm storage at 4 °C or 15 °C. We recently demonstrated that, INRA96® extender supplemented with egg yolk (EY) and glycerol (G), improves significantly the fertility rates of equine frozen sperm compared to INRA82 extender supplemented with EY and G (as a control): respectively 71% (n=30/42) versus 40% (n=17/42) ($P<0.01$). More sterilized EY-plasma afforded the same protection as EY: respectively 69% (n=24/35) versus 60% (21/35) ($P>0.05$). EY and more precisely LDL, composed mainly of phospholipids (PL), are considered since a long time as cryoprotective agents. In our analytical approach to develop a new freezing extender, we have tested the effect of EY-phospholipids (EY-PL) instead of EY. In a fertility trial, we compared INRA96® extender supplemented with 1) EY and glycerol (as a control) to 2) liposomes of EY–PL and glycerol. Semen from four stallions was frozen and 66 mares (n=80 cycles) were inseminated. Pregnancy rate per cycle showed no significant difference between EY and EY-PL: 68% (27/40) versus 55% (22/40) ($P=0.23$). Our results demonstrate that PL-liposomes as cryoprotective agents are a promising approach. However, the ratio of the different PL has to be optimized to reach the same fertility rate as with EY or EY-plasma.

Sudden deaths in captive zebras: a preliminary answer into elucidating atypical myopathy in equids

Benamou-Smith, A.E.M.[1], Plouzeau, E.[1], Bailly, S.[2], Bailly, J.D.[2], Belli, P.[1] and Gaiilard, Y.[1], [1]VetagroSup, Lyon Equine Research Center, 1 Avenue Bourgelat, 69280 Marcy Letoile, France, [2]Toulouse Veterinary School (ENVT), Laboratoires des mycotoxicologie, 23 Chemin des capelles, 31076 Toulouse Cedex 3, France; a.benamou@vet-lyon.fr

The cause of death due to an acute myopathy among captive zebras was investigated. Four out of five young healthy adult zebras (Grevy's zebras) living on the « African plain » of the Lyon zoological park, died within 24 hours of displaying very acute clinical signs: tachycardia and tachypnea, congested mucuous membranes, muscular weakness and stiffness, hindlimb paresis and lateral recumbency. Urine showed a very dark brownish colour, (+++ blood and proteins on Dipstick). Blood analyses showed marked neutrophilia with toxic neutrophils, elevation of plasma CK, GGT and fibrinogen. Necropsy showed acute muscle degeneration and necrosis, hepatic and renal tubular degeneration. Clinical and histological observations are compatible with a diagnosis of atypical seasonal myopathy. This is the first description of this disease made in zebras. Based on possible hypothesis on causal agents for atypical myopathy, feed was examined for mycotoxins and found to be negative. The area of the African Plain which had been recently recreated from new and carefully planned on botanical grounds, was scrutinized. Complementary investigations were undertaken.Small quantities of alfalfa hay normally restricted to giraffes living on the same paddocks, had been accidentally eaten by zebras. Analysis of the alflafa hay revealed significant contamination with Aspergillus fumigatus and Stachybotryxs atra. Specific mycotoxic analysis was conducted on liver samples, and very high concentrations of trichothecenes produced by Stachybotryx (verrucarol: 1,220 ng/g) and Trichoderma (60 ng/g) were found. These results strongly support the role of Stachybotryx atra as a causal agent for atypical myopathy in our captive zebras. The presence of these toxins need to be verified in horses affected with atypical myopathy.

Session 37 Poster 17

Out of sight, out of mind. human visual attention and positioning affect horses' behaviour during an interaction

Fureix, C., Vallet, A.-S., Henry, S. and Hausberger, M., Université de Rennes 1, Ethologie Animale et Humaine, Campus Beaulieu batiment 25 - 263 avenue Gl Leclerc, 35042 - Rennes cedex, France; carole.fureix@univ-rennes1.fr

Horses' reactions towards humans are a combined result of their temperament, their previous experience but also of the skills of the human they are interacting with. However, very little is known yet about the relevant elements that have to be considered when humans interact with horses. Here we investigated whether positions and/or humans' visual attention had an impact on horses' reactions and level of obedience in a simple leading task. Professionals (n=8) and non-professionals (n=10) were asked to lead a horse (n=15) along a given path. The humans' position and visual attention towards horses, and the horses' latencies to obey (to start walking and to stop) and ears positions were recorded. Here we show that 1) horses obeyed more quickly and held ears upright longer when a human gazed at them and was close to the anterior part of their body(spearman correlation, $P<0.05$); 2) the positions of humans either far ahead of a horse (associated with less time gazing) or behind its shoulder (i.e. near the middle of its body) are associated respectively with low latencies to start walking and to stop (spearman correlation, $P<0.05$); and 3) professionals spent less time gazing at horses, more time far ahead of them, and appeared to induce higher obedience latencies in horses than did non-professionals (Mann Whitney, $P<0.05$). These results suggest that horses are able to perceive humans' visual attention and confirm that some relative human-horse positions could be more appropriate than others during interactions with horses. Moreover, this study supports experimentally the idea that professionals may have a lower attention level when interacting with horses as a result of task repetition and habit. The results reinforce the idea that knowledge of horse behaviour and of observation methods must be given to improve observational skills and attention, which are key elements to prevent accidents.

Session 37 Poster 18

Proteomic analysis of equine follicular fluid during late follicle development

Fahiminiya, S., Labas, V., Dacheux, J.L. and Gérard, N., INRA UMR 85, CNRS UMR 6175, Université de Tours, Haras Nationaux, Nouzilly, 37380, France; nadine.gerard@tours.inra.fr

Ovarian follicular fluid is in part an exudate of serum, as surrounding cell layers permit the free diffusion of proteins of up to 500 kDa. This fluid also contains locally produced factors related to the metabolic activity of ovarian cells and it reflects the physiological status of the follicle. Furthermore, it contains essential substances implicated in oocyte maturation and fertilization, granulosa cells proliferation and differentiation, ovulation and luteinization of the follicle. Studies on its components may contribute to understand the mechanisms underlying these processes. The aim of this study was to determine for the first time, the equine follicular fluid protein composition by SDS-PAGE as the dominant follicle develops and maturates, and to compare follicular fluid and serum protein profiles in order to identify the proteins which may have an essential role in follicle development and maturation. For this purpose, fluids from ovarian follicles of three increasing diameters: early dominant, late dominant and preovulatory (34 h after an injection of LH) were recovered by ultrasound guided follicle aspiration. Follicular fluids and serum proteins were separated by 1D-PAGE after enrichment and 2D-PAGE, and analyzed by mass spectrometry. A total of 459 protein spots were visualized by 2D-PAGE in crude equine follicular fluid. No difference was observed between follicular fluids at the three developmental stages studied. Thirty proteins were observed as differentially expressed between serum and follicular fluid. The enrichement method increased the resolution of 2D-PAGE and many more proteins were visualized on 2D-PAGE (n=992). Our results of protein identification confirm the presence of several high abundance serum proteins in equine follicular fluid, and demonstrated that electrophoresis methods may be useful approaches to identify novel protein candidates that may have a special role in follicle development and maturation.

Placental expression of leptin and neonatal plasma concentrations in horses and ponies

Chavatte-Palmer, P.[1], Charlier, M.[2], Duchamp, G.[3], Guillaume, D.[4], Wimel, L.[5], Keisler, D.[6] and Camous, S.[1], [1]INRA, UMR 1198 Biologie du Développement et Reproduction, 78350 Jouy en Josas, France, [2]INRA, UR1196 Génomique et Physiologie de la Lactation, 78350 Jouy en Josas, France, [3]INRA, UE1297 Unité Expérimentale de Physiologie Animale de l'Orfrasière, 37380 Nouzilly, France, [4]INRA, UMR85 Physiologie de la Reproduction et des Comportements, 37380 Nouzilly, France, [5]Haras nationaux, Statio expérimentale de la Valade, 19370 Chamberet, France, [6]University of Missouri, 160 Animal Sciences Research Center, Columbia, Missouri 65211-5300, USA; pascale.chavatte@jouy.inra.fr

Leptin is produced by the white adipose tissue and regulates food intake in adults. The perinatal leptin peak programs the hypothalamic regulation of food intake. Leptin also modulates bone growth and glucose homeostasis. Osteochondrosis (OCD) in horses can occur in foals and has been correlated to hyperglycemia and hyperinsulinemia. Ponies are naturally resistant to OCD. Differences in plasma leptin and glycaemia may explain these breed differences. Plasma from 10 ponies and 10 horse mares and their foals were collected. Placental and milk samples were also obtained. An iv glucose tolerance test (IVGTT) was performed at 5 months of age. Plasma leptin concentrations were higher before than after foaling and higher in ponies ($P<0.001$). At birth, leptin was low and not different between breeds. It increased steeply in ponies to reach higher concentrations than in horses at 2 days ($P<0.001$). Milk leptin was higher in horses on day 10 of lactation ($P<0.01$). Fasting glycaemia was lower in ponies than in horses at 5 months of age ($P<0.01$). There were no breed differences for IVGTT. Leptin mRNA levels were very low in all placentas. The higher plasma leptin in ponies compared to horse mares is probably related to their higher adiposity. The low placental expression and neonatal plasma leptin indicates that fetal plasma leptin is of fetal origin. The higher fasting glucose and the lower leptin concentrations observed in horses may be related to their subsequent higher sensitivity to OCD.

Session 37 Poster 20

The maximum temperatures (Tmax) distribution on the body surface of sport horses

Jodkowska, E.[1], Dudek, K.[2] and Przewoźny, M.[3], [1]Wroclaw University of Environmental and Life Sciences, Department of Horse Breeding and Riding, Kożuchowska 5a, 51-631 Wrocław, Poland, [2]Technical University of Wroclaw, Institute of Machines Design and Operation, Łukasiewicza 7/9, 50-231 Wrocław, Poland, [3]Equi Vet Serwis, Wygoda 6, 64-320 Buk, Poland; ewa.jodkowska@up.wroc.pl

The objective of the study was to indicate usefulness of measurement of maximum temperatures in designated region of a healthy horse body surface. Thermographic investigations (Thermovision®550, FLIR) were carried out on 35 horses from 6 to 16 years old that participate in show jumping competitions. The rectal temperatures of horses were in the range 37.5 - 38.2 °C. The investigations were done in a stable, the ambient temperature (Tab) was 14 °C and humidity (j) was 60%. Thermograms of the left and right side of the horses were obtained. Each thermogram consisted of 36 regions on body surface (before competitions) and 22 regions (straight after competitions). The maximum temperature ranges in the rest state were 21.8 °C - 31.0 °C and symmetrical regions did not differ statistically ($P>0.05$). The highest temperatures were on a head, a neck and a trunk, the lowest – on limbs. Hind legs were warmer than forelegs in the analogous areas, with the exception of gaskin and forearm. The warmest body areas had the largest surface which indicates their crucial role in thermoregulation of a horse organism. The ranking of maximum temperatures in specific regions of horse bodies in the rest state was appointed. For this reason the study results can be useful in veterinary diagnosis. The range of maximum temperatures after the competitions was 24.2 °C - 34.2 °C. The highest increment was observed on breast, an elbow, a forearm and a gaskin, the smallest – on a head, a pastern and a hoof (fore- and hind limbs). The characteristic of the body surface maximum temperatures after effort can be useful to estimate of horse training but not from etiologic point of view.

Session 37 Poster 21

Linear evaluation system for type and qualifiers evaluation in the Spanish Arab Horse

Cervantes, I.[1], Gómez, M.D.[2], Bartolomé, E.[3], Gutiérrez, J.P.[1], Molina, A.[2] and Valera, M.[3], [1]University Complutense of Madrid, Avda. Puerta de Hierro s/n, E28040, Spain, [2]University of Córdoba, CU Rabanales, E14071, Spain, [3]University of Sevilla, Ctra Utrera km1, E41013, Spain; icervantes@vet.ucm.es

A linear evaluation system for type has been established to collect morphological information in the Spanish Arab horse Breeding Program. This is a system designed to obtain an objective analysis of morphological characters and to maximize the relationship between morphological traits and functional aptitudes. The evaluation system consists of 29 variables directly related with a body measurement and 19 subjective variables (but linear scored). Additionally, 9 traits of movements and 1 of temperament are included (also linear scored). After a training period, a test was made to verify the system and to choose qualifiers to the routine evaluation. Two parameters were used to evaluate the qualifiers, the reliability (probability that the score given by the qualifier is in accordance with the measured value) and the repeatability (probability of a qualifier giving the same score to a horse that is evaluated in two different moments) of their evaluation. A total of 165 sheets were filled during the evaluation of 15 qualifiers. The qualifiers attained a repeatability ranged from 0.89 to 0.98, a reliability for morphological traits ranged 0.92-0.97 and for movements traits, 0.93-0.99. Regarding the variables, the reliability values were from 0.89 (chest width) to 0.99 (scapula length). The highest repeatability (0.99) was for hoof variables and for lateral view of the knee and the lowest for the lateral view of the hock and for lateral view of forelimb (0.90). The movements' traits reliability was larger than 0.95. The final qualifiers' score was a combined index: 0.5*rel.morph+0.3*rep+0.2*rel.mov. Eleven qualifiers passed the evaluation (≥0.96), but during the next two years the system and the qualifiers will be tested and revised after that time to ensure the correct development of the system in this breed.

Session 38 Theatre 1

Genome-based statistical analysis of quantitative traits: is a new paradigm needed?

Gianola, D., De Los Campos, G., Gonzalez-Recio, O., Long, N., Okut, H., Rosa, G.J.M., Weigel, K.A. and Wu, X.L., University of Wisconsin-Madison, Animal Sciences and Dairy Science, 1675 Observatory Dr., WI 53706, USA; gianola@ansci.wisc.edu

Information from dense genomic markers is now available in livestock. This data may enhance accuracy of prediction of breeding value (and reduce generation interval), but may be also be useful for developing genome-assisted management or disease treatment. Animal breeders emphasize additive genetic models because the concept of breeding value is paramount and, also, because variance component models do not recover sizable amounts of non-additive variance. That this also happens when there is known epistasis is disturbing, and begs the questions of how useful the standard methods are for prediction of complex phenotypes or for shedding light on genetic architecture. Machine learning methods have been developed in computer science and artificial intelligence with the objective of extracting knowledge or patterns from complex, high-dimensional, data. These methods are tuned mainly on the basis of predictive ability and can take epistasis into account without explicit modeling. Some of the procedures are reviewed based on our experience in Wisconsin, and from the perspective of genome-enabled animal breeding. The objective includes of performance such that complex interactions can be taken into account, with a by-product being prediction of breeding value. Methods discussed include reproducing kernel Hilbert spaces regression, neural networks and radial basis functions, as universal approximators of quantitative traits. Methods for enhancing predictions, such as bagging (bootstrap averaging) and boosting (using an ensemble of methods) are discussed as well. The procedures can produce findings which would remain undiscovered with parametric methods.

Prediction of 305-day milk production with partial test-day records by artificial neural network

Hosseinnia, P., Tahmoorespur, M. and Teimurian, M., Ferdowsi University of Mashhad, Animal Science, Genetic and Breeding, Mashhad, 91775_1163 Khorasan, Iran; pouria_ho59@yahoo.com

Milk production in dairy cattle is influenced by interaction between genetic and environmental effects, and also controlled by various linear and non-linear factors. A number of current models uses of test-day records and linear models to estimate of 305-day milk yield. Linear models only consider linear relationships but artificial neural network (ANN) can be used to implement linear and non-linear relationships for prediction. 305-day records were used for estimation of genetic parameters like breeding value. In this study first period of test-day records were used for prediction of 305-day records by a back propagation artificial neural network (bpANN). Five data sets include various numbers of first period test-day (TD) records (100, 500, 1000, 2,000, 5,000 data sets) utilized for feeding to the bp ANN with 3 input, hidden and output layers for predicting of milk, fat percentage (fat%) and protein percentage (prot%) yield variables. The best results between all data sets were observed in data set that include 100 number of TD records for milk, fat% and prot% variables with coefficient of correlation (r_p) and determination of coefficient percentage(R^2%) of 0.87 and 77%; 0.75 and 55%; 0.85 and 72%, respectively. We concluded that the bpANN model has reasonably accurate prediction, especially about milk yield, due to consideration of both linear and non linear relationships between input and corresponding output variable during prediction procedure. Also the result of this study indicted that size of data set are important in the ANN prediction.

Genomic metrics of individual autozygosity, applied to a cattle population

Sölkner, J.[1], Ferencakovic, M.[2], Gredler, B.[1] and Curik, I.[2], [1]BOKU - University of Natural Resources and Applied Life Sciences Vienna, Departmet of Sustainable Agricultural Resources, Gregor Mendel Str. 33, A-1180 Vienna, Austria, [2]University of Zagreb, Faculty of Agriculture, Svetosimunska 25, 10000 Zagreb, Croatia (Hrvatska); johann.soelkner@boku.ac.at

High throughput genotyping technology provides information about a large number of genetic markers roughly evenly spread across the genome. Levels of individual homozygosity and homozygous segments (runs of homozygosity) provide information about levels of inbreeding of animals. We computed several indicators of autozygosity for a population of 919 Simmental/Fleckvieh bulls born 1998 or later, genotyped with the Illumina Bovine SNP50 Beadchip, and compared them to inbreeding coefficients calculated from pedigrees with a completeness of >7 complete generation equivalents. The genomic metrics of autozygosity, applied to all autosomal loci and pseudo-autosomal loci on the X chromosomes were: 1) Level of homozygosity: number of homozygous loci/number of all loci. 2) Runs of homozygosity: overall length of homozygous segments >3 (>1, >5) MB. 3) Diagonal element of the genomic relationship matrix as implemented in genomic selection procedures, considering average and base population allele frequencies. 4) Inbreeding coefficient (F), as calculated by FEstim software, after applying MASEL software for selecting unlinked markers 5) Inbreeding coefficient (F), as calculated by PLINK software, with and without pruning to include loci in approximate linkage disequilibrium. Results indicate that level of homozygosity (r=0.555) seems to be more closely correlated with pedigree inbreeding coefficients than other, more sophisticated metrics of autozygosity. Among the metrics compared, the approach as implemented in the diagonal elements of the genomic relationship matrix had the lowest correlation with pedigree inbreeding.

Forensic science technique applied for calculation of kinship index

Bömcke, E.[1,2] and Gengler, N.[1,3], [1]University of Liege, Gembloux Agro-Bio Tech, Passage des Déportés 2, 5030 Gembloux, Belgium, [2]FRIA, Rue d Egmont 5, 1000 Brussels, Belgium, [3]FNRS, Rue d Egmont 5, 1000 Brussels, Belgium; elisabeth.bomcke@ulg.ac.be

Implementing conservation strategies needs the knowledge of relationships inside the concerned population. The aim of this study was to find tools to help scientists and breeders to manage endangered populations or populations with missing pedigree information. The animal genetics literature often seems unaware of relevant developments in human genetics (and conversely). In this study, an approach called Familial Searching was tested. This is used in forensic science, in addition to matching DNA evidence directly to criminal profiles, to search for people (present in a database) who are related to an individual that left DNA evidence at a scene of crime. This method is based on the calculation of likelihood ratios (LR) between genotype of an individual and genotypes of each other individuals of the database. In order to decrease the number of comparisons, the available pedigree information was used as 'local' prior information, i.e. relating to specific pairs of individuals. General knowledge about the studied population (e.g., generation interval, sexual maturity) was considered as 'global' prior information. Including prior information reduced the number of comparisons from over 50%. Results showed that the parents were always classified into the 4 highest LR. This method simplified parentage verifications, it allowed the detection of 90% of false parentage (LR=0). It also allowed to create new links in the pedigree through detection of unregistered parents. The method was tested on the Skyros pony, an indigenous Greek breed. For this breed, partial pedigree information was available, and 99 individuals were genotyped at 16 microsatellite loci. The method allowed to check about 2500 possible parent-child combinations, three registered parentages were considered as incorrect and one non-recorded parentage was detected. The method will now be tested on other breed and with other markers, e.g. SNPs.

Fine decomposition of the inbreeding and the coancestry coefficients by using the tabular method

Garcia-Cortes, L.[1,2], Martínez-Avila, J.[2] and Toro, M.[1], [1]ETSIA (Universidad Politecnica de Madrid), Produccion Animal, Ciudad Universitaria s/n, 28040, Spain, [2]INIA, Mejora Genetica Animal, Carretera La Coruña km 7, 28040 Madrid, Spain; miguel.toro@upm.es

We describe a simple algorithm to decompose both the inbreeding and the coancestry coefficients. The decomposition is performed in pieces coming from each ancestor, including the founders and the Mendelian sampling terms of non-founders. This algorithm replaces the conventional tabular method formulae with a set of recursive formulas. We illustrate the procedure with two small examples, including the pedigree of the celebrated bull Comet. The procedure was also tested with simulated pedigrees and it succeeded in analyzing the impact of successive bottlenecks into the average coancestry of the current cohort.

Multigenerational estimation of relatedness from molecular markers

Fernández, J., Instituto Nacional de Investigación y Tecnología Agraria y Alimentaria, Mejora Genética Animal, Ctra. Coruña Km 7,5, 28040, Spain; jmj@inia.es

Beside the problems derived of their dependence on the knowledge of ancestral allelic frequencies, most relatedness estimators using molecular information are able to account only for a limited range of types of relationships at a time. For example, sometime they can just differentiate full-sibs from unrelated individuals. Moreover, only close relationships are to be detected (full- or half-sibs, parent-offspring). One of these methods, included into the group of pedigree reconstruction estimators, worked by maximising the correlation between the molecular coancestry matrix and the matrix calculated from virtual feasible pedigrees. Consequently, it allowed for more complex relationships than other estimators by creating more 'deep' genealogies of virtual ancestors. However, a drawback it presented was that individuals being studied had to belong to the same generation (i.e. no one could be ancestor of other). I have extended the previous method to be able to deal with a group of genotyped individuals belonging to several generations and, therefore, not all contemporary. Simulations have proven that this method is able to reconstruct medium size pedigrees comprising a few generations using an affordable amount of markers. Accuracy of the estimations improved when some demographic information was available (e.g. knowledge of the age or birth date of candidates allows for discarding some individuals as ancestor and/or descendant). The method could also take advantage of already known genealogical relationships. This estimator could be very useful to determine the structure of populations, especially when non regular as natural ones or populations without management.

Looking for a reliable effective population size as a measure of risk status in rare breeds: the case of three Spanish ruminant breeds

Pastor, J.M.[1], Cervantes, I.[2], Gutiérrez, J.P.[2] and Molina, A.[3], [1]Department of Animal Health, Provincial Branch of the Andalusian Regional Government, E11071 Cádiz, Spain, [2]University Complutense of Madrid, Avda. Puerta de Hierro s/n, E28040, Spain, [3]University of Córdoba, CU Rabanales, E14071, Spain; icervantes@vet.ucm.es

The conservation of genetic diversity is a priority factor in small populations as rare breeds. Genetic diversity of the population can be measured by genealogical or molecular analyses. Effective size (N_e) is one of the most important topics in population genetics, given its usefulness as a measure of the long-term performance of the population regarding both diversity and inbreeding and, therefore, to characterise the risk status of livestock breeds. The risk level of threatened in a breed mainly depends of N_e and this varies according to methodology used to asses it. Classical developed expressions to obtain N_e rarely match real populations. When genealogies are available the effective size can be computed using increases in inbreeding, but the reliability of the results depends on the pedigree depth. However, pedigree records are usually missing in endangered populations and only molecular marker information can be available. In this case, the use of methodologies based on linkage disequilibrium and temporal methods can be developed. But it should be taken into account the size and the number of markers of the samples that cause variation in the final value. All this, jointly with some specific requirements of livestock breeds like sanitary restriction increase the difficulty of assessing a reliable and useful effective size. Pedigree and molecular information from three Spanish autochthonous ruminant breeds were used (the Pajuna cattle, Payoya goat and Merino of Grazalema sheep breeds), as examples. Simulations using sanitary restriction were done to compute the effective size. The N_e values using molecular data ranged between 11.7 and 174.3 and those based on variance of family size from 11.2 to 26.1.

Overview of Holstein sperm importation in Iran

Joezy-Shekalgorabi, S.[1] and Shadparvar, A.A.[2], [1]Islamic Azad University, Shahre Qods Branch, Dep. Animal Science, Shahre Qods, 37515-374, Tehran, Iran, [2]University of Guilan, Dep. Animal Science, Tehran-Rasht Road, 416351314, Rasht, Iran; Joezy@modares.ac.ir

Status of Holstein semen importation to Iran from 1994 to 2008 is demonstrated. Average semen importation was 72 and 38 percent for Sire of Sire (SS) and Sire of Dam (SD) pathways, respectively. Contributions of donor countries in SS pathway were 60% from USA, 33% from Canada and 1% from European countries. For SD pathway the corresponding contributions were 58, 32 and 10%, respectively. Semen importation from USA had an increasing trend from 2000. Average age of imported semen was 12.98 years in SS pathway and 12.65 years in SD pathways. Average age of imported semen had a decreasing rate in the recent years. Use of locally proved semen has drastically decreased especially in SS pathway in the recent years. More genetic and economic response has led to more use of imported semen in spite of its higher price. The reason for higher use of USA semen might be more emphasis on milk production, which is one of the most important traits in breeding goal in Iran. International evaluation is necessary for choosing the best semen for importation to Iran. To have more contribution in the market, AI centers of Iran have to optimize their progeny testing and sire selection programs.

Longevity analysis based on virtual objective culling criteria

Tarres, J., Fina, M. and Piedrafita, J., Universitat Autonoma de Barcelona, Departament de Ciencia Animal i dels Aliments, Facultat de Veterinaria, 08193 Bellaterra, Spain; marta.fina@uab.cat

True length of productive life (TL) of a cow is usually defined as the difference between the date of culling and the date at first calving. Culling criteria are usually subjectively applied by each farmer. However, it is possible to define longevity in an objective manner to evaluate cows for longevity including the reproductive performance of the cow. We analyze two culling criteria by defining two virtual dates of culling. For strict reproductive longevity (SRL), a cow was culled after having a calving interval longer than 400 days (13 months), whereas for lax reproductive longevity (LRL) a cow was culled after having two calving intervals longer than 400 days. Database contained records from 1990 to 2008 registered in sixteen breeding herds participating in the Yield Recording Scheme of the Bruna dels Pirineus breed. After editing, our database included data of productive life of 2138 cows. The percentage of censored records decreased from 58% for TL to 41% and 23% for LRL and SRL respectively. The average value of longevity also decreased from 9.6 years to 6.7 and 4.4 years, respectively. For parameter estimation, a proportional hazards model was used including parity and year effect and extended with the random effect of the herd and the sire of the cow. As expected, the variance of herd effect decreased from 0.418 for TL to 0.174 and 0.136 for LRL and SRL. This decrease in herd variance can be explained by the homogenisation of culling criteria. The effective heritability was 0.243 for TL, 0.182 for LRL and 0.082 for SRL. When the percentages of censoring were taken into account, the equivalent heritabilities were 0.127 for TL, 0.116 for LRL longevity and 0.065 for SRL. As the interest of a breeding program should be to increase the longevity of the efficient cows, the definition of lax reproductive longevity can be a compromise between heritability and reproductive efficiency.

Genetic parameter estimates of ultrasound measurements in growing animals in Bruna dels Pirineus beef cattle

Fina, M., Tarrés, J., Esquivelzeta, C. and Piedrafita, J., Universitat Autònoma De Barcelona, Ciència Animal I Dels Aliments, Facultat De Veterinària, 08193 Bellaterra, Spain; marta.fina@uab.cat

The application of ultrasound technology as a research tool allows the evaluation of carcass traits in live animals. In addition, these measurements have potential to increase the rate of genetic progress including estimation of heritabilities and genetic correlations in genetic evaluation programs for carcass merits. The objective of this study was to estimate genetic parameters for real-time ultrasound measurements of loin area (LA), loin depth (LD) and subcutaneous fat thickness over the rump (SF) in Bruna dels Pirineus beef cattle. The measurements were obtained using a Sonovet 2000 ultrasound unit equipped with a 3.5-MHz 17 cm linear transducer. The same ultrasound technician preformed all measurements. Every animal (n=352) was scanned 2-5 times for each variable to estimate the accuracy of ultrasonic records. The weight at scanning ranged from 158 kg to 608 kg. Repeatabilities between measurements for LA, LD and SF were 0.964, 0.988 and 0.875, respectively. Heritabilities and genetic correlations were estimated using multiple-trait restricted maximal likelihood. The final animal model included only one ultrasonic measurement per animal (the closest to the mean for each animal), fixed effects for year-season (7 breeding year-seasons over a 2.5-yr period) and feedlot, and the interaction of weight by sex as a linear covariate. The pedigree included 936 animals. Heretabilities for LA, LD and SF were 0.37±0.13, 0.36±0.12, and 0.27±0.13, respectively. Genetic correlations between LA and LD, LA and SF, and LD and SF, were 0.63±0.17, 0.36±0.28, and -0.41±0.42, respectively. Both the estimates of heritabilities and correlations indicate that a relevant additive genetic variance exists for all three traits and supports the use of live animal ultrasonic measurements as a selection tool in breeding cattle.

Heterosis and reciprocal effects for crosses between divergently selected lines in Coturnix Japonica quail

Rezvannejad, E., Pakdel, A., Mirai Ashtiani, S.R. and Mehrabani, H., Tehran University, Animal Science, Karaj, Emamzade Hasan bldv., 98765, Iran; rezvannejad2002@yahoo.com

A common assumption underlying most crossbreeding trials is the absence of any reciprocal differences. The purpose of the present study was to measure heterosis and reciprocal effects in body weight (BW) in Japonica quail. The genetic lines used in experiment had been selected for 7 generations for high (H) and low (L) BW at 4 wk of age. H line and L line were mated reciprocally and within line for produce HH, LH, HL, LL groups. Body weight was measured at hatch, 1, 2, 3, 4 wk of age for all groups. Specific contrasts utilized were the heterotic effect and the reciprocal crosses. Line H was heavier at 4 wk than line L ($P<0.001$). The heterosis was negative for BW at all ages for males except 4 wk for males and at all ages except 3 and 4 wk for females. Heterosis was low and non-significant because heritability for these traits is high. Reciprocal effects were significant in HL and LH crosses for BW at all ages except 4 wk BW in males and 3 and 4wk BW in females, indicating the presence of maternal effects. In summary, non-additive genetic variation was an important source of variation in this cross of Japanese quail lines differing in BW.

COST Action FA0802 FEED FOR HEALTH: animal-derived foods and human health: an introduction

Pinotti, L.[1] and Givens, D.I.[2], [1]Università di Milano, Veterinary Science andTechnology for Food Safety, via Celoria, 10, 20133 Milano, Italy, [2]University of Reading, Faculty of Life Sciences, Earley Gate, Reading RG6 6AR, United Kingdom; luciano.pinotti@unimi.it

Animal-derived foods make important contributions to Western diets but In the EU this is set against increasing obesity and an ageing population, both increase the risk of chronic disease. It is therefore crucial to understand the role of these foods relative to chronic disease risk and whether animal nutrition can improve them. This symposium, organised by COST Action FA0802 Working Group 1 Feed & Food for Health, will address these topics with a focus on milk and milk products.: Relating consumption to chronic disease. This will explore recent data/meta-analyses examining the relative risk of stroke, ischaemic heart disease, diabetes and certain cancers in high consumers of milk/milk products compared with low consumers. The lack of good evidence in certain areas will also be explored. Enhancing the fatty acid composition of milk fat. Dairy foods supply much of dietary saturated fatty acids (SFA) in many countries and there is some evidence that replacing some of the SFA with mono- or polyunsaturated fatty acids will be beneficial. The role of animal nutrition for making such changes will be discussed. Trans fatty acids from ruminant foods. Reducing the degree of saturation of milk fat by animal nutrition can increase in the production of trans fatty acids (TFA). There remains uncertainly as to whether ruminant TFA have the same negative effects on health as do industrially produced TFA. Recent studies comparing these two sources will be discussed. Bioactive peptides from milk. During digestion of milk proteins release an array of peptides. Some of these peptides appear to have important bioactive properties including blood pressure lowering effects although the exact mode of action is not clear. New findings on this topic will be discussed. This abstract is supported by Feed for Health, COST Action FA0802 (www.feedforhealth.org).

Milk in the diet, good or bad for health?

Givens, D.I., University of Reading, Faculty of Life Sciences, Earley Gate, Reading RG6 6AR, United Kingdom; d.i.givens@reading.ac.uk

Increasing obesity and an ageing population increase the risk of chronic disease and its cost. Diet is a modifier of risk and since milk and its products are staple components of many diets it is important to understand if these foods increase or decrease the risk of chronic disease. Elwood *et al.* (2008) showed the risk of stroke and/or heart disease in subjects with high milk or dairy consumption was 0.84 (95% CI 0.76, 0.93) and 0.79 (0.75, 0.82) respectively, relative to those with low consumption. Four studies reported diabetes as an outcome, and the relative risk in the highest consumers was 0.92 (0.86, 0.97). The World Cancer Research Fund (2007) report confirmed that increased milk consumption will probably decrease the risk of colo-rectal cancer. Some studies have shown a positive association between increased milk consumption and prostate cancer but the increased risk is small and not consistent across studies. It should however not be ignored. Set against the proportion of total deaths attributable to the life-threatening diseases in the EU i.e. vascular disease, diabetes and cancer, the available data provide evidence of an overall survival advantage from the consumption of milk although the situation with other dairy foods is much less clear and needs urgent clarification. For milk in particular there appears to be an enormous mis-match between both the advice given on milk/dairy foods items by various authorities and public perceptions of harm from the consumption of milk and dairy products, and the evidence from long-term prospective cohort studies. These foods do however supply a sizeable proportion of dietary saturated fatty acids in many countries but simply reducing milk consumption to reduce saturated fatty acid intake is not likely to produce benefits overall, although the production of dairy foods with reduced saturated fatty acids is likely to be helpful. This abstract is supported by Feed for Health, COST Action FA0802 (www.feedforhealth.org)

Can animal nutrition improve the fatty acid profile of ruminant-derived foods?

Fievez, V., Van Ranst, G., Lourenço, M., Goel, G. and Vlaeminck, B., Ghent University, Laboratory of Animal Nutrition and Animal Product Quality, Proefhoevestraat 10, 9090 Melle, Belgium; Veerle.Fievez@UGent.be

Lowering dietary saturated and trans fat is a key concept in nutritional advice to reduce CVD risk, whereas cis unsaturated fatty acids improve blood lipid profiles. Hence, the milk fat has been criticized as it naturally contains trans mono unsaturated (MUFA) (3%) and high amounts of saturated fatty acids (SFA) (70-75%) and is low in poly unsaturated (PUFA) (5%). Animal dietary strategies to increase the amount of dairy cis PUFA include supplementation with plant oil (seeds) or marine products and (fresh) forage feeding. Despite a two to twenty fold increase in some of the milk PUFA, their concentration in modified milk remains low. Transfer is limited through rumen biohydrogenation and a limited affinity of the mammary lipoprotein lipase for plasma cholesterol esters and phospholipids, the main carriers of PUFA. Substantial elevation of dairy PUFA requires rumen bypass fat supplements. Although a number of technologies have been proposed, protection efficiencies vary, overprotection should be considered and an emerging need exists for 'natural' protection. Despite modest changes in dairy PUFA, dietary supplementation with plant oil (seeds) is beneficial in terms of their considerable reduction in milk SFA, particularly palmitic acid and increase in oleic acid. Unfortunately, increases in trans MUFA seem inevitable. Plant oils and seeds produce similar changes in milk SFA and PUFA, but oils show a higher risk of increasing trans MUFA. Linseed supplementation and fresh forage feeding lead to similar changes in milk SFA but the increase in oleic acid is smaller with forage feeding, whereas the potential of ensiled forages to modify milk fatty acid composition is limited. Supplementation with marine products seems the most effective way to increase dairy conjugated fatty acid content, but this is associated with a large increase in numerous isomers of trans MUFA. This abstract is supported by Feed for Health, COST Action FA0802.

Fatty acid composition of conventional milk related to the diet in seventeen farms in France

Rouille, B.[1] and Montourcy, M.[2], [1]Institut de l Elevage, Dairy cow nutrition, Monvoisin - BP 85225, 35652 Le Rheu Cedex, France, [2]ENITA Bordeaux, 1 Cours du Général De Gaulle - CS 40201, 33175 Gradignan Cedex, France; benoit.rouille@inst-elevage.asso.fr

The aim of this study was to identify different fatty acid (FA) composition of dairy cow milk in some French feeding systems. Seventeen farms from six regions were selected. Milk samples and a description of the diet were collected at five periods between May 2008 and February 2009 (May, July, September, January and February). Five milk classes were identified regarding milk FA profile ($P<0.001$). Diets associated to milk classes were described and analysed through percentages of all the forages and the concentrates. Fatty acid composition of milks produced by the grazed grass diets are statistically different ($P<0.001$) from milks from other diets [milk richer in polyunsaturated fatty acids (PUFA), in trans fatty acids (trans FA), and with a lower omega-6/omega-3 ratio]. Increasing the percentage of grass in the diet induces variations on (i) PUFA, monounsaturated fatty acids (MUFA) (increase), (ii) saturated fatty acids (SFA), omega-6/omega-3 ratio (decrease). According to their FA composition, the results show a strong link between the fatty acid profile and the diet. Milks obtained with hay based diets have a low omega-6/omega-3 ratio but a high SFA share. Grass-based diets produce milks with better nutritional profile. It is particularly obvious for pasture-based diets. The results are totally different when the diet is based on maize silage. This study shows the large diversity of milk fatty acid composition existing in France. It also underlines efficient, fast and reversible diet solutions to improve milk regarding health promoting FA. This leads to possible technical advices on feeding systems and strategies during a whole year if the main goal is to improve milk quality (at least FA).

Trans fatty acids from ruminants: a health issue or not?

Chardigny, J.-M., Malpuech-Brugere, C. and Morio, B., Clermont Université, Université dAuvernge, INRA, UMR 1019, UNH CRNH Auverne, Unité de Nutrition Humaine BP 10448, F-63000 Clermont-Ferrand, France; Jean-Michel.Chardigny@clermont.inra.fr

Trans fatty acids (TFA) result from both catalytic partial hydrogenation of vegetable oils and ruminal biohydrogenation. Since the 90s, industrially produced TFA (IP-TFA) have been described as deleterious in relation to cardiovascular risk. IP-TFA are associated with an increase in serum LDL-cholesterol and a decrease in HDL cholesterol. Ruminant TFA (R-TFA) differ from IP-TFA as their profile is different, vaccenic acid being the major isomer. The maximum R-TFA intake is also limited. Indeed, milk fat TFA content cannot reach high values. The difference of impact of R-TFA versus IP-TFA has been considered only recently in intervention studies, but the Nurses' Health study already suggested a lack of association between cardiovascular disease and R-TFA intake. In 2008, we (belonging to the 'TRANSFACT consortium) and others published intervention studies data which indicated that R-TFA did not alter cardiovascular risk factors by contrast with IP-TFA. This has also been confirmed in a recent Danish epidemiological study. More recently, we have assessed the impact of milk fat with different saturated (SFA)/R-TFA contents, as a result of cows' feeding modifications. We conclude that the benefit of the decrease of SFA intake is not altered by an increase in R-TFA intake. R-TFA intake at moderate (i.e. in a balanced diet) level cannot be considered as a health issue. This abstract is supported by Feed for Health, COST Action FA0802 (www.feedforhealth.org).

Biologically-active peptides from bovine milk and their effect on human health

Politis, I., Agricultural University of Athens, Animal Science and Aquaculture, 75 Iera Odos Street, 11855 Athens, Greece; i.politis@aua.gr

Biologically-active peptides encrypted within the primary structures of milk proteins can be released following inadvertent proteolysis of bovine milk or by the action of lactic acid bacteria during the manufacture of fermented dairy products. Furthermore, milk proteins can be subjected to hydrolytic breakdown during gastric processing and later upon exposure of milk proteins with indigenous or intestinal bacteria-derived enzymes in the gut. Certain peptides generated during the proteolytic events described above possess various biological properties including antihypertensive and immunoregulatory properties. The casein derived tripeptides, isoleucine-proline-proline (IPP) and valine-proline-proline (VPP) have been studied more extensively for their antihypertensive properties. Even though the evidence available is not unequivocal, the majority of the published human studies suggest that IPP and VPP act as angiotensin converting enzyme inhibitors thereby inhibiting the vasoconstrictory effect of angiotensin II and potentiating the vasodilatory effects of bradykinin with consequent reduction in blood pressure. The evidence concerning the functional role of milk-derived immunoregulatory peptides and their potential to influence human health is still debatable. It is likely that these peptides modulate mucosal immunity of the neonate possibly by guiding local immunity until it develops its full functionality. The future challenge is to identify antihypertensive and immunoregulatory peptides present in bovine milk that can resist degradation by intestinal or serum peptidases, their mechanisms of action and, thus, find more possibilities for using these constituents in the development of functional foods. This abstract is supported by Feed for Health, COST Action FA0802.

Iodine supplementation with iodinized salt in dairy cows: effects on iodine content in milk and in blood plasma

Robaye, V., Knapp, E., Dotreppe, O., Hornick, J.-L., Istasse, L. and Dufrasne, I., Liège University, DPA - Nutrition Unit, Bd de Colonster 20 B43, B-4000 Liège, Belgium; eknapp@ulg.ac.be

Iodine is a trace element essential in the production of the thyroid hormones which are involved in the basal metabolism. The iodine content of locally produced feedstuffs used for livestock is rather low. Supplementation is therefore necessary. Iodine is secreted in milk. So, milk and milk products could be considered as a source of iodine for the consumers. The aim of the present work was to assess the effects of iodinized salt on iodine content in milk and in blood plasma. Two groups of 24 dairy cows were offered a diet made of grass silage (5.8 kg DM), maize silage (5.8 kg DM) and 9 kg of a compound feedstuff. One group (iodine group) was offered per cow 100 g salt enriched with iodine as KI (500 mg/kg salt) and with selenium as selenite (50 mg/kg salt). The control group received also 100 g salt enriched with selenium only. Selenium was used since involved in the iodine metabolism. The supplements were offered for a period of 51 days after which selenium enriched salt was offered in both groups. Teat dip without iodine was used. There were no effects of the supplementation on the chemical composition of the milk. The iodine content was low at about 33 mg/l in the milk samples of the control group. It remained low during the whole period. By contrast it increased sharply in the iodine group during the first 21 days and then reached a plateau at 141 mg/l. It quickly decreased after the end of the supplementation. Blood samples were obtained from 10 cows in each group on 4 occasions. The plasma iodine contents were 47 and 74 mg/l ($P>0.05$) in the control and iodine groups respectively when iodine was offered. It can be concluded that the iodine supplementation at an intake of 50 mg/day largely increased the iodine content in milk which of interest to produce milk with added value.

Beyond production nutrition: can we be health-positive and product-positive

Knight, C.H., University of Copenhagen, IBHV, Grønnegårdsvej 7, DK 1870 Frederiksberg C, Denmark; chkn@life.ku.dk

This short review will examine the different factors that are likely to influence the way in which we will feed dairy cows in the future. Currently, economic pressure dictates that most producers in the developed European milk industries concentrate on reducing costs, either through low-input or increased efficiency (larger units). The former is more likely to make use of grazed grass and low technology, the latter is more likely to employ energy-dense diets and advanced technology. By definition, an equal number of producers adopting each strategy will inexorably result in an increased proportion of the community production coming from large units; this trend has been evident for some time and continues. There is a paradox. Changes in consumer preferences over the same time have been in favour of product quality, with emphasis on consumer and animal health and welfare. Focusing on the animal health aspect, if the dairy farmer of the future is to deliver on the requirement for improvement, they will require either a financial incentive or a political dictate combined with the knowledge to achieve the improvement. Do we have that knowledge? Can we exploit it? Can we beneficially combine the quantitative production-oriented skills of the animal nutrition researcher with the health-oriented skills of the human nutritionist to get the best of both worlds? Can we learn from the lessons of the past, where niche markets have required an individual approach to feeding in order to achieve a specific product? Other papers in the session will show that manipulation of general management and specifically of the cows endocrinology and immunology can achieve improvements, but at what cost, and will the consumer pay? What will be the impact of the global increase in demand for milk? As developing-nation production industries become more efficient, will strategies that seek to improve animal health become increasingly attractive for European producers? If so, which will have the greatest potential; the low-input unit or the large unit?

Health and welfare positive nutrition for the dairy herd

Logue, D.N.[1] and Mayne, S.[2], [1]University of Glasgow, Faculty of Veterinary Medicine, Glasgow, G61 1QH, United Kingdom, [2]Department of Agriculture and Rural Development Northern Ireland, Belfast, BT4 3SB, United Kingdom; d.logue@vet.gla.ac.uk

This review reflects the views of a veterinarian (DL) and a nutritionist (SM) on the future for nutritional strategies to improve animal welfare in dairy herds. The number of small family farms is decreasing, to be replaced by bigger more commercial units. The daily care of animals is increasingly given over to technology. The divergence between grass-based systems and housed herds is increasing. The advent of genomic technologies potentially offers greater opportunity to breed for health related traits. Environmental change is slowly bringing a more diverse range of pathogens, parasites and pests into Northern Europe. In developed countries, consumers attach considerable importance to animal health and welfare, whilst the global increase in demand for milk and meat drives ever more (economically) efficient production. In European countries and elsewhere legislative control over welfare and environmental impact issues is increasing. Despite all this change, the basics of production systems and the major health and welfare issues faced remain relatively unchanged, at least until now. The young calf typically does not spend more than a few hours with its mother but is spared the worst excesses of white veal production. Despite considerable research into optimum rearing rates for dairy heifers, there appears to be just as much variability in what is done on the ground as ever. Knowledge on the best way of introducing down-calving heifers into the dairy herd is good, but is not always applied. Housing aspects such as cubicle design is optimised, where economics allows. Most farmers manage to achieve good body condition at calving, and the incidence of overt metabolic disease in early lactation is probably less than it was, but remains a problem. Nevertheless, mastitis and lameness are still the biggest issues requiring of improvement, and are likely to remain so for some time.

Nutritional modification of the immune response in dairy livestock

Ceciliani, F., University of Milano, Department of Animal Pathology, Hygiene and Veterinary Public Health, Via Celoria 10, 20133 Milano, Italy; fabrizio.ceciliani@unimi.it

Milk and other dairy products are important components in human diet because of their nutritional value. Research has focused on the modification of milk composition, aiming to unbalance the fatty acid content toward polyunsaturated fatty acids (PUFA), which are believed to be beneficial for human health. Both n3 and n6 PUFA are now regarded as important part of dairy livestock diet due to their positive effects in animal productions, including, for example, increasing pregnancy rate and stimulating immune function. Whilst it is evident that the inclusion of these powerful immunomodulatory molecules in dairy ruminant diets can affect the immune system, as it does in humans and laboratory rodents, yet the information about how these molecules may interact with dairy livestock immune functions remains largely unexplored. Breeding programs of dairy animals has been driven mainly toward the increase of milk production, at least so far. Therefore, the immune system which results from this somehow 'short sighting' selection of high production dairy cattle, is particularly fragile and sensitive to external challenges. Therefore, the possibility to strengthen the immune system by including modulatory molecules in animal diets represents a tempting option. Focus of the first part of this presentation will be a brief overview on the dairy livestock immune system, its critical points, such as the transition period, and the go and stop signals which trigger, or dampen, the unleash of immune reactions. The complex relationship between the ruminant immune system and the modern farming techniques, such as overcrowding or excessive administration of drugs for example, will be also discussed. The modulatory activities of PUFA on the immune response of cells involved in the defence of the organism against pathogens will be presented in the second part, as well as the *in vivo* and *in vitro* models commonly utilized to study the interaction between immune cells and fatty acids.

Session 39 Theatre 11

Nutritional manipulation of the endocrine system in dairy cows: implications for health, fertility and milk fatty acids

Garnsworthy, P.C., University of Nottingham, School of Biosciences, Sutton Bonington Campus, Loughborough, LE12 5RD, United Kingdom; Phil.Garnsworthy@nottingham.ac.uk

The endocrine system regulates many important physiological processes in the dairy cow, including nutrient partitioning over the short-term (homeostasis) and long-term (homeorhesis), lactation and reproductive cycles. Dietary manipulation of metabolic hormones (e.g. insulin, glucagon, leptin and IGF-I) alters endocrine signals and metabolite supplies to major organs (brain, liver, mammary gland, ovary) leading to responses in milk synthesis, fertility and general health. Insulin and glucagon are most responsive to diet composition, particularly to dietary starch and fat concentrations; growth hormone, leptin and IGF-I are associated more with animal factors, such as milk yield and body fatness. Low insulin status after calving (induced by high milk yield, over-fat cows and high fat/protein diets) is associated with rapid mobilization of body reserves, fatty liver, ketosis, and oxidative stress. Research demonstrates that increasing the insulin status of cows immediately postpartum enhances recruitment of ovarian follicles and shortens the anoestrous period. High insulin has detrimental effects on oocyte quality, however, so switching diets from insulin-stimulating to insulin-suppressing at the start of the breeding period can increase pregnancy rate significantly. Insulin is negatively related to milk yield and positively associated with activities of mammary fatty acid synthetase and desaturase enzymes. This raises the possibility of using milk fatty acid profile as a metabolomic indicator of cow health status and likelihood of reproductive success. It is concluded that understanding interactions between diet composition, endocrine changes and physiological responses provides opportunities to design nutritional strategies that improve the health, fertility and milk composition of dairy cows.

Session 39 Theatre 12

Systematic analysis of choline supplementation in dairy cows

Pinotti, L. and Baldi, A., Università di Milano, Department of Veterinary Sciences and Technology for Food Safety, Via Celoria 10, Milan, 20133, Italy; luciano.pinotti@unimi.it

Although, the choline requirement of dairy cows is still unknown, higher choline availability (by feeding rumen-protected choline) can improve milk production, suggesting that this substance may be a limiting nutrient in transition dairy cows. Based on these assumptions, we investigated the effects of rumen protected choline (RPC) administration on milk production and selected plasma metabolites (non-esterified fatty acids and beta-hydroxybutyrate) in 11 and 9 different studies, respectively, carried out between 1991 and 2008. Accordingly, 42 and 28 experimental groups for milk and plasma metabolites respectively, have been considered in the dataset. Mean and standard error data have been used in a regression model in which milk production response to RPC supplementation has been investigated. Dataset analysis indicated that although most of variability among experiments was related to treatments schedule, dry matter intake and dietary composition, these factors were also highly correlated. For this reason, and in order to avoid any redundancy in the model, our regression analysis included only RPC supplementation (control/treated) as fixed effect, while all the other variables have been considered as experimental effects and treated as random components in the mixed model, according to the idea that results of different experiments are affected by different experimental conditions. The data reviewed in this analysis are consistent with the fact that choline supplementation (choline in a rumen protected form 5 to 20 g/d) significantly ($P<0.05$) increased milk yield in dairy cows. By contrast, effect of choline supplementation on plasma metabolites are inconclusive.

Session 39 Poster 13

The effects of essential oils of cinnamon, garlic and laurel on milk fatty acid in dairy cows

Sahan, Z., Gorgulu, M., Celik, L. and Cinli, H., cukurova university, agricultural faculty, animal science, cukurova university, agricultural faculty, animal science, 01330 Adana, Turkey; gorgulu@cu.edu.tr

The study was conducted to investigate the effects of some essential oil on milk fatty acid composition. A total 28 dairy cows having 501±9.12 kg live weight, 25.19±0.73 kg milk yield, 68.1±5.8 DIM, 2.99±0.06 BCS and 2.03±0.19 lactation number were used and cows were allocated into four experimental groups according to the live weight, milk yield, DIM, lactation number and body condition score. First group received no essential oil, the second group received TMR containing 60 ppm cinnamon oil, the third group received TMR containing 30 ppm garlic oil and the forth group received TMR containing 90 ppm laurel oil. The doses were determined by *in vitro* studies investigating the effects of these essential oils on *in vitro* true digestibility of nutrients of barley, soybean meal and wheat straw. TMR was formulated as isocaloric and isonitrogenic. The results showed that total omega-3, C18:3:n3 and C17:0 were affected by treatments and total Omega-3 and C18:3n3 were decreased by essential oil compared to control and garlic oil increased C17:0 level over control and other treatments. Neither polyunsaturated nor saturated fatty acids in milk were affected by essential oils. The results revealed that cinnamon, garlic and laurel essential oils did not have potential to alter ruminal biohydrogenation process and modify fatty acid profile of milk when dairy cows fed with TMR containing 30 ppm, garlic, 60 ppm cinnamon and 90 ppm laurel essential oils.

Session 39 Poster 14

Effect of sunflower oil with organic selenium and vitamin E inclusion in lactation cow's rations upon milk production, composition and effect on human health

Saran Netto, A., Zanetti, M.A. and Salles, M.S.V., University of Sao Paulo, Animal Science, Rua Duque de Caxias Norte, 225, CEP 13635-900, Brazil; mzanetti@usp.br

The heart diseases are the main problem of public health, mainly due to the decrease in life quality, to the great financial expense, besides the responsibility for 30% of total human dead in Brazil. The milk frequently is relate to the heart disease and some professional of human healthy have suggested to take off the milk from the diet due his proportion of saturated fatty acids. To aggregate quality to the milk, healthy to the animal and to the man, this research was carried out to study the effect of the sunflower oil with organic selenium and vitamin E inclusion in lactation cow's rations. The milk from the different treatments (500 ml/day) was ingested by children from an elementary school. Twenty four Jersey cows were allocated in four treatments: C (control); C + A (5 mg of organic selenium + 1000 IU of vitamin E/day): O (3% of sunflower oil in DM diet); O + A (3% of sunflower oil in DM diet + 5 mg of organic selenium + 1000 IU of vitamin E/day). The experimental period lasted 12 weeks with 14 days for adaptation to the diets, with two daily milking, daily evaluation of feed intake and milk production. Weekly were analyzed milk fat, protein, lactose, total solids and somatic cells counting (SCC). In the weeks 0, 4, 8 and 12 were analyzed the cows serum levels of vitamin E and leucocytes. This research was not concluded yet and are to bee analyzed cows' serum selenium; selenium, vitamin E, cholesterol, lipid profile and rate of lipid oxidation in milk. The children that ingested 500 ml milk/day are being evaluated for growth and his serum lipid fractions. As the results, selenium and vitamin E supplementation improved mammary gland healthy and decreased sub clinic mastitis ($P<0.05$), increased milk production and milk fat ($P<0.08$). The oil in the cow diet decreased the dray matter intake ($P<0.05$). It was no effect ($P>0.05$) in the leucocytes in blood serum and in the others milk components.

Milk fatty acids in relation to production conditions

Kirchnerová, K.[1], Foltys, V.[1] and Špička, J.[2], [1]Animal production researche center Nitra, Hlohovecká 2, 951 41 Lužianky, Slovakia (Slovak Republic), [2]Jihoceská univerzita, Branišovská 31a, 370 05 České Budějovice, Czech Republic; kirchnerova@cvzv.sk

The aim of the paper was to study current fatty acids (FAs) profile of cow milk fat at herds of cows held in mountain dairy farms in Slovakia. The milk samples were taken in total from 181 cows of 5 different breeds at summer pasture period, and from the same cows at winter feeding with grass haylage and maize silage ratio. The FAs composition of individual milk was determined by GC-MS, where 54 FAs were identified and expressed relatively in % and evaluated in segments in accord with their biosynthetic origine. Short-chain fatty acids (SCFA) from biosynthesis in mammary gland, long chain saturated fatty acids (LCFA) mainly from blood plasma, into which they came from feed and body fat. Middle chain FAs (MCFA) can be of both origins. Mono unsaturated fatty acids (MUFA) come from resorption of feed's fat, however, poly unsaturated fatty acids (PUFA) come from own synthesis by means of dehydrogenating enzymatic system, while unsaturated fatty acids from lipids of feeds are hydrogenated in rumen. The most interesting is the conjugated linoleic acid (CLA), which has been shown to possess a number of human health benefits. The highest value was at Pinzgau -0.82%, lower ($P=0.015$) at Slovak pied -0.62%, and significantly ($P<0.001$) lower at Holstein -0.44%, and cross-breed Holstein x Pinzgau -0.45%, and Holstein x Pinzgau x Red -0.43% in summer period. In Winter period was similarly the highest content of CLA at Pinzgau herd -0.74%, lower ($P=0.015$)at Slovak pied -0.60% and significantly ($P<0.001$) lower at Holstein -0.46%, and cross-breed Holstein x Pinzgau -0.44%, and Holstein x Pinzgau x Red -0.35%. These results confirm the assumption, that there is a basis for the genetic variation among breeds, which is related to rumen output of trans-11 C18:1 and to a lesser extent cis-9, trans-11 CLA, and to the amount and activity of 9-desaturase in the tissue of mammary gland.

Effect of chestnut tannins and vitamin E on oxidative status in broiler chickens

Voljč, M., Nemec, M., Frankič, T., Trebušak, T. and Salobir, J., University of Ljubljana, Department of Animal Science, Groblje 3, 1230 Domžale, Slovenia; mojca.voljc@bfro.uni-lj.si

Increased levels of PUFA and the lack of antioxidants in animal diets can enhance lipid oxidation. The objective of the present study was to establish the effect of chestnut tannins (Farmatan) and vitamin E on oxidative stress in broiler chickens. Fifty male broilers (ROSS 308) were divided into five groups fed isocaloric and isonutrient diets for 25 days. One group received diet containing 7.5% of palm fat (ContPalm). The diets of other groups contained the same amount of linseed oil and were either unsupplemented (ContLin) or supplemented with vitamin E or/and Farmatan (F) as follows: VitE (85 IU vitamin E), TAN (3 g F/kg) and VitE+TAN (85 IU vitamin E and 3 g F/kg). Lipid oxidation was studied by measuring malondialdehyde (MDA) in blood plasma, liver and fresh breast muscle and by analysing antioxidant status (TAS) of plasma, glutathione peroxidase (GPx), superoxide dismutase (SOD), glutathione reductase (GR) assays. DNA fragmentation of blood lymphocytes was measured by Comet assay. In comparison to ContPalm high PUFA intake in ContLin group increased MDA formation in plasma and both tissues and increased DNA fragmentation in blood lymphocytes. In comparison to ContLin group Vitamin E supplementation effectively reduced MDA content in fresh breast muscle but not in plasma and liver. Tannins and vitamin E together were able to reduce MDA content in plasma and breast muscle. No effect on MDA formation in TAN group was found. Values of TAS, GPx, SOD and GR were not changed in any of experimental groups. In comparison to ContLin group, DNA damage of blood lymphocytes was decreased in all supplemented groups to the level of ContPalm group. Concentrations of vitamin E in plasma and both tissues increased in groups where diets were supplemented with vitamin E, to greater extent in VitE+TAN group. In conclusion, our study shows that combination of both supplements is more effective then the sole use.

Effect of raw and extruded soybeans on fatty acid profile and CLA of sheep milk

Nitas, D.[1], Petridou, A.[2], Karalazos, V.[3], Mougios, V.[2], Michas, V.[1], Abas, Z.[4], Nita, S.[5] and Karalazos, A.[6], [1]Dept Animal Production, ATEI, Thessaloniki, Greece, [2]Dept Physical Education and Sport Science, AUTH, Thessaloniki, Greece, [3]Dept Ichthyology and Aquatic Environment, UTH, Volos, Greece, [4]Dept Agricultural Development, DUTH, Orestiada, Greece, [5]Ministry Education, Rethymno, Crete, Greece, [6]School of Agriculture, AUTH, Thessaloniki, Greece; dnitas@ap.teithe.gr

Raw (RSB) or extruded soybeans (ESB) were fed to 24 ewes to study the influence on milk yield, milk fatty acid composition and conjugated linoleic acid (CLA) content. Ewes were fed one of three diets A, B and C. The control diet containing soybean meal, maize grain, barley grain, alfalfa hay, wheat straw and concentrates compared with diets B and C containing 14% RSB or 14% ESB, respectively. Measurements were made during the last 2 days of each of the 6 periods of the experiment. The replacement of a part of maize grain and soybean meal by either raw or extruded soybean seeds in diets B and C, respectively, did not affect the body weight of ewes, the daily milk yield, milk fat, protein, lactose, SNF contents, the somatic cell content, the fat corrected milk and energy corrected milk. The proportions of short and medium-chain fatty acids C8:0, C10:0, C12:0, C14:0 and C16:0 decreased and the proportions of long-chain unsaturated fatty acids C18:1n9t, C18:1n11t, C18:1n9c, C18:1n7c, C18:2n6t, C18:2n6c, C18:4n3, C20:3n6, C22:6n3c (DHA) CLA(c-9, t-11) and CLA(c-9, c-11), the monounsaturated (MUFA), polyunsaturated (PUFA), n6, n3, total CLA and total unsaturated fatty acids were significantly increased in milk fat of ewes when RSB and ESB diets were fed compared with the proportions of fatty acids of milk fat of ewes fed the control diet. The results demonstrate that unsaturated fatty acids content of sheep milk can be substantially increased by the inclusion of raw or extruded soybeans in the diet at 14%, without any negative effects on animal performance.

Effects of *L. plantarum* PCA 236 in feed administration on dairy goat faecal microbiota, plasma immunoglobulins, antioxidant status and milk fatty acid composition

Mountzouris, K.C.[1], Maragkoudakis, P.A.[2], Dalaka, E.[1], Zoumpopoulou, G.[2], Rosu, C.[2], Hadjipetrou, A.[3], Theofanous, G.[3], Carlini, N.[4], Zervas, G.[1] and Tsakalidou, E.[2], [1]Agricultural University of Athens, Dep. of Nutritional Physiology and Feeding, Iera Odos 75, 11855, Athens, Greece, [2]Agricultural University of Athens, Dep. of Food Science and Technology, Iera Odos 75, 11855, Athens, Greece, [3]Pittas Dairy Industries, P.O. Box 21600, 1511 Nicosia, Cyprus, [4]Probiotical S.p.a., Via E. Mattei 3, 28100 Novara, Italy; kmountzouris@aua.gr

The aim of this work was to assess the potential application of L. plantarum PCA 236 as a probiotic feed additive in dairy goats. Twenty four goats of the Damascus breed were divided in two treatments of 12 goats each: control (C) no addition of L. plantarum and probiotic (PRO) with in feed administration of L. plantarum, during the milking time, so that the goats would intake 12 log cfu/day. The experiment lasted for 5 weeks and individual faecal, blood and milk samples were collected weekly. Faecal samples were examined for the presence of L. plantarum PCA 236 by PCR and population levels of total aerobes, coliforms lactic acid bacteria (LAB), Streptococcus, Enterococcus, total anaerobes, Clostridium and Bacteroides by culture techniques. Plasma immunoglobulins and antioxidant capacity were also determined. Experimental data were analysed by the GLM repeated measures ANOVA procedure using SPSS v.8.0. L. plantarum was recovered in the faeces of all PRO animals that also had significantly ($P \leq 0.05$) higher LAB (by 1.2 log CFU/g) and lower faecal clostridia (by 1.3 log CFU/g) levels, compared to C. The antioxidant capacity and immunoglobulin concentrations in goat plasma did not differ between treatments. A significantly higher content of linoleic, a-linolenic and rumenic acid, was noted in the milk fat of the PRO treatment. The results indicate that the L. plantarum strain has displayed an interesting probiotic potential that needs to be further researched. This work was performed under the E.U. funded project 'PathogenCombat, FP6-007081'

Growth in the global livestock sector and related greenhouse gas emissions: life cycle assessment and evaluation of mitigation options

Gerber, P.[1], Henderson, B.[1], Vellinga, T.[2] and Steinfeld, H.[1], [1]FAO, Animal production and health division, Viale delle terme di Caracalla, 00100 Rome, Italy, [2]Wageningen University, Livestock Research, P.O. Box 65, 8200 AB Lelystad, Netherlands; Pierre.Gerber@fao.org

The growth of the livestock sector is being achieved at substantial environmental costs. Today, livestock are a major stressor of the global environment, contributing close to a fifth of the anthropogenic greenhouse gas emissions. At the same time, livestock are also a crucial driver of rural growth and a tool for improving food security. Policies are required to guide the sector in achieving sometimes conflicting development objectives. Potential pathways include encouraging resource use efficiency, correcting for environmental externalities and accelerating technological change. The paper will introduce the approach and preliminary results of a wide programme of work currently carried out at FAO and aiming at identifying low emission development pathways for the livestock sector. The development of mitigation strategies, tailored to different development priorities and agro-ecological conditions, is the ultimate objective of this undertaking. This requires the analysis of technical information generated by life cycle assessment, combined with socio-economic data and economic modelling of cost effectiveness, distributional effects and trade implications of policies addressing the livestock sector. It also requires to understand the indirect impact of mitigation policies in other sectors (e.g. forestry and energy) on livestock sector.

Session 40 Theatre 2

Evaluating the positive and negative externalities of livestock systems for increasing their socio-economic and environmental sustainability

Herrero, M.[1], De Boer, I.[2], Garnett, T.[3], Staal, S.[1], Duncan, A.[1] and Grace, D.[1], [1]International Livestock Research Institute, P.O. Box 30709, Nairobi, Kenya, [2]Wageningen University, Animal Production Systems Group, Marijkeweg 40, 6709 PG Wageningen, Netherlands, [3]Centre for Environmental Strategy, University of Surrey, Food Climate Research Network, Guildford, Surrey, GU2 7XH, United Kingdom; m.herrero@cgiar.org

The demand for livestock products will increase significantly over the coming decades as a result of population growth, income increases, urbanisation and others. This, together with the required growth in other sectors to meet human demands will increase the pressure on land use and other resources. The livestock sector will need to increase its efficiency in order to use less natural, human and financial resources per unit of output produced. This will require improvements in the efficiency of livestock product life cycles and value chains. These efficiency gains can be obtained through improved management (improved input use), new technologies (production, processing, transport), better regulation and policies (emission caps, market development and incentives) and livestock product demand management (i.e. shifts in dietary choices, consumer preferences). This paper analyses the positive and negative aspects of livestock systems and also illustrates some of the main entry points for improvements using case studies from livestock systems and value chains from both the developed and the developing world.

Specificity of environmental issues related to livestock in the tropics

Doreau, M.[1] and Blanfort, V.[2], [1]INRA, Herbivore Research Unit, 63122 Saint-Genès Champanelle, France, [2]CIRAD, Livestock Systems Research Unit, BP 701, 97387 Kourou, French Guiana; michel.doreau@clermont.inra.fr

Tropical areas are sometimes considered as involved to a large extent in the degradation of the environment, but until now few attention has been paid to this concern, because the productive function of livestock was until now the main issue. Greenhouse gases emissions in tropical areas are not well known, because they are calculated from temperate references, or at large geographic scales, excluding biological processes.. Methane emissions may be modulated by the specificity of tropical forages (plants with C4 metabolism, rich in tannins or in saponins). Carbone dioxide and nitrous oxide are expected to be lower than in temperate areas due to low inputs in extensive systems. Emissions are partially compensated for by carbon sequestration in soils; the extent of this sequestration is not known for permanent grasslands but reconstitution of carbone storage progressively occurs on pastures after deforestation. Other negative impacts as energy consumption, eutrophication, acidification, mineral and chemicals, are very low in smallholder or pastoral systems, but will increase with the development of intensive systems, especially pig, poultry and fish production, and to a lesser extent milk production. These systems are located where the density of population is high, but also correspond to products which are exported (poultry, fish). Soil degradation in arid areas due to overgrazing may be limited by monitoring the current use of land and encouraging sustainable systems. There is a risk of loss in vegetal biodiversity, due to selection of the most palatable plants by grazing ruminants, and an increase in pasture weeds. At least, the increase in drought due to global warming, which requires pasture management and cropping practices for sparing water. Environmental issues are to be better known, because the development of livestock farming in the tropics allows food self-sufficiency, and is a pillar of rural societies.

Session 40 Theatre 4

multi-scale trade-offs between production and environment in grassland agroecosystems: application to livestock farming systems and biodiversity conservation

Tichit, M.[1], Sabatier, R.[1], Teillard D'evry, F.[1] and Doyen, L.[2], [1]INRA, SAD Science for Action and Development, AgroParistech, Paris, France, [2]CNRS, MNHN, rue Cuvier, Paris, France; muriel.tichit@agroparistech.fr

In European grasslands, grazing and mowing regimes are major drivers of the maintenance of many wild species, either through direct or indirect effects. To date, both types of effects were studied separately, mainly at the field scale and with no quantification of performance. Objective of this study was to model the trade-offs between ecological and productive performance of a grassland agro-ecosystem at two nested scales i.e. field and farm. We developed a dynamic model linking grass dynamics controlled by grazing or mowing to stochastic population dynamics of two grassland bird species. Bird dynamics were driven by both direct and indirect effects of management. Viable control framework was used to predict productive and ecological performance. At field scale, results showed that the best ecological performance was obtained at intermediate levels of productive performance (60 to 108 grazing days /ha /year). Above 108 grazing days, no grazing strategy was viable for either species due to the negative direct effects of management. Between 108 and 240 grazing days, the ecological-productive relationship showed a convex Pareto like frontier. Any improvement in productive performance entailed a strong decrease in ecological performance. As expected, at farm scale the best ecological performance was obtained with extensive farms. However, the ecological-productive relationship showed a concave Pareto like frontier indicating the occurrence of compensations among management regimes. In intensive farms, it was necessary to allocate 40% of farm area in low-intensity grazing in order to compensate for the negative effects of mowing and high-intensity grazing on birds. However this land allocation involved a 25% reduction in productive performance. Finally, we discuss several forms of complementarities among farm types to improve overall performance.

Effect of spice supplementation on *in vitro* degradability, total gas, methane and selected bacteria for rice straw as a substrate

Khan, M.M.H. and Chaudhry, A.S., Newcastle University, School of Agriculture, Food and Rural Development, NE1 7RU, United Kingdom; a.s.chaudhry@ncl.ac.uk

Methane is one of the major greenhouse gases that also contribute to the global warming. As methane is one of the end products of rumen digestion, reducing methane from ruminants can also be of direct economic benefit because it coincides with the greater energy-use and so feed efficiency by the animal. Spices which have long been safely used for human consumption could be tested as alternative supplements to reduce methane production from ruminant livestock. In the present study, the effects of adding three spices (Coriander, Cumin and Turmeric) each at a fixed level of 40 mg/g rice straw as the common substrate were compared with rice straw alone (Control) on the *in vitro* total gas for 0 to 120 h and degradability, ammonia and methane production for 120 h only. Gram positive and Gram negative bacteria were also monitored after 120 h of incubation of rice straw with or without spices. The IVD and ammonia of rice straw increased significantly at 120 h ($P<0.03$) in the presence than the absence (control) of spices. The effect of spices was not significant for total gas whereas it was significant for methane production ($P<0.05$) from rice straw. Concentration of methane was significantly ($P<0.05$) lower in the presence than the absence of spices. In the presence of coriander, cumin and turmeric after 120 h of incubation the population of Gram negative bacteria were increased but that of Gram positive bacteria were decreased than the control. More Gram negative bacteria were observed in the presence of coriander and turmeric than turmeric. As spices reduced methane production for rice straw without showing any detrimental effect on its degradability or fermentation profiles, these spices may be effectively used to reduce methane production from low quality forages in ruminants.

Session 40 Theatre 6

Beef cattle in a mobile outdoor system during winter: general and excretory behaviour

Wahlund, L.[1], Lindgren, K.[1], Lidfors, L.[2] and Salomon, E.[1], [1]JTI - Swedish Institute of Agricultural and Environmental Engineering, P.O. Box 7033, SE-750 07 Uppsala, Sweden, [2]Swedish University of Agricultural Sciences, Department of Animal Environment and Health, P.O. Box 234, SE-532 23 Skara, Sweden; lotten.wahlund@jti.se

The aim of the project was to develop an animal friendly beef cattle system suited to winter conditions with low investment costs and low risk of plant nutrient losses. The study was carried out on a commercial farm in West Sweden during November 2008. A group of 17 pregnant heifers were kept outdoors on an expandable area of 750-2250 m^2 of arable grassland with access to a tent with a canvas roof, netting windbreak on sides and deep straw bedding. Feed-area and feed-racks were moved weekly while the water-area was stationary. Behaviour observations were performed daytime (7 am to 5 pm) for a total of 10 hours per week for six weeks. Number of heifers in each defined sub-area and behaviour were recorded every 15 minutes. All defecations and urinations were recorded on a map. The number of heifers was highest in the feed-area (51%) and in the tent (42%). The majority of feces and urine were deposited around feed-racks (66%) and in the tent (18%). Fewer heifers and excretions were recorded around the water-tank, just outside the tent and on a corridor of synthetic matting placed towards the feed. The records from the map showed an even distribution of manure and urine over the grassland. Most time in the feed-area was spent eating and most time in the tent for lying down and all heifers chose to lie in the tent-area to 100%. The results indicate that all the heifers had unrestrained access to weather shelter and feed in this system. By moving the feed-area weekly defecation and urination were spread evenly, without point-loads and decreased risk of plant nutrient losses.

Emission of ammonia and hydrogen sulfide from fattening pig houses with natural ventilation in South-East Spain

Madrid, J.[1], Orengo, J.[1], Valera, L.[1], Martínez, S.[1], López, M.J.[1], Megías, M.D.[1], Pelegrín, A.F.[2] and Hernández, F.[1], [1]University of Murcia, Dept. of Animal Production, Campus of Espinardo, 30100, Spain, [2]IES Ing. de la Cierva, Patiño, 30012, Spain; jorengo@um.es

The emission of polluting gases from pig farms is nowadays considered an important factor that should be controlled. For estimating gas emissions, it is necessary to know the ventilation flow and concentration of pollutant gas. In this work we monitored the emission of NH_3 and H_2S for a period of 46 days in a pig fattening farm in the South-East of Spain. This experiment was developed under natural ventilation and intensive rearing conditions during the growing phase (30-60 kg BW). Pigs were raised on partly slatted floor, with a pit that was emptied regularly. The animal density was 1 pig/m^2, with maximum indoor temperature that ranged from 20 to 30 °C. The ventilation flow was indirectly determined by the tracer gas method (CO_2). The production of CO_2 was estimated considering different factors (liveweight, energy intake and temperature) and adjusted for animal activity. Indoor and outdoor CO_2 concentrations were measured. Gas concentrations (CO_2, NH_3 and H_2S) were continuously monitored by X-am 7000 Dragër multigas detectors. The average ventilation rate was above 30 m^3/animal/hour. The average indoor concentration of CO_2 was 1281 ppm; a maximum limit of 3000 ppm is suggested for pig houses. The concentration of NH_3 reached a high average value (19.30 ppm) compared to the threshold value recommended by the International Commission of Agricultural Engineering. By contrast, the H_2S did not exceed a problematic concentration (average value of 0.23 ppm). The high NH_3 concentration joined to ventilation flow led to high values of NH_3 emissions. The results of our study could be explained by some factors (livestock building design, manure handling and indoor environmental factors) with significant influence on emission of polluting gases. This study is part of a project (08776/PI/08) funded by the Seneca Foundation of Region of Murcia (Spain).

Animal activity measured by pixel difference image in pig fattening houses under natural illumination conditions of Mediterranean area

Pelegrín, A.F.[1], Hernández, F.[2], Valera, L.[2], Orengo, J.[2], López, M.J.[2], Martínez, S.[2], Megías, M.D.[2] and Madrid, J.[2], [1]IES Ing. de la Cierva, Patiño, 30012, Spain, [2]Univ. of Murcia, Dept. of Animal Production, Campus of Espinardo, 30100, Spain; jorengo@um.es

The pig's activity and its behaviour influence on daily variation patterns of gas emission and dust concentration. Nowadays, the estimation of the animal's activity is carried out through automatic recording methods. The aim of this study was to register the pig's activity under intensive rearing and natural illumination conditions of Mediterranean area. Our study was based on a computer vision system that measured the activity by pixel difference image. This experiment was conducted during Autumn 2009 (10 h light:14 h dark) in a pig fattening farm located in South-East Spain. The pigs were fed ad libitum and monitored during the growing phase (30-60 kg BW). Two pens of 20 pigs and 20 m^2 were provided with a wireless infrared camera. Both cameras recorded scenes covering: concrete and slated floor, feeder and drinker areas. Images were analyzed in real-time by a computer software, based on the AForge.NET Framework and developed for this purpose. The image processor system measured the difference of pixel intensity between two consecutives frames (10 frames/s). The analyzed data were the hourly averages, expressed as relative activity. The maximum period of activity was found from 10:00 to 18:00 h, decreasing at the end of the afternoon. The minimum activity time was at 02:00 h. The relative activity data were adjusted on a sinusoidal model by the following equation: $1-0.823 \times \sin[(2 \times \pi/24) \times (h + 4)]$ (R^2=0.66). Our model estimated greater amplitude on the average relative activity in comparison with other activity models for pigs in Northern Europe. These results suggest that the animal's activity should be specifically evaluated to obtain accurate information about its behaviour and the pollutants emission. This study is part of a project (08776/PI/08) supported by the Seneca Foundation of Region of Murcia (Spain).

Simultaneous emissions and dispersion of the ammonia and methane plume inside and around a dairy farm in Segovia (Spain)

Sanz, F.[1], Montalvo, G.[2], Gómez, J.L.[1], Bigeriego, M.[3], Pineiro, C.[4] and Sanz, M.J.[1], [1]CEAM, Darwin, 14, 46980, Valencia, Spain, [2]TRAGSEGA, GR Segov, 40195, Spain, [3]Spanish Ministry of Environment and Rural and Marine Affairs, Alfonso XII, Madrid, 28071, Spain, [4]PigCHAMP Pro Europa, GR Segov, 40195, Spain; gema.montalvo@pigchamp-pro.com

A large proportion of the NH_3 emitted locally is deposited in the immediate neighborhood of the source rather than transported over long distances. Quantitative information about the spatial location of emission sources, as well as estimations of the emissions, is crucial for target-oriented abatement.and for a realistic distribution of NH_3 sources and sinks. Few studies show the distribution of NH_3 emissions around point sources in Europe. In general ammonia gas deposits close to the source of emission, studies in the surrounding of a poultry farm in UK reported deposition fingerprints of less than 1 km. In Spain, concentration fields decreased in the laying hens farm to levels of 3-5 µg/m^3 (background levels for the area) within distances of less than 600 km, whereas in the case of broiler meat farm at 600 m concentrations were twice that. The aim of this paper is to perform NH_3 concentrations in the surroundings of a dairy farm, since no measurements around this type of farms are available, and combine the measurement with estimates of the fluxes of ammonia from the farm. Ammonia concentrations in the surroundings of the farm were determined by passive samplers located at 2 m above ground in 60 sampling points. To estimate the fluxes form the farm building continuous measurements an Innova (mod. 1412-5) were performed in different points. The concentrations nearby the stable were of maximum 150 µg/m^3 and decreased down 23 µg/m^3 within 200 m, the wind direction determined the shape of the plume. The background concentrations were estimated at the level of 5 µg/m^3. The estimated NH_3 emissions from the stable were of 10.5 kg of NH_3/cow/yr, those levels are lower that the ones used for the Spanish inventory. Also CH_4 emissions were estimated (79 kg of NH_3/cow/yr).

Greenhouse gases and ammonia emissions from intensive pig farm

Mihina, S.[1,2], Palkovicova, Z.[2], Knizatova, M.[2] and Broucek, J.[2], [1]Slovak University of Agriculture, Production Enginering, A. Hlinku 2, 94976 Nitra, Slovakia (Slovak Republic), [2]Animal Production Research Centre Nitra, Institute of Livestock Systems and Animal Welfare, Hlohovecka 2, 95141 Luzianky, Slovakia (Slovak Republic); stefan.mihina@uniag.sk

Concentrations of ammonia (NH_3) and greenhouse gases (CH_4, N_2O, CO_2, H_2O) were monitored in an intensive pig farm during two fattening periods in summer and winter seasons, respectively. Pigs were housed on slatted floor and manure was stored in pit under them and discharged after each fattening period. Barns were mechanically ventilated by under-pressure systems. Concentrations of gases were measured by Photoacoustic Multi-gas Monitor device 1312 based on the photoacoustic infrared detection method. Air samples for analysis of concentrations of gases were taken from air stream at three ceiling fans, animal level and from outdoor surroundings. Concentrations of NH_3 and CO_2 were significantly larger and were increased faster during summer fattening periods than in winter. Similar differences were noted for N_2O concentration, although, obtained values, similar to CH_4, increased during summer fattening period and declined in winter. Average values of CH_4 concentrations in both seasons were in balance. Total emissions of the gases measured had similar tendency corresponding to their concentrations.

Environmental impacts: sustainability indicators of the dairy stewardship alliance

Matthews, A.G., University of Vermont, Center for Sustainable Agriculture, P.O. Box 8062, Burlington, VT 05402, USA; allen.matthews@uvm.edu

Which sustainable practices contribute to increased environmental stewardship on dairy farms? The Dairy Stewardship Alliance (Alliance) study has developed and vetted sustainability indicators for dairy farming. To be sustainable, practices guided by the indicators must enhance the natural environment and herd health, support profitability and improve the quality of life for farmers and their communities. The Alliance's self-assessment provides measurable indicators for continuous improvement on the farm. The assessment includes modules focusing on biodiversity, animal husbandry, community health, on-farm energy, soil health, water quality, pest and nutrient management as well as a farm financial inventory. This research is collaborative effort between dairy farmers, University of Vermont, Ben & Jerry's, St. Albans Coop and Vermont's Agency of Agriculture. Participating farms improved sustainable farming practices and utilized the results to guide their management decisions. Farmers have developed a better understanding of their production practices, explored alternatives and implement changes to improve environmental quality and the sustainability of their farms. The Alliance coordinated with a similar effort in the EU. Researchers are examining measures for continuous improvement that might create financial opportunities for dairy farmers and create product value for their co-ops. The Alliance has now expanded into the Northeast Dairy Sustainability Collaborative of dairy cooperatives, processors, researchers and farmers. As a group, we have become involved in an industry wide effort to identify ways to reduce Green House Gas (GHG) emissions and carbon footprints throughout the dairy production and distribution system (Value Chain). A recent focus has been to develop a 'low carbon farming' matrix to measure improvements which decrease atmospheric concentrations of GHG by increasing carbon sequestration and/or by reducing GHG emissions from agricultural operations.

Life cycle assessment of organic milk production in Denmark

Mogensen, L., Knudsen, M.T., Hermansen, J.E., Kristensen, T. and Nguyen, T.L.T., Aarhus University, Faculty of Agricultural Sciences, P.O. Box 50, DK-8830 Tjele, Denmark; Lisbeth.Mogensen@agrsci.dk

Livestock farmers experience a growing pressure for a reduction of the environmental impact related to cattle production. Not least the effect on greenhouse gas emissions is considered important by the public and the policy makers. The purpose of the present study was to investigate the environmental load of Danish organic milk through a life-cycle assessment (LCA) at farm gate outlining a method that are based on nitrogen balances and data that typically are available at the farm. By investigating at what stages in the chain, the most important environmental load was created it was possible to suggest improvement options. LCA is a collection and evaluation of the input and outputs and the potential environmental impacts of a production system throughout its life cycle. The functional unit (FU) of this study was an organic dairy farm with 192 cows and 341 ha land. The main processes at the farm include methane emissions from enteric fermentation and from handling of manure in stable, storage and application. Furthermore, crop production cause an emission of laughing gas (N_2O) from application of manure and from crop residues left in the fields. N_2O is also released from manure deposit at pasture, and handling and storage. Import includes manure, concentrated feed (cereals as energy feed and soy meal as protein feed), straw, fixation, and seed. Energy use includes the direct on-farm use of diesel and electricity and the indirect use of diesel by the contractor. The total GHG emissions related to the production at the farm was 2348 t CO_2-eq or 1.36 kg CO_2-eq per kg milk produced. The on farm processes was responsible for 72% of the emission, the import for 16% and the energy use for the last 12%. The GHG emissions could be reduced by increasing the feed efficiency and/or by increasing the nitrogen efficiency in the field (lower nitrogen supply without crop yield reduction).

LCA of pig feed production: calculating the CO_2-emissions of different feed components

Ginneberge, C.[1], Meul, M.[1], Fremaut, D.[1] and De Smet, S.[2], [1]University College Ghent, Department Animal Production, Schoonmeersstraat 52, 9000 Ghent, Belgium, [2]Ghent University, Lanupro, Proefhoevestraat 10, 9090 Melle, Belgium; celine.ginneberge@hogent.be

Greenhouse gas emissions, and their impact on climate, is one of the key issues for agriculture's sustainability. Intensive animal production systems largely rely on the input of fossil energy, through which they contribute to the emission of CO_2. Energy is used on farms in a direct way as diesel fuel, electricity or gas, but also in an indirect way during the production of farm inputs such as mineral fertilizers, pesticides, machinery and feed. Hereby, feed production has a major contribution in the total energy use of intensive animal production systems; e.g. up to 70% of the total energy use on specialized pig farms in Flanders. Hence, energy consumption and CO_2-emission from intensive animal production systems could be substantially decreased by reducing the CO_2-emissions from the manufacturing of animal feed, resulting in compound feeds with a lower carbon emission compared to the currently used feeds. The aim of the present study is to formulate CO_2-low pig feeds that still meet zootechnical and economic requirements and can readily be used in farming practice. To achieve this, a life cycle analysis (LCA) is performed for different feed components. As a case study, we compare the CO_2-emission throughout the production chain of soybean meal with the emissions during the production of lucerne, a locally produced crop and possible alternative protein source for soybean meal. In a later phase of the research, these emission-values will be used as extra variables in the formulation of compound feeds in an attempt to formulate feeds with a low carbon emission. Critical issues that arise during the LCA relate to (i) the allocation of environmental impacts between co-products, (ii) the methodological choices that need to be made and (iii) the availability and reliability of data. In this study, we discuss these critical issues and simulate their impact on the LCA-outcome.

N-efficiency and ammonia emission reduction potential through dietary adaptations in Swiss pig production

Bracher, A. and Spring, P., Swiss College of Agriculture, Department of Animal Sciences, Länggasse 85, 3052 Zollikofen, Switzerland; annelies.bracher@bfh.ch

As most European countries, Switzerland has signed the Gothenburg-Protocol to abate acidification, eutrophication and ground-level ozone in 1999. Among other pollutants, the protocol sets emission ceilings for ammonia. The aim of the present study was to evaluate the current feeding strategies in an area with dense pig production and to estimate the potential to reduce N excretion from pigs by dietary modifications. The study was based on IMPEX-data from 887 grower-finisher pig farms in the state of Lucerne from 2008. The IMPEX-data contain information on N-input via feed and purchased pigs and N-output via sales of pigs for each farm. Based on this information, the N use efficiency and the N-excretion via feces and urine can be calculated. The influence of different feeding strategies was analyzed. The average dietary crude protein concentration over the entire grower-finishing phase was 159.0±6.9 g at 13.7 MJ DE. The feeding strategies (1, 2 and 3-phase feeding, on farm mixing of whey, other by-products or standard ingredients) had little effect (158-163 g/kg) on dietary CP concentrations. Interestingly, multi-phase feeding systems did not yield a lower average CP concentration in the overall diet compared to 1-phase feeding, as phase 2 and 3 diets contained no to only moderate reductions in CP concentrations. The average N use efficiency over all farms was 32%. Independent of the feeding strategy, N use efficiency correlated negatively with dietary crude protein content. However, at every dietary N-concentration, the N use efficiency was highly variable between farms. Based on the data summary and simulation of dietary formulations, the N-input reduction potential and its associated cost were estimated. The data show that the potential to reduce N-input and concomitantly ammonia emissions is mainly given for finishing diets. At least a part of this potential could be implemented at no or minimal extra cost.

Structural rearrangements as keys towards an integral sustainable pig husbandry

Van Weeghel, H.J.E., Van Eijk, O.N.M., Miedema, A.M. and Kaal-Lansbergen, L.M.T.E., Wageningen UR Livestock Research, P.O. box 65, 8200 AB Lelystad, Netherlands; ellen.vanweeghel@wur.nl

Production efficiency has been the master narrative of Dutch livestock industry for the past decades. However, it is increasingly acknowledged that this exclusive emphasis is a prime cause of a range of undesired side effects, for instance on animal welfare and ecology. This study identified four structural rearrangements in order to design an integrally sustainable pig production system that takes into account the interests of a broad set of stakeholders. The project Porkunities ('Pork Opportunities') conducted by Wageningen UR Livestock Research, designed systems for a sustainable pig husbandry. In its designs it strived to meet and unify the needs of the pig, the pig farmer, the planet and the consuming citizen. In the project the needs of these four stakeholders were identified and described and three interactive design rounds were held. The resulting pig husbandry designs were analysed and evaluated on realisation of the heterogeneous sets of needs. The results of the project show four recurring issues where a change in thinking is required to find sustainable solutions. These structural rearrangements seem to be crucial in realising the goal of integral sustainability, and are: 1) grant the pig the freedom to fulfil its own needs, rather than applying technique to do so for the pig; 2) make full use of the pig as food-waste converter, rather than feeding it foodstuffs that compete with human food 3) harvest minerals and energy as by-products of pig production, rather than treating these products as waste 4) connect the farm with its environment and invest in an active relation with nature, society and consumer, rather than just reducing negative effects. This study shows what the stakeholders want and need, the deeper content and implications of the structural rearrangements, the designed solutions to meet these rearrangements and the value of a different mindset in provoking the needed innovations for an integral sustainable pig production.

Supporting conversion processes to organic farming with models built in a participatory way with dairy farmers

Gouttenoire, L.[1,2,3,4], Cournut, S.[1,2,3,4] and Ingrand, S.[1,2,3,4], [1]INRA, UMR Métafort, 63100 Clermont-Ferrand, France, [2]Clermont Université, VetAgro Sup, UMR 1273, BP 10448, 63000 Clermont-Ferrand, France, [3]AgroParisTech, UMR 1273, BP 90054, 63172 Aubière, France, [4]Cemagref, UMR 1273, BP 50085, 63172 Aubière, France; lucie.gouttenoire@clermont.inra.fr

To improve sustainability, farmers may redesign their livestock farming systems in depth, e.g. by converting to organic farming. Assuming that modelling livestock farming systems can support such redesign processes, we built models of the operation of livestock farming systems in a participatory way with farmers. Farmers' viewpoints were formalised by drawing causal maps with different local groups of farmers converting or already converted to organic farming. In this communication, we will focus on the way such models can support individual farmers in their conversion processes. To that end, the content of the models was analysed so as to better structure the questions and issues raised by the farmers themselves. Then, the links between a model that was collectively built on the one hand and the individual questions and issues for participating farmers on the other hand were explored. Benefits for participating farmers can be seen at three levels: (i) Mapping and analysing the models can help farmers to gain a better understanding of the processes at stake during a conversion. (ii) Farmers can discover new ideas, analyse their weak points in the farm operation and identify where their neighbours' experience could help to overcome them. (iii) Farmers are made aware of the specificities of their objectives and strategies compared with their counterparts, and they can then analyse their consequences in a structured way. Concrete examples are given to illustrate each of those three points. The originality of our approach is to consider conversions to organic farming as individual processes within the larger context of the evolutions of a whole local professional group, which may foster both individual and collective innovation towards more sustainability.

New turnout to pasture's practices to improve grassland management in the ewe milk production

Thénard, V.[1], Vidal, A.[2], Lepetitcolin, E.[2] and Magne, M.A.[1], [1]INRA, UMR AGIR, BP 52627, 31320 Castanet Tolosan, France, [2]AVEM, BP 419, 12 104 Millau, France; vincent.thenard@toulouse.inra.fr

Grasslands are now being acknowledged for their multifunctional role. However the intensification of livestock production has led to a decrease in grassland use. In the production of the French cheese named Roquefort many farmers have abandoned natural grassland use to feed the milking ewes. This trend is less pronounced in few farmers' group wishing reduction in farm input and environemental impact while using the grassland. In order to maintain the level of the milk production the farmers have to innovate in new grazing systems and turnout to pasture practices. Designing new grazing practices requires understanding of links between the animal feeding and grassland management. To develop the necessary knowledge we have built a research project involving farmers. This project study has focused on the benefits from the large diversity of cultivated grasslands providing more flexibility in the farm management. We have conducted 20-farmers' interviews to have an insight into the grassland and herd management's practices, and milk production results. A Multivariate Analysis has permitted to define 5 patterns of turnout to pasture managements. For each pattern we have described a different milking curve for two-first months of pasture. The milking curves are based on the milk level and the milk persistency. The main results report on links between feeding practices and milk production. Further analysis of the different patterns could identify the predisposing factors of the milk persistency. We have showed that some practices such as large diversity of grazing grassland use are favorable for maintaining the high level of milk production. The two main conclusions are firstly the traditional feeding system can be used to the lower levels of milk production. Secondly, the livestock intensification needs protein supply use combined with the large diversity of cultivated grassland. These two points can be easily applied while advising farmers.

Predicting the location of dairy cow urinations

Clark, C.E.F., Levy, G., Beukes, P., Romera, A. and Gregorini, P., DairyNZ, Feed and Farm Systems, Corner Ruakura and Morrinsville Roads, Hamilton, 3240, New Zealand; cameron.clark@dairynz.co.nz

Capturing urine and spreading it evenly across a paddock reduces the risk of nitrogen loss to the environment in pasture-based dairy systems. This study determined the difference in urination frequency by location and then developed a model for predicting urination frequency. Forty-eight mixed age, multiparous Holstein-Friesian cows in early and late lactation were assigned to 6 groups and grazed according to the following treatments (2 groups per treatment): Pasture grazing offered for 4 h after each milking (2×4), 8 h between morning and afternoon milkings (1×8), or for 24 h excluding milking times (control). Between grazing periods, the 1×8 and 2×4 cows (restricted cows) were confined to a stand-off area. The distribution of urination events for each treatment were recorded using intra-vaginal sensors and geographic positioning systems. There was no difference (P>0.05) in urination frequency between the 1x8 and 2x4 h grazing access treatments or with stage of lactation. Restricted cow urination frequency was lower on the stand-off (0.4 urinations/cow/h) than the dairy, pasture or laneway (1.1, 0.7, 1.3 urinations/cow/h, respectively). Control cow urination frequency was lower on pasture (0.5 urinations/cow/h) than on the dairy or laneway (1.4 or 1.2 urinations/cow/h, respectively). On the assumption that cows do not urinate when resting and that cows rest for 8 h/d on soft surfaces, restricted cow rest time on pasture, back -calculated from urination frequency was 2 h in both early and late lactation. Thus, cows offered pasture for 0 h, 8 h (restricted) and 21 h (control) rest for 0 h, 2 h and 8 h on pasture. This was described by the following prediction equation: Rest time on pasture = 0.01*Total time on pasture2+0.17*Total time on pasture. This empirical model allows the location of dairy cow urination events to be predicted when time on pasture is varied, according to the proportion of total active time on each surface.

Green house gas emission from commercial milk production units in Denmark by life cycle assessment

Kristensen, T. and Mogensen, L., Aarhus University, Faculty of Agricultural Sciences, Dept. of Agroecology and Environment, Blicher Alle 20, DK 8830 Tjele, Denmark; troels.kristensen@agrsci.dk

Agriculture is believed to have a significant role in the total emissions of green house gas (GHG), but there are a limited number of investigations of the actual level of GHG emissions at farm level. The purpose of the present study was to quantify the GHG load of Danish milk production through a life-cycle assessment (LCA) at farm gate outlining a method that are based on data that typically are available at commercial farms and to analyses the reasons to variation between farms in order to find promising ways to reduce the emission at system and management level. Annual data from 37 conventional (135 cows, 137 ha) and 32 organic (115 cows, 178 ha) dairy farms were collected and the potential environmental impact of the production of one kg energy corrected milk (ECM; 4.10% fat; 3.30% protein)) was evaluated throughout its life cycle. The main processes at the farm include methane (CH_4) emissions from enteric fermentation and from handling of manure. Furthermore, crop production cause an emission of laughing gas (N_2O) from application of manure and from crop residues left in the fields. N_2O is also released from manure deposit at pasture, and from other parts of the N turnover at farm level. Secondary emissions from import include manure, fertilizer, concentrated feed and use of diesel and electricity and the indirect use of diesel by the contractor. Allocation between milk and meat was made on either economic value or biological importance based upon energy required. The total GHG emissions, included the secondary emission was 1.28 kg CO_2-eq per kg milk. After economic allocation the emission was 1.09 kg CO_2-eq, and after biological allocation 0.90 CO_2-eq per kg milk. A preliminary analysis showed no effect of production system on the GHG emission per kg milk, while it seems that it could be reduced by increasing the roughage production or milk yield.

Reliefs and shrub covers which may be used to design pastoral management

Lecrivain, E.[1], Garde, L.[2] and Dormagen, E.[1], [1]INRA, SAD, Ecodéveloppement, 84914 Avignon, France, [2]CERPAM, Route de Durance, 04100 Manosque, France; lecriv@avignon.inra.fr

The aim is to stand out husbandry systems where flock behavioural abilities participate to the maintenance of pastoral areas in shrubby hillsides. The objective is to characterize from flock behaviour, what is the respective role of the relief and the degree of shrub cover which contribute to the stabilisation of group grazing activity when forage resources are scattered. The study focuses on two productive flocks shepherded on two pastoral units in the winter period for 4 years.We monitored flocks behaviour during 17 daily grazing circuits. 54 grazing sites and 239 segments forming limits to the grazing sites have been analysed. The results reveal that the relief has a major influence on the flock's movements against the shrub cover and that the flocks spread out an intense grazing activity into particular areas. Some natural configurations are favorable to group grazing activity. Slopes, plateaux and small valleys favour an intense grazing. The average area of the grazing sites is 3 ha for 800 to 900 heads. Shrub-covered garrigue, is accessible for intense grazing activity until 60 up to 80% of shrub cover if the shrubs'average height do not exceed 40 cm, and until 40 up to 60% if the shrubs'average height do not exceed 60 cm. The relief is also more limiting grazing sites than shrub cover. Alone it forms a limit in 54% of cases. The relief/vegetation association is limiting in 12% of cases. In all, the relief participates in 66% of the limits. Two third are impassable or passable marked reliefs. The last third is secondary reliefs, not very marked and easy to cross by the flocks: change of incline, change of direction which break up any stable grazing activity. In conclusion, this series of criteria lead us to better identify areas which facilitate stabilized grazing sequence in such heterogeneous type of environment and to reconsider the evaluation of their grazing potential. They become essential descriptors for designing new pastoral management and developing sustainable pastoral systems.

Session 40 Poster 11

Survey of current Swiss pig feeding practices and potential for ammonia emission reduction

Bracher, A. and Spring, P., Swiss College of Agriculture, Department of Animal Sciences, Länggasse 85, 3052 Zollikofen, Switzerland; peter.spring@bfh.ch

To achieve the goals on ammonia emission, mitigation programs focus on measures that diminish ammonia losses from animal housings, manure storage and spreading. Dietary modifications as a begin-of-pipe measure will affect all emission stages from barn to field. An effective way to reduce N-excretion, from livestock is achieved by the reduction of dietary CP, while maintaining amino acid levels. The aim of the present study was to gain an overview of the current pig feeding practices in Switzerland. A survey was conducted in feed manufactures, comprising 80% of the Swiss pig feed market. Based on diet specifications and sales volumes, the usage of NPr-feed (N and P reduced feed) and the average nutrient content of different feeds were calculated. To verify if declared diet specifications corresponded with the actual concentrations, declared and analyzed data from 108 diets were compared. The survey showed that 70% of the feed available on the market are sold as NPr-feed. The percentage is highly variable within regions and between feed mills, reaching over 90% in animal dense regions. The average CP concentrations for grower-finisher pig diets were 172.95±4.83 g (at 13.57 MJ DE) and 158.04±6.17 g (at 13.72 MJ DE) for standard and NPr diets, respectively. The average CP concentrations for dry sow diets were 144.97±8.1 g (at 12.05 MJ DE) and 139.12±9.9 g (at 12.26 MJ DE) and for lactating sow diets 178.85±7.1 g (at 13.68 MJ DE) and 164.81±8.4 g (at 13.73 MJ DE) for standard and NPr diets, respectively. The analyses revealed no protein over-formulation compared to the declared values. The analyzed energy values surpassed declared values on average by +0.4 MJ DE/kg (+3.05%) and analyzed crude protein contents deviated from declared contents on average by -0.46 g/kg (- 0.25%). The 30% of total feed still being sold at standard CP concentrations offers a considerable potential to reduce N-input and concomitantly ammonia emissions in Swiss pig production.

Session 41 Theatre 1

Consumers' perceptions of food quality products: Greece's experiences

Vlachos, I and Kleanthous, P., Agricultural University of Athens, Iera Odos 75, 118 55, Greece; ivlachos@gmail.com

Purpose: The aim of this research was to understand Greek consumer's perceptions regarding price, quality, and value as determinants of product choice. Quality perception is linked to food choice and consumer demand. We address value as key determinant of food choice. We also address questions of price perception and the validity of willingness-to-pay measurements. Design/methodology/approach: We conduct an extensive literature review of food product marketing in Greece. Then, we developed a questionnaire and made a survey using questionnaire interviews in 2009. Questions were focused on meat and meat products, as well as traditional products, such as PDOs. We employed two-step cluster analysis, multinomial logistic regression and stepwise descriminant analysis as the most appropriate methods for our analysis. We assessed homogeneity within clusters in order to add validity to our results. Findings: The findings revealed different clusters associated to shopping behavior and product choice. The groups differ in terms of demographic characteristics as well as price, quality, and value perceptions. It is concluded that value and food quality are central issues in today's food marketing. Research limitations/implications: We focus on one European Union member state. Implications/Future Research: This study presents the current status of consumers' perceptions of food quality products. The results of our survey can positively contribute to food managers as well as to any food marketing researchers.

Consumer perceptions of home made, organic, EU certified, and traditional local products in Slovenia

Klopcic, M.[1], Verhees, F.[2] and Kuipers, A.[3], [1]University of Ljubljana, Biotechnical Faculty, Department of Animal Science, Groblje 3, 1230 Domzale, Slovenia, [2]Wageningen University, Marketing and Consumer Behaviour Group, Postbus 8130, 6700 EW, Wageningen, Netherlands, [3]Wageningen University & Research Centre, Expertise Centre for Farm Management and Knowledge Transfer, De Leeuwenborch, Postbox 35, NL-6700 AA, Wageningen, Netherlands; Marija.Klopcic@bfro.uni-lj.si

Perceptions of food characteristics were measured for cheeses and sausages: regular, organic, PDO, mountain and farm-made. Questionnaires were sent to 2,300 consumers and 340 were returned: 220 respondents rated 4 or 5 cheeses and 120 rated 4 sausages. As variables were asked: age, education, stage in career, and region. Results did not indicate a serious bias in the responses compared with Slovene population. Respondents were questioned about their perceptions of cheeses and sausages. For example, to measure consumer's perception of a cheese's effect on health 4 attributes were included: nutritious, health, healthy product, and safety of product. This resulted in an order of importance of the various food attributes. Excellent taste, healthy, produced in Slovenia, enjoyment and environmental friendly produced are considered as important, while traditionally produced, easy to prepare, nutritional value, regionally produced and low price are not so important. Respondents were also questioned about their intentions to buy cheeses and sausages. Coefficients for intentions to buy on perceived food characteristics were calculated. High convenience and indulgence have a positive influence on consumer's intention to buy cheeses, but health, low price and sustainable production not. Low price and indulgence determine consumer's intention to buy sausages. This research shows opportunities for processors of quality products to improve their market position. To identify market segments marketers should not use only consumer's stated perceptions. The consumer seems to act differently in expressing a public opinion and his behaviour in the market place.

Overview Consumer research in Western Balkan countries

Kuipers, A., Wageningen University and Research Centre, Expertise Centre for Farm Management and Knowledge Transfer, P.O Box 35, 6700 AA Wageningen, Netherlands; abele.kuipers@wur.nl

A set of 225 publications from Western Balkan countries was gathered and systematically ordered as action of the EU FP7 project 'Focus Balkans - Food Consumer Sciences in the Balkans'. Six country reports provided the base data. Methodologically, there is in general a lack of primary data, and a disproportion between qualitative and quantitative studies is observed in favour of quantitative. This is probably due to the perception in Western Balkans that quantitative research is 'more valuable'. The majority of publications gathered are qua research concept 'environment' oriented. Part of the publications' contents is said to be either too general or too theoretical. Quite some papers give general information like 'consumers request a certain quality from the food producers' or 'product quality can be achieved by implementing ISO series of standards'. There are not enough data on consumers' attitudes, knowledge and habits regarding food in general, and especially regarding different food types chosen for case studies. Somewhat less focus on fruit than on other product groups (organic, traditional and health claimed) is observed. There may be a perception that research in such 'soft' science has a rather low 'scientific value' compared to 'pure' science. The 'house' of scientific expertise in this field has therefore, perhaps, a poor foundation, while 'the pillars behind food consumer science' are not interrelated and fastened together. A multidisciplinary approach is rare. It is observed that the accessibility of the scientific work is limited, while there is a lack of interaction with the international scientific community.This is illustrated by the fact that by far the most publications are published in the local languages and nearly all written by local authors. Some of the countries conclude that there is a huge knowledge gap on food consumers´ science and that should be used as a stimulation for developing the base for further research.

Health claimed products and consumer attitudes in Balkan countries

Stojanovic, Z., Ognjanov, G. and Dragutinovic-Mitrovic, R., Faculty of Economics, University of Belgrade, Department of Economic Policy and Development, Kamenicka 6, Belgrade, 11000, Serbia; zaklina@ekof.bg.ac.rs

Generally, in WBC a little attention was paid to the products with health claims, both from consumer and producer point of view. There is no data regarding consumers' knowledge about functional food and attitudes towards this type of food. This paper will present research findings of EU founded Project FOCUS-Balkans related to WBC consumers' demand towards products with health claims in dairy sector - milk and yoghurt. Interviewing was a key method of data collection. Data were collected from all WBC. Totally 29 out of 45 identified producers and 26 retailers were interviewed. Additionally, seven EU exporters were interviewed. Totally 52 out of 120 observations are dedicated to the dairy products – milk, yoghurt/kefir and cheese. The price of the products with H&N claims is absolutely the most important factor influencing the local consumers' choice of food. However, they also see health benefits of food as of growing importance. Restraints are economic difficulties, price, lack of nutritional knowledge and eating habits. Availability has been the solemn restraint not given a great importance by either processors or retailers in almost all WBC. Perception of consumers of products with health claims is quite identical: female belonging to the age groups of 15 to 40 years, or elder (40-64), with higher or middle income, secondary or high education, with or without health problems, living in urban areas. Consumers are perceived mostly as practicing a healthy life style, following modern trends and fashion in food consumption, active (sportsmen, businessmen) or mothers who are expected to provide healthy food for their families. Retailers demand in WBC seems to be dissatisfactory. Processors mostly complain about the limited shelf space for which 'they have to fight'. Majority of experts think that food labelling is not satisfactory and not understandable for consumers.

Food quality products and consumer attitudes in Egypt

Khallaf, M.F., Ain Shams University, food scince, 68 Hadayeq Shoubra, Cairo, 11241, Egypt; dr.khallaf86@yahoo.com

There are many forms for consuming meat and meat products in Egypt and other Arabian countries. Fresh meats are consumed as 'cooked' meats, i.e. roasted, broiled, fried, braised, stewed, micro -waved meats... etc. On the other hand, there are many Egyptian meat products that are widely consumed in Egypt. These meat products have a special taste and are rich in its flavor, because of its processing method, ingredients and method of cooking.This presentation highlights the principles and processing steps of some Egyptian meat products, named 'Basturma', 'Oriental sausage', 'Hawawshy' and 'Shawerma'. The first product 'Basturma' is a cured product, the second one 'Oriental sausage' is a frozen product, the third product 'Hawawshy' is a minced product an the last one 'Shawerma' is a barbecued product. After this presentation, you will be able to: 1) list factors affecting the selection of meats; 2) describe how to properly store meats to maintain their quality; 3) describe the principles and methods of preparing and cooking some Egyptian meat products; 4) detect critical control points (CCPs) during the processing steps of such meat products; 5) know adulteration ways of mentioned products.

Consumer attitudes to animal food quality products in Italy: a survey

Lazzaroni, C.[1], Gigli, S.[2], Iacurto, M.[2], Vincenti, F.[2] and Biagini, D.[1], [1]University of Torino, Department of Animal Science, Via Leonardo da Vinci, 44, 10095 Grugliasco (TO), Italy, [2]Agricultural Research Council, Research centre for meat production and genetic improvement, Via Salaria, 31, 00015, Monterotondo (RM), Italy; federico.vincenti@entecra.it

Food quality is composed by characteristics of food that are acceptable to consumers, identifiable through labelling, marketing and quality rules. According to the different kind of consumers, a lot of different product characteristics affect their choices. Nowadays consumers expect their food to be safe, wholesome, tasty, and typical (linked to tradition and land). The latest requirement has a special feeling in Southern Europe and in Italy, where the tradition for typical products is higher than in the Central and Northern European Countries. In fact, Italy has 21.1% of the registered PDO, PGI and TSG products in EU (fresh meat 2.7%, processed meat 32.0%, fishery 11.1%, cheese 20.5%, other animal products 4.2%), and the remaining Southern European Countries (France, Spain, Portugal, Greece, Slovenia, Malta and Cyprus) 56.9% (fresh meat 84.7%, processed meat 51.6%, fishery 38.9%, cheese 55.6%, other animal products 83.3%), leaving only 22.1% to the other 20 EU Countries. Moreover, other typical labels are widely spread in Italy, such as Slow Food Presidia (58.0% of world list). The consumption trend of animal food products is a good indicator of consumers' attitudes and in Italy in the last years we are facing ups and downs. I.e., from 2003 to 2007 the consumption of fresh meat is decreased to 97.7% (beef to 99.7% - but the PGI Vitellone Bianco dell'Appennino Centrale raised to 149.2%, pork to 99.6%, sheep meat to 94.8%, poultry to 94.8%, while rabbit meat raised to 100.6%), that of processed meat products to 97.9% (while the PDO products raised to 102.6%, even noting that the Prosciutto di Parma is decreased to 97.8%), that of milk is raised to 107.3%, that of cheese is decreased to 99.8% (and again PDO products raised to 106.3%, and Parmigiano Reggiano to 113.8%).

Food quality products and consumer attitudes in France

Hocquette, J.F.[1], Jacquet, A.[2], Giraud, G.[3], Legrand, I.[4], Sans, P.[5], Mainsant, P.[6] and Verbecke, W.[7], [1]INRA, UR1213, Theix, 63122 St-Genes Champanelle, France, [2]INAO, 6, rue Fresnel, 14000 Caen, France, [3]VetAgroSup, 63370, Lempdes, France, [4]Institut de l Elevage, 14310, Villers-Bocage, France, [5]ENVT, 31076, Toulouse Cedex 3, France, [6]INRA, 94200, Ivry-sur-seine, France, [7]Ghent University, B-900, Ghent, Belgium; jfhocquette@clermont.inra.fr

The French market of animal products is nowadays very segmented, due to the raising of many quality signals: i) official label identifying a superior quality (Label Rouge), environmental quality (Organic Farming), or quality linked to origin (PDO) or provenance (PGI), ii) product descriptions highlighting a specific feature: 'farm processed', 'mountain produce', etc. (iii) certification of products aimed at applying normative standards. For official signs, professionals voluntarily undertake to set up and monitor a quality-focused approach individually (organic farming) or collectively independent and competent bodies carry out regular checks, and the public authorities supervise the system. Beef and lamb fresh meat under quality schemes represent about 6% and 15% of the French meat production. Geographical indications identify a product as originating from a region, where a given and unique trait is attributable to this region. Protected Denomination of Origin (PDO) was mainly developed for dairy products. The integration of local breeds into PDO concerns all ewe cheese PDO, the main part of meat PDO, about half of cow PDO cheeses and a third of goat PDO cheeses. The breed may contribute to the product's specificity or may only be a marketing claim. The evolution of demand results from changes (i) in the demographic composition and way of life of the consumers, (ii) in the characteristics of products, and (ii) in preferences of consumers for specific attributes (taste, health, traceability, etc). For beef, the French consumers seem to be favourable to a beef eating-quality guarantee. The market outcome of certification programmes depends upon consumer understanding of and confidence in high quality labels.

Are there connections between animal welfare in traditional Portuguese beef production systems and beef quality?

Viegas, I.[1], Vieira, A.[2], Stilwell, G.[1], Lima Santos, J.[3] and Aguiar Fontes, M.[1], [1]FMV-TUL, CIISA, Av. Univ. Técnica, 1300-477 Lisboa, Portugal, [2]Rafael Filho SA, R.Henriques Lagoa Furadouro, 2490-403 Santarém, Portugal, [3]ISA-TUL, Tapada Ajuda, 1349-017 Lisboa, Portugal; inesviegas@gmail.com

Farm animal welfare is a growing concern for consumers' in Europe and in spite of different interpretations within different parties, there has been a profound evolution in animal welfare perception in Europe. This, together with pressure from different society groups and growing evidence that animal welfare has direct and indirect impact on food safety and quality, led to significant changes in the CAP, and to new legislation for an animal friendlier production sector. Also, new support measures for farmers complying with standards above legally imposed ones aid those who supply cost increased products with differentiated quality, as animal friendlier products. As part of a research project on Portuguese consumers' willingness to pay for beef with credence quality attributes, this article is centred in unveiling connections between traditional production systems and animal welfare. Therefore, two common beef production systems in Portugal are described, as well as their geographical distribution. A descriptive analysis is relevant for understanding why local breeds are preferred by producers, and why are (semi) extensive systems better adapted to large part of the country. Portuguese beef production systems are also described in terms of their animal welfare status and probable critical points. Again, native breeds are more adapted to local conditions. Having understood the welfare status of beef cattle in Portugal, relationships with beef quality can be suggested. A twofold link between Portuguese native breeds and beef quality is proposed. First, welfare may be translated into beef intrinsic quality resulting in an increased experienced quality by consumers. Second, as animal welfare is a credence quality attribute, it can be part of market strategies to be explored together with the products' intrinsic quality.

Session 41 Theatre 9

Consumer attitudes toward meat comsumption in Spain. Special reference to quality marks and kid meat

Alcalde, M.J.[1], Ripoll, G.[2] and Panea, B.[2], [1]University of Seville, Ctra Utrera Km.1, 41013 Seville, Spain, [2]CITA, Av Montañana 930, 50059 Zaragoza, Spain; aldea@us.es

The compsumtion of meat in Spain in relationship with the average EU is higher. In 2008 the meat compsumtion was 65.3 kg/per capita. But fresh meat was 48.9 kg, of which only 5.6 kg was certified meat. In the north of Spain the meat compsumtion is higher than in the south. Consumption of poultry meat (16.1 kg) and pork (14.1 kg) focuses on the fresh meat while the compsumtion of lamb/kid meat (lamb and kid come together in the statistics) was only 3.3 kg/per capita. The lamb/kid meat is considered expensive (6.8 out of 10), with a family and festive character. The main reason for its purchase is its taste. Consumers consider (53.4%) that the Spanish production of lamb/kid meat is linked to the environment. In the particular case of kid meat, the estimated compsumtion is the 0,5 kg/per capita. The Spanish goat effectives (third country of the EU, after Greece and France) were in the South but the higher consumption of kid meat was in the North-East. But it doesn't exist a Protected Geographical Indication (PGI), only a regional quality mark is registered. So a consumers test was made in Sevilla (Southern Spain) to check their appraisal where no habit of compsumtion meat exits. The effect of breed, carcass weight (heavy and light) or lactation (natural or artificial) on meat quality of suckling kids from seven Spanish breed were investigated. A home test was carried out. Consumers tested all types of samples in a balanced designed. There were some differences as breed for light kids but not for heavy kids. But neither weight inside breed nor lactation system inside breed had effect on any of studied variables. At the end of the test, some consumers wanted more meat and they asked where to buy it and others no more long wanted.

Quality and safety of products of animal origin and consumers attitudes: Cyprus perspective

Kalli, N., Kyriakidou, P. and Kyriakides, G., Veterinary Services of Cyprus, Veterinary Public Health Division, Athalassa, 1417 Lefkosia, Cyprus; nkalli@vs.moa.gov.cy

Cyprus is an island in the South-East Mediterranean Sea, at the crossroads of three continents, Asia, Africa and Europe. Throughout its history, different civilizations dominated the island, leaving their marks on the culture, food and peoples' habits. To this day, a vast variety of traditional products comprise common ingredients of the cypriots' diet. Cyprus-produced traditional products include zalatina, wine sausages, tsamarella, lountza, halloumi and anari. 1058740 animals (pigs, sheep, goats, bovine) are slaughtered each year to meet the demands of a human population of 796,900. This highlights the high consumption of meat in Cyprus. In 2004, Cyprus joined the EU therefore adopting and complying with the relevant EU legislation. As a result, the production of food and the use of methods that were not in accordance with the provisions of EU legislation were terminated. Public announcements were made for the consumers and scientific advice had been provided to the producers while implementing the EU standards, leading consumers in becoming more aware and better informed on the importance of food safety and quality. The official controls became systematic and strict, in order to meet the consumers' expectations on the quality and safety of food. Inspections, sampling and reports were rearranged accordingly and a new evaluation and control system which ensures Public Health and promotes the production of a variety of products was developed. The consumers can now choose from an array of local traditional products with ensured quality and safety. One of the missions of the Veterinary Services of Cyprus is to foster Excellence in the evaluation and supervision of the quality and safety of products of animal origin for the benefit of Public Health. This also pertains to the production of traditional products which are an integral part of the Cypriot cuisine and culture.

Session 41 Theatre 11

Consumer attitudes towards red meat in Turkey recently

Çelik, K. and Uzatıcı, A., Çanakkale Onsekiz Mart University, Faculty of Agriculture, Depart. of Animal Sci., Çanakkale, Çanakkale, Turkey; urartular@gmail.com

World feed prices have increased significantly in 2007. This increase 71% in barley, 57% in wheat, 44% in soybean, 71%, in sunflower seed meal, in wheat bran 62%, while in corn by 39% was realized. However, world red meat prices, especially in countries which pasture-based training did not increase too much is seen. Similarly the results of high increases and instability of feed, red meat consumers' purchasing power has weakened in last ten years in Turkey. When we are looking into purchasing power of the Turkish people that they have had trouble reaching to meat and meat products recently. People has also significantly towards poultry products, pasta and rice because of the serious decline in purchasing power due to general and agricultural policy. The pasta consumption increased at a rate of % 53.8 clearly shows this trend. However, per capita annual consumption of poultry products in 1994 was 2,5 kg in 2009 with an increase of % 240 increase to 11 kg is observed. While the case in Turkey the EU beef carcass prices are lower than Turkey. If we look at prices between 2002-2009 are considered, beef prices in the EU is 270-332 kg Avro/100, whereas prices in Turkey in the same period, is 302-438 kg Avro/100. On the basis of consumer prices for red meat prices has increased more than 50% in the last 2 years and it means Turkish people will consume more grains and vegetables.

Transforming consumer's healthy eating perceptions into product development

Ranilović, J., Podravka food company, Quality Control, Ante Starčevića 32, 48000 Koprivnica, Croatia (Hrvatska); jasmina.ranilovic@podravka.hr

A recently published survey on a national sample of Croatian adults aged 15 years and over was performed relating to consumer healthy eating interpretations. This study showed that behind the most frequently mentioned concepts of healthy eating being 'fresh, natural foods' (50%), various interpretations were lay down like food without additives, pesticides, fertilizers and/or GMO, organic or bio produced food, safe packed food, domestic and traditional food products, good and safe for health, and from Dalmatian, Croatian or Mediterranean origin. It indicates that for Croatian believes regarding to 'healthy eating', the way a food product is manufactured, packed or originated might be more relevant than safe and healthy eating aspects. This is probably the case, not because of lack of safety concerns among Croatian adults, but because they assume the local or regional originated, domestic and traditional food to be the 'food you can trust' in terms of honesty, quality and food safety. For product developers, these perceptions give wide opportunities to develop a variety of food products that are 'safe by design', taking into account the complexity with which food is currently being viewed by the Croatian consumers. It implies developing effective communication strategies for increasing the competitiveness in the market and at the same time utilizing labeling and consumer education to enhance healthy dietary practices.

Session 41 Theatre 13

New horizons for maximising the value of the beef carcase

Pethick, D.W.[1], Polkinghorne, R.[2] and Thompson, J.M.[3], [1]Murdoch University, 90 South St, 6150 Murdoch WA, Australia, [2]Marrinya Pty Ltd, 70 Vigilantis Rd, 3875 Wuk Wuk VIC, Australia, [3]University of New England, 11/3 Stewart St, 2444 Port Macquarie NSW, Australia; d.pethick@murdoch.edu.au

The Meat Standards Australia (MSA) beef grading scheme can predict the eating quality grade of 40 muscles by 5 cooking methods. The quality grades, namely, unsatisfactory (2*), good every day (3*), better than every day (4*) and premium (5*), represent the predicted rating of untrained consumers. This voluntary scheme has been operating for 10 years in Australia and in 2009 the number of carcases graded reached 1 million with price premiums for farmers, wholesalers and retailers. Despite this success most product in Australia is sold as MSA beef with only a few supply chains marketing higher quality grades. The true potential of the MSA system will be discussed in a scenario where by yearling beef carcases (18mo, *Bos taurus*), which is optimally processed (tenderstretch hanging), are then deboned such that all muscles are allocated to an optimal quality grade by cooking method. Accordingly traditional cut nomenclature would no longer be used and instead beef would be sold as 4* steaks, 4* roasts, 3* stir fry etc. For example 4* grilling steak would be selected from 4-5 different muscles which conventionally retail at widely different prices (e.g. oyster blade at $AUS 10/kg and striploin at $AUS 25/kg). Based on both willingness to pay data gathered from 5 different countries (n=9,840 consumers) and actual prices received in an Australian retail business, the value of the new system can be estimated. Thus if 3* beef has a retail value of 100% then 4* and 5* star beef can be sold at 150% and 200% of the 3* price respectively. The final scenario is to then overlay muscle yield with eating quality grade as a system that would deliver transparent value based payment back to beef farmers. The differences in the value of the carcases, corrected to constant weight, can range up to $AUS 700/carcase at retail, which is variably apportioned to differences in eating quality and yield.

Milk fatty acid composition in France: perception by the consumers and relations with dairy cow diets

Rouillé, B., Institut de l Elevage, Dairy cow nutrition, Monvoisin - BP 85225, 35652 Le Rheu Cedex, France; benoit.rouille@inst-elevage.asso.fr

The diversity of milk fatty acids (FA) induces a high interest of milk fat for human nutritional issues. However, milk fat does not have a good reputation in human nutrition due to its relative richness in satured FA (SFA). Increased mono-unsatured FA (MUFA), poly-unsatured FA (PUFA) and decreased omega-6/omega-3 ratio (<5) are goals to improve milk quality according to nutritionists. In fact, consumers do not realize that milk fat is not the only quantitative target and is not the good qualitative target in order to improve human diet. Moreover, people do not even know what a FA is, except maybe for omega-3, which are very often used in marketing. The aim of this study was to identify different fatty acid (FA) composition of dairy cow milk in some French feeding systems in order to show that cow milk can be modulated to fit human nutrition. Seventeen farms from six regions were selected. Milk samples and a description of the diet were collected at five periods. Five milk classes were identified regarding the milk FA profile ($P<0.001$). Diets associated to milk classes were described and analysed. Fatty acid composition of milk produced by the grazed grass diets are statistically different ($P<0.001$) from milk from other diets [milk richer in polyunsaturated fatty acids (PUFA), in trans fatty acids (trans FA), and with a lower omega-6/omega-3 ratio]. Grass-based diets produce milk with better nutritional profile. This study shows the large diversity of milk fatty acid composition existing in France but it also shows that milk is produced by very different feeding systems. This leads to possible technical advices on feeding systems and strategies during a whole year so as to fit human nutritional requirements. Three goals must be reached to improve the image of milk products for the consumers: (i) explaining the diversity of milk FA, (ii) communicating on the nutritional and sensorial benefits of milk fat, and (iii) showing the diversity of milk and the technical advices available to modify the FA profile.

Sensory evaluation of beef eating quality in France and UK at two cooking temperatures

Micol, D.[1], Jailler, R.[1], Jurie, C.[1], Meteau, K.[2], Juin, H.[2], Nute, G.R.[3], Richardson, R.I.[3] and Hocquette, J.F.[1], [1]INRA, URH, INRA Theix, F63122 Saint Genes Champanelle, France, [2]INRA, EASM, Le Magneraud, F17700 Saint Pierre d Amilly, France, [3]DFAS, University of Bristol, Langford, BS40 5 DU, United Kingdom; hocquette@clermont.inra.fr

This study is part of the EU ProSafeBeef project on producing safe beef and beef products with enhanced quality characteristics. The aim of this study is to compare the eating quality of beef by sensory evaluation in France (F) and United Kingdom (UK) according two endpoints grilled cooking temperatures, 55 °C usual in France or 74 °C usual in UK. Muscle (Longissimus thoracis) samples came from 33 young bulls of three breeds of different earliness (Angus, Limousin, Blond d'Aquitaine). Each loin sample was aged 14 days, frozen and stored (-20 °C) until sensory evaluation at 55 or 74 °C in France and UK. The taste panel comprised 10 (UK), 12 (F) trained assessors. Panel members rated tenderness, juiciness, beef flavour intensity, abnormal flavour intensity, residue and overall liking on a continuous scale from 0 to 10. All data were analyzed by using the SAS GLM procedure. The model included the fixed effects: country (C), cooking temperature (T°), CxT° interaction and animal. Tenderness scores were significantly different between F and UK; and at 74 °C endpoint temperature steaks were tougher in both countries. Juiciness was the highest in UK at 55 °C and the lowest in France at 74 °C. Beef flavour was higher in France than in UK and higher at 74 °C than at 55 °C in UK. Abnormal flavour was the highest in France at 74 °C and in UK at 55 °C perhaps according to different consumer preferences associated with different cooking habits in the two countries. Residue evaluation was significantly higher in UK than in France, it was also higher at 74 °C than at 55 °C. Nevertheless overall liking scores were low in both countries (2.3-3.3), significantly different between France and UK, and the lowest in France at 74 °C.

Session 41 Poster 16

Dietary factors affecting butyric spores content in dairy cattle faeces in the Parmigiano Reggiano district

Bani, P., Minuti, A., Calamari, L. and Bertoni, G., Istituto di Zootecnica, Universita Cattolica del Sacro Cuore, Via Emilia Parmense, 84, 29122 PIACENZA PC, Italy; giuseppe.bertoni@unicatt.it

Trends in consumers preferences indicate an increasing attention to food quality and safety, as well as to sustainable production and ethical working practices. Consumers attractiveness of PDO label seems variable through countries and in some cases consumers demonstrate to pay more attention to foods brand, felt as guarantee of higher quality, than to PDO label. In both cases manufacturers are induced to pay attention to the quality of their products. For long ripening cheese, like Parmigiano-Reggiano, late blowing caused by butyric clostridia represents a major negative issue in the production process. To avoid them, in the Parmigano Reggiano productive chain dairy cows are fed dry or fresh forages only, without any possibility to use (cheaper) silages, as they are considered a main source of butyric spores. Furthermore, no anti butyric agents like lysozyme foreseen for other PDO cheeses production, are allowed. This suggests a particular attention to monitor any change in the milk-cheese productive chain that could impair this old but fragile equilibrium. With this background a research was carried out to assess the impact of some dietary aspects on the presence of butyric spores in faeces – vehicle to milk contamination – of dairy cattle bred in the Parmigiano Reggiano district: farms feeding moderate vs. high quantities of concentrate and farms with traditional feeding technique vs. TMR. Chemical characteristics of rations and feces were controlled, as well as butyric contamination of feeds and feces. As result, spores excreted in the feces exceeded those ingested, but the increase was neither significantly influenced by dose of concentrate used nor by the adoption of TMR. Dietary starch content, however, appeared as a risk factor that must be further studied. Modern evolution in dairy cows' feeding strategy is worth to be monitored to assess its possible effects on risk of butyric milk contamination.

Session 41 Poster 17

Effects of caponization on growth performance and carcass composition of broilers and male-layers

Symeon, G.K.[1], Mantis, F.[2], Bizelis, I.[1], Kominakis, A.[1] and Rogdakis, E.[1], [1]Agricultural University of Athens, Animal Breeding and Husbandry, 75, Iera Odos, GR 118 55, Athens, Greece, [2]Technological Educational Institution of Athens, Food Technology, EGALEO, GR 122 10, Athens, Greece; symeon@aua.gr

The aim of the present study was to evaluate the effect of caponization on growth performance and carcass composition of broilers and male layers. For the purpose of this study three experiments were conducted. In the first two, a medium growth broiler hybrid (Redbro) was used and the birds were reared until either the 18th (exp 1) or the 24th (exp 2) week of age. In the third experiment, male-layers (Lohmann Silver) were used and they were reared until the 34th week. For the evaluation of growth performance live weights and feed intake were recorded weekly. Five slaughters were performed in experiment 1 (6, 9, 12, 15 and 18 weeks of age), one slaughter in experiment 2 (24 weeks) and three slaughters in experiment 3 (26, 30 and 34 weeks). At slaughter, cold carcass weight and carcass yield was recorded along with edible viscera weights, carcass parts weights and tissue weights (muscles, bones and skin plus visible fat) of the thigh, drumstick and breast. The data were analyzed using GLM procedures fitting the treatment as the fixed factor. Caponization did not affect final live weight and feed intake in both hybrids. On the contrary, caponization resulted in an increase of fat deposition in terms of abdominal fat and of skin plus any visible fat of the commercial parts. Also, a significant decrease of foot muscle weight was recorded ($P<0.05$) along with a tendency for increased breast muscle weight. These differences were dependent on the hybrid and the age at slaughter. Specifically, they were more established on the male-layers compared to broilers while they were more pronounced with increasing age. Conclusively, caponization could be used for the production of 'special quality' chicken meat without harming production efficiency

Session 42 Theatre 1

The role of horses in Greek antiquity: myth, normal life, religion and art

Karakitsou, E.[1], Tahas, S.[2], Fragkiadaki, E.[3] and Xylouri, E.[3], [1]Archeologist, Kapandriti Attikis, 19014, Attiki, Greece, [2]Aristotle University of Thessaloniki, School of Veterinary Medicine and Surgery, Thessaloniki, 54124 Thessaloniki, Greece, [3]Agricultural University of Athens, Faculty of Animal Science and Aquaculture, Iera Odos 75, 11855, Athens, Greece; efxil@aua.gr

The aim of the present paper is to present the role of horses in Greek antiquity regarding myth, everyday life, religion and art. Horses have been well known as sym-bols of power, wealth status and social image from antiquity up to our days. Horse domestication started in South Russia as early on as the fourth millennium B.C. be-cause the horse proved a faithful companion for man in his life and, sometimes, in his death as well as a symbol of Olympian Divine. The first attempt of domestication served the target of meat and milk production. Sequentially, the horse's presence was directly associated with the chariot, its supporting people's movements and contribut-ing to conquest of new regions. The burial of horses next to their owners is a practice that was found sporadically in many areas around the Greek World, even since the Mycenaean era. The significance of horses in every-day human life is recognized by the fact that important authors in Ancient Greece dedicated monographs to horses. Two major Gods of Ancient Years, Athena and Poseidon, in their mythic course, were associated with horses. Athena's relation with her horse was close and multi-valued, as she domesticated horses and discovered the headstall. She was wor-shiped as the Horse and Headstall Goddess. In Thessaly, Ancient Greeks considered that Poseidon was responsible for the first appearance of horses after hitting a rock with his trident and this is why he is also referred to as 'Rocky'. According to Homer, in T 400-424, Achilles, the famous Greek Hero of Troy, is presented to have talked lovingly to his horse Xanthos that predicted his owner's death.

Session 42 Theatre 2

The role and potential of equines in a sustainable rural development in Europe: social aspects

Miraglia, N.[1], Saastamoinen, M.[2] and Martin-Rosset, W.[3], [1]Molise University, Department of Animals, Vegetables, Environmental Sciences, Via De Sanctis, 86100 Campobasso, Italy, [2]2MTT Agrifood Research, Equine Research, Opistontie 10, 32100 Ypäjä, Finland, [3]INRA, Department of Animal Husbandry and Nutrition, Theix, 63122 Saint Genès-Champanelle, France; miraglia@unimol.it

Equines are currently playing a considerable role in the context of rural sustainable development. Their multi-purpose uses represent a strong advantage in the territory management if compared to other species. Horse husbandry contribute the diversification of agricultural activities and the utilization and preservation of extensively cultivated and natural areas, as well as development of agritourism on horse farms. Such activities are more and more linked to the maintenance of population in rural areas, to new relationships between citizen and cultural rural life and consequent preservation of traditional socio-cultural life. Another important aspect concerns the socio-economic issues arising by this agricultural context where equines are involved. This report highlights many of these issues referred to different European countries, from Northern to Central and Mediterranean countries emphasizing the diversity of the 'equine culture' and the equine-related activities focused to the leisure and tourism activities, the relationship between urban citizen and cultural rural life, the preservation of rural socio-cultural life and the most relevant socio-economic issues.

Rural Sociology with a focus on human/horse relationship: a contribution of the Equine Landscapes Research Network

Evans, R.[1], Kjäldman, R.[2] and Couzy, C.[3], [1]Integrate Consulting; University of Highlands and Islands., 67 Metal Street, CF24 0LA, United Kingdom, [2]Sosiaalipedagoginen Hevostoiminta, Espoo, Espoo, Finland, [3]Institut de l elevage, Agrapole, Lyon, France; ritva.kjaldman@gmail.com

The first ever Rural Sociology Working Group with a focus on human/horse relations was held at the European Society for Rural Sociology in Vaasa Finland August 2009. The topics, which were presented from 10 countries, ranged from highly practical grounded policy issues to explorations of human-animal relations through the lens of the human/horse partnership. Over the course of five Workshops, the papers explored five themes: 1) Performing Equine Landscapes: the symbolic and material production of time and space. 2) Equine Landscapes of economic and social development: consumption of and in rural spaces. 3) Gender & meaning: construction of the gendered self through equine activities. 4) Human-animal interaction: Embodiment; Learning, Discipline. 5) Equine Landscapes: therapeutic landscapes and landscapes of recovery and recuperation As human geographers, anthropologists, rural sociologists, agricultural economists and development specialists, it was agreed that the study of human/horse relations can contribute something special and unique to our fields of study and out of this, came the determination to design the Equine Research Network (www.integrateconsulting.co.uk/eqrn) to carry on the collaborations and networking forward. This Report will detail some of the topics covered and update the progress of the Equine Research Network.

Horse and territory: Equidae holders' relationship with space

Vial, C.[1], Soulard, C.[2] and Perrier-Cornet, P.[2], [1]Institut Français du Cheval et de l Equitation, Campus Supagro-INRA, bâtiment 26, UMR MOISA, 2 place Pierre Viala, 34060 Montpellier, France, [2]Institut National de Recherche Agronomique, Campus Supagro-INRA, 2 place Pierre Viala, 34060 Montpellier, France; vialc@supagro.inra.fr

In France, equestrian activities have been growing since the 1990s. This development takes place in rural and suburban areas, while these spaces are changing. These modifications are creating an ongoing space need for city expansion and leisure activities whereas agricultural functions are retreating. In this context, equestrian activities appear to be a good indicator of the new territorial dynamics. As a result we wonder what role equestrian activities play in rural development and which relationships Equidae holders implement in the space? Two suburban and two rural areas were studied in France. A logistic regression based on existing databases was carried out in order to characterize the presence of Equidae. Then local exhaustive inventories of Equidae, their owners and the spaces they occupy were achieved. Afterward 27 Equidae holders were investigated so as to understand the different strategies they implement to use spaces. The statistical analysis provides evidence that the Equidae presence would be mainly linked with the residential development. At the local level, the findings show that Equidae generally occupy between three and four percent of the total surface of the studied areas. The importance of equestrian leisure activities among the horse chain is pointed out as well. The qualitative analysis of inquiries highlights different strategies implemented by Equidae holders, which range from a minimum domestic insertion to that of a rural entrepreneur according to a territorial insertion gradient. The use of agricultural lands by non-farmers residents is well supported by the horse example. This issue is a growing concern in France. This research program is relevant to enable all stakeholders of the horse chain and of land management to understand the impact of equestrian activities growth on territorial development.

Mismatch between breeding elite sport horses and rider amateurs expectations

Couzy, C.[1] *and Godet, J.*[2], [1]*Institut de l elevage, Agrapole, Lyon, France,* [2]*ISARA, Agrapole, Lyon, France; christele.couzy@inst-elevage.asso.fr*

One million and a half people are routinely riding in France; and 4.5 millions when occasional riders are included. Eighty percent of routine riders are amateurs. That raises yearly a market for 50,000 horses. Seventy percents are horses for leisure and education. Only 5% are bought for professional sportive competition (mostly jumping show) and 15% for 'amateur' competition. The challenge to be matched by new horse owners is to cope with breeders who are keen to breed elite horses as additional value is expected of course. And there are twice more breeders devoted to sport horse production than breeders oriented to leisure horse production. The aim of our study, carried out using breeders' interviews, raises the main reasons which could highlight the situation. The production of elite sport horses is based on an expensive breeding management, long term profitability only... but also on a lot of dreams related to hazard of the breeding systems. The breeders are in a vicious circle of excellence, for economical, cultural and psychological reasons. However few examples show that other production strategies are relevant and possible as far as the initial aim of the production is definitely fixed in the early stage in respect of the demand, and the management is based on the new knowledge drawn from equines sciences to reduce the cost of production. For sure these new breeding systems might raise more regular gain, less frustration but perhaps less pride?

The REFErences network, an actor in the economic knowledge of the french equine industry

Dornier, X.[1], *Heydemann, P.*[1] *and Morhain, B.*[2], [1]*Institut français du cheval et de l équitation, Observatoire économique et social du cheval, BP3, 19231 Arnac Pompadour Cedex, France,* [2]*Institut de l élevage, 9 Rue de la Vologne, 54520 Laxou, France; xavier.dornier@haras-nationaux.fr*

In order to grow up, the French equine industry needs technical and economic information as diagnosis tools and to support public and private actors. The French Institute of Horse and Horse Riding, the French Livestock institute, Horse Councils and Chambers of Agriculture have partnered in 2006 to improve economic knowledge in forming the Economic Network of the Equine Industry (REFErences) which combines three complementary approaches. In macroeconomics, the network aims to produce indicators for measuring the impact of activities related to horses in terms of employment and occupation of territory. It works by economic modelling based on chronological series consolidated with data from nearly 50 partners. This database is supplemented with regional surveys conducted every five years on a range of agricultural and non-agricultural enterprises of the equine industry and some complementary punctual surveys. In microeconomics, the goal is to better understand the functioning of markets and equine enterprises. The network is developing techno-economic referentials based on a system of 265 farms, mostly breeding farms (sport, leisure, meat) and equestrian establishments followed several times a year. In parallel, complementary studies and research are led to define modelling methods and to improve sociological understanding about current trends. A research program focuses on explaining in particular the presence of horses in a given area and on understanding the organization of related actors. Another program focuses on the structural complexity of the saddle horse market in a view to improve its efficiency. The aim is to produce objective socioeconomic data promoting a common diagnosis and exchanges between actors in order to raise up the awareness and recognition of the importance of the equine industry within the agricultural world.

Resources center for work in the french horse industry

Doaré, S. and Mahon, G., French National Stud, Institut Français du Cheval et de l Equitation, équiressources, Haras National du Pin Le Tournebride, 61310 Le Pin au Haras, France; sylvie.doare@haras-nationaux.fr

In France, more than 70,000 people work in the horse industry. In spite of a well developed educational system, covering the entire territory, the observations of the French National Stud and the French Equine Economic Cluster show the difficulties to find employees. In order to help the horse industry to reduce this gap, the project 'Equi-ressources' was etablished in 2006. The partnership between public and private organisations led to create the structure based on a national professional network, in May 2007. 'Equi-ressources' main goal is to link demand and offer in job and internship's matter, and to inform the public about trades, diplomas and vocational courses. All the services are available on-line: www.equiressources.fr. For 3 years, the activities of 'équi-ressources' are growing regularly. During this period, the job centre had worked for 623 different employers, publishing 1,400 job opportunities in all sectors. The jobs treated are covering a wide range of qualifications from stable jobs to stud managers. At the end of 2009, we received more than 4,000 CV. These persons can be soon employed, looking for a new place or actually without job. Especially for these kinds of demands, the 'Equiressources' team of 4 job and training advisors are able to give personalized recommendations. The student's requests are more about trades, vocational courses and internships. A first level of information is given by the website (450 connections/day). If needed, they can ask for further details to the advisors who had answered to 2,000 requests since the beginning of the activity. Whole 'Equi-ressources' missions will lead it to develop in 2010 a national observatory for trade, job and courses in the horse industry. The production of these datas would be particularly important for the decision-makers.

Carving out a new role for equine companions: a theory for efficacy of equine supported therapy

Perkins, A., Carroll College, Psychology, 1601 North Benton, 59625 Helena, Montana, USA; draperkins@carroll.edu

Horses have played many important roles throughout the history of humankind. They have been a food source, provided important transportation, and assisted warriors in conquering or settling new lands. In modern times, horses moved into positions of recreation and competition. Many horses now occupy the role of family pet. A new and more novel role for the horse is sweeping the globe. Horses are being incorporated into both mental and physical health treatment plans or utilized in educational settings to support educational goals. Why the horse? What is it about the horse that is proving valuable for this new industry? This talk will pull together several well developed theories in order to present a new hypothesis that may provide a foundational perspective for this phenomenon. Drawing on concepts in biophilia, co-evolution, theory and mind, and finally the more recent discovery of mirror neurons this presentation will demonstrate how horses may be capable of interpreting human intentions and how humans interpret equine body language. The audience will have an opportunity to use 'audience response technology' (clickers) to participate in a non-verbal exercise reading horse facial expressions or body language. Responses from people with a great deal of horse experience will be compared to people with little or no horse experience.

Session 42 Theatre 9

The use of animal in therapies: a critical review

Grandgeorge, M. and Hausberger, M., Université Rennes 1, Laboratoire Ethologie Animale et Humaine ETHOS, Bat 25, Campus de Beaulieu, 263 avenue du Général Leclerc, 35000 Rennes, France; marine.grandgeorge@univ-rennes1.fr

A central issue of human-animal interactions has been, and continues, to be mainly focused on the emotional and positive aspects that fixe the imagination of both public and researchers. It is particularly highlighted in the use of animals in therapeutic situations that involved 'weaker' people. After a historical report and a clarification of the term 'animal assisted therapy', we propose to develop a critical review of these particular therapies using horses as well as other species (i.e. dog, dolphin). It is firstly required to establish why animals are widely considered as potential therapists. We examine the theoretical framework included the relaxing effect and the social substitute status that animals can have, as well as the simple non verbal interactions between human and animal developed. We secondly describe several empirical researches to illustrate the heterogeneity (e.g. people, animal used) as well as the weakness of these studies (e.g. methodological devices, subjectivity, case studies) with a focused on autistic disorders. Thus, clear scientific evidence is still lacking and a need arises to better understand the relations between people and animals especially in a reproducible therapeutic ways. A project is currently led by different scientific teams to explore finely the relations between children, both with typical and autistic development- and animals, according to different contexts. We here present some results that reveal, on the one hand, the importance of owning a pet in the development of children with autism. On the other hand, the practice of horse riding is explored both in a typical practice in an equestrian club and in a therapeutic way. The results show an impact on children behaviors with unfamiliar animals. Other results are also discussed in the light of animal assisted therapies, providing us some ways of thinking about methodology and studies in research and the applications of animals in assisted therapies.

Session 42 Theatre 10

The impact of socio-pedagogic equine-activities intervention on special education pupils with neurological disorders

Kjäldman, R.K., University of Helsinki, Behavioural Sciences, Hakjärventie 11 a, 02820 Espoo, Finland; ritva.kjaldman@helsinki.fi

The purpose of this study is to discover if socio-pedagogic equine-activity has an impact on special childrens' well-being and what is the role of the horse in that social healing process. In Finland there are a growing number of at-risk children. New models of rehabilitation are needed to work with these children in supporting their schooling, and mental and physical health. Previous studies have shown promising signs that horses have a considerable part when used in therapeutic process with humans. The frame of reference of socio-pedagogic equine-activity intervention is based on social education and the human-equine bond. The purpose of the method is to support the social growth and welfare of those children, youth and adults, who were at-risk, through activities in cooperation with their peers and horses. Human-equine bond research needs to include several requirements such as standardized activities, to be a longitudinal study, to have control groups, and to measure outcomes such as cognitive, emotional, and behavior. In this research there are two test groups receiving equine-activities, and four comparison groups. All the pupils in this study have neurological disorders. The measurements are done three times for each group, their parents and teachers. The questionnaires used in this research measure human-pet relationship, empathy, aggression, locus of control, and loneliness. Test groups participated in socio-pedagogic equine-activities program during eight weeks. They had standardized activities from EPIC training manual. During eight weeks of time two of the comparison groups receive ART-intervention (Aggression Replacement Training), which consists of three elements: behavior (social skills and new alternatives for unsocial behavior), feelings (self-control) and values of life (moral reasoning). The last two comparison groups receive no intervention, but they fill in the same questionnaires as many times as do the other groups.

The impact of equine assisted learning as an intervention aimed at promoting psychological and social well-being amongst young people with behavioural issues

Carey, J.V., Festina Lente Foundation, Old Connaught Avenue, Bray, Co Wicklow, Ireland; jillcarey@festinalente.ie

Equine Assisted Learning is a learning-centred experiential education process used to assist young people, adolescents, and adults in learning about themselves. It also allows people to develop tools and strategies for making good choices, facilitating new insights and awareness, develop problem solving and communication skills, build healthier relationships and improve self-confidence. This project examines the impact of equine assisted learning as an intervention to promote the psychological and social well-being of young people between 11-17 years of age with behavioural issues. Each programme consists of 8 sessions with each session lasting approximately one and a half hours. Sessions are facilitated by a facilitator and an equine specialist. Specifically the equine assisted service focuses on facilitating participants to develop communications, self-awareness, relationships and social skills. The project used a mixed methods research design including quantitative and qualitative approaches. Questionnaires assessing the key variables have been completed pre and post each programme. There is a follow up 3 months after completion of the programme, which involves questionnaires. Interviews exploring these issues are completed three months after the end of the programme with a random selection of participants. Participants are male and female and between the ages of 11 and 17, the parents/guardians and teachers of the young people. Participants are referred by schools, family resource centres and support groups with responsibility for young people. All the young people referred have presented with behaviours that currently exclude them or run the risk of excluding them from mainstream education and/or mainstream settings and/or whose behaviours have interfered with their ability to develop and/or maintain healthy relationships with their peers.

Session 42 Theatre 12

The Pindar Project: a psychiatric and veterinary multicentric research in Therapeutic Riding

Cerino, S.[1], Bergero, D.[2], Gagliardi, G.[3] and Miraglia, N.[4], [1]Federazione Italiana Sport Equestri, Therapeutic Riding Department, Viale Tiziano 74, 00196 Rome, Italy, [2]University Ofveterinary Medicine, Animal Production, Eipemiology And Ecology, Via L. Da Vinci 44, 10095 Grugliasco (To), Italy, [3]Sammarco Veterinari Associati, C. Da Coluonni 1, Benevento, Italy, [4]University Of Molise, Animal, Vegetables And Environmental Sciences, Via De Sanctis, 86100 Campobasso, Italy; s.cerino@alice.it

The Pindar Project idea was born because of in TR international field there are only a few indexed papers, based on a really scientific recognition. On the other side, anybody having something to do with TR, knows that the positive results of rehabilitative activities with horses are specifically proved. So, starting from some very interesting results achieved by the psychiatrists' ROME F Mental Health Department pilot project, who have been working with young schizophrenic patiens since 2008, the Italian Equestrian Federation Therapeutic Riding Department has begun the Multicentric Pindar Project research on national basis. Such a project concerns 100 psychotic patients, aged 18-40, diagnosed with starting schizophrenia, fitted in a program of psychiatric rehabilitation by hippotherapy lasting one year. This program, started in February 2010, expects to be monitored on Time 0 and Time F. At the same time, a veterinary research concerning the fifty horses used by these patients has started, too. Its aim is to verify the horses' behavior so to estimate their fundamental neurophysiologic parameters. This paper aims to point out the operational mode and the preliminary results available till now.

Session 42 Poster 13

The numbers of horses and the jobs they generate in France: methods and results

Dornier, X. and Heydemann, P., Institut Français du Cheval et de l Equitation, Observatoire économique et social du cheval, BP3, 19231 Arnac Pompadour, France; xavier.dornier@haras-nationaux.fr

The socio-professionals of the equine industry show a growing interest for indicators measuring the importance of the sector: the REFErences network (Economic Network of the Equine Industry) produces indicators on the number of horses and jobs in the French industry at national and regional levels. It works from the development of databases, supplemented by surveys. The SIRE database, regulatory basis of horse identification in France, has allowed for a systemic approach to horse numbering. Additionally, the spreading use of chipping, gradually made compulsory in France for the entire equine population between 2003 and 2008, identified nearly all horses present on the territory. Modelling of survival curves from a sample of registered deaths and exported chipped horses were used to estimate the number of horses out of the model after being chipped. As a result, at the end of 2008, 1,118,000 horses had been chipped and we estimate there are still 850,000 of them present in France, but also about 50,000 that have not been chipped yet. This number was distributed by age, activity and region. Besides, equine activities are regarded as agricultural since 2003. This sector generates many direct agricultural jobs but also many non-farm employment as indirect or induced activities. Agricultural activities are relatively well informed by the equine or agricultural databases, only breeding as a secondary activity requires further investigations. As for non-agricultural activities, jobs are more diverse and range from the management of horse-racing betting to the equipment suppliers. As the sources are less complete, estimations are based on methods of national or regional surveys. Thus, the French equine industry approximately amounts to 40,000 agricultural jobs and 35,000 non-agricultural FTEs.

Session 43 Theatre 1

Genetic improvement of dairy sheep in France: results and prospects

Astruc, J.M.[1], Lagriffoul, G.[1], Larroque, H.[2] and Barillet, F.[2], [1]Institut Elevage, BP 42118, 31321 Castanet-Tolosan, France, [2]INRA, UR631, 31320 Castanet-Tolosan, France; jean-michel.astruc@toulouse.inra.fr

The selection of the 5 French dairy sheep breeds is presently based on the usual quantitative genetics approach, both for milk production and udder functional traits, in the framework of open nuclei totalizing, in 2009, 302,112 ewes in AC official milk recording in 781 nucleus flocks and 531,661 ewes in D recording in 1,369 commercial flocks of the base population, with 480,905 animal inseminations (AI), and 715 AI young rams progeny tested. Moreover since 2002, a gene-assisted selection (GAS) has also been implemented to select for the ARR allele of the PrP gene in the aim to increase genetic resistance against classical scrapie and bovine spongiform encephalopathy (BSE). In a first part, we summarize genetic parameters for milk and functional traits estimated in the French dairy sheep breeds, and their evolutions in the last 20 years. In a second part, we describe the breeding objectives and their evolutions for each breed, as well as phenotypic and genetic gains focusing mainly on Lacaune and Manech red faced breeds to emphasise key points regarding nucleus and commercial flocks during the last 25 years. In a third part, organization and results of the PrP gene GAS are presented. Then, given different on-going projects, we describe new traits which could be selected for in the future for French dairy sheep breeds such as milk fatty acid composition, milk persistency and once daily milking ability, milk flow kinetics, or nematode resistance. Finally we present an overview of different genomic on-going projects in French dairy sheep breeds, using the Illumina Ovine SNP50 BeadChip, for testing marker-assisted selection (MAS)/GAS or genomic selection (GS). GS seems feasible at least for the Lacaune and Manech red faced breeds, thanks to the size of their nuclei using extensively AI and to the storage of the DNA/blood of the AI rams since the middle of the 90's, providing actual reference populations in the range of 2,000 to 5,000 progeny tested AI rams per breed.

Breeding and recording strategies in small ruminants in the USA

Notter, D.R., Virginia Polytechnic Institute & State University, Animal and Poultry Sciences, Dept. of Animal and Poultry Sciences, Virginia Tech, Blacksburg, VA 24061, USA; drnotter@vt.edu

The U.S. National Sheep Improvement Program (NSIP) provides animal recording and genetic evaluation services for sheep, meat goats, and alpaca. The program began in 1987 to provide within-flock genetic evaluations and expanded to across-flock evaluations in 1995. The system is supported exclusively by user fees. Breed coordinators support each of 8 major sheep breeds (Targhee, Katahdin, Polypay, Suffolk, Hampshire, Rambouillet, Dorset, and Columbia; ranked by number of ewes evaluated), a consortium of breeders representing a few flocks of several minor sheep breeds, Kiko meat goats, and a group of breeders of Huacaya and Suri alpaca. Designations of 'major' or 'minor' breeds reflect participation in NSIP, not impact on the industry. In 2009, NSIP processed sheep and goat records from 160 flocks and 9,800 adult breeding females. Data are transferred from breeders to the breed coordinators (who perform initial data editing) and then to the Genetic Evaluation Center at Virginia Tech and breeding value predictions are returned once or twice per year using Excel spreadsheets. Major trait categories include direct and maternal effects on growth (weights at birth, weaning, postweaning, and yearling ages), wool production (fleece weight, fiber diameter, and staple length), and reproduction (numbers of lambs born and weaned per ewe lambing); reporting is mandatory only for number of lambs born and weaning weight. NSIP is not involved in milk recording. Research and development activities include development of breeding objectives in Targhee, a ewe productivity index in Polypay and Katahdin, and breeding value predictions for carcass traits (based on ultrasonic measurements in live animals) in Suffolk, fecal egg counts (an indicator of resistance to gastrointestinal parasites) in Katahdin, and detailed measures of fleece quality in alpaca. Strategies for long-term sustainability of programs such as NSIP in the absence of public funding will be discussed.

Using test-day models for the estimation of breeding values for milk quality traits in the Latxa sheep breed

Ugarte, E.[1], Afonso, L.[2] and Goñi, A.[2], [1]Neiker-Tecnalia, Animal Production, Campus Agroalimentario. Arkaute, 31006, Spain, [2]Universidad Pública de Navarra, Animal Production, Pamplona, 31006, Spain; eugarte@neiker.net

In the breeding program of the Latxa dairy sheep breed, estimated breeding values (EBVs) are calculated for production traits, both quantity and quality, using lactation models, i.e. lactation average values. Due to the high cost, milk quality is recorded only during the central part of lactation. In order to ensure good estimations of the average values, records have to meet certain requirements regarding the number of controls, as well as their distribution over lactation. As a result of this editing process, the number of lactations finally used for genetic evaluation for quality traits is around 70% of the total number of recorded lactations for these traits. Moreover, only records of herds in which the AC control method is applied are used representing 40% of the flocks under quantitative recording. The use of test-day models could allow obtaining higher technical returns to quality recording, as it would allow increasing the amount of valid records. The objective of this study was to analyze whether the application of test-day models allows evaluating more animals and improving the accuracy of the estimates. Univariate animal models using repeated records were applied. The results show that, although the application of test-day models enabled an increase in the number of records being used, the estimates of variance components were lower than those obtained when using lactation models. The estimates of heritability for fat and protein percentage were 0.10 and 0.30, versus 0.14 and 0.38 in lactation models. Pearson and Spearmen correlation coefficients between the EBVs obtained with both models were very low (around 50%) if males and females were considered in the analysis. The correlation coefficients were higher when comparing only males (between 0.68 and 0.80 depending on the trait and male type). In the case of milk production, changes do not seem to be important.

Session 43 Theatre 4

Estimation of genetic parameters for body weights of Kurdish Sheep in different ages using multivariate animal models

Shokrollahi, B. and Zandieh, M., Islamic Azad University, Department of Animal Science, Sanandaj, Iran; Borhansh@yahoo.com

Kurdish sheep are indigenous to the Kurdistan and Khorasan provinces of Iran, and highly valued for their meat production. Genetic parameters and (Co) Variance components were estimated by restricted maximum likelihood (REML) procedure, using 6 different animal models for body weight at birth, 3, 6, 9 and 12 months of age in a Kurdish sheep flock. Direct and maternal breeding values were estimated using the best linear unbiased prediction (BLUP) by DF-REML software. The data used in the present study, collected from one research flock of Kurdistan Kurdish sheep, included 2476 animals during the period of 1992 to 2002 (11 years). The Log L were estimated for all of the six models. Direct heritability values for the traits of this study based on the model with the highest Log L (best model) were 0.161, 0.233, 0.260, 0.091, and 0.122, respectively. Furthermore, maternal heritabilities (m^2) based on the best model were 0.238, 0.023, 0.014, 0.005, 0.004, respectively. Genetic correlations between direct and maternal effects for body weight at birth, 3, 6, 9 and 12 months of age were 0.0351, 0.5911, 0.7427, 0.7968 and 0.5735, respectively. Direct additive genetic correlations between birth weight and body weight at later ages were positive and relatively high, ranging from 0.32 to 0.79, whereas they were positive and high between weaning and later weights, and 6 months with yearling weight, ranging from 0.39 to 0.76. Genetic progress for body weight atn birth, 3, 6, 9 and 12 months of age during 11 years period were 12.42, 223.54, 466.42, 89.67 and 102.67 g, respectively.

Session 43 Theatre 5

Growth performances and carcass characteristics of black goats in the rural commune of Aït Bazza, Morocco

El Amiri, B.[1], Nassif, F.[1], Araba, A.[2] and Chriyaa, A.[1], [1]INRA, Centre régional de Settat, Productions Animales, BP 589, 26000, Settat, Morocco, [2]IAV Hassan II, Productions et biotechnologies animales, BP 6202, Rabat-Instituts, 10101, Rabat, Morocco; bouchraelamiri@hotmail.com

Aït Bazza is one of the middle Atlas rural communes located in Boulemane province in the northern central part of Morocco. In this area, goats are an important source of animal proteins. However, the goat production is based mainly on extensive systems, characterized by low productivity. In addition, the greatest percentage of goat populations is maintained by small farmers living in extreme poverty. In these conditions, animal nutrition depends on pasturing of natural prairies and forests. Yet very little is known about the animal growth and carcass characteristics. Thus, the objective of the present study was to investigate growth performances and carcass characteristics of black goats under traditional raising systems in Aït Bazza. Two parallel experiments were performed. The first one concerned the monitoring of growth performances under extensive management. Two farms were chose randomly and a flock of 20 adult animals were monitored. Additionally a flock of 18 Barcha population was introduced as a control. The second was carried out in the small slaughterhouse of Aït Bazza (Marmusha) from July 2007 to July 2009. Goats are slaughtered over a wide range of body weights and ages, from kids to old bucks and does. A total of 168 carcasses were studied. Hot carcass and non carcass components were weighed immediately and carcass measurements (carcass length, breast width, long leg length and long leg width) were recorded. Then, ratios such as carcass compactness, dressing percentage were calculated. The results showed that the growth performances from birth to 90 days are very low varying from 1.8 kg at birth to 10.5 kg at 90 days. Female kids were significantly lighter than male kids. The dressing percentage varies from 45.5 to 50%. The analysis still in progress to evaluate meat quality.

Session 43 Theatre 6

Performance of Awassi lines in Bedouin sheep flocks in the Negev, Israel

Al Baqain, A.[1], Herold, P.[1], Abu Siam, M.[2], Gootwine, E.[3], Leibovich, H.[4] and Valle Zárate, A.[1], [1]Animal production in the Tropics and Subtropics, Garbenstr. 17, 70599 Stuttgart, Germany, [2]Siam Veterinary Clinic, POB 519, 85357 Rahat, Israel, [3]ARO, The Volcani Center, POB 6, 50250 Bet Dagan, Israel, [4]Extension Small Ruminant Production, POB 67, 42905 Moshav Udim, Israel; Brigitte.Anna.Al.Baqain@uni-hohenheim.de

Crossbreeding with new genotypes is widely used by Bedouins in the Negev in an attempt to increase productivity of their flocks. But the suitability of these new lines under the harsh environment and different production conditions is not known. An on-farm-performance testing in 14 Bedouin farms in the Negev in Israel was therefore started in 2007. A total of 2416 ewes were ear-tagged and recording was done up to 2 years. Afec Awassi (BB) rams were distributed in 7 Bedouin flocks with recording of birth and weaning weights of lambs. Lambing rate of the local Awassi ranged by farm from 63% to 100%, with Assaf and crossbred ewes showing similar results. Prolificacy was significantly affected by farm, breed and parity. Afec Awassi (B+) ewes had a significantly higher ($P<0.05$) number of lambs born per ewe lambing (LB/EL) and lambs born alive per ewe lambing (LBA/EL) with 1.74 LB/EL and 1.57 LBA/EL, than Assaf ewes (1.34 LB/EL; 1.24 LBA/EL), Assaf crossbred ewes (1.27 LB/EL; 1.17 LBA/EL) or local Awassi ewes (1.18 LB/EL; 1.08 LBA/EL). Highest prolificacies and lowest lamb mortalities at birth were found in farms with good feeding and high labor input. The mortality rate of lambs at birth ranged by farm from 1.3% to 17.2% and from birth to weaning between 5% and 23.3% and was similar between breeds. Significant differences ($P<0.05$) in litter weight of lambs at birth and at weaning were found between B+ ewes and local Awassi ewes with an average of 7.8, 31.3 kg and 5.3, 18.2 kg respectively. Thus, B+ ewes outperformed the local Awassi in weaned kg per ewe. Due to the very restricted number of farms with B+ ewes in this survey, further studies of the performance of B+ ewes under all production conditions are suggested.

Session 43 Theatre 7

EC Reg no. 852/2004: use of check list in ewes dairy farm in Tuscany

Brajon, G.[1], Dal Prà, A.[1], Ragona, G.[1], Bacci, M.[2], Baroni, N.[2], Funghi, R.[2], Vittori, M.[2], Bendinelli, R.[2] and Capelli, V.[2], [1]Istituto Zooprofilattico Sperimentale delle Regioni Lazio e Toscana, Sezione di Firenze, Via Castelpulci, 43, 50018 - Scandicci, Italy, [2]Associazione Produttori Pastorizia Toscana, Via della Villa Demidoff, 64/E, 50144 - Firenze, Italy; giovanni.brajon@izslt.it

The authors verified if the check lists are applicable that are adopted in Tuscany for the milk production of ewes implementing Annex I part A of the EC Reg. N. 852/2004. Every six month check lists are filled out by the farmers that outline the dairy farm towards the fulfillment of the above mentioned EC rules. The framework of the check list is based on a series of questions that form 'thematic blocks' dividing the productive process of the farm in four parts: 1 health and animal welfare, 2 nutrition and feeding, 3 environment, 4 milking hygiene and milk cold storage. Every question of the check list that defines the conformity of the farm to a specific rule that can be objectively checked can be answered with 'yes', 'no' or 'not found'. The check lists were tested on 67 dairy sheep farms in Tuscany with the help of six technicians of the sheep breeder association (A.P.P.T.) in order to find out if the questions are relevant. The collected data has been analyzed by student's T test. The results showed that questions concerning the structure of the plants were satisfactory. Some difficulties were found concerning mandatory recording and farm management (Annex I part A of the EC Reg. N. 852/2004); there has been a remarkable amount of 'not found' answers (37%). The check lists were subsequently revised to allow an easier use on the farm.

Session 43 Theatre 8

Community-based breeding: a promising approach for genetic improvement of small ruminants in developing countries

Tibbo, M.[1], Sölkner, J.[2], Wurzinger, M.[2], Iñiguez, L.[1], Okeyo, A.M.[3], Haile, A.[4], Duguma, G.[2,4], Mirkena, T.[2,4] and Rischkowsky, B.[1], [1]ICARDA, P.O. Box 5466, 5466 Aleppo, Syrian Arab Republic, [2]BOKU, Gregor-Mendel-Str. 33, A-1180 Vienna, Austria, [3]ILRI, P.O. Box 30709, 0108 Nairobi, Kenya, [4]ILRI, P.O. Box 5689, 5689 Addis Ababa, Ethiopia; m.tibbo@cgiar.org

Within-breed selection programs based on developed country approaches, and importation of exotic breeds for crossbreeding, have generally failed in developing countries. Main reasons for the failures were lack of involving or inadequate participation of livestock keepers in the design and implementation of the programs in addition to due considerations for infrastructural and institutional arrangements. This paper summarizes important steps in designing community-based breeding program that suits the conditions of communities and the needs of the poor livestock keepers, and discusses preliminary findings from a pilot sheep breeding project in Ethiopia. The steps include characterization of the production system and the breeds, assessment of constraints to access to inputs and market services, definition of breeding goals, assessment of alternative breeding strategies, implementing breeding programs, and assessing impacts of genetic change. Methods such as participatory rural appraisal (PRA), formal surveys, choice experiments, ranking of animals from own flock and ranking of animals of other farmers, and simulation studies were used. The approach ensuring full participation of livestock keepers allowed the identification of breeding goals and selection criteria, and the most appropriate and acceptable breeding strategy for four breeds in Ethiopia. At present 500 smallholder households owning about 8,000 sheep are registered in the program. The first indication of success has been the implementation of participatory selection of the first sets of best replacement rams based on performance records and dam history. Capacity building to improve the ability of communities and researchers to manage the breeding programs was embedded in the project.

Session 43 Theatre 9

Sequencing of prion protein gene (PRNP) in East Adriatic sheep populations revealed new non-synonymous polymorphisms linked to ARQ allele

Cubric-Curik, V.[1], Kostelic, A.[1], Feligini, M.[2], Ferencakovic, M.[1], Ambriovic-Ristov, A.[3], Cetkovic, H.[3] and Curik, I.[1], [1]Faculty of Agriculture University of Zagreb, Department of Animal Science, Svetosimunska 25, 10000 Zagreb, Croatia (Hrvatska), [2]Istituto Sperimentale Italiano Lazzaro Spallanzani, Laboratorio di Epigenomica Applicata, Viale Giovanni XXIII 7, 26900 Lodi, Italy, [3]Rudjer Boskovic Institute, Division of Molecular Biology, Bijenicka 54, 10001 Zagreb, Croatia (Hrvatska); vcubric@agr.hr

Polymorphisms of PRNP are known for scrapie susceptibility in sheep populations. The genotypes of 780 sheep, sampled along all Eastern Adriatic, were determined for a 628 bp long fragment in PrP exon 3 gene. In the scrapie susceptible locus (codons 136-154-171), alleles ARQ (0.566±0.013), ARR (0.260±0.012), AHQ (0.106±0.008), VRQ (0.057±0.006), ARH (0.011±0.003) and ARK (0.001±0.001) were found. Overall, the frequencies of ARR/ARR, VRQ/VRQ and ARQ/VRQ were 0.097±0.007, 0.051±0.002 and 0.064±0.003, respectively. We also observed three new non-synonymous alleles caused by mutations at codons 145 (G145GGC→V145GTC), 160 (Y160TAC→D160GAC) and 185 (I185ATC→T185ACC) as well as at previously reported mutations at codons 101 (Q101CAG→R101CGG), 112 (M112ATG→T112ACG), 127 (G127GGC→S127AGC), 137 (M137ATG→T137ACG), 141 (L141CTT→F141TTT), 143 (H143CAT→R143CGT), 175 (Q175CAG→Q175CAA), 180 (H180CAT→Y180TAT), 231 (R231AGG→R231CGG) and 237 (L237CTC→L237CTG). The most frequent were alleles R231CGG (0.159), L237CTG (0.159) and T112ACG (0.041), while other mutations were rare (<0.040). All three new mutations were confirmed by cloning. Linkage disequilibrium, composite genotypes and cloning suggested that additional non-synonymous polymorphisms are linked only to ARQ allele. Thus, decrease of ARQ alleles would lead to elimination of non-synonymous variability at PRNP (exon 3) which potentially might be beneficial.

Fertility and prolificacy traits of Chios and Farafra sheep under Subtropical conditions in Egypt

Hamdon, H.[1], Abd El Ati, M.N.[2] and Allam, F.[2], [1]Faculty of Agriculture, Animal Production & Poultry Dept., Sohag University, Sohag, 82786, Egypt, [2]Faculty of Agriculture, Animal Production & Poultry Dept., Assiut University, Assiut, 71526, Egypt; hamdon9@yahoo.com

Eight hundred and twenty five Farafra and two hundred and five Chios ewes were used for comparison during the experimental period (two years including six mating seasons). Ewes lambed per ewe joined (EL/EJ), estimates of lambs born per ewe joined (LB/EJ), lambs weaned per ewe joined (LW/EJ), kilogram born per ewe joined (KB/EJ) and kilogram weaned per ewe joined (KW/EJ) were recorded and calculated. Results showed that EL/EJ in Farafra ewes was significantly (P<0.01) higher than Chios ewes (0.67 vs. 0.49). In Farafra ewes, LB/EJ, LW/EJ, KB/EJ and KW/EJ were 0.86, 0.72, 2.94 and 8.86, respectively. The corresponding values in Chios ewes were 0.63, 0.43, 2.36 and 5.71, respectively. The best reproductive performance was observed in September (0.71, 0.96, 0.78, 3.43 kg and 9.58 kg) followed by May (0.65, 0.80, 0.66, 2.72 kg and 8.12 kg), then in January (0.50, 0.62, 0.51, 2.10 kg and 6.46 kg) for EL/EJ, LB/EJ, LW/EJ, KB/EJ, and KW/EJ, respectively. It was observed that fertility traits increased significantly (P<0.01) up to 6 - <8 years old then decreased with advancing age. Chios ewes had slightly higher LB/EL (1.30) and KB/EL (4.83 kg) than Farafra ewes (1.28 and 4.39 kg). Meantime, Farafra ewes had higher LW/EL (1.08) and KW/EL (13.25 kg) than Chios ewes (0.89 and 11.71 kg). Litter size at birth and at weaning was better in February compared to October and June lambing seasons. February lambing season showed the best values of KW/EL (4.83 kg), and KW/EL (13.49 kg), while, October and June lambing seasons had 4.21 kg & 4.17 kg for KB/EL and 12.60 kg &12.81 kg for KW/EL, respectively. It can be concluded that fertility and prolificacy traits of Chios flock should be improved. Moreover, the selection program for Farafra flock should be continued.

Session 43 Poster 11

Morphological characterisation of Spanish Assaf dairy sheep breed

Cervantes, I.[1], Legaz, E.[2], Pérez-Cabal, M.A.[1], De La Fuente, L.F.[3], Mártinez, R.[4], Goyache, F.[5] and Gutiérrez, J.P.[1], [1]Facultad de Veterinaria, UCM, Ava. Puerta de Hierro, E-28040 Madrid, Spain, [2]CGSC, Campo Real, E-28510 Madrid, Spain, [3]University León, Universidad 25, E-24071 León, Spain, [4]Assaf, Zamora, E-49010 Zamora, Spain, [5]SERIDA, Camino de Rioseco 1225, E-33392 Asturias, Spain; gutgar@vet.ucm.es

Body measurements of Spanish Assaf dairy sheep breed (Assaf.E) have been analysed in order to define its morphological standard. Data consisted of 16 body measurements and weight from 341 individuals (61 males and 280 females) ranging from 2-5 years old and from 9 different regions (provinces). Additionally 8 udder measurements were collected in the period of maximum level of lactation. The body measurements included 7 widths, 5 lengths, 2 heights and 2 perimeters. Regarding the udder traits, 7 variables were measured and 1 was described using a linear scale (1 to 5). A general linear model was performed separately by gender using age and flock as fixed effects. The effect of the flock was significant in all variables in females except for the width of the ear and for the teat placement. Flock effect was also significant in 9 variables for the males. The age effect was significant for 13 variables in females but for only 5 measurements in males. The 66.9% and 75.7% of total phenotypic correlations between body variables for females and males were respectively significant, their absolute value ranging from 0.25 to 0.84. The udder variables showed little significant correlation with body measurements and only 25% of the correlations among udder traits were significant. The factor analysis grouped traits regarding animal size in the first factor, and the second and third factors grouped mainly udder traits. The Mahalanobis distance showed the flocks in Madrid where the Assaf was introduced via absorption with a different breed as the most differentiated. This analysis is expected to be useful in describing the morphological type in the Spanish Assaf population.

Evaluation of the effect of group of genetic resistance against scrapie on reproduction traits in chosen Polish sheep breeds

Niżnikowski, R., Czub, G., Strzelec, E., Głowacz, K., Popielarczyk, D. and Świątek, M., Warsaw University of Life Sciences, Sheep and Goat Breeding, Ciszewskiego 8, 02-786 Warsaw, Poland; roman_niznikowski@acn.waw.pl

The research was carried out on two Polish sheep breeds: Polish Heath (134 hds) and Żelaźnieńska Sheep (69 hds), bred at the Warsaw University of Life Sciences' Research Farm in Żelazna. The individual reproduction data of ewes were collected basing upon the breeding information books and allowed to calculate the individual indicators of fertility, prolificacy, lambs' survivability, average reared litter size and average reared litter size per ewe. The genetic polymorphism in the PRNP gene was established due to the single nucleotide polymorphism methods at 4 codons (136, 141, 154 and 171) and allowed to classify animals into the three groups of genetic resistance against scrapie: G1 (ALRR/ALRR), G2 (ALRR/ALRQ, ALRR/AFRQ, ALRR/ALRH, ALRR/ALHQ) and G3 (all alleles' combinations excluding the ALRR allele). The VLRQ and VFRQ alleles were not detected. The results were calculated using the LSM method in SPSS v.12 software. The effects of group of genetic resistance, number of reproduction seasons and ewe's birth type as well as the interaction: group of genetic resistance x ewe's birth type were used in the statistical model. The effect of chosen factors were evaluated with the F-test. Generally, the group of genetic resistance did not affected the reproduction traits in both sheep breeds. Among the other factors, only the effects of number of reproduction season on almost all examined traits (excluding the fertility indicator in both breeds and the average reared litter size per ewe in Żelaźnieńska sheep) was observed. The results led to the conclusion that it is possible to use the independent selection both in the aim of improving the reproduction traits and increasing the frequencies of beneficial PRNP alleles to obtain the genetic resistance against scrapie disease in flocks of Polish Heath Sheep and Żelaźnieńska Sheep without negative effect on reproduction traits.

Genetic parameters and breeding values for the four major sheep breeds in Switzerland

Burren, A., Hagger, C., Schneeberger, M. and Rieder, S., ETH Zurich, Institute of Plant, Animal and Agroecosystem Sciences (IPAS) TAN D 2, Tannenstrasse 1, 8092 Zurich, Switzerland; markus.schneeberger@inw.agrl.ethz.ch

Four major sheep breeds – White Alpine (WAS), Brown Headed Meat (BFS), Black Brown Mountain (SBS) and Valais Black Nose (SN) – with population sizes of more than 10,000 animals, are kept in Switzerland. During 2009, population parameters, direct and maternal breeding values were estimated for average daily gain up to 45 days of age, using REML and BLUP methodology. The data comprised 405,233 litters and 526,257 ancestors in the pedigree for WAS, 116,991 litters and 171,994 ancestors for BFS, 103,298 litters and 157,712 ancestors for SBS,123,087 litters and 152,896 ancestors for the SN breed. Trait records were taken from 1995 upwards, pedigree records from 1985 upwards. The statistical model accounted for fixed effects of sex of lamb, litter size, litter number*age of dam, age at weighing, lambing season and random effects of herd*year, permanent environment, direct component of the animal, maternal component and residual effect. Litter size was taken as a combination of total lambs born and number of lambs alive at 45 days of age. Litter size and age of dam were the two fixed effects with the largest influence on growth of lamb up to 45 days. Estimated heritabilities for the direct component were 0.21, 0.20, 0.22 and 0.27 for the WAS, BFS, SBS and SN breed, respectively. For the maternal component, estimated heritabilities were 0.17, 0.19, 0.16 and 0.15 for the WAS, BFS, SBS and SN breed, respectively. The genetic correlations between direct and maternal components were negative and differed among breeds. Estimated values for this parameter were -0.47, -0.38, -0.45 and -0.66 for the WAS, BFS, SBS and SN breed. The results seem plausible and correspond to estimates found in the literature for similar traits. The average breeding values per year of birth gave indications on the genetic trends of the two components. The genetic trends differed between breeds. They provide information on the selection practised within a breed.

Ultrasound measurements of lamb loin muscle and back-fat depth in the Czech Republic

Milerski, M., Research Institute of Animal Science, Přátelství 815, 104 00 Prague - Uhříněves, Czech Republic; milerski.michal@vuzv.cz

Genetic improvement of lamb carcass quality characteristics is one of the main goals of breeding programs for meat breeds of sheep in the Czech Republic. For that reason the *in vivo* ultrasound measurements of loin muscle depth (UMD) and fat and skin layer thickness (UFD) between 13th thoracic and 1st lumbar vertebra are carried out in lambs at 100 (±20) days of their age. At the same time the live weights of lambs are collected. Real-time ultrasound scanners with 5 MHz linear probes are used for ultrasound measurements. During years 1999-2009 totally 50362 lambs of Suffolk (SF), Charollais(CH), Oxford Down (OD), Texel (T), German Blackhead (GB) and Romney (RM) breeds were scanned in nucleus flocks. Breeding values for UMD and UFD are estimated using multitrait animal model. The models for both traits include heard-year, sex, number of lambs reared in the litter, age category of the dam, age and live weight of the lamb at the scanning and additive genetic effects. Relative weights of the traits measured by ultrasound in the selection indexes ranged between 16-27% for UMD and between 5-9% for UFD. In RM the traits measured by ultrasound are not included into selection index, nevertheless the EBV for UMD and UFD are published separately and might be also used by breeders for selection. Average UMD genetic gains per year during the period 2003-2009 was +0.12 mm/year (+0.14s.d.) for SF, +0.11 mm/year (+0.12 s.d.) for OD, +0.10 mm/year (+0.10 s.d.) for RM, +0.04 mm/year (+0.05 s.d.) for T and +0.02 mm/year (+0.03 s.d.) for BG. Average UFD genetic gains per year ranged from +0.055 mm/year (+0.04 s.d.) in RM to -0.015 mm/year (-0.02 s.d.) in OD.

Impact of selection for scrapie resistance on genetic diversity of the Sambucana sheep breed

Sartore, S., veterinary faculty, animal production, via leonardo da vinci 44, 10095 grugliasco, Italy; stefano.sartore@unito.it

The aim of the present investigation was to evaluate the consequences on the molecular diversity of Sambucana sheep breed (Piedmont, Italy) due to selection on the PRNP gene since 2005. The Sambucana breed had 0.319 and 0.544 allele frequencies for the ARR and ARQ alleles, respectively, before the PRNP selection programme started. It has a limited size (around 3000 animals), and the management of genetic variation is based on a rotational mating scheme. Two groups of 78 and 66 young rams were chosen to be representative of the genetic diversity. The first group included animals born before 2005, the second group included animals born in 2008-2009, after 4 years of selection. The PRNP gene and 14 microsatellite markers were genotyped for all the rams. The microsatellite were recommended by the FAO-ISAG and were considered as selectively neutral. For the PRNP gene, the selection for scrapie resistance increased the ARR frequency to 0.567 and decreased the ARQ frequency to 0.366. The impact of selection on neutral diversity was very low, according to information on the 14 microsatellite markers. Number of alleles, allelic richness, and expected heterozygosity showed no significant differences (t-test) between the two successive groups. The between-group diversity as estimated by the genetic distance of Reynold, appropriated for livestock populations with short-term divergence, was very small. The present investigation provides evidence of no overall genetic differentiation in a short-term perspective. Besides, the change of allele frequencies at the PRNP locus shows that the selection for scrapie resistance was less effective than in other breeds, over a few years.

Session 44 Theatre 1

Pre-requisites for genomic selection

Woolliams, J.A., University of Edinburgh, The Roslin Institute & R(D)SVS, Roslin, Midlothian EH25 9PS, United Kingdom; john.woolliams@roslin.ed.ac.uk

The community has begun to quantify from direct experience what is required for the successful application of genomic selection in dairy cattle, for example, how much accuracy in milk yield can be achieved from SNP chips of a particular density. However, can we predict what the future outcomes may be as more records are accumulated, the SNP chip changes, more traits are addressed, and selection based on genomic predictors progresses across generations in the population. Moreover to what degree and in what way can other livestock sectors - or silviculture, or aquaculture - learn from the experience of dairy cattle? There are key areas of structure and technology of the dairy industry that have underpinned success. For example: (i) improvement schemes generating highly accurate estimates of breeding values through progeny testing to allow for the testing, assessment and refinement of genomic techniques; (ii) the structure of the bovine genome, shaped in part by the selection history of dairy cattle, leading to linkage disequilibrium extending over relatively long distances across the genome; and (iii) the large community of scientists driving advances in bovine genomic technology developing large numbers of SNPs from relevant populations. Therefore to map the pre-requisites for genomic selection in other species there is a need to develop a set of scaling rules accounting for these different attributes of improvement structure, traits to be improved, the species genome and the properties of the SNP chips to be used. Theory exists to generate this scaling.

Session 44 Theatre 2

Accuracy of genomic selection using multi-breed reference populations

Gredler, B.[1], Pryce, J.[2], Bolormaa, S.[2], Egger-Danner, C.[3], Fuerst, C.[3], Emmerling, R.[4], Sölkner, J.[1] and Hayes, B.J.[2], [1]BOKU University Vienna, Department of Sustainable Agricultural Systems, Gregor Mendel Str. 33, 1180 Vienna, Austria, [2]Department of Primary Industries Victoria, Biosciences Research Division, 1 Park Drive, 3083 Bundoora, Australia, [3]ZuchtData EDV-Dienstleistungen GmbH, Dresdner Str. 89/19, 1200 Vienna, Austria, [4]Bavarian State Research Center for Agriculture, Institute of Animal Breeding, Prof.-Dürrwaechter-Platz 1, 85580 Grub, Germany; birgit.gredler@boku.ac.at

The objective of this study was to compare the accuracy of direct genomic breeding values (DGV) using single-breed, dual-breed, and three-breed reference populations. Animals included in the analyses were 1,141 Australian Holstein (HF), 364 Australian Jersey (JE), and 1,596 dual purpose Fleckvieh (FV) bulls. All bulls were genotyped for 54,001 Single Nucleotide Polymorphism (SNP) using the Illumina Bovine SNP50TM Bead-Chip. SNP-data quality checks were applied within each breed resulting in 36,986 SNPs. The HF and FV populations were split in a reference and validation set including 755 and 386 bulls for HF and 1,247 and 349 bulls for FV, respectively. Daughter yield deviations (DYD) for milk yield, protein yield, fat yield and fertility were used as response variables. For FV the 5 fertility traits non-return-rate after 56 days in heifers and cows, days to first service, and interval between first and last insemination in heifers and cows were available, whereas for HF and JE bulls only information on calving interval was assessed. To account for breed differences DYDs were corrected for breed effects. Methods used for genomic breeding value estimation were a BLUP approach with a genomic relationship matrix and BayesA which estimates single SNP-effects. First results show that there is no clear difference in accuracies using a multi-breed reference population compared to using a single-breed reference set for HF and FV. The JE bulls can be best predicted using a HF reference set. Accuracies of DGV for FV using a HF reference set are low, DGV for HF using FV reference set are slightly higher.

Session 44 Theatre 3

Including non-additive effects in Bayesian methods for the prediction of genetic values from genome-wide SNP data

Wittenburg, D., Melzer, N. and Reinsch, N., Leibniz Institute for Farm Animal Biology, Genetics and Biometry, Wilhelm-Stahl-Allee 2, 18196 Dummerstorf, Germany; wittenburg@fbn-dummerstorf.de

It is a challenging task to predict genetic values for traits in dairy cattle on the basis of genome-wide SNP markers. Methods including additive genetic effects have already been studied, but the importance of non-additive effects for the genetic variation is not fully understood. A better prediction of genetic values is intended, when additive, dominance and pairwise epistatic effects are jointly involved in fitting a model to a trait. In order to avoid the estimation of additional covariance components, the genotypic effects have to be appropriately re-parameterised in advance. Three methods of orthogonalisation were taken from the literature (e.g. Cockerham's model). The methods differ in whether they assume Hardy-Weinberg equilibrium or not. The re-parameterisation directly affects the design matrices of the additive and dominance effects. In case of including epistatic effects, it is necessary to standardise the genotypic effects to reach numerical stability. So far the Bayesian methods used for marker assisted prediction applied Gibbs sampling steps which require a lot of computing time. These sampling methods collapse for 50K (or more) SNP markers, if further non-additive effects are included. For some time an approximate Bayesian approach is available which applies the analytically derived posterior density for a marker effect rather than samples thereof. This approach is known to be slightly less accurate but much faster than the conventional Bayesian methods (e.g. BayesB). The fast Bayesian method is used to estimate genetic effects on the basis of simulated datasets. Different scenarios are simulated to study the loss of accuracy of prediction, if epistatic effects are not simulated but modelled and vice versa. This study shows that the fast Bayesian method is convenient for genetic value prediction, but it is rather sensitive concerning variation in default values for the hyper-parameters involved.

Session 44 Theatre 4

Evaluation of selection strategies including genomic breeding values in pigs

Haberland, A.M.[1], Ytournel, F.[1], Luther, H.[2] and Simianer, H.[1], [1]Georg-August-University Goettingen, Department of Animal Sciences, Albrecht-Thaer-Weg 3, 37075 Goettingen, Germany, [2]SUISAG, Allmend, 6204 Sempach, Switzerland; ahaberl@gwdg.de

The recent developments in animal breeding through the increasing amount of genomic data should also influence the breeding programs. In order to optimize the design of a pig breeding program, we thus investigated different selection strategies, particularly those including genomic breeding values (GEBVs) as selection criteria. A breeding population of 450 pigs of the Swiss terminal sire-line was modeled using ZPLAN+, a computer program based on selection index theory and gene flow method. This software deterministically quantifies parameters like genetic gain, generation intervals, and accuracies of the estimated breeding values as well as breeding costs for an investigated breeding program. GEBVs were assumed to be available for most of the 14 traits and were fitted into the selection index considering them as auxiliary traits correlated to the traits in the breeding goal. Heritabilities of the genomic traits were calculated depending on the reliability of their estimation for two sizes of the calibration group, namely 500 or 1000 animals. Because of the small population size of the breeding nucleus, a major proportion of very young boars is used as sires in order to control the risks of inbreeding. Hence, the selection decision is based only on information on own performance (daily gain and backfat thickness at a weight of 95 kg), performance of ancestors and performance of two full sibs tested on station. The accuracy (rAI) of the estimated breeding value (EBV) in this case is 0.56. This conventional selection scheme was compared to a scheme where selection was based exclusively on genomic information and to a scheme combining conventional and genomic information sources. Due to a higher accuracy of EBV (rAI(500)=0.72 and rAI(1000)=0.83), adding genomic information as information source to the selection index increases the monetary genetic gain by 27 to 47 percent, respectively.

Potential application of marker assisted selection to increase disease resistance in Canadian swine herds

Jafarikia, M.[1], Lillie, B.[2], Maignel, L.[1], Hayes, T.[2], Adler, M.[3], Wimmers, K.[3], Wyss, S.[1] and Sullivan, B.[1], [1]Canadian Centre for Swine Improvement, Ottawa, ON, Canada, [2]University of Guelph, Guelph, ON, Canada, [3]Research Institute for the Biology of Farm Animals, 18196 Dummerstorf, Mecklenburg-Vorpommern, Germany; mohsen@ccsi.ca

This study explored the potential impact of selecting for disease resistance genetic markers on economically important performance traits in Canadian pigs. Thirty-eight single nucleotide polymorphisms (SNPs) located in genes associated with disease resistance were studied. The numbers of animals sampled from across Canada were 302, 245, 245 and 64 for Yorkshire, Duroc, Landrace and crossbreds, respectively. The minor allele frequency of SNPs ranged from 0.03 to 0.48 with an average and standard deviation of 0.22±0.15. Most of the SNPs did not show any significant association with the performance traits included in this study. The level of genetic variability in the SNPs and lack of association with performance traits are important for potential application of Marker Assisted Selection (MAS) for disease resistance. The A allele in a SNP located in the Toll-like Receptor 4 gene (TLR4 G(962)A) and the G allele located in the Mannan-binding lectin A gene (MBL1 G(271)T) were associated with higher number born in Duroc and Yorkshire, respectively. Previous research has shown that these two SNPs are less frequent in diseased subgroups of animals. In contrast, while alleles T and C of the MBL2 G(-1636)T and MBL2 T(-2148)C SNPs were associated with faster growth in Landrace, they were more frequent in diseased subgroups. Allele T of the MBL1 C(273)T and MBL1 C(687)T SNPs were also more frequent in diseased subgroups and positively associated with number born in Yorkshire. Selection of alleles that have a positive effect on resistance to disease and also a positive or no effect on performance could potentially be beneficial. Further study is necessary to determine the magnitude of the economic effect of each SNP allele where the favourable allele for disease resistance is unfavourable for performance.

First screening of three major Canadian swine breeds using a high density SNP panel

Jafarikia, M., Maignel, L., Wyss, S. and Sullivan, B., Canadian Centre for Swine Improvement, Central Experimental Farm, Building #54, 960 Carling Avenue, Ottawa, ON, K1A 0C6, Canada; mohsen@ccsi.ca

The high density porcine SNP (single nucleotide polymorphism) panel made available to the swine industry in the past year has provided an opportunity to carry out genome-wide selection and quantitative trait loci mapping at closer marker distances. The quality of the SNP data depends on factors such as the distribution of SNPs across the genome, minor allele frequency (MAF), Mendelian segregation and Hardy-Weinberg Equilibrium (HWE). A total of 643 pigs were sampled from herds across Canada and genotyped using the Illumina porcine 60K SNP chip. From 64,232 genotyped SNPs, 14,555 (23%) were not yet mapped to specific chromosomes and 1,402 (2%) were located on sex chromosomes. About 4% of the autosomal SNPs were not in HWE ($P<0.01$) in at least one of the three breeds under study. Porcine chromosome 1, the largest autosomal chromosome, had the greatest number of SNPs (6,798 SNPs or about 10%). The rest of SNPs, 41,477 (65%), were distributed across other autosomal chromosomes with an average of 2,304 per chromosome, a minimum of 1,129 SNPs for chromosome 12 and maximum of 4,168 SNPs for chromosome 14. About 8% of SNPs had a MAF of zero, while 11% of SNPs had MAF greater than zero and below 0.10, 59% were between 0.10 and 0.40 and 22% had MAF above 0.40. The average MAF of autosomal SNPs within the chromosomes ranged from 0.23 to 0.28 with an average of 0.26. The current variation with relatively high MAF is promising for potential benefit from genomic selection.

Genotyping panels available in cattle and their properties

Druet, T.[1], Zhang, Z.[1], Coppieters, W.[1], Mulder, H.A.[2], Calus, M.P.L.[2], Mullaart, E.[3], Schrooten, C.[3], De Roos, A.P.W.[3] and Georges, M.[1], [1]Unit of Animal Genomics, University of Liège, 1 avenue de l Hôpital - B34 (+1), 4000 Liège, Belgium, [2]Animal Breeding and Genomics Centre, Wageningen UR Livestock Research, Edelhertweg 15, 8200 AB Lelystad, Netherlands, [3]CRV, Wassenaargweg 20, 6800 AL Arnhem, Netherlands; tom.druet@ulg.ac.be

Three years ago, genotyping arrays of approximately 50 thousands (50K) markers were developed in cattle. These tools allowed the breeders to apply genomic selection but also to implement fine-mapping studies much more rapidly. In 2010, chips having much more markers (600K-1000K) are expected to be launched for cattle, which will likely impact the cost of older 50K chips. In parallel, whole genome sequencing costs are steadily decreasing. Further, low density marker panels for screening larger portions of the cattle population at lower costs are under development. This paper will compare the properties of these different marker panels in terms of cost and genome coverage (measured as the linkage disequilibrium with ungenotyped SNPs). Both costs and prediction accuracy of missing markers (imputation) when using lower density panels determine which marker panel would be the most cost effective in genomic selection. With current marker panels (60K) it is possible to obtain an allelic imputation error rate close to 0.5% and results show that error rates further decrease with higher marker densities. On the contrary, allelic imputation error rates increase when using lower marker density panels: ~3% with 3,000 SNPs. The error rate is a function of the relationship between the animals genotyped on the lower density panels and the animals genotyped for all the markers. With both parents genotyped, allelic imputation error rates are close to 0.5%, even with only 3,000 SNPs (and still lower at increasing densities). Finally, the precision of genomic selection for animal genotyped on different chips will also be presented.

Mutation rate and number of QTL for genomic selection simulations

Casellas, J., Universitat Autònoma de Barcelona, Departament de Ciència Animal i dels Aliments, Facultat de Veterinària, 08193 Bellaterra, Spain; joaquim.casellas@uab.cat

Genomic selection has revolutionized animal breeding, although its performance has not almost been evaluated in real populations but in simulated scenarios. The values assumed for mutation rate (ρ) and the initial number of QTL (ξ) influence subsequent results, although a wide range of magnitudes has been used. This research was an attempt to define optimum values for ρ and ξ on the basis of the ratio between additive genetic (σ_a^2) and mutational (σ_m^2) variances, and the number of polymorphic QTL. Five different ρ (10^{-2}, 10^{-3}, 10^{-4}, 10^{-5} and 10^{-6}) were evaluated against 500, 1000 and 1,500 QTL spread in a 2500-cM genome with 25 chromosomes of equal size. For each combination of ρ and ξ, 100 populations with $N_e$modified σ =100 were simulated along 1000 non-overlapping generations, they expanding to 1000 individuals up to generation 1,003. New mutations were sampled from a gamma process with parameter 0.4 and multiplied by 1 or -1 with probability 0.5. Both ρ and ξ_a^2/σ_m^2 and the number of polymorphic QTL, although largest differences were due to ρ. Indeed, differences between 1000 and 1,500 QTL were not significant for σ_a^2/σ_m^2 ($P>0.05$). The expected σ_a^2/σ_m^2 under a neutral model with only mutation and drift operating is $2N_e$. The average value for σ_a^2/σ_m^2 under $\rho=10^{-4}$ and $\rho=10^{-5}$ agreed with this assumption ($P>0.05$), whereas smaller and larger ρ departed from this value ($P<0.05$). A moderate advantage was suggested for $\rho=10^{-4}$ due to the smaller dispersion of σ_a^2/σ_m^2 estimates. The average number of polymorphic QTL significantly ($P<0.001$) increased with $\rho=10^{-4}$values agreed with the expected number of loci reported in literature for quantitative traits (, although both ρ>50-100 loci). In conclusion, a $\rho=10^{-4}$ with ~1000 QTL could be advocated, although plausible scenarios may also be generated under alternative assumptions.

Choice of parameters for a single-step genomic evaluation

Misztal, I.[1], *Aguilar, I.*[1], *Legarra, A.*[2] *and Lawlor, T.J.*[3], [1]*University of Georgia, Animal and Dairy Science, Athens GA 30602, USA,* [2]*INRA, 32326, Castanet Tolosan, France,* [3]*Holstein Association, Brattleboro, VT 05302, USA; ignacy@uga.edu*

In a single step procedure, the pedigree-based matrix A is replaced by matrix H that blends pedigree and genomic relationships. The inverse of matrix H involves an expression $G^{-1} - A_{22}^{-1}$, where G is genomic relationship matrix and A_{22} is pedigree relationship matrix for genotyped animals. Two modifications to that expression: $(\alpha\, G + \beta\, A_{22})^{-1} - A_{22}^{-1}$ and $\tau\,(0.95\, G + 0.05\, A_{22})^{-1} - \omega\, A_{22}^{-1}$ were investigated with regard to accuracy and scale of genomic predictions. While the first one is equivalent to assuming a genomic and polygenic effect for genotyped animals, the second one is equivalent to assuming double prior for the additive effect. Data included final scores recorded from 1955 to 2009 for 6.2 million Holsteins, pedigrees for 10.5 millions animals, and SNP50k genotypes for 6,508 bulls. Analyses used a repeatability animal model. Comparisons involved R^2 and regression coefficients (REG) based on 2004 predictions of young bulls and their 2009 daughter deviations. REG below 1.0 indicates inflation of genomic predictions. The initial expression yielded R^2=0.41 and REG=0.75. With the first modification, varying α from 0.6 to 1.2 decreased R^2 less than 0.01 and decreased REG from 0.81 to 0.71. Increasing β from 0 to 0.6 decreased the R^2 and REG by 0.02 or less. With the second modifications, varying τ from 0.6 to 1.5 increased R^2 by about 0.02 and increased REG by 0.02 (ω=0) to 0.15 (ω=1.0). Decreasing ω from 1.0 to 0 decreased the R^2 by 0.03 and increased REG from 0.2 (τ=1) to 0.3 (τ=0). Parameters τ=1.5 and ω=0.4 yielded R^2=0.40 and REG=1.0. While the scale of G (parameters α and τ) has a small effect on R^2 and REG, matrix G as used here is about 50% too large. The scale of A_{22}^{-1} (parameter ω), which is associated with parental index based on genotyped bulls, has a large impact on inflation of genomic predictions.

The pig as a model for the metabolic syndrome in humans

Litten-Brown, J.C., University of Reading, School of Agriculture, Policy and Development, Earley Gate, P.O. Box 327, Reading RG6 6AR, United Kingdom; j.c.litten-brown@reading.ac.uk

The Metabolic Syndrome consists of a cluster of risk factors for type 2 diabetes and cardiovascular disease which are generally classified as a combination of insulin resistance, central obesity, raised plasma triacyleglycerol (TAG) concentrations, reduced high-density lipoprotein (HDL) cholesterol and hypertension. Due to the varying definitions the global prevalence is hard to determine effectively but is currently estimated at 16% however the figure is growing at an alarming rate. This multi-factorial scenario has historically been studied in the rat model and much has been learnt about the prevention and treatment of the disease. Humans suffering from the metabolic syndrome regularly exhibit three or more of the clinical signs simultaneously, however it is rare that the rat model exhibits this many. Hence the differences in the metabolism and physiology between rats and humans are too great and an alternative model has been sought. Pigs demonstrate anatomical, physiological and metabolic similarities to humans, in addition they exhibit three or more of the clinical signs of the metabolic syndrome. Pigs are an appropriate biomedical model for many medical conditions and are now considered to be the optimum non-primate model for the metabolic syndrome in humans. Moreover as pigs produce litters of multiple offspring from the same pregnancy they are useful for developmental programming studies, enabling greater understanding of the mechanisms behind the metabolic syndrome and not just the treatment of the disease. This paper will discuss the pig as a model for the metabolic syndrome, covering the areas of developmental programming, cardiovascular disease and diabetes in particular.

Effects of a neonatal diet enriched in proteins on growth and adipose tissue gene expression in a porcine model of intra uterine growth restriction

Sarr, O.[1,2], Gondret, F.[1,2], Jamin, A.[1,2], Le Huërou-Luron, I.[1,2] and Louveau, I.[1,2], [1]INRA, UMR1079, Domaine de la Prise, 35590 Saint-Gilles, France, [2]Agrocampus Ouest, 65 rue de Saint-Brieuc, 35042 Rennes, France; Isabelle.Louveau@rennes.inra.fr

Neonatal nutrition has been suggested to influence growth and development in later life. The high protein content of formula offered to babies born with small weights is suspected to increase the risk of later obesity. This study was undertaken to examine the immediate and long term effects of neonatal diets differing in protein content on growth, plasma hormonal and metabolic status and gene expression in adipose tissue of pigs born with small birth weights. Piglets (10th percentile) were fed milk-replacers formulated to mimic sow milk (AP, 4.4 g of protein/100 kcal) or provide an excess of proteins (HP, 6.2 g of protein/100 kcal) from day 2 to weaning (day 28). Ten piglets were killed at day 28. After weaning, other animals (n=25) were offered ad libitum a similar high fat diet (12% fat) up to day 160. From birth to weaning, HP piglets had a greater ($P<0.05$) daily weight gain than AP piglets. Lipid contents in carcass as well as in adipose tissues were lower ($P<0.05$) in HP than in AP piglets. Leptinemia was reduced ($P<0.05$) in HP piglets whereas plasma concentrations of metabolites did not differ between the two groups. In subcutaneous adipose tissue, mRNA levels of glucose transporter 1, adipose triglyceride lipase and stearoyl-CoA desaturase 1 tended ($P\leq0.1$) to be lower in HP than in AP piglets. Beyond weaning, growth performance did not differ between HP and AP groups. At day 160, carcass composition and plasma concentrations of leptin, insulin and metabolites were similar in HP and AP pigs. However, insulin receptor, adipose triglyceride lipase and leptin mRNA levels in subcutaneous adipose tissue tended to be greater ($P\leq0.1$) in HP than in AP pigs. In summary, our data clearly show that a high protein formula induced a transitory reduction in adipose tissue development and support an alteration of adipose tissue physiology in the long-term.

Behavioural tests to screen satiating properties of dietary fibre sources in adult pigs

Souza Da Silva, C.[1,2], Van Den Borne, J.J.G.C.[2], Gerrits, W.J.J.[2], Kemp, B.[1] and Bolhuis, J.E.[1], [1]Wageningen University, Adaptation Physiology Group, Marijkeweg 40, 6709 PG Wageningen, Netherlands, [2]Wageningen University, Animal Nutrition Group, Marijkeweg 40, 6709 PG Wageningen, Netherlands; carol.souza@wur.nl

Dietary fibres (DF) are known to affect satiety in different ways. Underlying mechanisms have not been studied in detail yet, which complicates the application of DF in human and animal diets. The project Fermentation in the gut to prolong satiety focuses on identifying the working mechanisms by which different types of DF affect satiety, using the pig both as a target animal and as a model for humans. The aim of the present (pilot) study was to develop reliable behavioural tests for assessing satiety in adult pigs. Ten adult sows fed a standard commercial diet were used in a cross-over design. The study consisted of two parts. In part 1, voluntary feed intake (VFI) was determined and sows were trained to perform two feeding motivation tests, an operant consumer-demand (CD) test and a runway (RW) test. For obtaining a feed reward, in the CD test the sow had to turn a wheel on a progressive ratio schedule with a step size of one (PR1); whereas in the RW test, the sow had to walk a fixed route. In part 2, an experimental contrast in feeding level was created (75% vs. 60% of VFI) and responses in feeding motivation were measured by the tests at 1, 3 and 7 h post morning meal on three different days. In the CD test, sows on a low feeding level (60% of VFI) obtained more rewards and turned the wheel more times than sows on a high feeding level (75% of VFI) at all test times. The effect of feeding level was significantly different ($P<0.05$) at 1 and 3 h after feeding. In the RW test, there was no effect of feeding level on time spent walking to the feed reward ($P>0.05$) at all test times. Thus, in the CD test changes in satiety could be reliably measured, as sows on a low feeding level had a higher feeding motivation (cf. lower satiety) than sows on a high feeding level.

What are the prerequisites for the establishment of a pig model to investigate the mechanisms of conditioned food aversion in humans?

Meunier-Salaün, M.C., Gaultier, A., Malbert, C.H. and Val-Laillet, D., INRA, UMR1079, SENAH, 35590 St Gilles, France; Marie-Christine.Salaun@rennes.inra.fr

Conditioned food aversion (e.g. after chemotherapy) can have deleterious effects in humans. The pig is an excellent model in human nutrition and neurosciences. The aim of our study was to validate a pig model to describe the behavioral and neurobiological consequences of conditioned flavor aversion. We identified four prerequisites and used 36 pigs to validate our model: 1) identify a substance, injection site and concentration leading to visceral illness: we compared the effects of apomorphin, veratrin, erythrocin and LiCl injections after a meal (N=8); 2) select different flavors and concentrations to avoid spontaneous preference/aversion: we added cinnamon (C), thyme (T), or orange (O) essential oils to food (10 ml/kg, 1, 2, or 5%) and performed food-choice tests (N=20); 3) verify the establishment of a flavor aversion after conditioning (when the flavor is in the food or in the ambient air above the trough), and 4) elaborate a strategy to describe brain activations during flavor exposure: in anaesthetized pigs, we diffused a taste on the tongue and an odorized air in the nostril to evaluate brain activation via tomography (N=8). Only LiCl injections induced emesis and stereotypies in pigs. Intraduodenal injection was more effective than gastric injection (1st vomiting: 6.0±1.6 min; nb of vomiting: 10±1; 17% of time in stereotypies, $P<0.05$ vs. control). There was no spontaneous preference between C, T and O when dilutions were 2, 5 and 5% respectively. After conditioning, pigs completely avoided the food with a negatively-associated flavor ($P<0.01$). They were also able to select their food on the basis of the ambient air (orthonasal perception is sufficient for food discrimination). Negative flavors induced different brain activations compared to control. Anaesthetized pigs can perceive flavors and their brain responses differ according to their experience with flavors.

The use of sheep as biomedical models

Chavatte-Palmer, P., Morel, O., Heyman, Y., Pailhoux, E. and Laporte, B., INRA, UMR 1198 Biologie du Développement et Reproduction, Domaine de Vilvert, F-78350 Jouy en Josas, France; pascale.chavatte@jouy.inra.fr

Due to obvious ethical considerations in the field of human biomedical research, animal models are of critical importance for medical research. Rodents and lagomorphs are certainly the most commonly used, essentially because of their low cost, their handling and rearing facilities, their limited ethical impact, and the availability of a wide range of genetic research tools in these species. Nevertheless, these animal models present some limitations. The physiological mechanisms observed in these species might be far from those in humans. For example, rodents and lagomorphs are not suitable for the evaluation of several topics in the field of perinatal research because of their large number of embryos and short length of gestation. Large animal species are needed when surgical approaches or new medical devices have to be evaluated. The pig is widely used in these situations, as well as small ruminants. Concerning physiological, anatomical and genetic considerations, large primates could be preferred because of their important similarities with humans but their use is greatly limited by their behavioral and social organization, raising important ethical questions, and their elevated cost. A very large number of experiments using ruminants as animal models have been published. A rapid bibliometric analysis performed using the Pubmed database from 1969 to January 2010 allowed to find 1108 reviews of the bibliography using the Mesh keywords 'ruminant & animal model', with sheep being the most widely used model. As a consequence, it appears neither possible nor suitable to carry out an exhaustive overview of all research possibilities offered by the different ruminant models. This presentation will focus on the most outstanding examples of great biomedical advances carried out with ruminant in the field of perinatal research, which is our field of expertise, based on a limited number of examples covering the major fields of biomedical research.

Bilateral internal thoracic artery resection. Influence in sternal wound healing after median sternotomy: early experience in the growing sheep model

Beltrán De Heredia, I.[1], Gallo, I.[2], Sáenz, A.[2], Artiñano, E.[3], Martínez-Peñuela, A.[3], Pérez-Moreiras, I.[2], Esquide, J.[2], Larrabide, I.[2] and Camacho, I.[2], [1]Neiker-Tecnalia, P.O. Box 46, 01080 Vitoria-Gasteiz, Spain, [2]Policlínica Gipuzkoa, Division of Cardiac Surgery, Paseo Miramón 174, 20009 Donostia, Spain, [3]Policlínica Gipuzkoa, Department of Pathology, Paseo Miramón 174, 20009 Donostia, Spain; ibeltran@neiker.net

It has been clearly demonstrated that the use of 1 or 2 internal thoracic arteries (ITA) during coronary artery bypass grafting results in excellent long-term graft patency. Unfortunatelly, this technique is limited by the increased risk of deep sternal wound infection and dehiscences associated with ITA harvesting because of devascularization of the sternum. Median sternotomy is the most common access for open heart surgery. Sternal wound complications, including dehiscence and infection, remain challenging. They occur in 1 to 3% of patients undergoing cardiac surgery, leading to a variable mortality rate ranging from 14 to 47%. The aim of this experimental study in the growing sheep model is to evaluate the influence of resection of both internal mammary arteries in the healing of the sternal wound. We looked also at the effects of the plasma rich in growth factors (PRGF) as an agent on bone healing. In 24 female sheep, a median sternotomy was surgically created and both internal mamary arteries were resected. In 12 of them (group control) the sternum was closed with three figure-of-eight wires. In 12 (group PRGF) three cloths of autologous PRGF were applied over the sternum after its closure in the same manner as the control group. All sheep were put to death 3 months followup. The sternum was removed and fixed. In the control group we found extensive cartilaginous areas. In the PRGF group, the presence of trabecular bone tissue was common, with formation of hematopoyetic medullary tissue. The process of new bone formation was accelerated in the PRGF group and resection of both mammary arteries does not have influence in sternal wound healing.

Performance of calves fed ration containing 2-hydroxy-4-(methylthio) butanoic acid

El-Bordeny, N.E.[1] and Abedo, A.A.[2], [1]Ain Shams University, Animal production, 68 Hadayeq Shoubra, Cairo, Egypt, 11241, Egypt, [2]N, Animal production, Dokki, Giza, Egypt., 11111, Egypt; abedoaa@hotmail.com

Ruminal escape of methionine hydroxy analog [D,L-2-hydroxy-4-(methylthio)-butanoic acid (HMB) can be enhance calves performance. So this study was conducted to evaluate effect of HMB supplementation on growth and economic efficiency of calves. Thirty-sex crossbred male calf with mean initial live body weight 251.52±3.16 kg were divided into two similar groups. Diets were identical except for HMB supplementation and the animals were fed total mixed ration containing 28.17% corn silage, 32.32% yellow corn, 10.06% soybean meal, 10.77%wheat bran 10.77% rice bran, 5.03% rice hulls 1.44% limestone, 0.72% salt, 0.36% mineral and vitamin mixture and 0.36% buffering agent of DM basis without or with 10 g HMB for G1 and G2 respectively. All nutrients digestibility and feeding value were did not affected ($P<0.05$) by HMB supplementation,. Supplementation with HMB significantly ($P<0.05$) increased both serum total protein and albumin, but decreased ($P<0.05$) urea nitrogen. Globulin values and albumin/globulin ratio were did not significant different. Average body weight gain was significantly increased ($P<0.01$) (1.145 kg) for calves fed diet supplemented with HMB than 0.995 kg/h/d for calves fed control diet. Also, feed conversion as DM, TDN and DCP were ($P<0.01$) improved (6.52, 4.09 and 0.52) for calves fed HMB compared with 7.51, 4.98 and 0.59 kg/kg gain, respectively for calves fed diet without supplementation. It could be concluded that supplementation of calves' diet with HMB improved efficiency of protein utilization, body weight gain and feed conversion.

Session 46 Poster 2

The effects of dry matter content and hay particle size of total mixed ration on eating and lying behaviours of dairy cows

Hosseinkhani, A.[1]*, Daghigh Kia, H.*[1] *and Valizadeh, R.*[2]*,* [1]*University of Tabriz, Faculty of Agriculture, Department of Animal Science, 29 Bahman Bolvard, 5166614766, Iran,* [2]*University of Ferdosi Mashhad, Department of Animal Science, Mashhad, 51666, Iran; hoseinkhani2000@yahoo.com*

Water addition to total mixed ration (TMR) and particle size reduction of hay are techniques used to minimize diet selection of dairy cows but these changes in physical form of diet may alter some eating behaviours of cows. There is evidence that the longer the animal stands after milking, the lower the risk for bacterial penetration of the teats when the cow eventually lies down. The objective of this study was to understand whether the particle size and dry matter content of the diet affects DMI, eating and lying behaviours of dairy cows. Eight multiparous Holstein cows were allocated to the treatments in a change over design with periods of 21 d in early lactation period. The average day in milk and milk production of the cows were 28±12 d (mean ± SD) and 43±3.5 kg/d, respectively. They were fed ad-libitum twice daily and had free access to drinking water. The balanced diets had the same chemical composition. Diet main ingredients were as follows (g/kg): Lucerne hay (200), maize silage (150), barley grain (310), cottonseed (90), soybean meal (120), safflower meal (60), wheat bran (30) and protected fat (20). Two particle size of Alfalfa hay (5 and 20 mm) and two levels of TMR dry matter (without and with water addition up to 50% of DM) were applied in the treatments. Water was sprinkled to diet during every day diet preparation. All animals were milked three times a day. Eating, standing and lying behavioural activities were observed manually by a team of observers and recorded for 24 h (5 min intervals) during 14-15th days of each experimental period. The data were analyzed using mixed model procedure of SAS. Experimental treatments had no effect on DMI, eating behaviours (eating time (min/d), eating rate, meal number and meal duration (min/meal)) and lying behaviour (Total lying time (min/d) and Latency to lie down (min) after comeback from milking.

Session 46 Poster 3

Effects of different levels of dried citrus pulp and urea on performance of fattening calves

Foroughi, A.R., Ahooei, G.H.R., Shahdadi, A.R. and Tahmasbi, G.H.,; afroghi@yahoo.com

Twenty fattening Brown Swiss male calves were used in this study. At the start of the experiment, calves averaged 192.15±30 kg live weight and 196.3±24 days age were housed in individual tie stalls and randomly allocated to four experimental treatments include: 1) without urea and dried citrus pulp (DCP) (Control), 2) 12% DCP + 0% urea, 3) 0% DCP + 0.65% urea, and 4) 12% DCP + 0.65% urea. The experimental design was 2×2 factorial. The experimental diets were consisted of 35% forage (corn silage) and 65% concentrate. The length of the experiment was 100 days (10 days for adaptation period and 90 days for experimental period). Feed offered and orts were measured and recorded daily to calculate feed intake. Weighting of calves was done each one month. Rumen liquid samples were obtained 2 hours after morning feeding at the end of the study. For determining of nutrients digestibility, faeces were collected in days 95-100. Data showed that treatments had no significant effect on dry matter intake, average daily gain and feed to gain ratio, although there was a numerical increase in treatment 4 than other treatments for dry matter intake and average daily gain. There isn't significant difference between rumen pH of calves, but rumen N-NH_3 was significantly differ between treatments ($P<0.05$), which calves receiving diet 3 significantly had the highest rumen N-NH_3. The nutrients digestibility significantly affected by experimental treatments ($P<0.05$). The dry matter and crude protein digestibility in treatment 4 was higher than other treatments, but the organic matter digestibility in treatment 3 was higher than other treatments.

Effects of replacing barley grain with sorghum grain on fattening performance of Holsteins male calves

Foroughi, A.R.[1], Ahmadi, V.[2], Vakili, R.[3], Nejad Mohammad Nameghi, H.[4] and Shahdadi, A.R.[5], [1]Institute of Scientific Applied Higher Education of Jihad-e-Agriculture, Shahid kalantry street, Mashhad, Iran, [2]Jihad Keshavarzy office of gain, Eslamabad Dairy Farm, gain, Iran, [3]Azad University, university street, kashmar, Iran, [4]Education and culture Office of Khorasan Razavi, Vakilabad street, mashhad, Iran, [5]University of agricultural sciences and natural resources of gorgan, University Street, Gorgan, Iran; Afroghi@yahoo.com

In order to investigating performance of male calves fed different levels of sorghum (Sorghum bicolor) instead of barley grain, fifteen Holstein bull calves with 209.46±21.06 kg live weights were used in Randomized Block Design. The calves were classified and randomly allocated to three experimental treatments include: 1) the concentrate based on sorghum grain; 2) the concentrate containing of same levels of sorghum and barley grain; and 3) the concentrate based on barley grain. Rations were adjusted to NRC (1996) recommendations. The experimental diets were consisted of 30% forage (corn silage and alfalfa hay) and 70% concentrate. The length of this study was 120 days (30 days for adaptation period and 90 days for experimental period). Feed was made as TMR and fed adlibitum to calves. Intake was measured daily during design period. Blood and rumen liquid samples were at the end of study 2 hours after morning feeding. Weighting of calves was done each one month. Data were analyzed in a completely randomized design. Data showed that treatments had no significant effect on dry matter intake, daily weight gain and feed to gain ratio, although treatment 2 had better results than other treatments. Blood parameters (except cholesterol) were significantly different between treatments ($P<0.05$). Blood glucose in treat 3 was significantly higher than treats 1 and 2 ($P<0.05$). Blood triglyceride and urea nitrogen in treat 2 were significantly higher than other treatments ($P<0.05$). Treatments had no significant effect on blood cholesterol and rumen pH.

Yeast culture as probiotic in young calves

Petkova, M.[1] and Levic, J.[2], [1]Institute of Animal Science, Pochivka 1, 2232 Kostinbrod, Bulgaria, [2]Institute for Food Technology, bul. Cara Lazara 1, 21000 Novi Sad, Serbia; m_petkova2002@abv.bg

Although ruminants are particularly well adapted for feed digestion, owing to their ruminal microflora, at times, biochemical conditions prevailing in the rumen can prevent optimum feed utilization. Certain microorganisms (probiotics) can therefore act as rumen stabilizers following a meal rich in readily fermentable carbohydrates. A popular group of probiotics are the live yeast products, based on S. cerevisiae. The aim of this study was to evaluate the effects of yeast culture (YC), and administrated per os, on growth performance and blood biochemical changes in calves. A total of twenty four calves, Holstein breed were separated into two equivalent treatments (per six male and female calves) for 148 days, including 48 days preliminary period, 70 days experimental period and 30 days post experimental period. Calves were fed whole mother milk with in additional TMR with 55:45 ratio of compound feed and roughage with protein level 0.220 and 0.113 kg/kg DM respectively at the beginning and by the end of the experiment. Experimental calves received in addition to the mother milk YC in liquid form. Body weight changes (two times per month), total intake of dry matter (DMI), biochemical changes in the blood of calves (levels of protein, glucose, urea and amino-N, as well as activities of both GPT-ase and GOT-ase, two time per month) were determined. Average daily gain (ADG) and utilization of feed (ADG: DMI) were calculated. The obtained results allow to make the next conclusions: there was a tendency towards positive impact of YC on average daily gain (4%), more clear express in male calves (9%); the effect of yeast culture on FCR was in the frame of 9,8%; no effect was found on the level of glucose, amino-N and urea under the influence of YC as well as GOT-ase activity; the observed levels of protein in the blood were increased by 3% as well as activity of GPT-ase by 5.4%. This experiment did not show expected positive impact of yeast culture added to daily rations of young calves.

Session 46 Poster 6

Using different levels of sorghum in finishing Ghezel×Arkhar-Merino crossbred lambs diets and its effects on animal performance

Hosseinkhani, A., Taghizadeh, A., Daghigh Kia, H., Rostami, A. and Shoja, J., University of Tabriz, Animal science, 29 bahman bolvard, faculty of agriculture, 5166614766, Iran; hoseinkhani2000@yahoo.com

Reducing feeding costs is one of the major problems in animal production. In most of the semi-arid regions of the world such as Iran, producing of feed for animal is difficult and farmers have to use some alternative feedstuffs such as sorghum to reduce feeding costs. In many cases the main problem in utilization of these alternative feeds is the presence of anti-nutritional factors such as tannins found in them. Adverse effects of tannins are associated with their ability to bind with dietary protein, carbohydrate and minerals and restrict intake and reduce overall weight gains by livestock. The aim of the present study was to investigate the effects of replacing dietary barely with different levels of sorghum on lamb performances. Sixteen Ghezel×Arkhar-Merino crossbred male lambs with live weights 46±5.8 (mean ± SD) were used in the experiment. Experimental animals were kept at research farm of university of Tabriz. All dietary treatments had alfalfa hay (20% total DM), and remain part of diets (80% total DM) were contain different levels of barley grain substituted with sorghum grain. Lambs were randomly assigned to one of the four dietary treatments in a completely randomized design assignment, in which sorghum grain was used in the levels of 0, 60, 70 and 80 percent of total ration. Experimental diets were containing 0, 0.588, 0.686, 0.784 percent of total tannin/kg of diet DM respectively. Feed intake and weight gain data were analyzed using the general linear model of the Statistical Analysis Systems. The results of experiment showed that different levels of tannin in the diet had no significant effect on dry mater intake, average daily gain and feed conversion ratio of lambs. May be the low number of experimental units affected the accuracy of the results so it is recommended that experiment done with more animals in each treatment.

Session 46 Poster 7

Composition of the whole plant and silage from maize hybrids with the cry1Ab trait versus its nonbiotech counterpart

Balieiro, G.N., Possenti, R., Branco, R.B.F., Nogueira, J.R., Ferrari, E.J., Monteiro Dos Santo Cividanes, T. and Paulino, V.I., Agência Paulista de Tecnologia dos Agronegócios, APTA/SAA, Bandeirantes Avenue, 2419, Ribeirão Preto, São Paulo, 14030670, Brazil; geraldobalieiro@apta.sp.gov.br

This study aimed evaluating the effect of cry1Ab gene on the nutrient composition maize hybrids with the cry1Ab trait versus its nonbiotech counterpart. The hybrids DKB 390 from Dekalb and AG 8088 from Agroceres that contains cry1Ab gene from Bacillus thuringiensis (Bt) and yours isogenics that not contains the cry1Ab gene were used. The experimental design was the randomized block with five replications, in factorial arrangement 2 x 2. The crop as silage was harvested when the whole plant had 69 to 73% moisture and 95 days old. Acid detergent fiber, lignin, cellulose, hemicellulose contents in the whole plant did not differ between hybrids or MGO. The values of dry matter (DM), crude protein (CP), no fiber carbohydrates, *in vitro* digestibility, Ca, P and K of whole plant from hybrids that contains cry1Ab gene were higher and neutral detergent fiber (NDF) were smaller than conventional hybrids. The values of DM, total carbohydrates, CP, crude fat, Ca and P of maize silage from hybrids that contains cry1Ab gene were higher than conventional hybrids. The neutral detergent insoluble nitrogen value in whole plant from hybrid DKB 390 that contains cry1Ab gene was smaller than hybrid DKB 390 conventional. The results demonstrated that gene Bt avoids the mechanisms of resistance expression of the plant against the infesting, including lignin and protein accumulation in the cellular wall. The effect of cry1Ab gene allowed continuous growth, anticipating maturity and grain filling, changing the nutrient composition. The effect the cry1Ab gene can vary significantly due to genetic differences of maturity stages between hybrids in harvest.

Comparison difference levels of phytase enzyme in diet of Japanese quail (*Coturnix japonica*) and some blood factors

Vali, N., University, Animal Science, Islamic Azad University, Shahrekord branch, Iran, P.B. 166, Iran; nasrollah.vali@gmail.com

A total of 210, 3-d-old Japanese quail chicks, (*Coturnix japonica*) allocated to 21 cages, each cage containing 10 chicks, that were received seven diets with three replication of each diet. Experimental diets were arranged with five levels 150, 300, 600, 1,200 and 2,400 phytase enzyme unit (FTU/Kg) with commercial name (Natuphos 500) as treatments1, 2, 3, 4 and 5 respectively, and two control groups (positive and negative controls). At the end of each experimental period (each week to 45 days of ages), birds were weighed and feed consumption was recorded for feed efficiency computation. At the age of 45 days, four chicks were randomly selected from each treatment blood samples for subsequent determination of minerals (Ca, P and Mg) and alkaline phosphates (ALP) in serum. Data were subjected to analysis of variance using the general linear models (GLM) procedures of SAS software. Liver weight in treatment 5 (4.68±0.23 g) at 45 days of age was more than other groups that had significant difference ($P<0.05$). Quails of treatment 4 had calcium (Ca) and phosphorus (P) 10.19±0.80, 7.11±0.46 mg/dl respectively that were more than other groups. The amount of magnesium on treatment 1(3.27±0.15 g), positive control (3.27±0.15 g) and negative control (3.56±0.15g) which were not significantly different ($P<0.05$), whereas treatment1 and control groups had significant differences with other treatments ($P>0.05$). The Alkaline phosphates of treatment 5 (366.83±38.07 IU/L) in blood serum at 45 days of age represents the most increase among other treatments, which had significant differences with treatment 1 (124.99±50.80 IU/L) and control groups, although weren't significant difference with the treatments 2 (260.26±0.01 IU/L) and treatments 3(288.37±40.64 IU/L).

Metabolizable energy and amino acid bioavailability of field pea seeds in broilers' diets

Dotas, V., Hatzipanagiotou, A. and Papanikolaou, K., Aristotle University, Faculty of Agriculture, Thessaloniki, 54124, Greece; vdotas@agro.auth.gr

The aim of this study was to determine the apparent (AME) & true (TMEn) metabolizable energy as well as the crude protein (CP) & amino acid (AA) bioavailability of broilers' diets and field pea seeds (FPS) of the Greek cultivar 'Olympos'. Forty eight broilers were used. Birds were placed in individual cages in vertical decks of a digestibility chamber and randomly allocated into 4 dietary treatments. Each group was comprised of 6 male and 6 female broilers. Birds consumed 80 g/d of either a typical commercial diet or the same diet in which 10, 20 or 30% had been substituted by ground FPS. The experiment lasted 15d (counting from the 28th d of age). The first 4 days needed for the adaptation of birds in cages, 7 days was the duration of preliminary period and 4 days needed for the excreta collection. The endogenous nitrogen losses were determined by using an indigestible marker and this procedure lasted 28 h. Final LW of birds was significantly ($P<0.05$) reduced only in the treatment containing the highest rate of FPS inclusion. Apparent & true CP bioavailabilities of the diets were not significantly affected by FPS inclusion up to the level of 30%. Apparent and true CP bioavailability of FPS showed a linear trend of reduction by increasing inclusion rates of FPS and this reduction was significant ($P<0.05$) at the level of 30%. Amino acid bioavailability as determined by the difference between the untreated diet's one and that containing 10% FPS remained at high levels (~80%), with the exception of methionine and valine and was similar to AA mean. The means AME and TMEn of FPS were estimated equal to 10.8 and 11.0 MJ ME/kg, respectively. FPS 'Olympos' are a valuable energy and protein source for broilers and could be included in their diets up to a level of 20%, without any adverse effect on the diets' digestibility and broilers' performance, contributing to natural feed resources exploitation.

Session 46 Poster 10

The effects of dietary hesperidin supplementation on broilers performance and meat quality

Simitzis, P.E., Symeon, G.K., Charismiadou, M.A., Ayoutanti, A. and Deligeorgis, S.G., Agricultural University of Athens, Animal Science, 75 Iera Odos, 11855 Athens, Greece; pansimitzis@yahoo.gr

Large quantities of citrus fruits are produced in Greece that are usually used for juice extraction creating problems in disposing the peels. Hesperidin is a bioflavonoid, which is contained in citrus cultivation by-products and possesses intense antimicrobial, antifungal and antioxidant properties. Natural antioxidants are currently receiving considerable attention in animal nutrition because of changes in legislation, consumer trends, isolation of antibiotic resistant pathogens and finally their association with food quality. In this study, the dietary effects of hesperidin on broiler growth performance and meat quality were investigated. Eighty Cobb male chickens were randomly allocated into 4 equal groups. One of the groups served as a control and was given a basal diet, whereas the other three groups were given the same diet further supplemented with hesperidin at 1.5 g/kg, or hesperidin at 3.0 g/kg, or a-tocopheryl acetate at 0.2 g/kg. Feed and water were offered ad libitum. Feed intake was recorded and all chickens were weighed for estimation of body weight gain and feed conversion ratio at 21 and 40 days of age. At 40 days of age, the chickens were slaughtered and samples of breast muscle were further examined. Colour (L, a* and b* parameters), pH, tenderness (shear force value), cook loss, total lipids and oxidative stability of lipids (on the basis of the malondialdehyde content) were measured. Growth performance was unaffected by the experimental diets. Colour, pH, tenderness, cook loss and intramuscular fat were not significantly influenced by supplementation. Lipid oxidation showed that as hesperidin increased in the diet, malondialdehyde values decreased in tissue samples, suggesting that hesperidin particularly at 3 g/kg exerted an antioxidant effect on chicken tissues. However, dietary a-tocopheryl acetate supplementation at 0.2 g/kg of feed displayed greater antioxidant activity than hesperidin at either supplementation rate.

Session 46 Poster 11

Effect of β-D-glucanase and β-D-mannanase with two various levels of metabolisable energy on performance of broiler chicks fed barley-soybean diets

Karimi, K., Sadeghi, A.A. and Forudi, F., Islamic Azad University(I.A.U)- Varamin branch, Animal Science, School of Agriculture- Choobbori-Varamin-Tehran, 3371857554, Iran; ka_egh@yahoo.com

Effects of two Metabolisable Energy (ME) levels with or without ones or combination of two commercial feed enzyme supplements containing ß-D-glucanase and ß-D-mannanase on feed intake (FI), gain (G), feed conversion ratio (FCR) and abdominal fat (AF) were evaluated. This experiment carried out in 2×2×2 factorial design with 2 ME levels (equal of broilers need or 100Kcal/kg Lower than it) and 2-glucanase levels(0 and 0.1%) and 2 ß-mannanase levels(0 and 0.1%). Results indicate that the interactions between ME levels and enzyme treatments in some times on G, FI, FCR were significant ($P<0.01$). Some times, high ME levels had no significant affects on performance traits ($P>0.05$). In 41 day-old male broiler chicks, effect of enzyme treatments on AF was significant ($P<0.05$). In these broilers glucanase increased AF from other enzyme treatment. Combination of glucanase and mannanase in soybean barley based diets some times had better effects on performance traits. In each periods of age (except week 4) the best of G in week's 1-3 and 6 and total period FCR improvement were seen in the combination of two enzymes.

Meta-analysis of phosphorus utilisation by broilers receiving corn-soybean meal diets: influence of dietary calcium and microbial phytase

Letourneau Montminy, M.P.[1], Narcy, A.[2], Lescoat, P.[2], Bernier, J.F.[3], Magnin, M.[4], Pomar, C.[1], Nys, Y.[2], Sauvant, D.[5] and Jondreville, C.[6], [1]Agriculture and Agri-Food Canada, Sherbrooke, Qc, J1M 1Z3, Canada, [2]INRA UR83, Nouzilly, 37380, France, [3]Laval University, Qc, G1V0A6, Canada, [4]BNA Animal Nutrition, Chateau-Gontier, 53200, France, [5]INRA Agroparistech, Paris, 75231, France, [6]INRA, Nancy-University, Vandoeuvre-lès-Nancy, 54500, France; Marie-Pierre.Letourneau@agr.gc.ca

A good understanding of bird's response regarding dietary phosphorus (P) is a prerequisite for the reduction of this pollutant and expensive essential element in broilers' diets, with productivity remaining of major concern. A database including information from 15 experiments and 203 treatments was used to predict the response of 21-day old broilers to dietary non-phytate P (NPP), taking into account main factors of variation, calcium (Ca) and microbial phytase derived from A. niger, in terms of average daily feed intake (ADFI), average daily gain (ADG), gain to feed (G:F) and tibia ash concentration. The response of these criteria to dietary NPP concentration is curvilinear. Dietary Ca affected the intercept and the linear component for ADG ($P<0.01$), G:F ($P<0.05$) and tibia ash concentration ($P<0.001$), whereas for ADFI, it affected only the intercept ($P<0.01$). Microbial phytase addition impacted on the intercept, the linear and the quadratic coefficient for ADFI ($P<0.01$), ADG ($P<0.001$) and G:F ($P<0.05$), and on the intercept and the linear component ($P<0.001$) for tibia ash concentration. The increase in dietary Ca concentration aggravated P deficiency while phytase addition had a positive effect. The more P deficiency was marked, the more bird response to phytase was exacerbated. However, even if the decrease in dietary Ca may improve P utilisation, it could in turn become limiting for bone mineralization. In conclusion, this meta-analysis provides ways to reduce dietary P in broilers' diets without impairing performance by taking into account dietary Ca and microbial phytase.

The effect of an increasing dietary energy level on the feed intake and production of breeding ostriches

Brand, T.S.[1,2], Olivier, T.R.[2] and Gous, R.M.[3], [1]Elsenburg Animal Production Institute, Department of Agriculture: Western Cape, Private Bag X1, Elsenburg 7607, South Africa, [2]Stellenbosch University, Department of Animal Science, Stellenbosch, 7600, South Africa, [3]University of KwaZulu Natal, Department of Poultry Science, Pietermaritzburg, 3200, South Africa; tersb@elsenburg.com

The experiment was conducted to determine to what extent dietary energy content will affect different production parameters of breeding ostriches. Ninety pairs of breeding ostriches were divided into six groups, consisting of 15 breeding pairs per group. Six diets with increased metabolisable energy contents (8.0, 8.7, 9.4, 10.1, 10.8 and 11.5 MJ ME/kg feed) were provided ad libitum to birds during the breeding season. All the other nutrients were balanced according to the nutrient requirements of the birds. Production data was analysed by one-way analysis of variance. Average feed intake/bird/day (kg) amongst the different diets were not statistically different ($P>0.05$), with an average feed intake value of 3.7±0.2 kg per bird per day. The increase in live masses of both the female (respectively 5.3, 2.8, 5.1, 10.8, 11.3 and 15.7 kg) and male (respectively-2.1, 1.6, 3.2, 4.9, 0.9 and 10.0 kg), however indicated that both the male and female breeders over-consumed energy on the diets with the higher energy values. No significant differences ($P>0.05$) were observed for other production parameters measured, like total eggs produced per female per season (mean value of 45.6±5.8), number of chicks hatched (mean value of 21.3±4.5), number of infertile eggs (mean value 11.6±3.6), number of dead-in-shell eggs (mean value of 7.5±1.8) or egg mass (1405±30.5 g). The present study revealed that breeding ostriches were unable to regulate their feed intake due to dietary energy level. Results from the present study were important for the determination of the maximum feed intake of breeding ostriches as well as the determination of the required concentration of the other accompanied essential dietary nutrients.

Session 46 Poster 14

The effect of supplementation to irrigated pastures and stocking rate on the production of finishing ostriches

Brand, T.S.[1,2], Strydom, M.[2] and Van Heerden, J.M.[3], [1]Elsenburg Animal Production Institute, Agriculture: Western Cape, Private Bag X1, Elsenburg 7607, South Africa, [2]Stellenbosch University, Animal Science, Stellenbosch, 7600, South Africa, [3]Agricultural Research Council, Pasture Science, Stellenbosch, 7600, South Africa; tersb@elsenburg.com

The experiment was conducted to determine the effect of stocking rate and level of supplementation on the production of finishing ostriches while grazing irrigated lucerne pasture. A lucerne pasture was divided into 16 paddocks of approximately 0.85 ha each. Two hundred ostriches (±6 months old) were randomly allocated into four groups. The birds rotationally grazed lucerne pasture with a stocking rate of either 15 birds/ha or 10 birds/ha while received either 0 g or 800 g supplementary feed/day. Four lucerne paddocks were randomly allocated to each group. Birds were weighed every 14 days, while pasture production and utilization were recorded with exclosure cages. Data was analyzed with analysis of variance. Ostriches receiving supplementary feed of 800 g/bird/day had significantly higher mean live weight at 54 weeks and significantly better feed conversion ratio's (FCR) than ostriches receiving no supplementation. Ostriches receiving 800 g supplementary feed/bird/day reaching target weight (±95 kg) within the period of the trial, while ostriches receiving no supplementation did not reach slaughter weight. Stocking rate did not have an influence on either mean live weight at 54 weeks or FCR. A significant interaction was found between level of supplementary feed and stocking rate regarding average daily gain (ADG) of the birds. Stocking rate influenced ADG only for birds receiving no supplementary feed. As stocking rate increases, ADG of birds receiving no supplementary feed declined. At a stocking rate of 10 birds/ha, supplementary feeding did not influence lucerne DM intakes of the ostriches, but when birds receive no supplementation at this stocking rate, lucerne DM intakes gradually increased.

Session 46 Poster 15

Dietary evaluation of faba bean seeds in broilers

Gourdouvelis, D.[1], Dotas, V.[1], Mitsopoulos, I.[2], Nikolakakis, I.[3] and Dotas, D.[1], [1]Aristotle University, Faculty of Agriculture, Thessaloniki, 54124, Greece, [2]TEI of Thessaloniki, Sindos, 57400, Greece, [3]TEI of W. Macedonia, Florina, 53100, Greece; topidaias@gmail.com

The aim of this study was to estimate the apparent (AME) and true (TMEn) metabolizable energy and the digestibility of crude protein (CP) of faba bean seeds (FBS) of the Greek cultivar 'Polikarpi' in broilers' diets. Forty eight broilers were randomly allocated into 4 dietary treatments (M, B_{10}, B_{20} & B_{30}) and placed in individual cages. Birds in treatment M were fed on a commercial diet based on wheat, maize and soybean meal, while those in treatments B_{10}, B_{20} & B_{30} were fed on the same diet with a proportion of 10, 20 and 30% of diet M replaced by FBS, respectively. The equilibrium in Ca, P and methionine-cystine of the test diets was achieved by the addition of monocalcium phosphate and methionine. The experiment lasted 15 days (counting from the 28th d of age). The first 4 days needed for the adaptation of birds in cages, the next 7 days was the duration of the pro-experimental period and the last 4 days needed for the excreta collection. Birds consumed 80 g of feed per day. Metabolic N determination (using Fe_2O_3 as marker) lasted 28 hours. The AME & TMEn and the digestibility of CP of both diets and FBS were determined by the method of difference. Final LW of birds significantly ($P<0.05$) decreased in B_{30} treatment compared to the other treatments. Apparent and true digestibilities of CP in diets were not significantly affected by the replacement with FBS up to the level of 30%. Apparent and true digestibility of CP in FBS significantly ($P<0.05$) decreased at the inclusion rate of 30%. AME & TMEn of FBS were estimated equal to 11.2 & 11.5 MJ ME/kg, respectively. FBS 'Polikarpi' are a worthy alternative protein & energy source for broilers without having any adverse effects in diets' digestibility and broilers' performance. Feeding trials in large scale units are needed to verify the results of current study.

Session 46 Poster 16

The effect of dietary selenium sources on growth performance and oxidative stability in broiler chickens

Heindl, J., Ledvinka, Z. and Zita, L., Czech University of Life Sciences Prague; FAFNR, Department of Animal Husbandry, Kamycka 129, 16521, Prague 6 - Suchdol, Czech Republic; zita@af.czu.cz

This study examined the effect of supplementation of dietary sodium selenite (SS), selenium enriched yeast (Sel-Plex) and selenium enriched alga Chlorella (SCH) on growth performance, carcass composition, feed conversion and oxidative stability of meat in broiler chickens. Sexed cockerels Ross 308 were allotted to 4 dietary groups after 105 broiler chickens per group. Commercial pelleted type feed mixture contained (22.67% crude protein, 12.87 MJ/kg ME and 50 mg/kg α-tocopherol). The basal diet was supplemented with 0 (control) or 0.15 mg/kg Se from SS, Sel-Plex or SCH. Water and feed was available ad libitum. The broiler chickens were slaughtered at 35 days of age. Results of the experiment show that selenium supplement caused significant differences ($P \leq 0.001$) in live weight at 21 days of age but at 35 days of age did not increased body weight (min. 2010 g for SCH group; max. 2123 g for Sel-Plex group). There were no significant differences between the groups in feed conversion and mortality. Carcass weight was significantly ($P \leq 0.05$) affected by supplements (min. 1331 g for SCH group, max. 1473 g for SS group). Carcass composition was not influenced by feeding regime influencing gizzard weight, abdominal fat. Selenium supplements did not significantly affect the liver weight, hearts and dressing percentage. The inclusion of selenium in the diet did not enhance the oxidative stability of meat expressed as reduced malondialdehyde (MDA) values in breast meat after 0; 3 and 5 days of storage in a refrigerator at 3 to 5 °C. Best results after all measuring reached the control group. The study was supported by Ministry of Education, Youth and Sports of the Czech Republic (Project No. MSM 6046070901).

Session 46 Poster 17

Effect on growth performance of restricting xylanase supplementation of wheat based diets to specific periods of the broiler's life

Ponte, P.I.P.[1], Figueiredo, A.A.[2], Correia, B.A.[2], Ribeiro, T.[1], Falcão, L.[2], Freire, J.P.[2], Prates, J.A.M.[1], Ferreira, L.M.A.[1], Fontes, C.M.G.A.[1] and Lordelo, M.M.[2], [1]Faculdade de Medicina Veterinária - UTL, Av. Universidade Técnica, Alto da Ajuda, 1300-477 Lisboa, Portugal, [2]Instituto Superior de Agronomia - UTL, Tapada da Ajuda, 1349-017 Lisboa, Portugal; pponte@fmv.utl.pt

Endo-β-1,4-xylanases can effectively attenuate the anti-nutritive properties expressed by soluble arabinoxylans found in wheat-based diets for poultry. However, in previous studies, the use of xylanase supplementation to improve the nutritive value of wheat-based diets was explored throughout the entire experimental period. Here, we report 2 experiments where the effect of restricting xylanase supplementation to specific periods on broiler performance was evaluated. In experiment 1, 1-d-old chicks were fed wheat based diets supplemented with a commercial xylanase for the entire duration of the experiment (28 d), the last 21 d, the last 14 d, the last 7 d of the experiment or a non-supplemented basal diet. In experiment 2, birds were fed enzyme supplemented diets for the entire duration of the experiment (36 d), the last 27 d, the last 18 d, the last 9 d of the experiment and a group was fed a non-supplemented basal diet. The data revealed that broilers fed diets supplemented with xylanase throughout their entire life had similar growth performances, intestinal viscosity, gastrointestinal enzyme activity and organ sizes to broilers fed supplemented diets only in the last 14 d (experiment 1) or the last 18 d (experiment 2) of the experimental period. These results indicate that supplementation of wheat based diets for broilers with xylanases during the entire production cycle may be technically unnecessary.

Effects of dietary glycerol addition on the intramuscular fatty acid composition in broilers

Papadomichelakis, G., Mountzouris, K.C., Zoidis, E., Pappas, A.C., Arvaniti, A. and Fegeros, K., Agricultural University of Athens, Nutritional Physiology and Feeding, 75 Iera Odos Str., 118 55, Athens, Greece; cfeg@aua.gr

The present work sought to determine the effects of dietary glycerol content on the intramuscular fatty acid profile in broilers. Eighty Cobb chicks (one-day old) were fed a control (C) or 3 diets (G7, G14 and G21) with increasing levels of crude glycerol (7, 14 and 21%, respectively; n=20 per group). Glycerol had a 0.5% total fatty acids (FA) content with 13.0% palmitic, 42.6% oleic, 31.6% linoleic and 5.9% α-linolenic acid and originated from a mixture of different oils (palm, soybean, cotton and rape seed oil). At 42 days of age, broilers were weighed and slaughtered. After chilling at 4 °C for 24 h, carcass weight and dressing percentage were determined and the left half of the breast muscle was excised and stored at -20 °C until FA analysis. Data were analysed by one-way (diet) ANOVA using SPSS v.16.0. Body weight at slaughter and carcass weight were significantly higher ($P<0.05$) in G7 compared to C, G14 and G21 broilers. Total weights of FA (mg/100 g fresh tissue) decreased linearly ($P<0.01$), indicating lower muscle lipid content with increasing dietary glycerol. On the other hand, saturated (SFA) and polyunsaturated (PUFA) FA increased linearly ($P<0.01$ and $P<0.001$, respectively); however PUFA:SFA ratio was higher ($P<0.05$) for G21 compared to the other diets. Changes in the intramuscular fatty acids were not always consistent with the dietary FA profile. In conclusion, muscle FA composition in the present work was affected by the impact of glycerol on other parameters (e.g. reduced feed intake and body weight), less so by its FA profile. An influence of the glycerol FA residues cannot be dismissed and is expected to vary, depending primarily on the efficiency of FA extraction, the origin of crude glycerol (type of seed oil) and the feeding regime.

Improving n-3 fatty acids in poultry meat through dietary supplementation with extruded linseed and DHA gold

Ribeiro, T.[1], Lordelo, M.[2], Prates, J.A.M.[1] and Fontes, C.M.G.A.[1], [1]FMV-UTL, CIISA, Av. Universidade Técnica, 1300 Lisboa, Portugal, [2]ISA-UTL, Tapada da Ajuda, 1349-071 Lisboa, Portugal; amtribeiro@fmv.utl.pt

Long chain n-3 fatty acids have important therapeutic and nutritional benefits in humans. Poultry meat is a good target for modification by means of nutritional strategies, particularly for modifying lipid content and composition by improving long chain n-3 fatty acids and decreasing n-6 fatty acids. In addition, poultry meat has not been associated with risk of cancers nor cardiovascular diseases. In this trial, we used two different polyunsaturated fatty acids sources: extruded linseed (SL), which is a good source of α-linolenic acid (18:3n-3) and a product derived from the algae Schizochytrium (DHA gold™) that contains a high content of docosahexaenoic acid (22:6n-3). These products were added to poultry diets with the aim of assess the impact on poultry meat. One-day old chicks (120 birds) were divided into 40 cages and fed with a basal diet in the first 3 weeks. In the final 2 weeks of the experiment, animals were fed on 4 different diets. One group was fed with a corn-based diet without supplementation (CN), while the 3 others were fed with the same corn based diet but with inclusion of 7.4% of DHA gold™ (DHA) or 13.4% of extruded linseed (SL) or 3.7% of DHA gold™ and 7.7% of extruded linseed (DHASL). All diets contained 8% of crude fat, and 50% of that was derived from the respectively fat dietary source. With this formulation, a 2% supplementation with n-3 fatty acids was assured. Birds submitted to the SL treatment displayed lower performance, while animals submitted to the DHA treatment had the best performance. Breast pH was affected by the inclusion of SL while thigh color was different between DHA and SL treatments. In addition, the results of meat fatty acids, vitamins and cholesterol will be presented and discussed. This work was supported by the SFRH/BD/32321/2006 (to T.R.) from Fundação para a Ciência e Tecnologia (Portugal).

Combination of citric acid and microbial phytase on digestibility of calcium and phoshphorus and mineralization parameters of tibia bone in broilers

Ebrahim Nezhad, Y., Ghyasi Gale-Kandi, J. and Farahvash, T., Islamic Azad University-Shabestar Branch, Animal and Poultry Science, Islamic Azad University-Shabestar Branch, Faculty of agriculture, Shabestar, East azerbaijan, 5381637181, Iran; ebrahimnezhad@gmail.com

This experiment was conducted to evaluate the combined effects of citric acid (CA) and microbial phytase (MP) on digestibility of calcium and phosphorus and mineralization parameters of tibia bone in broilers chicks. This experiment was conducted using 360 Ross-308 male broiler chicks in a completely randomized design with a 3×2 factorial arrangement (0, 2.5 and 5% CA and 0 and 500 FTU MP). Four replicate of 15 chicks per each were fed dietary treatments including (i) P-deficient basal diet [0.2% available phosphorus (aP)] (NC); (ii) NC + 500 FTU MP per kilogram of diet; (iii) NC + 2.5% CA per kilogram of diet; (iv) NC + 2.5% CA + 500 FTU MP per kilogram of diet; (v) NC + 5% CA per kilogram; and (vi) NC + 5% CA + 500 FTU MP per kilogram of diet. The content of calcium, phosphorus and length of tibia bone and digestibility of calcium and phosphorus was evaluated. The results showed that interaction effect of CA×MP on tibia calcium content to low available phosphorus diets was significant ($P<0.01$). Adding of CA to P-deficient diets, was increased tibia phosphorus content of broilers in compared with control group ($P<0.01$). Adding MP to P-deficient diets, based on corn-soybean meal, cause increased digestibility of calcium, phosphorus and also, length of tibia in broilers ($P<0.01$). Also, adding of CA to broiler diets deficient in available phosphorus, significantly increased digestibility of phosphorus ($P<0.01$) and length of tibia ($P<0.001$) in compared with control group. From this study it could be deduced that adding of MP to low available phosphorus diets can cause improvement of utilization of phytate phosphorus. Also adding CA as a chelator to diet can improve tibia mineralization parameters in broilers.

Session 46 Poster 21

Efficacy of Pulsatilla nigricans like growth promoter in broiler chickens

Morfin-Loyden, L., Camacho-Morfin, D., Hernández P., S.I. and Vicente A., P., Facultad de Estudios Superiores Cuautitlán. Campo 4. Universidad Nacional Autónoma de México., Dpto. of Animal Sci., Km 2.5 Carretera Cuautitlán-Teoloyucan, 54500, Mexico; morfinde@yahoo.com

A study was conducted to investigate the efficacy of Pulsatilla like growth promoter in broiler chickens. The test was conducted on 248 female and 248 male COBB chickens of one day old; the animals were allocated in eight groups for each sex and distributed in a completely randomized design, consisting of a 2 x 2 factorial arrangement with four repetitions of 31 birds each one. The factors were the sex and treatment. It was used alcohol 720 (A) like control and Pulsatilla 200 C (P), both products came from a commercial homeopathic laboratory. The treatments were administered on the drinking water, 0.2 ml/kg of living weigh of alcohol or Pulsatilla one time every week. The experiment lasted six weeks. The animals were weighted every seven days and every day was register dry matter feed intake. Average total body gain (ABG) and feed conversion ratio (FCR) were calculated. The total intake values for female A, P were 5,144.9+87.5 and 5,163.8+181.0 g respectively, for males in the same order 5,288.2+63.2 and 5,280.3+106.8, g. Female ABG, A and P 2,316.5+85.1, 2,351.5+28.9 and males was in the same order 2,558.7+31.4 y 2,595.8+30.4 g. FCR were for female 2.2+0.1 and 2.2+0.1 and for males 2.1+0 and 2.0+0.1. There were statistic differences between sexes but there were not differences between treatments or interactions. It was concluded that Pulsatilla 200 C didn't acts like grown promoter in COBB broiler chicken.

Session 46 Poster 22

Effects of dietary fat, vitamin E and Zinc on immune response and blood parameters of broiler reared under heat stress

Vakili, R.[1], Gofrani Ivar, Y.[2], Rashidi, A.[3] and Khatibjoo, A.[4], [1]Islamic Azad University-kashmar branch, Animal science, Ahmad Abad st.,2nd kholahdoz, no.22,mashhad, 9183816811, Iran, [2]Tarbit modaress University-, Animal science, Tehran, 532, Iran, [3]Islamic Azad University-kashmar branch, Animal science, Kashmar, 532, Iran, [4]Ferdowsi university, Animal science, mashhad, 91838, Iran; rezavakili2010@yahoo.com

The purpose of this experiment was to determine the effect of corn–soybean basal diets containing (5% tallow, 5% fish oil and 5% fish oil with vitamin E and zinc) each one in two temperatures (21 °C and 32 °C) on immune response and blood parameters of Arbor Acres (AA) broiler growing from 1 to 42 days of age. The results demonstrated that heat stress decreased bursa of fabricius weight and spleen weight; also high ambient temperature decreased humoral and cell-mediate immune responses and increased hetrophil to lymphocyte ratio (H/L). Fish oil plus vitamin E and zinc diet improved humoral and cell-mediate immune responses, decreased H/L ratio and increased bursa of fabricius and spleen weight at two temperatures. The diets containing tallow decreased bursa of fabricius weight and spleen weight. In chicks fed by fish oil, the response to 2, 4-dinitrochlorobenzene (DNCB) was improved. It did not have significant influence on phytohemagglutinin M (PHA-M) and sheep red blood cells (SRBC). The results of experimental temperatures on blood parameters demonstrated that high ambient temperature increased cholesterol, triglyceride, glucose, hematocrite percentage and serum malondialdehyde (MDA) although it did not affect total protein. Higher cholesterol and triglyceride observed in chicks fed with diet containing tallow and higher MDA in chicks fed with diet containing fish oil. Fish oil plus vitamin E and zinc decreased cholesterol, triglyceride, MDA and glucose significantly. Fat type did not affect total serum protein and hematocrite percentage.

Session 46 Poster 23

The effect of vegetable oil on the performance and carcass value in broiler rabbits weaned at 21 days of age

Zita, L., Czech University of Life Sciences Prague; FAFNR, Department of Animal Husbandry, Kamycka 129, 16521 Prague 6 - Suchdol, Czech Republic; zita@af.czu.cz

The objective of the present study was to investigate the effect of commercially available oil Akomed R®, lipase addition on growth, feed consumption, mortality and carcass value in early weaned broiler rabbits. As vegetable oil we used Akomed R®, commercial oil (AarhusKarlshamn AB, Sweeden) which contains 60.8% caprylic acid, 38.7% capric acid and 0.5% lauric acid. Forty two Hyplus rabbits were assigned to three groups. Group 1 was fed a control feed mixture, the second group received an experimental feed mixture containing 1% of Akomed R® and 0.5% lipase (PES–S, Inotex, Czech Republic) from 21 to 42 days of age. The third group received the same feed as the latter group from 21 to 77 days of age. Temperature of 16 oC and relative humidity about 65% were maintained during the whole fattening period. Water and feed were available ad libitum. Results of the experiment show that there was no growth promotion effect of Akomed R® and lipase supplement. Daily feed intake for the whole fattening period was significantly higher in the control group. Rabbits in the second group which received experimental feed till 42 days of age had the feed intake about 9% lower and in the third group approximately 15% lower in comparison with the control group. Mortality was not influenced. The effect of Akomed R® and lipase on dressing percentage was not observed. However, the renal fat was non significantly reduced from 2.99% in the control group to 2.59% or 2.10% in the experimental groups. This study was supported by the Ministry of Education, Youth and Sports of the Czech Republic (Project No. MSM 6046070901).

Replacement of barley grain for corn grain impacts on growth performance and carcass composition in New Zealand rabbits

El-Adawy, M.M.[1], Mohsen, M.K.[2], Salem, A.Z.M.[1], El-Santiel, G.S.[2] and Dakron, M.Z.[2], [1]Alexandria University, Faculty of Agriculture, Department of Animal Production, El-Shaty, 21519, Egypt, [2]Kafr el Sheikh University, Faculty of Agriculture, Department of Animal Production, Kafr el Sheikh, 33717, Egypt; dr.eladawy@yahoo.com

This work was aimed to evaluate the barley grains replacement with corn in rabbit's diets at 0 (B0), 20 (B20), 40 (B40), 60 (B60), 80 (B80) and 100 (B100) % of the corn grains diet until the 13 weeks of rabbit's age. Forty eight weanling's male White New Zealand rabbits (875±28.3 g LW, of 6 weeks of age), were allocated randomly in six experimental groups. Average daily gains (ADG), feed conversion (FC), cecal turnover rate (CTR) and carcass composition (CC) were determined. Rabbits of B20 had a highest ($P<0.05$) ADG and FC with a lower feed intake than the other groups. CTR was significantly increased in rabbits B60 and B80 than those fed the other experimental diets. Perslaughter weight, hot carcass and cold carcass weight; and dressing percentage (i.e. CC) were varied ($P<0.05$) among the experiment groups and the rabbits B20 had the better values than others. Carcass drip loss weight, fur, liver, kidneys, spleen, gallbladder, heart, lings + trachea, digestive tract (full and empty) and secum + appendix (full and empty) were not affected by the barley replacements. Highest improvement ($P<0.05$) in economical efficiency was in B20 rabbits. It was concluded that replacement of barely at 20% of the corn grains diet had better nutritive impacts on body rabbit's performance and carcass composition. This work was supported by CNCSIS-UEFISCSU, project number1055/2009 PNII-IDEI code 898/2008.

The effects of probiotic and herbal coccidiostatic on fattening performance in rabbit broilers

Majzlik, I., Mach, K., Vostrý, L. and Hofmanová, B., Czech University of Life Sciences, Animal Science and Ethology, Kamýcká 129, 16521 Prague 6, Czech Republic; majzlik@af.czu.cz

Several trials were conducted with hybrid genotype HYPLUS (♂PS 59 x ♀PS 19) to evaluace the performance parameters under feed additives PROBIOSTAN (probiotic product Czech origin) and herbal coccidiostatic EMANOX (France origin). Animals were fed ad libitum with commercial granulated feed mixture supplied by different concentrations of both additives. The mixture for control group was supplied by chemical coccidiostatic ROBENIDIN. Generally, the testing period was covering 35-78 days of life. The average daily gains were 43.91-47.70 g with feed consumption 3.42-3.57 kg per 1 kg of live weight gain and the daily feed consumption ranged 152.72-163.92 g. The final body weight ranged between 2,653.75-2,718.67 g and dressing percentage 56.54-58.62% resp. The best results of fattenig trials showed groups fed with both experimental additives PROBIOSTAN (0.2%) and EMANOX (500 g in 1,000 kg feed mixture) with the highest total body weight gain and finishing body weight at the lowest feed consumptions. The health status was positively influenced by both experimental additives, but the chemical coccidiostatic ROBENIDIN did not influenced the incidence of coccidiosis. This project was supported by grant MSM 604 607 09 01.

Session 34 Poster 25

Study the effects of age at first calving and mother age on milk productive traits of Iranian Holstein cows

Ghazi Khani Shad, A.[1], Heidari, M.[1] and Sayyad Nejad, M.B.[2], [1]Islamic Azad University,Saveh Branch, Islamic Azad University, Saveh, Iran, 3145526852, Iran, [2]Iranian Animal Breeding Center, Karaj, Iran, 3175245621, Iran; a.ghazikhani@iau-saveh.ac.ir

The aim of this paper was studying the effects of age at the first calving (AFC) and mother age (MA) on the productive traits in Iranian Holstein cows. Study was carried out using collected data during 25 years since 1981. Data included production records have been recorded for each animal during six lactations. The records were milk, fat and protein percent and ECM (Energy Corrected Milk) were corrected based on 305 days of production and two milking times. The AFE and MA were classified in six groups (between 18 to 30 months) and has been considered in statistical model. Also this item has been considered as a covariable in the model. Analysis of variance and regression analysis was carried out for the effect of AFC and MA as classified and covariable. The means of production traits in different lactations including the milk, fat, protein production and ECM showed an increasing trend from first to third lactation and decreased gradually. The fat percent revealed an increasing trend from the first to sixth lactation but protein percent was approximately constant. The effect of MA on most of the production traits in five first lactation and longevity was significant ($P<0.01$) but for the protein percent from second to sixth lactation and lifetime was not significant ($P>0.01$). There was a significant regression relationship between MA and most production traits ($P<0.01$). The notable effect were obtained for the effects of AFC on the milk, protein and ECM in first, second and fourth lactation and fat production in the first lactation and fat percent in the four first lactation ($P<0.01$). On the other hand, there was a significant relation between AFC and all traits ($P<0.01$). The phenotypic correlation between studied traits showed that there is a significant and medium correlation between most traits ($P<0.01$).

Session 41 Theatre 18

Slovene consumers' attitude regarding quality of eggs from different laying hens housing systems

Pohar, J. and Perc, V., University of Ljubljana, Biotechnical Faculty, Department of Animal Science, Groblje 3, 1230 Domzale, Slovenia; jure.pohar@bf.uni-lj.si

In order to find out what is the comprehension of the type of housing system for laying hens by Slovene consumers, face-to-face interviews were conducted with 151 female consumers at the point of purchase. In addition to the questions opt to find out what is the general level of recognition and comprehension of laying hens housing systems used in Slovenia, the interviewees were also asked to express their perception of the quality of eggs from free-range, barn and cage rearing systems by giving scores for each system on a 7 point Likert scale. The highest average score for quality was given for the eggs from free-range system (over 6 points) and the lowest for cage system (below 2 points). Interviewees' perception was that the free-range system is the most appropriate also from the standpoint of animal welfare, followed by barn and cage system. The percentage of consumers who know from which housing systems come the eggs they normally purchase is very low. In spite of the general perception that the free-range housing system is the best regarding the quality of eggs and animal welfare, they mainly buy eggs from the barn housing system.

Changes in some physicochemical properties of heat-processed soybean meal

Caprita, A. and Caprita, R., Banat University of Agricultural Sciences and Veterinary Medicine , Calea Aradului 119, 300645 Timisoara, Romania; rodi.caprita@gmail.com

Mild heat treatment of soybeans destroys or reduces heat-labile antinutritional factors (including trypsin inhibitors, hemagglutinins, antivitamins, goitrogens and phytates), and improves digestibility. Protein quality of soybean meal is linked to both the reduction of antinutritional factors, and the optimization of protein digestibility. Direct analysis of both specifications is difficult in routine operations. The objective of our study was to study the effect of heat processing on some physicochemical properties of soybean meal (SBM). Commercial SBM (43.7% crude protein) was ground to pass the 200 μ sieve and was heated in a forced air oven at 120 °C for varying periods of time: 5, 10, 15, 20, 25 and 30 minutes. The following parameters were determined: urease index (UI), KOH protein solubility (PS), Protein Dispersibility Index (PDI), Nitrogen Solubility Index (NSI), refractive index (Abbe refractometer), and dynamic viscosity (Brookfield viscometer Model DVIII Cone CP-40). UI, PDI and NSI decreased with the heating time. Experimental data revealed a positive correlation between UI and both PDI (r=0.9857) and NSI (r=0.9713). The refractive index and the dynamic viscosity of dilute KOH solution extracts also decreased as the heating time increased, and the values are highly correlated with PS: r=0.9382 for refractive index and r=0.8943 for dynamic viscosity.

Dietary water extract viscosity in relation to the content of non-starch polysaccharides

Caprita, R. and Caprita, A., Banat University of Agricultural Sciences and Veterinary Medicine , Calea Aradului 119, 300645 Timisoara, Romania; rodi.caprita@gmail.com

Non-starch polysaccharides (NSP) from various cereal grains are considered to have antinutritive effect in poultry nutrition if they are present in diets at high concentrations. The detrimental effect of soluble NSP is mainly associated with the viscous nature of these polysaccharides and their physiological effects on the digestive medium. Soluble NSP increases the viscosity of the small intestinal chime, generally hampering the digestion process, whereas insoluble NSP impedes the access of endogenous enzymes to their substrates by physical entrapping. Almost all water-soluble polysaccharides produce viscous solution. The viscous properties of NSP depend on several factors, including their chemical composition, molecular size, and composition of the extraction media. Wheat and barley contain substantial amounts of both soluble and insoluble NSP. The main water soluble NSP are arabinoxylans in wheat and β-glucans in barley. Extract viscosity values of grains could be used as predictors of anti-nutritional properties of NSP in cereals. Experiments were carried out to investigate the relationship between water extract viscosity and wheat and barley containing feeds. Feeds with different wheat and barley content (0 to 40%) were milled by a laboratory grinder at a 600 μm sieve. Two extraction procedures were experimented, without and with inactivation of endogenous enzymes. The samples were treated in boiling aqueous ethanol and both the ethanol-treated and untreated flours were extracted with water. The water extract viscosity was determined using a cone/plate viscometer (Brookfield Model DVIII Cone CP-40) at 100 rpm and 25 °C. Aqueous extract viscosities correlated well with the wheat and barley concentrations of the feeds. Water extract viscosity in wheat containing feeds is lower than in barley containing feeds. This work was supported by CNCSIS-UEFISCSU, project number1054/2009 PNII-IDEI code 894/2008.

Authors index

B

C

D

G

K

M

Q

R

T

U

V

Z

Printed in the United States
by Baker & Taylor Publisher Services